# A Textbook of Two Dimensional Geometry

A Textbook of

# Two Dimensional Geometry

(SECOND EDITION)

**SAT PAL**
**HARBANS LAL**
Department of Mathematics
Hans Raj College
University of Delhi, Delhi
INDIA

PUBLISHING FOR ONE WORLD

**NEW AGE INTERNATIONAL (P) LIMITED, PUBLISHERS**
New Delhi • Bangalore • Chennai • Cochin • Guwahati • Hyderabad
Jalandhar • Kolkata • Lucknow • Mumbai • Ranchi
Visit us at **www.newagepublishers.com**

Published by New Age International (P) Ltd., Publishers
First Edition: 1984
Second Edition: 1996
Reprint: 2008

---



***Branches:***
- 36, Malikarjuna Temple Street, Opp. ICWA, Basavanagudi, **Bangalore**. © (080) 26677815
- 26, Damodaran Street, T. Nagar, **Chennai**. © (044) 24353401
- Hemsen Complex, Mohd. Shah Road, Paltan Bazar, Near Starline Hotel, **Guwahati**. © (0361) 2543669
- No. 105, 1st Floor, Madhiray Kaveri Tower, 3-2-19, Azam Jahi Road, Nimboliadda, **Hyderabad**. © (040) 24652456
- RDB Chambers (Formerly Lotus Cinema) 106A, 1st Floor, S.N. Banerjee Road, **Kolkata**. © (033) 22275247
- 18, Madan Mohan Malviya Marg, **Lucknow**. © (0522) 2209578
- 142C, Victor House, Ground Floor, N.M. Joshi Marg, Lower Parel, **Mumbai**. © (022) 24927869
- 22, Golden House, Daryaganj, **New Delhi**. © (011) 23262370, 23262368

---

**ISBN : 978-81-224-0925-3**

**Rs. 185.00**

**C-08-03-2366**

**2 3 4 5 6 7 8 9 10**

Printed in India at Ram Printograph, Delhi.
Typeset at Datatime Associates, New Delhi.

PUBLISHING FOR ONE WORLD
**NEW AGE INTERNATIONAL (P) LIMITED, PUBLISHERS**
4835/24, Ansari Road, Daryaganj, New Delhi-110002
Visit us at **www.newagepublishers.com**

# PREFACE TO THE SECOND EDITION

The authors take pleasure in presenting this second edition of the book which has been thoroughly revised in the light of the revised syllabi of B.A. (Pass), B.Sc. (General and Honours) students of Delhi and other Indian Universities.

Some salient features of the present edition are:

- The subject matter has been presented in a rigorous manner and the language used is simple and effective, so as to make the exposition clear and lucid even to the weakest student.
- The subject matter has been developed in a very smooth and coherent manner so as to provide an orderly presentation to the reader.
- About one hundred solved examples and about four hundred unsolved problems have been added in this edition. Most of these have been selected from the latest examination papers of various universities and the competitive examinations.
- Each concept has been illustrated with the help of a sufficient number of illustrative examples.
- The answers to the unsolved problems have been thoroughly checked.
- There is an index of technical words from two dimensional geometry at the end of the book.
- Every effort has been made to make the book free of errors.

We would be failing in our duty if we do not place on record the valuable suggestions given to us by our colleagues in Hans Raj College, Delhi and from other friends in various colleges of this university, which have been duly incorporated. Our sincere thanks are for the publishers for bringing out in time this edition of the book.

We hope, the book in its present form will prove more useful for the students it is intended for.

Suggestions for the improvement of the book will be gratefully acknowledged and taken care of in the future editions of the book.

S. Pal and H. Lal

# PREFACE TO THE FIRST EDITION

Though several books on Two Dimensional Geometry are available, most of them were written many years ago. The concepts of sets and modulus of a number were not taught at the school level, then. The present-day under-graduate student, however, is familiar with these concepts and the authors have not hesitated to utilize them, as and when needed, in the present book. Also, in the aforesaid books, a number of theorems were stated and proved, in the form of necessary conditions only, but the same were also used, quite frequently, as sufficient conditions. Here, in this book, we have tried to bridge this gap, as far as possible, by properly stating and proving them in the form of necessary as well as sufficient conditions.

The idea of locus, in Chapter 2, has been explained with the help of sets and the authors urge upon the reader to go through its contents very carefully, as it would be used in the sequel, quite often.

In Chapter 8, entitled, "A line and a curve", we have dealt with all the common aspects of the standard curves, *viz.*, circle, parabola, ellipse and hyperbola. This has been presented in a very simple and lucid way, so as to enable the reader to become familiar with the general properties of these curves, simultaneously.

Throughout the book, a large number of solved examples have been given to illustrate the various concepts and the techniques. On a number of places, more than one method has been indicated to solve the same question. The problems have been carefully selected (from the examination papers of our Universities and of the various Competitive Examinations) and graded properly. Also the answers have been thoroughly checked.

Decimal notation has been used in the book to denote articles, theorems, examples and exercises, which run independently of each other, e.g. § 6.1.2 refers to sub-section 2, in article 1 of Chapter 6.

We hope this book will be useful for all the B.A./B.Sc., Honours, pass course as well as Engineering Students of our Universities and also for the students preparing for the various Competitive Examinations.

During the preparation of this book, we have consulted many texts on the subject and we feel indebted to their authors and the publishers.

We express our gratitude to all our friends who helped us in the preparation of this book and especially to our senior colleague, Prof. I.K. Malhotra, who very kindly went through the manuscript with great care and keen interest and suggested a large number of valuable comments and also to Dr Qazi Zameeruddin, of Kirori Mal College, Delhi. We also thank the publishers for their unfailing courtesy and co-operation.

Intimation of errors, suggestions and comments shall be gratefully received and acknowledged.

*July, 1984*

S. PAL
H. LAL

# CONTENTS

*Preface to the Second Edition* iii
*Preface to the First Edition* v

**1. CO-ORDINATES** **1**

1.1 Co-ordinate System on a Line 1
1.2 Cartesian Co-ordinates of a Point in a Plane 3
1.3 Area of a Triangle 11
1.4 Choice of Axes 14
1.5 Polar Co-ordinates 17

**2. EQUATION OF LOCUS** **25**

2.1 Locus of a Point and its Equation 25
2.2 Method to find the Equation of the Locus of a Point 27
2.3 Intersection of Loci 32
2.4 The Locus Represented by the Equation $S_1 . S_2 = 0$ 34
2.5 The Locus Represented by the Equation $S_1 + kS_2 = 0$ 34

**3. CHANGE OF AXES** **37**

3.1 Shifting of the Origin 37
3.2 Rotation of Axes without Changing the Origin 39
3.3 Rotation of Axes along with Change of Origin 41
3.4 Removal of First Degree Terms 42
3.5 Removal of the Term $xy$, from $f(x, y) = ax^2 + 2hxy + by^2$ without Changing the Origin 45
3.6 Invariants 49

**4. STRAIGHT LINE** **54**

4.1 General Equation of a Line 54
4.2 Particular Cases of the Equation of a Line 55
4.3 Reduction of the General Equation $ax + by + c = 0$, of a Line to Special Forms 65
4.4 To find the Perpendicular Distance from the Point $(x_1, y_1)$ on the Line : $ax + by + c = 0$ 67

4.5 To find the Angle Between the Lines whose Slopes are given 71
4.6 To find the General Equations of Lines Parallel and Perpendicular to the Line $ax + by + c = 0$ 72
4.7 To find the Point of Intersection of Two Lines 76
4.8 Line Through the Intersection of Two Lines 77
4.9 System (or Family) of Lines 84
4.10 Concurrence of Lines 86
4.11 Collinearity of Three Points 92
4.12 Position of a Point Relative to a Line 93
4.13 Two Sides of a Line 94
4.14 To find the Bisectors of the Angles between the Lines : $a_1x + b_1y + c_1 = 0$, $a_2x + b_2y + c_2 = 0$ 95

**5. PAIR OF STRAIGHT LINES** **110**

5.1 Homogeneous Equations 110
5.2 To find the Angle between the Lines given by $ax^2 + 2hxy + by^2 = 0$ 118
5.3 To find the Equation of the Bisectors of the Angles between the Lines given by $ax^2 + 2hxy + by^2 = 0$ 119
5.4 General Equation of Second Degree 123
5.5 To find the Equation of the Lines Joining the Origin to the Points of Intersection of a Line and a Curve 136

**6. CIRCLE** **151**

6.1 To find the Equation of a Circle whose Centre and Radius are given 151
6.2 To find the Position of a Point $P(x_1, y_1)$ w.r.t. a Given Circle: $x^2 + y^2 + 2gx + 2fy + c = 0$ 160
6.3 Circle through the Points of Intersection of Two Given Circles 161
6.4 Tangents and Normals to a Circle 169
6.5 Angle of Intersection of Two Circles 180
6.6 Radical Axis of Two Circles 185
6.7 Co-axal System of Circles 193

**7. SOME STANDARD CURVES** **211**

7.1 Tracing of Curves 211
7.2 Parametric Equations of the Standard Curves 218
7.3 Position of a Point w.r.t. Standard Curves 219

8. A LINE AND A CURVE 224

8.1 To find the Points of Intersection of a Line and a Curve 224
8.2 To find the Equation of the Tangent at $(x', y')$ to the Curve, $S = 0$ 232
8.3 Normal 239
8.4 Chord of Contact 244
8.5 Pole and Polar 249
8.6 To find the Equation of the Chord of a Curve, $S = 0$, in terms of its Middle Point 259
8.7 Pair of Tangents from a Point to a Curve 269

9. CONIC SECTIONS 278

9.1 Conic 278
9.2 Conic Sections 281

10. PARABOLA 284

10.1 Standard Form of the Parabola 284
10.2 Summary of the Results from Chapter 8 294
10.3 Some Propositions on the Parabola 296
10.4 To find the Equation of the Normal to the Parabola, in Terms of its Slope, $m$ 304

11. ELLIPSE 325

11.1 To find the Equation of the Ellipse in the Standard Form 325
11.2 Summary of the Results from Chapter 8 333
11.3 Auxiliary Circle 334
11.4 To find the Locus of the Middle Points of a System of Parallel Chords of an Ellipse 360

12. HYPERBOLA 389

12.1 To find the Equation of a Hyperbola in the Standard Form 389
12.2 Summary of the Results from Chapter 8 399
12.3 Geometrical Interpretation of the Parameter $\theta$, when the Equations of the Hyperbola are taken as $x = a \sec \theta$, $y = b \tan \theta$ 401
12.4 Four Normals from a Point to a Hyperbola 407
12.5 Asymptotes 414
12.6 To find the Locus of the Middle Points of a System of Parallel Chords, having Slope $m$, of the Hyperbola 426

12.7 Rectangular Hyperbola 434

13. TRACING OF THE GENERAL CONIC 447

13.1 The Nature of Conic Represented by $S = 0$ 447
13.2 To find the Co-ordinates of the Centre of the Conic, $S = 0$ 450
13.3 To find the Direction and Lengths of the Axes of a Central Conic, $S = 0$ 452
13.4 To find the Eccentricity of a Central Conic 455
13.5 To find the Co-ordinates of the Focus (or Foci) of a Conic, $S = 0$ 462
13.6 To find the Equation of the Directrix (or Directrices) of a Conic, $S = 0$ 463
13.7 Tracing of an Ellipse 464
13.8 Tracing of a Parbola 468
13.9 Tracing of a Hyperbola 477

14. POLAR CO-ORDINATES 490

14.1 To find the Distance between Two Points $(r_1, \theta_1)$ and $(r_2, \theta_2)$ 490
14.2 To find the Equation of a Line in Polar Co-ordinates 492
14.3 To find the Equation of a Circle 500
14.4 To find the Equation of the Conic when the Focus is taken as a Pole 502

*Answers* 527
*Index* 569

# 1. CO-ORDINATES

## INTRODUCTION

In two dimensional Analytical Geometry or Co-ordinate Geometry, as it is popularly known, we study geometry, by using algebraic methods. This is done by associating a pair of real numbers, with every point in a plane. The pair of real numbers associated with any point in the plane is called its co-ordinates. One type of such co-ordinates are called *Cartesian Co-ordinates,* in honour of its inventor, René Des Cartes (1596–1650), a French Mathematician, who had a dream one night, of associating a pair of real numbers to every point in the plane. We shall utilise algebraic techniques to associate equations with the geometric curves and thus study their properties. In this chapter, we first, establish a co-ordinate system on a line and then in a plane.

## 1.1. CO-ORDINATE SYSTEM ON A LINE

Consider any line $x'x$, choose any two distinct points on it and label them 0 (zero) and 1. The point 0 (zero) is called the *origin,* and we denote it by $O$. We define the positive direction on the line to be from the point 0 to 1, and the length of the line segment between them to be the *unit of length,* that is, we define the length of this segment to be 1.

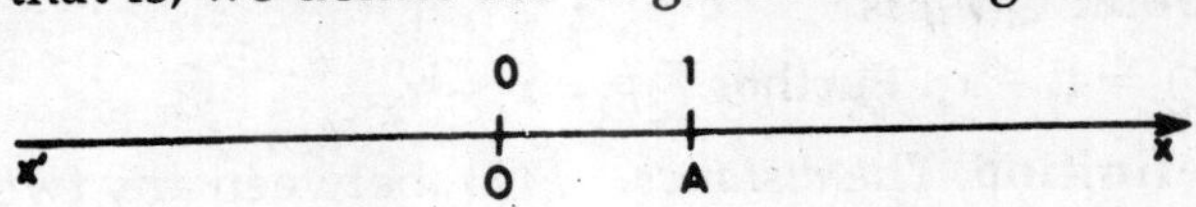


The line can be in any position and the points 0 and 1 can be any distinct points on it. However, this line is usually taken to be horizontal and its positive direction to the right, as indicated by the arrow head.

We make the following assumption which is, in fact, the geometric equivalent of the *Axiom of Completeness:*

Corresponding to each point of the line, there is a unique real number and to each real number, there corresponds a unique point on the line; we assume further that if $A$ and $B$ are any two points on the line, and $a$ and $b$ are the corresponding real numbers, then $A$ is to the right of $B$ if and only if $a > b$.

A line with its points in correspondence with the real numbers is called a *co-ordinate line* and the real number corresponding to a particular point on the line is called the (one-dimensional) *co-ordinate* of that point. We indicate that a point $P$ has the co-ordinate $x$ by: $P(x)$. Thus $A$ and $O$ have co-ordinates (1) and (0) respectively.

### 1.1.1. Direction of a line segment and distance between two points

Let $P(x_1)$ and $Q(x_2)$ be any two points on the co-ordinate line as shown below:

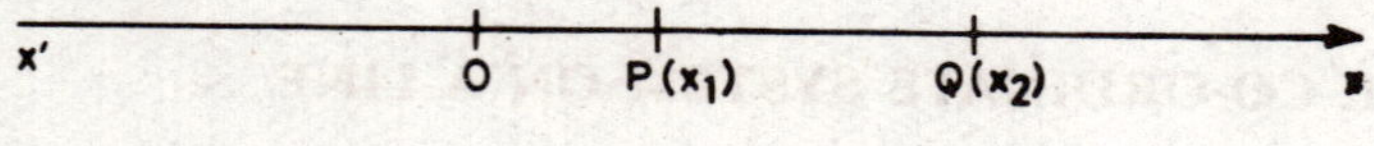


The line segment $PQ$ can be looked upon in two ways; from $P$ to $Q$ or from $Q$ to $P$. We define the direction of $PQ$ to be positive or negative, according as $Q$ is on the right or on the left of $P$.

**Definition.** The *directed distance*, $\overline{PQ}$ , from a point $P(x_1)$ to a point $Q(x_2)$ is

$\overline{PQ} = x_2 - x_1$. Further $\overline{QP} = x_1 - x_2$.

**Definition.** The *distance*, $|PQ|$, between any two points, $P(x_1)$ and $Q(x_2)$, is

$|PQ| = |x_2 - x_1|$, where $|x_2 - x_1|$

denotes the modulus (or the absolute value) of the real number, $x_2 - x_1$.

Thus the distance between any two points on the co-ordinate line is always non-negative and also $|PQ| = |QP|$.

**Example 1.1.** Find the distance between the points: $P(-2)$ and $Q(3)$. Also find $\overline{PQ}$ and $\overline{PQ}$.

*Solution.* $|PQ| = |3 - (-2)| = 5.$

$$\overline{PQ} = 3 - (-2) = 5 \text{ and } \overline{QP} = (-2) - 3 = -5.$$

## 1.2 CARTESIAN CO-ORDINATES OF A POINT IN A PLANE

Consider, in a plane, two intersecting co-ordinate lines, inclined at any angle (not necessarily a right angle), which intersect at the zero point on each line. Let them be $x'Ox$ and $y'Oy$.

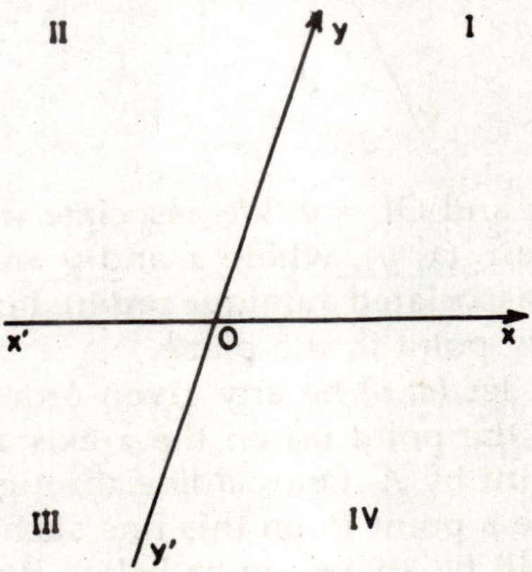


The line $x'Ox$ is called the *x-axis* and the line $y'Oy$ is called the *y-axis*; together, they are called the *co-ordinate axes* and their point of intersection, $O$, is called the *origin*. We take $Ox$ as the positive side of the *x-axis* and $Ox'$ as its negative side. Also $Oy$ and $Oy'$ are respectively taken as the positive and negative sides of the $y$-axis. Clearly the co-ordinate axes divide the plane into four sections. These four sections are called the *four quadrants of the co-ordinate plane,* as shown above.

Now we fix the position of any point $P$ of this plane w.r.t. the above axes.

Draw, through $P$, two lines, $PL$ and $PM$, parallel to the co-ordinate axes, meeting them at $M$ and $L$ respectively (see the figure below).

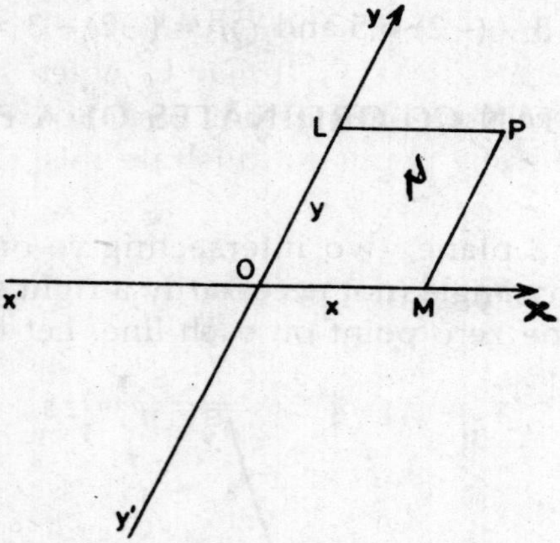


Let $OM = x$ and $OL = y$. We associate with the point $P$, the ordered pair, $(x; y)$, where $x$ and $y$ are real numbers. Thus we have associated a unique ordered pair of real numbers with every point in the plane.

Conversely, let $(a, b)$ be any given ordered pair of real numbers. Plot the point $(a)$ on the $x$-axis as in § 1.1 and denote this point by $A$. Draw a line through $A$, parallel to the $y$-axis. Take a point $P$ on this line such that $AP = |b|$. The point $P$ will be above, on or below the $x$-axis according as $b$ is $>$, $=$ or $< 0$. The figure below, shows the point $P$ when $a > 0$, $b > 0$. This $P$ corresponds to the ordered pair $(a, b)$.

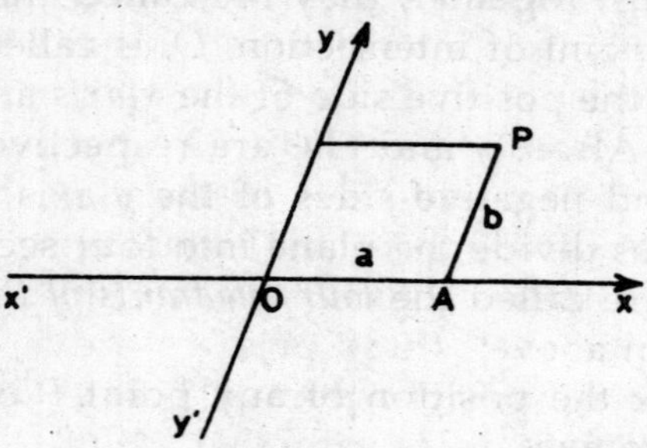

Thus we have a one-to-one correspondence between the points in a plane and the ordered pairs of real numbers. The unique ordered pair $(x, y)$ associated with a point $P$ in the plane, is called its *co-ordinates*; $x$ is called its abscissa or the *x-co-ordinate* and $y$ is called its *ordinate* or the *y-co-ordinate*. We write it as: $P(x, y)$. It may be noted that the signs of $x$ and $y$, determine the position of the point.

We exhibit some points to illustrate this.

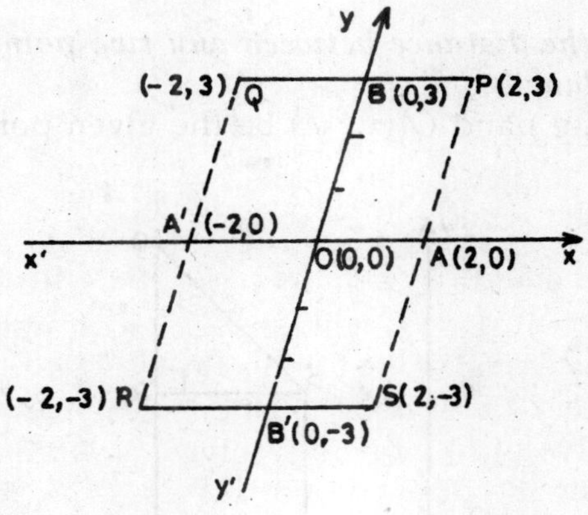


The afore-said system of co-ordinates is called the *Oblique Cartesian Co-Ordinate System*, whenever $x'Ox$ and $y'Oy$ are inclined at an angle other than a right angle. However, if the angle between them is a right angle, the co-ordinate axes, in this case, are called *Rectangular* and this system of co-ordinates is, then, said to be *Orthogonal* or *the Rectangular Cartesian Co-ordinate System*. Throughout this text, we shall confine to this system only, except in the concluding Chapter where we shall study polar co-ordinates, as well.

Thus the co-ordinates of a point $P(x, y)$ in the Rectangular Cartesian System are as exhibited below:

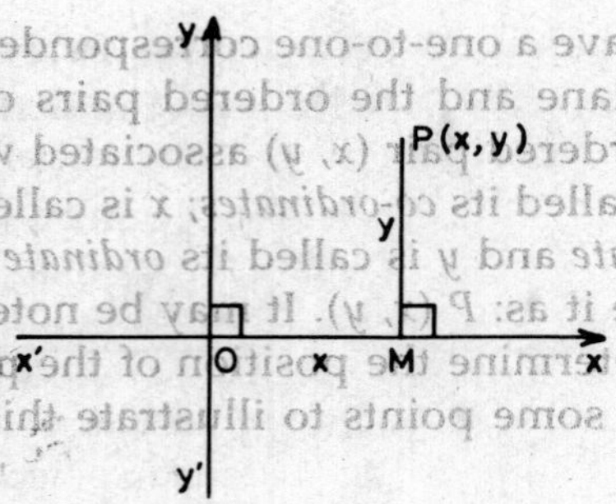


### 1.2.1. Distance Formula

***To find the distance between any two points in the co-ordinate plane.***

Let $P(x_1, y_1)$ and $Q(x_2, y_2)$ be the given points. To find: $|PQ|$.

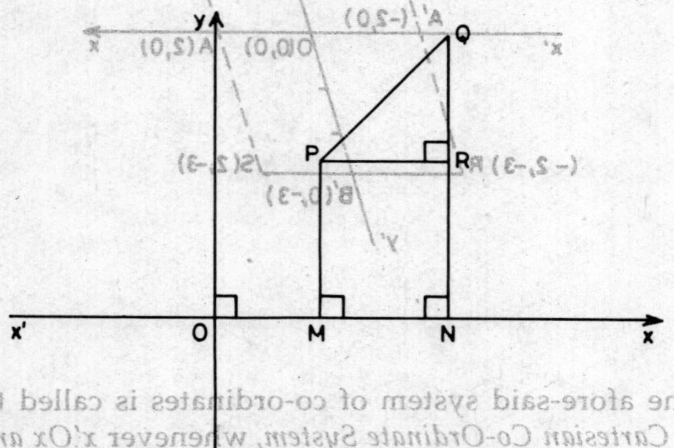


Draw $PM$ and $QN$, perpendiculars on the axis of $x$ and $PR \perp NQ$. From $\Delta PRQ$,

$$PQ^2 = PR^2 + RQ^2 = MN^2 + (NQ - NR)^2$$

$$= (ON - OM)^2 + (NQ - MP)^2 = (x_2 - x_1)^2 + (y_2 - y_1)^2$$

$$\therefore \quad |PQ| = \sqrt{(x_2 - x_1)^2 + (y_2 - y_1)^2}\,.$$

**Remark.** The above formula has been proved when $P$ and $Q$ are in the first quadrant. However, it remains to be true for all positions of the points, keeping in mind, the proper signs of their co-ordinates.

**Example 1.2.** Show that the following points are collinear (i.e. they lie on a line): (1, 2), (3, 6) and (6, 12).

*Solution.* Call these points by $A$, $B$ and $C$ respectively. They are collinear iff sum of the distances between two pairs of points is equal to the distance between the third pair. Now

$$|AB| = \sqrt{(3-1)^2 + (6-2)^2} = 2\sqrt{5}$$

$$|BC| = \sqrt{(6-3)^2 + (12-6)^2} = 3\sqrt{5},$$

and $$|AC| = \sqrt{(6-1)^2 + (12-2)^2} = 5\sqrt{5};$$

whence $|AB| + |BC| = |AC|$, i.e. $A$, $B$, $C$ are collinear.

*Notation.* We shall denote the distance between two points $P$ and $Q$ of the co-ordinate plane, either by $|PQ|$ or by $\overline{PQ}$.

### 1.2.2. Section Formulae

***To find the co-ordinates of a point which divides the line joining two points in a given ratio.***

Let $P(x_1, y_1)$ and $Q(x_2, y_2)$ be the given points and let $R(x, y)$ be a point which divides $PQ$ in the ratio $m : n$, internally.

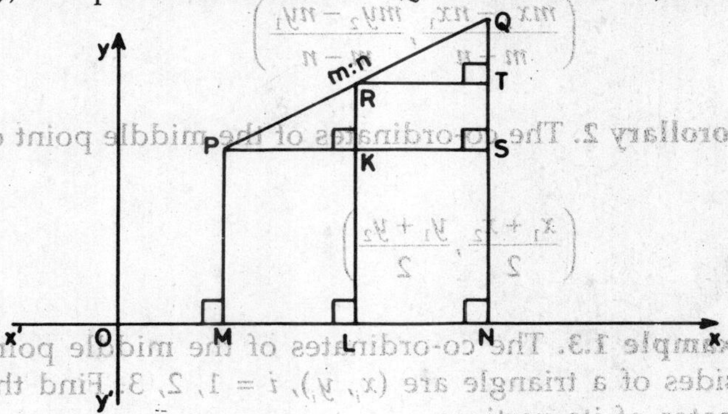

Draw *PM, RL QN* perpendiculars on the *x*-axis, and *PS, RT* parallel to it. Then by the similar triangles, *PKR* and *RTQ*,

$$\frac{PK}{RT} = \frac{KR}{TQ} = \frac{PR}{RQ} = \frac{m}{n}$$

First consider,

$$\frac{PK}{RT} = \frac{m}{n}$$

$$\frac{x - x_1}{x_2 - x} = \frac{m}{n}$$

$$\Rightarrow x = \frac{mx_2 + nx_1}{m + n}$$

Similarly, $\dfrac{KR}{TQ} = \dfrac{m}{n} \Rightarrow y \Rightarrow \dfrac{my_2 + ny_1}{m + n}$

Thus the co-ordinates of *R* are

$$\left(\frac{mx_2 + nx_1}{m + n}, \frac{my_2 + ny_1}{m + n}\right)$$

**Corollary 1.** If *K* divides *PQ* externally in the ratio $m : n$ $(m \neq n)$, then the co-ordinates of *R* are

$$\left(\frac{mx_2 - nx_1}{m - n}, \frac{my_2 - ny_1}{m - n}\right)$$

**Corollary 2.** The co-ordinates of the middle point of *PQ* are

$$\left(\frac{x_1 + x_2}{2}, \frac{y_1 + y_2}{2}\right)$$

**Example 1.3.** The co-ordinates of the middle points of the sides of a triangle are $(x_i, y_i)$, $i = 1, 2, 3$. Find the co-ordinates of its vertices.

*Solution*. Let $ABC$ be the triangle with vertices at $(a, b)$, $i = 1, 2, 3$: and let $L$, $M$, $N$ be the middle points of the sides as shown. Then,

$$\left(\frac{a_2+a_3}{2}, \frac{b_2+b_3}{2}\right) = (x_1, y_1), \left(\frac{a_3+a_1}{2}, \frac{b_3+b_1}{2}\right) = (x_2, y_2)$$

and $$\left(\frac{a_1+a_2}{2}, \frac{b_1+b_2}{2}\right) = (x_3, y_3)$$

whence $a_2 + a_3 = 2x_1,\ a_3 + a_1 = 2x_2,\ a_1 + a_2 = 2x_3$ (1)

and $b_2 + b_3 = 2y_1,\ b_3 + b_1 = 2y_2,\ b_1 + b_2 = 2y_3$ (2)

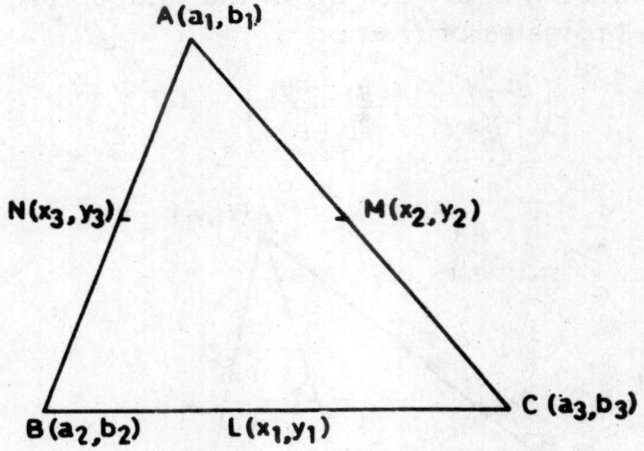


Equations (1) imply on addition,

$$a_1 + a_2 + a_3 = x_1 + x_2 + x_3$$

But $a_2 + a_3 = 2x_1$

$\therefore$ $a_1 = -x_1 + x_2 + x_3$.

Similarly, $a_2 = x_1 - x_2 + x_3$

and $a_3 = x_1 + x_2 - x_3$.

Equations (2) yield

$$b_1 = -y_1 + y_2 + y_3,\ b_2 = y_1 - y_2 + y_3,$$

and $b_3 = y_1 + y_2 - y_3$, similarly.

Thus the co-ordinates of the vertices of the triangle are

$(-x_1 + x_2 + x_3, -y_1 + y_2 + y_3)$,
$(x_1 - x_2 + x_3, y_1 - y_2 + y_3)$
and $(x_1 + x_2 - x_3, y_1 + y_2 - y_3)$.

**Example 1.4.** Show that the internal bisectors of the angles of a triangle are concurrent and hence find the co-ordinates of its *incentre*.

*Solution*. Let $ABC$ be any triangle with vertices at $(x_i, y_i)$, $i = 1, 2, 3$, and let $a, b, c$, denote the lengths of its sides. Let $AD$, $BE$ and $CF$ be the internal bisectors of the angles of this triangle.

By Geometry, $D$ divides $BC$ in the ratio $c : b$.
$\therefore$ the co-ordinates of $D$ are

$$\left(\frac{bx_2 + cx_3}{b+c}, \frac{by_2 + cy_3}{b+c}\right)$$

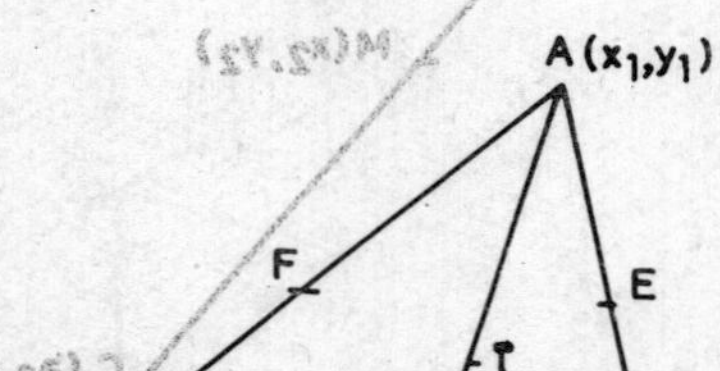


Let $I$ be the point on $AD$ dividing it in the ratio $b + c : a$.
$\therefore$ the co-ordinates of $I$ are

$$\left(\frac{ax_1 + bx_2 + cx_3}{a+b+c}, \frac{ay_1 + by_2 + cy_3}{a+b+c}\right) \quad (1)$$

By symmetry, this point $I$ lies on the bisectors, $BE$ and $CF$, also.

Thus the internal bisectors of the angles of a triangle meet at $I$, whose co-ordinates are given by (1). This point, $I$, is the *incentre* of the $\Delta ABC$.

**Example 1.5.** One end of a diameter of a circle is at (2, 3) and its centre is at (–2, 5). Find the co-ordinates of the other end of this diameter.

*Solution*. Let $A$ and $C$ be the points (2, 3) and (- 2, 5).

Let $B$ $(x, y)$ be the other end of the diameter. Since $C$ is the mid-point of $AB$,

$$\frac{x+2}{2} = -2, \quad \frac{y+3}{2} = 5$$

$\Rightarrow x = -6, \; y = 7$

$\Rightarrow$ Coordinates of $B$ are (–6, 7)

## § 1.3 AREA OF A TRIANGLE

**To find the area of a ΔPQR whose vertices are $(x_i, y_i)$ i = 1, 2, 3.**

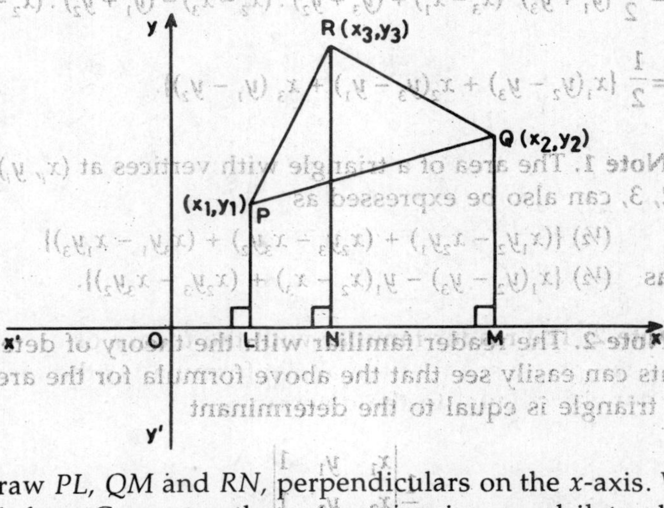


Draw $PL$, $QM$ and $RN$, perpendiculars on the $x$-axis. We recall, from Geometry, that a *trapezium* is a quadrilateral in which a pair of opposite sides is parallel and if $ABCD$ is a trapezium having $AB \parallel DC$, then its area is

$$\frac{1}{2}\,(|AB| + |DC|)\,.\,|DE|$$

where $DE \perp AB$.

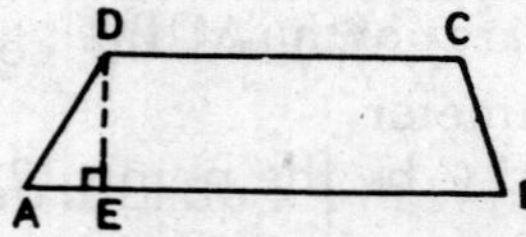


$$\therefore \text{ Area of } \Delta\, PQR = \begin{cases} \text{area of trapezium } PLNR \\ + \text{ area of trapezium } RNMQ \\ - \text{ area of trapezium } PLMQ \end{cases}$$

$$= \begin{cases} \frac{1}{2}(|LP|+|NR|).|LN|+\frac{1}{2}(|NR|+|MQ|).|NM| \\ -\frac{1}{2}(|LP|+|MQ|.|LM| \end{cases}$$

$$= \frac{1}{2}\{y_1 + y_3)\ (x_3 - x_1) + (y_3 + y_2) . (x_2 - x_3) - (y_1 + y_2) . (x_2 - x_1)\}$$

$$= \frac{1}{2}\{x_1(y_2 - y_3) + x_2(y_3 - y_1) + x_3\,(y_1 - y_2)\}.$$

**Note 1.** The area of a triangle with vertices at $(x_i, y_i)$, i = 1, 2, 3, can also be expressed as

$$(½)\ \{(x_1y_2 - x_2y_1) + (x_2y_3 - x_3y_2) + (x_3y_1 - x_1y_3)\}$$

or as $\quad (½)\ \{x_1(y_2 - y_3) - y_1(x_2 - x_3) + (x_2y_3 - x_3y_2)\}.$

**Note 2.** The reader familiar with the theory of determinants can easily see that the above formula for the area of the triangle is equal to the determinant

$$\frac{1}{2}\begin{vmatrix} x_1 & y_1 & 1 \\ x_2 & y_2 & 1 \\ x_3 & y_3 & 1 \end{vmatrix}$$

**Note 3.** The areas of the triangles *PQR* and *PRQ* are opposite in sign. In numerical cases, we usually ignore the sign of the area and take the absolute value of the area of the triangle as the final result.

**Corollary 1.** The area of the $\Delta OPQ$ is $\frac{1}{2}(x_1y_2 - x_2y_1)$, where $O$ is the origin.

**Corollary 2.** The area of a quadrilateral $PQRS$ is $|\text{area } \Delta PQR| + |\text{area } \Delta PRS|$.

**Corollary 3.** The condition of collinearity of the points $(x_i, y_i)$, $i = 1, 2, 3$ is given by

$$\begin{vmatrix} x_1 & y_1 & 1 \\ x_2 & y_2 & 1 \\ x_3 & y_3 & 1 \end{vmatrix} = 0$$

**Example 1.6.** Find the area of the quadrilateral with vertices at (1, 2), (– 4, 1), (–3, 0) and (5, 2).

*Solution.* Call the given points by $P, Q, R$ and $S$ respectively.

Join $PR$.

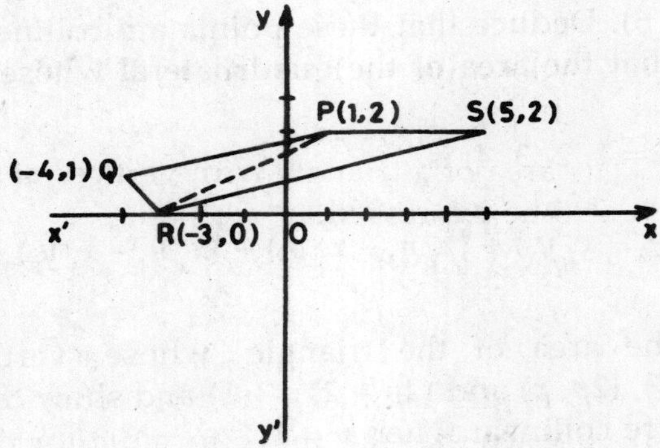


Now area of $\Delta PQR$

$$= \frac{1}{2}\{x_1(y_2 - y_3) + x_2(y_3 - y_1) + x_3(y_1 - y_2)\}|$$

$$= \frac{1}{2}|1(1-0) - 4(0-2) + (-3)(2-1)|$$

$$= 3$$

and area of $\Delta PRS$

$= \frac{1}{2} \mid 1(0-2) + 3(2-2) + 5(2-0) \mid = 4.$

$\therefore$ area of $PQRS = 3 + 4 = 7$.

## Exercise 1.1

1. Show that the triangle with vertices at (0, 6), (–5, 3) and (3, 1) is isosceles.
2. Find $x$ given that $|PQ| = |QR|$ and $P(2, 3)$, $Q(3, 5)$ and $R(x, 4)$.
3. Find the co-ordinates of the points which divide the line joining $P(2, 3)$ and $Q(3, 5)$ both internally and externally in the ratio 5 : 4.
4. Prove that the medians of a triangle are concurrent and hence find the co-ordinates of its *centroid*.
5. Find the area of the triangle with vertices at (4, 4), (3, 5) and (2, 6). Deduce that these points are collinear.
6. Prove that the area of the quadrilateral whose vertices are $(x_i, y_i)$, $i = 1, 2, 3, 4$; is
$$= \frac{1}{2}\{(x_1 y_2 - x_2 y_1) + (x_2 y_3 - x_3 y_2) + (x_3 y_4 - x_4 y_3) + (x_4 y_1 - x_1 y_4)\}.$$
7. Find the area of the triangle whose vertices are $(p + 1, 1)$, $(2p, p)$ and $(2p + 2, p + 1)$ and show that these points are collinear when $p = 1$.

## 1.4 CHOICE OF AXES

Sometimes a careful choice of the axes or of the origin, leads to a great deal of simplification, in certain problems. We mention, here, a few such situations:

(1) In dealing with problems regarding two perpendicular lines, we usually take their point of intersection as the origin and these lines as the co-ordinate axes.

(2) In dealing with questions having two fixed points,

say $A$ and $B$, we take $|AB| = 2a$, middle point $O$ of $AB$ as the origin, line $AOB$ as the $x$-axis and a line through $O$, perpendicular to it as the $y$-axis. Clearly, then, the co-ordinates of the fixed points are $(\pm a, 0)$.

(3) In dealing with sums, where some kind of symmetry exists, it is sometimes useful to take the co-ordinates of the points in a general way i.e. $(x_i, y_i)$, $i = 1, 2, ..$ etc.

(See: Questions 6 to 10 in the Miscellaneous Exercise).

We illustrate these situations with the help of the following examples.

**Examples 1.7.** In any triangle $ABC$, prove that

$$AB^2 + AC^2 = 2\,(AD^2 + DC^2)$$

where $D$ is the middle point of $BC$.

*Solution.* Let the co-ordinates of $A$, $B$, $C$ and $D$ be as shown in the figure.

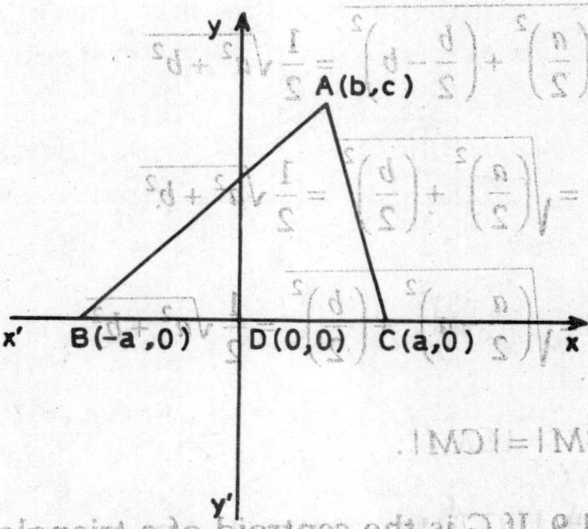


Now $\quad AB^2 + AC^2 = [(b + a)^2 + c^2] + [(b - a)^2 + c^2]$

$$= 2(a^2 + b^2 + c^2) \quad (1)$$

and $\quad 2(AD^2 + DC^2) = 2[b^2 + c^2) + a^2] \quad (2)$

The result follows from (1) and (2).

**Examples 1.8.** Prove that in a right angled triangle, the mid-point of the hypotenuse is equidistant from its vertices.

*Solution.* Let the given right angled triangle be $ABC$, with right angle at $B$. We take $B$ as the origin and $BC$, $BA$ as the co-ordinate axes. Let $M$ be the mid-point of the hypotenuse, $AC$. Taking the co-ordinates of the various points as shown in the figure, we have.

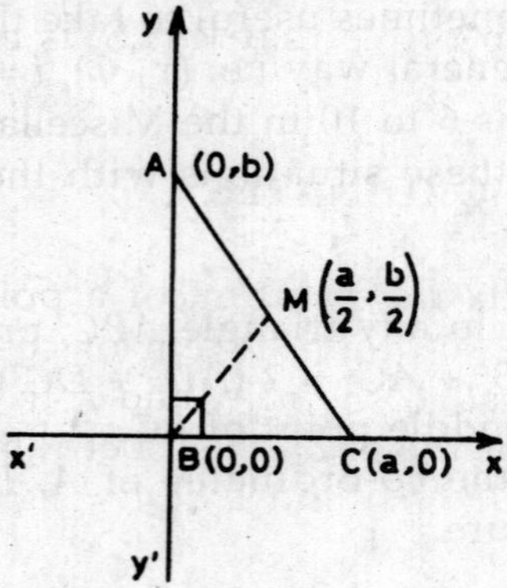


$$|AM| = \sqrt{\left(\frac{a}{2}\right)^2 + \left(\frac{b}{2} - b\right)^2} = \frac{1}{2}\sqrt{a^2 + b^2}$$

$$|BM| = \sqrt{\left(\frac{a}{2}\right)^2 + \left(\frac{b}{2}\right)^2} = \frac{1}{2}\sqrt{a^2 + b^2}$$

and $$|CM| = \sqrt{\left(\frac{a}{2} - a\right)^2 + \left(\frac{b}{2}\right)^2} = \frac{1}{2}\sqrt{a^2 + b^2}$$

$\therefore \quad |AM| = |BM| = |CM|.$

**Example 1.9.** If $G$ is the centroid of a triangle $ABC$, then prove that $3(GA^2 + GB^2 + GC^2) = AB^2 + BC^2 + CA^2$.

*Solution.* Taking $G$, the centroid of the given triangle as the origin, and the vertices $A$, $B$, $C$ respectively as $(x_i, y_i)$ $(i = 1, 2, 3)$; then

$$\frac{x_1 + x_2 + x_3}{3} = 0 = \frac{y_1 + y_2 + y_3}{3}$$

$\Rightarrow \quad x_3 = -(x_1 + x_2);\ y_3 = -(y_1 + y_2).$

Now L.H.S. $= 3\,(GA^2 + GB^2 + GC^2)$

$= 3\,\{(x_1^2 + y_1^2) + (x_2^2 + y_2^2) + (x_1 + x_2)^2 + (y_1 + y_2)^2]$

$= 6\,[(x_1^2 + x_2^2 + y_1^2 + y_2^2 + x_1x_2 + y_1y_2]$

Similarly we can prove that R.H.S. is equal to the above expression.

## 1.5 POLAR CO-ORDINATES

In order to fix the position of a point in a plane, we introduce, here, another system of co-ordinates. Take a fixed point $O$, called the *pole*, and a fixed line OX, called the *initial line* (or the *polar axis*). Let $P$ be any point in the plane, so that $|OP| = r$ and $\angle XOP = \theta$ ; $\theta$ measured in the

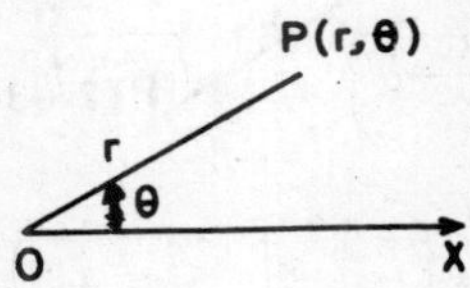


counter-clockwise sense, then $(r, \theta)$ is one set of *polar coordinates* of the point $P$; $r$ is called the *radius vector* and $\theta$ is called the *vectorial angle* of the point $P$. The polar co-ordinates of the pole, $O$ are taken as $(0, \theta)$, where $\theta$ is arbitrary.

**Remark.** In what we have said above, both $r$ and $\theta$ have been positive or zero. If $\theta < 0$, the angle $\theta$ is measured in the clock-wise sense and if $r < 0$, we first measure the angle $\theta$, and then, measure the distance, $|r|$, in the opposite direction, as explained below.

**Example 1.10.** Plot the following points:

(i) $(2, 30°)$, (ii) $(2, -30°)$, (iii) $(-2, 30°)$,

(iv) $(-2, -30°)$, (v) $(2, 210°)$.

*Solution.* (*i*) Draw line $OP'$ making an angle 30° with the initial line, $OX$, in the counter-clockwise sense and measure $OP = 2$, on this line. Then $P$ represents (2, 30°).

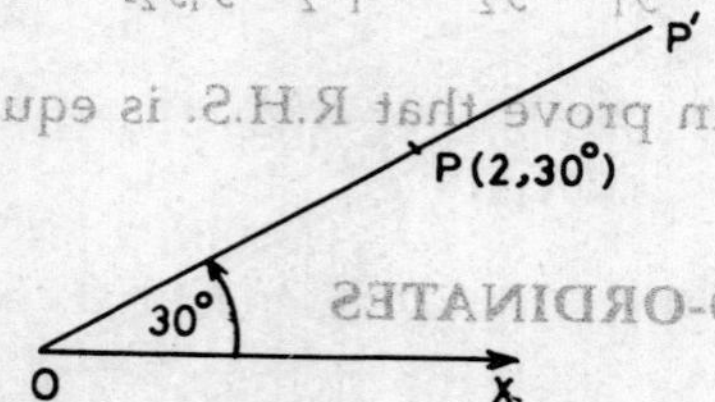


(*ii*) As in part (*i*), but angle to be taken in the clockwise sense.

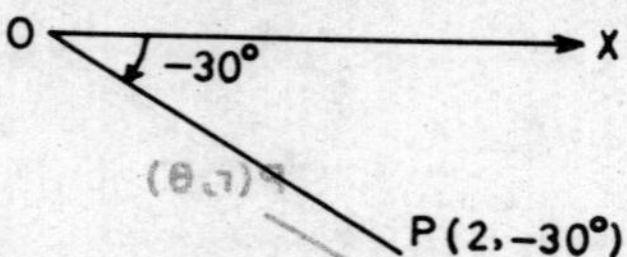


(iii)

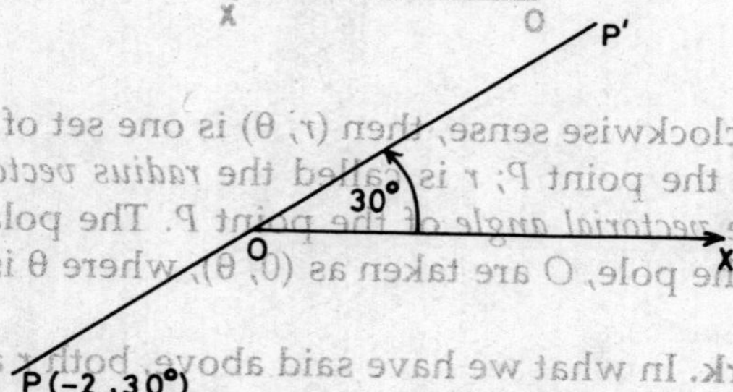


Make $\angle XOP' = 30°$. Produce $P'O$ and measure $OP = 2$ on it. Then $P$ represents (–2, 30°).

(iv) The point P whose polar co-ordinates are (– 2, – 30°) is shown below:

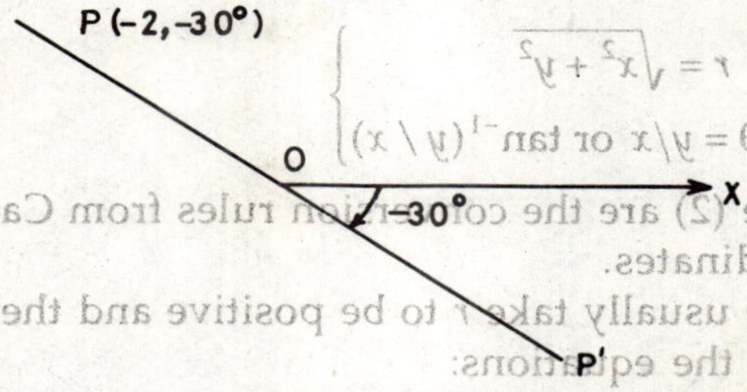


(v) It is easy to see that the point represented by (2, 210°) is the same as (–2, 30°).

**Remark.** In contrast with Cartesian co-ordinates of a point, it may be noted that whereas each pair $(r, \theta)$ represents a unique point, the polar co-ordinates of a point are not unique, since the point $P(r, \theta)$ can also be written as $(r, \theta + 2k\pi)$ or as $(-r, \theta + \pi + 2k\pi)$, where $k$ is any integer.

### 1.5.1. To find the relations between the Cartesian and the polar co-ordinates of a point.

Let $(x, y)$ and $(r, \theta)$ be the Cartesian and the polar co-ordinates of a point $P$ respectively. Then, from $\Delta OMP$,

$$\left.\begin{aligned} x &= r\cos\theta \\ y &= r\sin\theta \end{aligned}\right\} \quad (1)$$

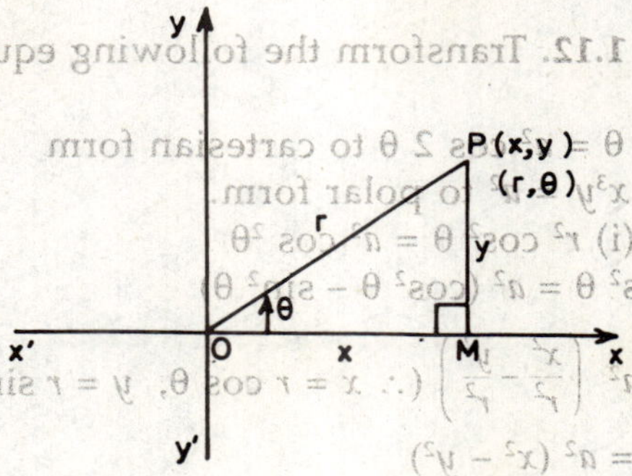

Formulae (1) are the conversion rules from polar to Cartesian co-ordinates. They imply

$$\left.\begin{aligned} r &= \sqrt{x^2 + y^2} \\ \tan\theta &= y/x \text{ or } \tan^{-1}(y/x) \end{aligned}\right\} \quad (2)$$

Formulae (2) are the conversion rules from Cartesian to polar co-ordinates.

**Note.** We usually take $r$ to be positive and then $\theta$ is obtained from the equations:

$$\cos\theta = x/r \text{ and } \sin\theta = y/r.$$

**Example 1.11.** Find polar co-ordinates of the point whose Cartesian co-ordinates are $(-1, -1)$.

*Solution.* Here $x = -1, y = -1$.

Let $(r, \theta)$ be polar co-ordinates of the point $(-1, -1)$. Then

$$r = \sqrt{x^2 + y^2} = \sqrt{2}$$

Now $\quad \cos\theta = x/r$ and $\sin\theta = y/r$ imply

$$\cos\theta = -1/\sqrt{2} = \sin\theta,$$

whence $\quad \theta = \pi + \pi/4 = 5\pi/4$

$\therefore$ One set of polar co-ordinates of the point $(-1, -1)$ is $(\sqrt{2}, 5\pi/4)$.

**Remark.** If we find $\theta$, from the equation $\tan\theta = y/x = 1$; then $\theta = \pi/4$ and then $(x,y) = (r\cos\theta, r\sin\theta) = (1, 1) \neq (-1. -1)$.

**Example 1.12.** Transform the following equations as indicated:

(i) $r^2 \cos^2\theta = a^2 \cos 2\theta$ to cartesian form

(ii) $xy^3 + x^3y = a^2$ to polar form.

*Solution.* (i) $r^2 \cos^2\theta = a^2 \cos {}^2\theta$

$\Rightarrow \quad r^2 \cos^2\theta = a^2 (\cos^2\theta - \sin^2\theta)$

$\Rightarrow \quad x^2 = a^2 \left(\frac{x^2}{r^2} - \frac{y^2}{r^2}\right)$ $(\because x = r\cos\theta,\ y = r\sin\theta)$

$\Rightarrow \quad r^2 x^2 = a^2 (x^2 - y^2)$

$\Rightarrow \quad (x^2 + y^2)\, x^2 = a^2\, (x^2 - y^2).$

(ii) By putting $x = r\cos\theta$, $y = r\sin\theta$

in $xy^3 + x^3y = a^2$, we get

$$r\cos\theta.\ r^3\sin^3\theta + r^3\cos^3\theta.\ r\sin\theta = a^2$$

$\Rightarrow \quad r^4 \sin\theta\cos\theta\,(\sin^2\theta + \cos^2\theta) = a^2$

$\Rightarrow \quad r^4 = a^2/\sin\theta\cos\theta$

$\Rightarrow \quad r^4 = 2a^2 \operatorname{cosec} 2\theta,$

which is the required polar transform.

**Example 1.13.** If the points whose polar co-ordinates are $(r_1, \theta_1)$, $(r_2, \theta_2)$ and $(r_3, \theta_3)$ are collinear, then prove that

$$\begin{vmatrix} r_1\cos\theta_1 & r_1\sin\theta_1 & 1 \\ r_2\cos\theta_2 & r_2\sin\theta_2 & 1 \\ r_3\cos\theta_3 & r_3\sin\theta_3 & 1 \end{vmatrix} = 0$$

*Solution*. Let $(r_i, \theta_i) \equiv (x_i, y_i)$ ; $i = 1, 2, 3$.

$\therefore\ x_i = r_i\cos\theta_i$ and $y_i = r_i\sin\theta_i$.

The condition of collinearity of the points $(x_i, y_i)$ ; $i = 1, 2, 3$; is

$$\begin{vmatrix} x_1 & y_1 & 1 \\ x_2 & y_2 & 1 \\ x_3 & y_3 & 1 \end{vmatrix} = 0$$

whence follows the required condition.

## Exercise 1.2

1. Plot the following points whose polar co-ordinates are, (2, 180°), (– 2, 0°), (– 3, 45°), (– 3, – 45°), ( 2, – 30°).
2. Give two more equivalent polar co-ordinates of each of the points given in Question 1.
3. Find the Cartesian co-ordinates of the following points whose polar co-ordinates are:
   (i) (2, 60°), (2, 180°), (3, $\pi/2$).

(ii) $(-2, 30°)$, $(-3, -30°)$
(*Hint:* First find equivalent polar co-ordinates in which $r$ is positive.)

4. Find the polar co-ordinates of the following points whose Cartesian co-ordinates are:
(i) $(2, 2)$, $(2, -2)$, $(-2, 2)$, $(-2, -2)$.
(ii) $(2, 0)$, $(0, 2)$.
5. Find the polar transforms of the following equations:
(i) $Ax + By + C = 0$.
(ii) $x^2 + y^2 = a^2$.
(iii) $(x^2 + y^2)^2 = a^2 (x^2 - y^2)$
6. Find the Cartesian transforms of the following equations:
(i) $r = \cos \theta$.
(ii) $r = \sin \theta$.
(iii) $r^2 = \cos 2\theta$.
(iv) $r^2 = \sin 2\theta$.
7. Taking a vertex of a regular hexagon as pole and a side (length = $a$) as the initial line, find a set of co-ordinates for each of the remaining vertices.
8. The vectorial angle of a point $P$ on the join of points $(r_1, \theta_1)$, and $(r_2, \theta_2)$ is $\frac{1}{2}(\theta_1 + \theta_2)$. Show that the length of the radius vector of $P$ is $[2r_1 r_2 \cos \frac{1}{2}(\theta_1 - \theta_2)/(r_1 + r_2)]$.

**Miscellaneous Exercise on Chapter 1.**

1. Prove that the middle point of the line joining (4, 5) and (2, 3) satisfies the equation $3x - 2y = 1$.
2. Prove that (2, 4), (2, 6) and $(2 + \sqrt{3}, 5)$ are the vertices of an equilateral triangle, whose side is 2.
3. Find the co-ordinates of the centre of the circle passing through the points $(7, -1)$, $(-2, 8)$ and $(1, 2)$. (See also Example 6.1).

4. Let the vertices of the triangle $ABC$ be $A(4, -2)$, $B(-2, 4)$ and $C(5, 5)$. Find the co-ordinates of its centroid and the incentre.
5. Find the points on the line joining $P(-5, 1)$ and $Q(3, 5)$, which form a triangle of area 2 with the points $B(1, 5)$ and $C(7, -2)$. (Note. The area = $\pm$ 2).

   $\left[ \textit{Hint: } \text{Any point on the line } PQ \text{ is } \left( \frac{-5k+3}{k+1}, \frac{k+5}{k+1} \right) \right]$
6. If $G$ is the centroid of a triangle $ABC$ and $O$ any other point, prove that
   $OA^2 + OB^2 + OC^2 = GA^2 + GB^2 + GC^2 + 3GO^2$
   (*Hint:* Let the co-ordinates of $A$, $B$, $C$ and $O$ be $(x_i, y_i)$, $i = 1, 2, 3$ and $(a, b)$ respectively and take $G$ as the origin). See also example 1.9.
7. Show that the line joining the mid-points of any two sides of a triangle is half the third side.
8. Let $ABCD$ be a rectangle and $P$ be any point in its plane. Show that
   $$PA^2 + PC^2 = PB^2 + PD^2.$$
9. Let $ABCD$ be a parallelogram, whose diagonals meet at $O$. If $P$ be any point in its plane, prove that
   $$PA^2 + PB^2 + PC^2 + PD^2 = AB^2 + BC^2 + 4PO^2.$$
10. Prove that the lines joining the mid-points of the opposite sides of a quadrilateral and the line joining the mid-points of diagonals are concurrent.
    (*Hint:* Show that the mid-points of all these lines are same).
11. Find the co-ordinates of the circumcentre of the triangle whose vertices are (5, 5), (4, 2) and (3, 4).
12. Find the area of the triangle formed by the points $(at_i^2, 2at_i)$, $i = 1, 2, 3$ and hence show that the points can never be collinear.
13. If $\Delta_1$, $\Delta_2$ denote the areas of the triangles whose vertices are
    $(a, b)$, $(b, c)$, $(c, a)$ and $(bc - a^2, ca - b^2)$, $(ca - b^2, ab - c^2)$, $(ab - c^2, bc - a^2)$ respectively then show that
    $\Delta_2 = (a + b + c)^2 \Delta_1$.

14. Show that if the points

$$\left(\frac{x_i^3}{x_i - 1}, \frac{x_i^2 - 3}{x_i - 1}\right), \text{ where } i = 1, 2, 3$$

are collinear, then

$$3(x_1 + x_2 + x_3) - (x_1x_2 + x_2x_3 + x_3x_1) + x_1x_2x_3 = 0,$$

given that no two $x$ s are equal.

# 2. EQUATION OF LOCUS

## INTRODUCTION

In the previous chapter, it was stated that Co-ordinate Geometry is that branch of Mathematics, in which we study geometry by employing algebraic techniques. This is done by introducing the concepts of locus of a point and its equation. We shall define these concepts, in this chapter and explain how the equation of the locus of a point can be arrived at.

## 2.1 LOCUS OF A POINT AND ITS EQUATION

*Locus of a point*, in the plane, moving under a given set of geometric conditions, is the set of all those points in it, which satisfy these conditions. Thus, if $T$ is the locus of a point, moving under a set of conditions $H$, then

a point $P \in T$ iff $P$ satisfies the conditions $H$.

We can also say that the locus of a point is the path traced by a moving point under the given set of conditions.

If $(x, y)$ be the co-ordinates of a point, moving under a given set of geometric conditions, then a relation, between $xy$, $y$ and the given entities satisfying these conditions, is called *the equation of its locus*. Clearly every point on the locus will satisfy this equation and every point satisfying this equation, will lie on the locus.

**Example 2.1** Find the equation of the locus of a point which moves in such a way that its distance from the point (1, 2) is 3.

*Solution*. Let $(\alpha, \beta)$ be one of the positions of the moving point. Then, by hypothesis

$$\sqrt{(\alpha-1)^2+(\beta-2)^2}=3$$
$$\Rightarrow \quad \alpha^2+\beta^2-2\alpha-4\beta-4=0$$

Thus the point $(\alpha, \beta)$ satisfies the equation

$$x^2+y^2-2x-4y-4=0.$$

Clearly any point on this equation satisfies the given condition. This is, therefore, the required equation of the locus. The locus represented by this equation is a circle.

**Note.** In the sequel, we shall use "find the locus of a point" and "find the equation of the locus of a point" as synonyms. In fact, we restrict ourselves at the moment to finding the equation of the required locus, and what this equation represents, in the plane, will be taken up in the subsequent chapters.

**Example 2.2.** Explain the relationship between the locus and its equation obtained in Example 2.1.

*Solution*. Let the set $T$ be the locus of the point which moves in such a way that its distance from the point (1, 2) is 3.

Then the following are equivalent:

(i) a point $P(x, y) \in T$.

(ii) $P(x, y)$ is at a distance 3 from the point (1, 2).

(iii) $x, y$ satisfy the equation $x^2+y^2-2x-4y-4=0$

For example, $x = 1$, $y = -1$ satisfy (iii).

$\Leftrightarrow$ the point $(1, -1)$ is at a distance 3 from the point (1, 2).

$\Leftrightarrow$ Point $(1, -1) \in T$.

Again $x = 5$, $y = 1$ does not satisfy (iii)

$\Leftrightarrow$ the distance between the points (5, 1) and (1, 2) is not 3.

$\Leftrightarrow$ Point $(5, 1) \notin T$.

We exhibit this set in the following figure:

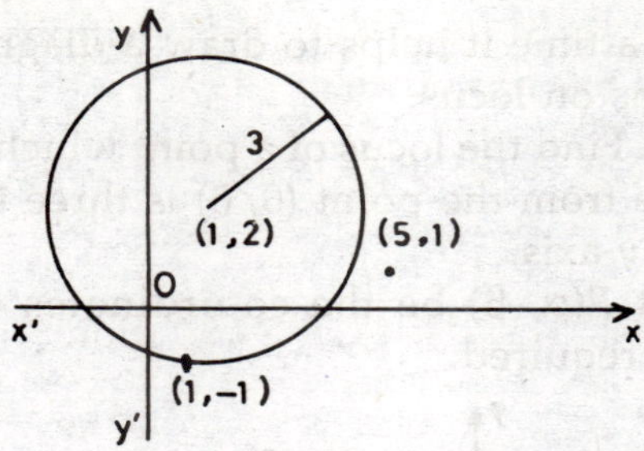


Thus we arrive at the following two concepts:

(1) A set $T$, usually, a geometrical figure (known as the locus of a point).

(2) An algebraic equation in $x$ and $y$ (known as the equation of the locus).

Then the relationship between (1) and (2) is that

a point $(x, y) \in T$ iff $x$, $y$ satisfy the algebraic equation (2).

## 2.2 METHOD TO FIND THE EQUATION OF THE LOCUS OF A POINT

We now solve some examples to illustrate how the equation of the locus of a point can be found. Generally, we proceed as under:

**Step I.** Assume $(\alpha, \beta)$ [or any other symbol say $(x', y')$ or $(x_1, y_1)$ or $(h, k)$] as the co ordinates of a moving point on the locus.

**Step II.** Assume the necessary quantities, if needed.

**Step III.** Obtain a relation/relations between $\alpha$, $\beta$; the given quantities and the assumed quantities (if any) in Step II, by applying the given set of geometric conditions in the question.

**Step IV.** Eliminate all the assumed quantities, to get a relation in $\alpha$, $\beta$ and the given quantities only.

**Step V.** Replace $\alpha$, $\beta$ by $x$, $y$ respectively to get the required equation of the locus.

**Note:** Many a time it helps to draw a diagram for solving the problems on locus.

**Example 2.3.** Find the locus of a point which moves such that its distance from the point (0, 0) is three times its distance from the $y$-axis,

*Solution*. Let $P(\alpha, \beta)$ be the co-ordinates of the point whose locus is required.

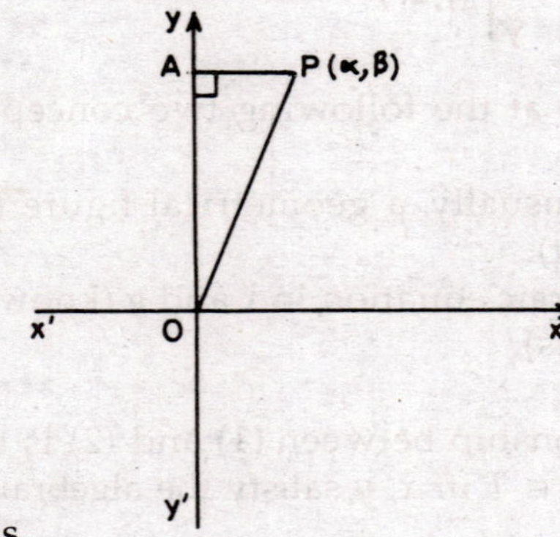


By hypothesis,

$$|OP| = 3.\ |AP|$$

$$\Rightarrow \quad \sqrt{(\alpha^2 + \beta^2} = 3\ |\ \alpha\ |$$

$$\Rightarrow \quad \alpha^2 + \beta^2 = 9\ \alpha^2$$

$$\Rightarrow \quad 8\alpha^2 - \beta^2 = 0.$$

Changing $(\alpha, \beta)$ to $(x, y)$, we get

$$8x^2 - y^2 = 0$$

Any point on this equation, clearly satisfies the hypothesis. Thus this is the desired locus of the point in question.

**Example 2.4.** Show that the equation of the locus of a point, such that the sum of its distances from two fixed points $(ae, 0)$ and $(-ae, 0)$ is $2a$, is given by

$$\frac{x^2}{a^2} + \frac{y^2}{a^2(1-e^2)} = 1, \text{ (given: } e \neq 1).$$

*Solution*. Let $P(\alpha, \beta)$ be the point whose locus is required.

By hypothesis,

$$\sqrt{(\alpha - ae)^2 + \beta^2} + \sqrt{(\alpha + ae)^2 + \beta^2} = 2a \qquad (1)$$

Also

$$[(\alpha - ae)^2 + \beta^2] - [(\alpha + ae)^2 + \beta^2] = -4aex \qquad (2)$$

Dividing (2) by (1), we get

$$\sqrt{(\alpha - ae)^2 + \beta^2} - \sqrt{(\alpha + ae)^2 + \beta^2} = -2e\alpha \qquad (3)$$

Adding (1) and (3), we have,

$$2\sqrt{(\alpha - ae)^2 + \beta^2} = 2(a - e\alpha)$$

$$\Rightarrow \quad (\alpha - ae)^2 + \beta^2 = (a - e\alpha)^2$$

$$\Rightarrow \quad \alpha^2 (1 - e^2) + \beta^2 = a^2(1 - e^2)$$

$$\Rightarrow \quad \frac{\alpha^2}{a^2} + \frac{\beta^2}{a^2(1 - e^2)} = 1,$$

whence the locus of $(\alpha, \beta)$ is

$$\frac{x^2}{a^2} + \frac{y^2}{a^2(1 - e^2)} = 1$$

**Example 2.5.** A stick of length $l$ rests against the floor and a wall of a room. If the stick begins to slide on the floor, find the locus of its middle point.

*Solution*. Let a cross-section of the floor and the wall be taken as the co-ordinate axes and let $AB$ be one of the positions of the stick.

Let $A$ be $(a, 0)$ and $B$ be $(0, b)$. Let $P(\alpha, \beta)$ be the middle point of the stick $AB$, whose locus is to be found.

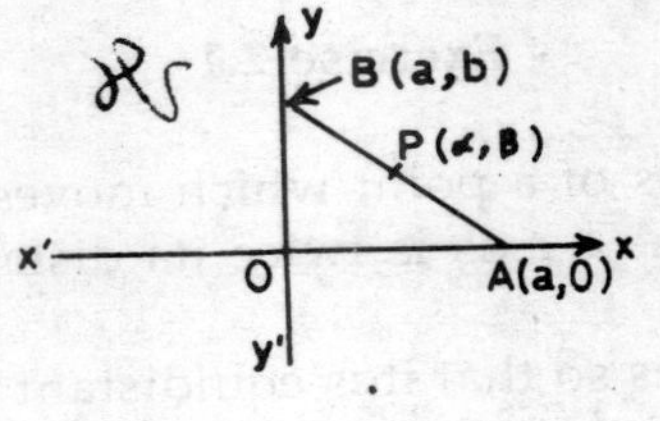

Also, the middle point $AB$ is $\left(\frac{a}{2}, \frac{b}{2}\right)$

$$\therefore \qquad \alpha = \frac{a}{2}, \ \beta = \frac{b}{2} \qquad (1)$$

Now $$AB = l$$

$$\Rightarrow \qquad \sqrt{(a-0)^2 + (0-b)^2} = l$$

$$\Rightarrow \qquad a^2 + b^2 = l^2 \qquad (2)$$

Eliminating $a$ and $b$ from (1) and (2), we get

$$4\alpha^2 + 4\beta^2 = l^2.$$

Thus the locus of $(\alpha, \beta)$ is given by

$$4(x^2 + y^2) = l^2.$$

**Example 2.6.** Find the equation to the path of a point which moves so that it remains equidistant from two fixed points.

*Solution.* Let $A$ and $B$ be the two fixed points. Join $AB$ and take $AB$ as the $x$-axis. Let the mid. *point*, $O$, of $AB$ be taken as the origin and a line through $O$, perpendicular to $AB$ as the $y$-axis. If $AB = 2a$, then $A$ is $(-a, 0)$ and $B$ is $(a, 0)$. Let $P$ $(x,y)$ be the point whose locus is required.

By hypothesis,

$$PA = PB \quad \Rightarrow PA^2 = PB^2$$

$$\Rightarrow \quad (x+a)^2 + (y-o)^2 = (x-a)^2 + (y-o)^2$$

$$\Rightarrow \quad 4xa = 0 \quad \Rightarrow x = 0 \quad (\text{as } a \neq 0),$$

which is the required locus (It may be noted that this is nothing but $y$-axis).

## Exercise 2.1

1. Find the locus of a point which moves so that its distance from the $x$-axis is twice its distance from the $y$-axis.
2. A point moves so that it is equidistant from the points

(3, 4) and (5, – 2). Show that the equation of its locus is $x - 3y = 1$.

3. A point moves so that its distance from (3, 0) is twice its distance from (– 3, 0). Find the equation of its locus.

4. Two fixed points $A$ and $B$ have co-ordinates (5, 3) and (3, – 2) respectively. Find the equation of the locus of a point, $P$:
   (i) such that area of $\Delta\ PAB$ is 5 units.
   (ii) such that the points $P$, $A$, $B$ are collinear.
   (*Hint:* Let the area be $5k$ where $k = 1$ or $-1$ for part (i) and $k = 0$ for part (ii).

5. Find the equation to the locus of a point whose distance from the origin exceeds its distance from the $y$-axis by 3.

6. $A$ and $B$ are two fixed points. Find the locus of a point $P$ such that $APB$ is a right angled triangle.

7. (a) Show that the point with co-ordinates (2, 2) lies on the locus whose equation is

$$2x + 3y = 10.$$

(b)(i) Show that $(a \cos \theta, a \sin \theta)$ lies on the locus whose equation is

$$x^2 + y^2 = a^2, \text{ for all values of } \theta.$$

(ii) Do the same for

$$(a \cos \theta, b \sin \theta) \text{ and } \frac{x^2}{a^2} + \frac{y^2}{b^2} = 1$$

(iii) Do the same for

$$(a \sec \theta, b \tan \theta), \text{ and } \frac{x^2}{a^2} - \frac{y^2}{b^2} = 1$$

(iv) Do the same for

$$\left(ct, \frac{c}{t}\right) \text{ and } xy = c^2.$$

for all values of $t \neq 0$.

(v) Do the same for

$(at^2, 2at)$ and $y^2 = 4ax$ for all $t$.

8. $P$ and $Q$ are two variable points on the axes of $x$ and $y$

respectively, such that $|OP| + |OQ| = a$. Find the equation of the locus of the foot of the perpendicular from the origin on $PQ$.

9. A point moves so that the difference of its distances from the point $(a, 0)$ and $(-a, 0)$ is $2c$. Find the locus of this point.

10. A rod of length $(a + b)$ slides between two perpendicular straight lines. Find the locus of a point which divides the extremities in the ratio $a : b$.

## 2.3 INTERSECTION OF LOCI

Let $S_1 = 0, S_2 = 0, \ldots, S_n = 0$ be $n$ given equations and let $T_1, T_2, \ldots, T_n$ be the loci represented by them respectively. We wish to find the locus of a point whose co-ordinates satisfy all the equations:

$$S_1 = 0, S_2 = 0, \ldots, S_n = 0$$

Now, $T$ consists precisely of those points whose co-ordinates satisfy

$$S_i = 0\ (i = 1, 2, \ldots, n).$$

Thus $x$ and $y$ satisfy all the equations

$$S_1 = 0, S_2 = 0, \ldots, S_n = 0$$

$\Leftrightarrow$ The point $(x, y) \in T_i$ for all $i = 1, 2, \ldots, n$

$\Leftrightarrow$ The point $(x, y) \in T_1 \cap T_2 \cap \ldots \cap T_n$.

Hence, the required locus is $T_1 \cap T_2 \cap \ldots \cap T_n$.

**Example 2.7.** Find the intersection of the loci whose equations are $x + 2y - 5 = 0$ and $3x - 2y + 1 = 0$.

*Solution.* Solving the two equations simultaneously, we see that these equations have the only solution, $x = 1, y = 2$.

$$\therefore \quad T_1 \cap T_2 = \{(1, 2)\}.$$

**Example 2.8.** Find the points of intersection of the loci given by the equations $x^2 + y^2 = 25$ and (i) $x + y = 7$, (ii) $4x + 3y = 25$, (iii) $x = 7$.

*Solution.* (i) We have two equations $x^2 + y^2 = 25$ and $x + y = 7$. Let $T_1$ and $T_2$ respectively denote their loci. Then $x = 7 - y$.

$$\Rightarrow \quad (7 - y)^2 + y^2 = 25$$

$$\Rightarrow \quad y^2 - 7y + 12 = 0$$

$$\Rightarrow \quad y = 3 \text{ or } y = 4.$$

Then $x = 7 - y \quad \Rightarrow \quad x = 4$ or $x = 3$.

Hence, the points of intersection are (4, 3) and (3, 4).

$$\therefore \quad T_1 \cap T_2 = \{(4, 3), (3, 4)\}.$$

(ii) If we now solve $x^2 + y^2 = 25$ and $4x + 3y = 25$, the quadratic in $y$, is $y^2 - 6y + 9 = 0$.

$$\therefore \quad y = 3, 3 \text{ and so } x = 4, 4.$$

Hence, the two loci meet in a pair of coincident points (4, 3) and (4, 3),

whence $\quad T_1 \cap T_2 = \{(4, 3)\}$.

**Note.** In this case, we say that the two loci touch each other and the point (4 , 3), then, is called their *point of contact*.

(iii) $x^2 + y^2 = 25$, $x = 7$. The quadratic in y, in this case is $y^2 = -24$, which has no real roots.

Thus, the loci do not have any point in common.

$$\therefore \quad T_1 \cap T_2 = \phi \text{ (empty set)}.$$

**Note.** In Co-ordinate geometry, we generally say that the two loci meet in a pair of imaginary points, whose co-ordinates are $(7, \pm i2\sqrt{6})$.

**Exercise 2.2.**

1. Find the set of points of intersection of the following loci:

   (i) $2x + y = 5$
   $x + 2y = 4$.

   (ii) $x - y = 0$
   $y + 2x = 0$

$3x - 2y = 0.$

(iii) $x + 2y = 5$

$2x + 4y = 10.$

(iv) $x + 2y = 5$

$2x + 4y = 9.$

(v) $x^2 + 2y^2 = 6$

$x - 2y = 0.$

2. Find the condition that the locus represented by $y = mx + c$ may have real intersection with the curves:

(i) $x^2 + y^2 = a^2$,

(ii) $\frac{x^2}{a^2} + \frac{y^2}{b^2} = 1$

(iii) $\frac{x^2}{a^2} - \frac{y^2}{b^2} = 1$

(iv) $y^2 = 4ax$.

Also find the condition, in each case, if the loci

(a) touch each other,

(b) have imaginary points of intersection.

## 2.4. THE LOCUS REPRESENTED BY THE EQUATION, $S_1 . S_2 = 0$

Let $T_1$ and $T_2$ be the loci represented by the equations $S_1 = 0$ and $S_1 = 0$, and $T$ be the locus represented by $S_1 . S_2 = 0$.

The co-ordinates $(x, y)$ of a point $P$, satisfy $S_1 . S_2 = 0$ iff they satisfy either $S_1 = 0$ or $S_2 = 0$.

whence $(x, y) \in T$ iff $(x, y) \in T_1$ or $(x, y) \in T_2$.

$$\Rightarrow \quad T = T_1 \cup T_2.$$

## 2.5. THE LOCUS REPRESENTED BY THE EQUATION, $S_1 + k S_2 = 0$

Let $T_1$ and $T_2$ be the loci represented by the equations, $S_1 = 0$ and $S_2 = 0$ respectively, and $k$ any real number.

Let the set $T$ be the locus represented by the equation,

$$S_1 + kS_2 = 0$$

Now $(x, y) \in T_1$, and $(x, y) \in T_2$

$\Rightarrow$ $(x, y)$ satisfies the equations $S_1 = 0$ and $S_2 = 0$

$\Rightarrow$ $(x, y)$ satisfies the equation $S_1 + kS_2 = 0$

$\Rightarrow$ $(x, y) \in T$

Thus $T_1 \cap T_2 \cap T$

i.e. the locus represented by the equation $S_1 + kS_2 = 0$ contains all the points which satisfy both $S_1 = 0$ and $S_2 = 0$.

**Remark.** For the illustration and the limitations of this article, see Example 4.11 and the following remark.

**Example 2.9.** What locus is represented by the equation $xy = 0$?

*Solution.* Let $S_1 = 0$ and $S_2 = 0$ be $x = 0$ and $y = 0$ and, $T_1$ and $T_1$ be the loci represented by them respectively.

Further let $T$ be the locus represented by $xy = 0$.

Then $T_1 = \{(x, y) \mid x = 0\}$ = the *axis* of $y$.

$T_2 = \{(x, y) \mid y = 0\}$ = the *axis* of $x$.

As $xy = 0 \Rightarrow x = 0$ or $y = 0$, we get

$$T = T_1 \cup T_2 = \text{the co-ordinate axes}$$

## Exercise 2.3.

1. What loci are represented by
   (i) $(x + y) + k(x - y) = 0$, $k$ any real number.
   (ii) $(2x + 3y - 1)(x + 2y + 5) = 0$.
   (iii) $x^2 + y^2 = 5$, $y = 2$.
2. Let $T_1$ be the locus of a point satisfying a condition $A$, $T_2$ be the locus of a point satisfying conditions $A$ and $B$, and $T_3$ be the locus of a point satisfying conditions $A$ or $B$. What is the relation between $T_1$, $T_2$ and $T_3$?

## Miscellaneous Exercise on Chapter 2.

1. Show that the loci represented by $x \cos \theta + y \sin \theta - a$ and $x^2 + y^2 = a^2$ interest in two coincident points for all values of $\theta$.

2. Show that the loci represented by $\frac{x\cos\phi}{a}+\frac{y\sin\phi}{b}=1$ and $\frac{x^2}{a^2}+\frac{y^2}{b^2}=1$ intersect in two coincident points for all values of θ.
3. Find the condition that the loci represented by $lx + my + n = 0$ and $\frac{x^2}{a^2}-\frac{y^2}{b^2}=1$ may intersect in two coincident points.
4. Let $P(x, y)$ be a variable point and $S(a, 0)$ be a fixed point. Suppose $P$ moves so that $SP = |x + a|$. Prove that the locus of $P$ is given by $y^2 = 4ax$.
5. Find in the simplest form, the equation to the locus of a point which moves such that the sum of its distances from the points (1, 0) and (–1, 0) is 3.
6. A line of length $2a$, has its extremities on two fixed perpendicular lines. Find the locus of the foot of the perpendicular from the intersection of the two fixed lines to the given line.
7. $A$ and $B$ are two fixed points $(a, 0)$ and $(-a, 0)$ respectively. Find the equation of the locus of $P$, when
   (i) $PA^2 - PB^2 = 4a^2$
   (ii) $PB^2 + PC^2 = 2PA^2$, where $C$ is the point $(c, 0)$
8. If the join of the points (3, 2) and (2, 3) subtend a right angle at the point $P$, find the locus of $P$.
9. Find the equation to the locus of a point $P$, which moves so that its distance from a fixed point $(\alpha, \beta)$ remains always constant $(= k)$.
10. The base of a triangle is fixed and its vertex moves along a fixed line. Prove that the locus of its centroid is a line parallel to the fixed line.
    (*Hint:* Take the base of the triangle as *x-axis* and its midpoint as the origin. Let its vertex be (a, b), moving on a line given by, $y = mx + c$).

# 3. CHANGE OF AXES

## INTRODUCTION

It is found convenient, sometimes, while solving certain problems, to shift the origin to another point, whereas the axes remain parallel to the original axes or to rotate the axes through a certain angle without changing the origin or a combination of both these processes. These processes, in Co-ordinate Geometry, are known as *transformations or change of axes*. By a mere shifting of the origin to a suitable point or a rotation of axes through a suitable angle, in a certain situation, simplifies it to a great extent.

### 3.1 SHIFTING OF THE ORIGIN

Let $x'Ox$ and $y'Oy$ be a system of co-ordinate axes at $O$ and let $P(h, k)$ be a given point. Draw $X'PX$ and $Y'PY$ parallel to the axes at $O$, as shown in the figure:

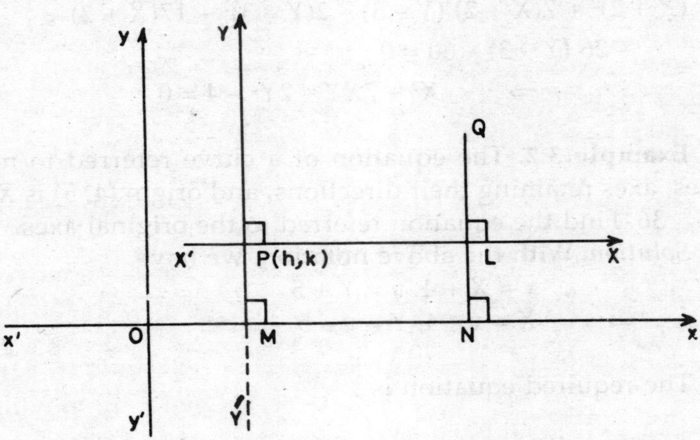


Let $(x, y)$ and $(X, Y)$ be the co-ordinates of a point $Q$ referred to the axes at $O$ and $P$ respectively ($P$ is called the new origin). Draw $PM$ and $QN$, perpendiculars on $x'Ox$, where $QN$ meets $XP'X$ at $L$.

Then $$x = ON = OM + MN = OM + PL = h + X$$
$$y = NQ = NL + LQ = MP + LQ = k + Y$$

$$\Rightarrow \quad x = X + h \text{ and } y = Y + k.$$

These are the required relations for changing the original system of co-ordinate axes at $O$ to the new system of co-ordinate axes at $P$ and vice-versa.

This transformation is known as a *Translation*.

Also $X = x - h$ and $Y = y - k$ is called the *inverse translation*.

**Example 3.1.** Find the equation to which the equation

$$x^2 + 7xy - 2y^2 + 17x - 26y - 60 = 0$$

is transformed if the origin is shifted to the point $(2, -3)$, the axes remaining parallel to the original axes.

*Solution.* Here the new origin is $(2, -3)$.

Then, $x = X + 2$, $y = Y - 3$.

and the given equation transforms to

$$(X + 2)^2 + 7(X + 2)(Y - 3) - 2(Y - 3)^2 + 17(X + 2) - 26(Y - 3) - 60 = 0$$

$$\Rightarrow \quad X^2 + 7XY - 2Y^2 - 4 = 0$$

**Example 3.2.** The equation of a curve referred to new axes, axes retaining their directions, and origin $(4, 5)$ is $X^2 + Y^2 = 36$. Find the equation referred to the original axes.

*Solution.* With the above notation, we have

$$x = X + 4, \; y = Y + 5$$
$$\Rightarrow \quad X = x - 4, \; Y = y - 5$$

$\therefore$ The required equation is

$$(x - 4)^2 + (y - 5)^2 = 36$$
$$\Rightarrow \quad x^2 + y^2 - 8x - 10y + 5 = 0.$$

## 3.2. ROTATION OF AXES WITHOUT CHANGING THE ORIGIN

Let the axes $x'Ox$ and $y'Oy$ be rotated through an angle $\theta$ to obtain the new axes $X'OX$ and $Y'OY$; (origin in this case is not being changed). Let $P$ be any point whose co-ordinates referred to the original and the rotated axes be $(x, y)$ and $(X, Y)$ respectively.

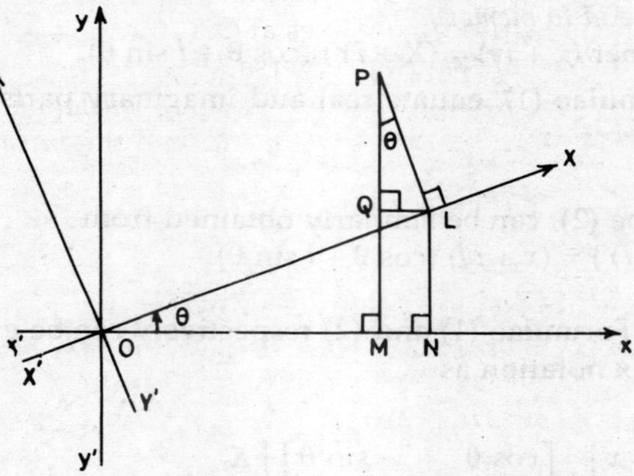


Draw $PM \perp x'Ox$ and $PL \perp X'OX$. Also draw $LQ \perp MP$ and $LN \perp x'Ox$. Then

$x = OM, y = MP, X = OL, Y = LP$ and $\angle MPL = \theta$.

$\Rightarrow x = OM = ON - MN = ON - QL = QL \cos\theta - LP \sin\theta$

$= X \cos\theta - Y \sin\theta$.

and $y = MP = MQ + QP = NL + QP = OL \sin\theta + LP \cos\theta$

$= X \sin\theta + Y \cos\theta$.

Thus the required transformation formulae are

$$\left.\begin{aligned} x &= X\cos\theta - Y\sin\theta, \\ y &= X\sin\theta + Y\cos\theta \end{aligned}\right\} \quad (1)$$

This transformation is known as a *Rotation*.

**Note 1.** Formulae (1) can also be written as

$$\left.\begin{aligned} X &= x\cos\theta + y \text{ in } \theta \\ Y &= -x\sin\theta + y\cos\theta \end{aligned}\right\} \quad (2)$$

One can obtain them directly also, from the above figure.

**Note 2.** *Aid to memory:*
Remember $(x + iy) = (X + iY)(\cos\theta + i\sin\theta)$.

For formulae (1), equate real and imaginary parts on both sides.

Formulae (2), can be similarly obtained from
$(X + iY) = (x + iy)(\cos\theta - i\sin\theta)$.

**Note 3.** Formulae (1) and (2) respectively can be expressed in matrix notation as

$$\begin{bmatrix} x \\ y \end{bmatrix} = \begin{bmatrix} \cos\theta & -\sin\theta \\ \sin\theta & \cos\theta \end{bmatrix} \begin{bmatrix} X \\ Y \end{bmatrix} \quad (1)'$$

$$\begin{bmatrix} X \\ Y \end{bmatrix} = \begin{bmatrix} \cos\theta & \sin\theta \\ -\sin\theta & \cos\theta \end{bmatrix} \begin{bmatrix} x \\ y \end{bmatrix} \quad (2)'$$

**Example 3.3.** The equation of a curve referred to a given system of axes is $3x^2 + 2xy + 3y^2 = 10$. Find its equation if the axes are rotated through an angle 45°, the origin remaining unchanged.

*Solution.* With the above notation, we have

$$x = X\cos 45^\circ - Y\sin 45^\circ = \frac{X - Y}{\sqrt{2}}$$

and $$y = X\sin 45^\circ + Y\cos 45^\circ = \frac{X + Y}{\sqrt{2}}$$

Thus, $3x^2 + 2xy + 3y^2 = 10$ transforms to

$$3\left(\frac{X-Y}{\sqrt{2}}\right)^2 + 2\left(\frac{X-Y}{\sqrt{2}}\right)\left(\frac{X+Y}{\sqrt{2}}\right) + 3\left(\frac{X+Y}{\sqrt{2}}\right)^2 = 10$$

$$\Rightarrow \quad 2X^2 + Y^2 = 5.$$

## 3.3. ROTATION OF AXES ALONG WITH CHANGE OF ORIGIN

Let the origin be shifted to a point $P(h, k)$ and also the axes be rotated through an angle $\theta$.

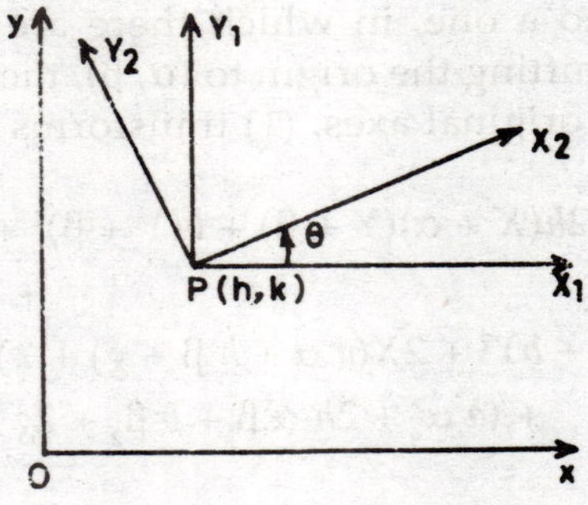


Then, we have, as in the figure

$$x = X_1 + h.$$
$$y = Y_1 + k.$$

Also
$$X_1 = X_2 \cos\theta - Y_2 \sin\theta,$$
$$Y_1 = X_2 \sin\theta + Y_2 \cos\theta,$$

where $(x, y)$, $(X_1, Y_1)$ and $(X_2, Y_2)$ are the co-ordinates of an arbitrary point respectively w.r.t. the axes $ox$, $oy$; $PX_1$, $PY_1$ and $PX_2$, $PY_2$ and hence

$$\left.\begin{aligned} x &= X_2 \cos\theta - Y_2 \sin\theta + h \\ y &= X_2 \sin\theta + Y_2 \cos\theta + k \end{aligned}\right\} \quad (1)$$

are the required formulae connecting the co-ordinates of a point w.r.t. the co-ordinate axes at $O$ and $P$ respectively.

## 3.4. REMOVAL OF FIRST DEGREE TERMS

It is desired to shift the origin to a new point, axes remaining parallel, so that the equation

$$ax^2 + 2hxy + by^2 + 2gx + 2fy + c = 0 \quad (1)$$

transforms into a one, in which there are no first degree terms. Thus, shifting the origin to $(\alpha, \beta)$, the axes remaining parallel to the original axes, (1) transforms to

$$a(X + \alpha)^2 + 2h(X + \alpha)(Y + \beta) + b(Y + \beta)^2 + 2g(X + \alpha) + 2f(Y + \beta) + c = 0$$

$$\Rightarrow aX^2 + 2hXY + bY^2 + 2X(a\,\alpha + h\,\beta + g) + 2Y\,(h\,\alpha + b\,B + f) + (a\,\alpha^2 + 2h\,\alpha\,\beta + b\,\beta_2 + 2g\,\alpha + 2f\beta + c) = 0.$$

This equation will be free of first degree terms,

iff $a\,\alpha + h\,\beta + g = 0,\ h\,\alpha + b\,\beta + f = 0$ (2)

whence $(\alpha, \beta) = \left(\dfrac{hf - bg}{ab - h^2}, \dfrac{gh - af}{ab - h^2}\right)$

Thus we can remove the first degree terms in (1) iff $ab - h^2 \neq 0$.

Also, then, the constant term

$= a\,\alpha^2 + 2h\,\alpha\,\beta + b\,\beta^2 + 2g\,\alpha + 2f\,\beta + c$

$= \alpha\,(a\,\alpha + h\,\beta + g) + \beta\,(h\alpha + b\beta + f) + (g\,\alpha + f\,\beta + c)$

$= g\,\alpha + f\,\beta + c$ (3)

$$= \frac{g(hf - bg) + f(gh - af) + c(ab - h^2)}{ab - h^2}$$

$$= \frac{abc + 2fgh - af^2 - bg^2 - ch^2}{ab - h^2} = \frac{\Delta}{ab - h^2}$$

where $\Delta = abc + 2fgh - af^2 - bg^2 - ch^2$.

Hence, the transformed equation is

$$aX^2 + 2hXY + bY^2 + (g\alpha + f\beta + c) = 0 \tag{4}$$

$$\Rightarrow aX^2 + 2hXY + bY^2 + \frac{\Delta}{ab - h^2} = 0 \tag{5}$$

Here, $\Delta = abc + 2fgh - af^2 - bg^2 - ch^2$

$$= \begin{vmatrix} a & h & g \\ h & b & f \\ g & f & c \end{vmatrix}$$

is called the *discriminate* of the equation,

$S = ax^2 + 2hxy + by^2 + 2gx + 2fy + c = 0$.

**Note 1.** Clearly

$$\frac{\partial S}{\partial x} = 2\,(ax + hy + g) \text{ and } \frac{\partial S}{\partial y} = 2\,(hx + by + f).$$

The point $(\alpha, \beta)$ is the point of intersection of $\frac{1}{2}\frac{\partial S}{\partial x} = 0$ and $\frac{1}{2}\frac{\partial S}{\partial y} = 0$.

**Note 2.** $\alpha = G/C$, $\beta = F/C$ where $C$, $F$, $G$ are respectively the cofactors of $c$, $f$, $g$ in $\Delta$.

Also the constant in equation (4) is

$$g\,\alpha + f\,\beta + c + (gG + fF + cC)/C = \Delta/c$$

**Note 3.** If $C = 0$ i.e. if $h^2 = ab$, the above transformation fails, as the point $(\alpha, \beta)$ to which the origin is to be shifted, recedes to infinity.

**Example 3.4.** Transform $12x^2 + 7xy - 12y^2 - 17x - 31y - 7 = 0$ to rectangular axes through the point $(1, -1)$ inclined at an angle $\tan^{-1}(4/3)$ to the original axes.

*Solution*. ***Method I***. Let us first shift the origin to the point $(1, -1)$ by setting

$$x = x' + 1, \quad y = y' - 1.$$

$\therefore$ The given equation becomes

$$12(x' + 1)^2 + 7(x' + 1)(y' - 1) - 12(y' - 1)^2 - 17(x' + 1) - 31(y' - 1) - 7 = 0$$

$$\Rightarrow 12x'^2 + 7x'y' - 12y'^2 = 0 \qquad (1)$$

Now we shall rotate the axes through $\tan^{-1}(4/3)$

Let $\theta = \tan^{-1}(4/3) \Rightarrow \tan\theta = 4/3$ $(0 < \theta < \pi/2)$

$\Rightarrow \sin\theta = 4/5$; $\cos\theta = 3/5$

$$\text{set: } x' = X\cos\theta - Y\sin\theta = \frac{3}{5}X - \frac{4}{5}Y$$

$$y' = X\sin\theta + Y\cos\theta = \frac{4}{5}X + \frac{3}{5}Y$$

$\therefore$ Equation (1) transforms to

$$12\left(\frac{3}{5}X - \frac{4}{5}Y\right)^2 + 7\left(\frac{3}{5}X - \frac{4}{5}Y\right)\left(\frac{4}{5}X + \frac{3}{5}Y\right) - 12\left(\frac{4}{5}X + \frac{3}{5}Y\right)^2 = 0$$

$$\Rightarrow \quad XY = 0$$

***IInd Method***. We solve the equations

$$a\alpha + h\beta + g = 0$$

$$h\alpha + b\beta + f = 0$$

$$\text{i.e.} 12\,\alpha + \frac{7}{2}\,\beta - \frac{17}{2} = 0$$

$$\text{and} \quad \frac{7}{2}\,\alpha - 12\,\beta - \frac{31}{2} = 0$$

and find $(\alpha, \beta) = (.\ 1, -1)$

So, if we transform it to the point (1, – 1), by § 3.4 the transformed equation is

$$aX^2 + 2hXY + bY^2 + (g\alpha + f\beta + c) = 0$$

Now $g\alpha + f\beta + c = \left(\dfrac{-17}{2}\right)(1) - \dfrac{31}{2}(-1) - 7$

$\therefore$ the new equation is $12X^2 + 7XY - 12Y^2 = 0$

Now proceed as in method I.

## 3.5. REMOVAL OF THE TERM xy, FROM f(x, y) = ax² + 2hxy + by² WITHOUT CHANGING THE ORIGIN

Clearly, $h \neq 0$.

Rotating the axes through an angle θ, we have

$$x = X\cos\theta - Y\sin\theta$$

$$y = X\sin\theta + Y\cos\theta.$$

$$\therefore\ f(x, y) = ax^2 + 2hxy = by^2$$

$$= a(X\cos\theta - Y\sin\theta)^2 + 2h(X\cos - \theta\, Y\sin\theta).$$

$$(X\sin\theta + Y\cos\theta) + b(X\sin\theta + Y\cos\theta)^2$$

$$= (a\cos^2\theta + 2h\cos\theta\sin\theta + b\sin^2\theta)X^2$$

$$+ 2[(b - a)\cos\theta\sin\theta + h(\cos^2\theta - \sin^2\theta)]XY$$

$$+ (a\sin^2\theta - 2h\cos\theta\sin\theta + b\cos^2\theta)Y^2$$

$$= F(X, Y), \text{ (say)}.$$

In $F(X, Y)$, we require that the coefficient of the $XY$-term to be zero.

$$\therefore\ 2[(b - a)\cos\theta\sin\theta + h(\cos^2\theta - \sin^2\theta)] = 0.$$

$$\Rightarrow (a - b)\sin 2\theta = 2h\cos 2\theta.$$

$$\Rightarrow \tan 2\theta = \frac{2h}{a - b} \text{ or } \cot 2\theta = \frac{a - b}{2h}$$

We use the former or the later equation according as $a \neq b$ or $a = b$. These yield $\theta$, the angle through which the axes are to be rotated (the origin remaining unchanged) in order to remove the $xy$-term from $f(x, y)$.

**Example 3.5.** Given the equation $= 4x^2 + 2\sqrt{3}xy + 2y^2 = 1$. Through what angle should the axes be rotated so that the term in $xy$ be wanting from the transformed equation.

*Solution.* Comparing the given equation, with

$ax^2 + 2hxy + by^2$, we get $a = 4$, $h = \sqrt{3}$, $b = 2$.

Let $\theta$ be the angle through which the axes are to be rotated.

$$\text{Then } \tan 2\theta = \frac{2h}{a-b}$$

$$\Rightarrow \quad \tan 2\theta = \frac{2\sqrt{3}}{4-2} = \sqrt{3} = \tan\frac{\pi}{3}$$

$$\Rightarrow \quad 2\theta = \frac{\pi}{3}, \pi + \frac{\pi}{3}$$

$$\Rightarrow \quad \theta = \frac{\pi}{6}, \frac{2\pi}{3}$$

**Example 3.6.** By a suitable translation and rotation of axes, remove the terms $x$, $y$ and $xy$ from the equation
$S \equiv 14x^2 - 4xy + 11y^2 - 36x + 48y + 41 = 0$

*Solution.* We first remove the terms in $x$ and $y$ by a suitable translation.

$$\text{We solve } = \frac{\partial S}{\partial x} = 0 = \frac{\partial S}{\partial y}$$

$28x - 4y - 36 = 0$ and $-4x + 22y + 28 = 0$

$\Rightarrow 7x - y - 9 = 0$ and $2x - 11y - 24 = 0$

$\Rightarrow x = 1, y = -2$

Shift the origin to $(1, -2)$ by setting

$x = x' + 1, y = y' - 2.$

The given equation becomes

$$14x'^2 - 4x'y' + 11y'^2 - 25 = 0 \qquad (1)$$

We now remove the product term by rotating the axes through an angle $\theta$ given by

$$\tan 2\theta = 2h/(a - b) = -4/(14 - 11) = -4/3.$$

$$\Rightarrow \frac{2\tan\theta}{1-\tan^2\theta} = -\frac{4}{3}$$

$\Rightarrow 2\tan^2\theta - 3\tan\theta - 2 = 0.$

$\Rightarrow \tan\theta = 2, -1/2$

Taking $\tan\theta = 2$, we have

$$\sin\theta = 2/\sqrt{5} \text{ and } \cos\theta = 1/\sqrt{5} \ (0 < \theta < \pi/2)$$

$\therefore\ x' = X\cos\theta - Y\sin\theta = (X - 2Y)/\sqrt{5}$

and $y' = X\sin\theta + Y\cos\theta = (2X + Y)/\sqrt{5}$.

Substituting these in (1), we get

$$\frac{14}{5}(X-2Y)^2 - \frac{4}{5}(X-2Y)(2X+Y) + \frac{11}{5}(2X+Y)^2 = 25$$

$\Rightarrow 2X^2 + 3Y^2 = 5.$

**Note.** If we take $\tan\theta = -\frac{1}{2}$ and $\theta$ such that $-\frac{\pi}{2} < \theta < 0$; then $\sin\theta = -1/\sqrt{5}$ and $\cos\theta = 2/\sqrt{5}$ and (1) simplifies to $3X^2 + 2Y^2 = 5$.

**Example 3.7.** Transform the equation

$$x^2 + 4xy + y^2 - 2x + 2y + 4 = 0$$

into the form

$$\frac{Y^2}{b^2} - \frac{X^2}{a^2} = 1$$

*Solution.* To remove the first degree terms, we shift the origin to the point $(\alpha, \beta)$. Then by **§3.4,**

$a\,\alpha + h\,\beta + g = 0,\ h\,\alpha + b\,\beta + f = 0$

$\Rightarrow \alpha + 2\,\beta - 1 = 0,\ 2\alpha + \beta + 1 = 0$

whence $(\alpha, \beta) = (-1, 1)$.

Then, the transformed equation is

$aX^2 + 2hXY + bY^2 + (g\,\alpha + f\,\beta + c) = 0$. (By **§ 3.4**)

$$\Rightarrow X^2 + 4XY + Y^2 + 6 = 0. \qquad (1)$$

Now to remove the $XY$ term, from (1), we rotate the axes through an angle $\theta$ given by,

$$\cot 2\theta = \frac{a-b}{2h} = 0 \qquad \text{(By § 3.5)}$$

$$\Rightarrow \cot 2\theta = 0 \;\Rightarrow\; 2\theta = \frac{\pi}{2} \text{ or } \frac{3\pi}{2} \;\Rightarrow \theta = \frac{\pi}{4} \text{ or } \frac{3\pi}{4}$$

Taking $\theta = \frac{\pi}{4}$ $\left(\text{we can also take } \theta = \frac{3\pi}{4}\right)$

$$\Rightarrow \cos\theta = \frac{1}{\sqrt{2}},\ \sin\theta = \frac{1}{\sqrt{2}}.$$

Now, if $(X', Y')$ be the co-ordinates of the point when the axes are rotated through $\theta = \frac{\pi}{4}$, we have

$$X = \frac{X'-Y'}{\sqrt{2}}, Y' = \frac{X'+Y'}{\sqrt{2}}$$

thus, equation (1) becomes

$$\left(\frac{X'-Y'}{\sqrt{2}}\right)^2 + 4\left(\frac{X'-Y'}{\sqrt{2}}\right)\left(\frac{X'+Y'}{\sqrt{2}}\right) + \left(\frac{X'+Y'}{\sqrt{2}}\right)^2 + 6 = 0$$

$$\frac{Y'^2}{6} - \frac{X'^2}{2} = 1 \Rightarrow \frac{Y'^2}{b^2} - \frac{X'^2}{a^2} = 1, \text{ where } a^2 = 2 \text{ and } b^2 = 6$$

**Note.** Usually, we use the formula, $\tan 2\theta = \frac{2h}{a-b}$ for finding the angle of rotation, θ. However if $a = b$, we use $\cot 2\theta = \frac{a-b}{2h}$ as in this case $\tan 2\theta$ is not defined.

## 3.6. INVARIANTS

We rotate the axes by an angle θ, without changing the origin, and suppose

$$ax^2 + 2hxy + by^2$$

changes to $\quad AX^2 + 2HXY + BY^2$.

Then by **§ 3.5,**

$$A = a\cos^2\theta + 2h\cos\theta\sin\theta + b\sin^2\theta,$$

$$B = a\sin^2\theta - 2h\cos\theta\sin\theta + b\cos^2\theta,$$

and $\quad H = h(\cos^2\theta - \sin^2\theta) + (b-a)\cos\theta\sin\theta$.

Now, clearly

(i) $\quad A + B = a + b$

and (ii) it is straight forward to verify that

$$AB - H^2 = ab - h^2.$$

Hence, we see that the quantities $a + b$ and $ab - h^2$ remain unchanged by any rotation of the axes without changing the origin. Such quantities $a + b$ and $ab - h^2$ are called *invariants* of the quandratic form $ax^2 + 2hxy + by^2$.

**Example 3.8.** Prove that the transformation of rectangular axes, which converts $\frac{X^2}{p} + \frac{Y^2}{q}$ into $ax^2 + 2hxy + by^2$

will convert $\frac{X^2}{p-\lambda}+\frac{Y^2}{q-\lambda}$ into

$$\frac{(ax^2+2hxy+by^2)-\lambda(ab-h^2)(x^2+y^2)}{1-\lambda(a+b)+\lambda^2(ab-h^2)}.$$

*Solution*. Since both the systems are rectangular, by invariants,

$$\frac{1}{p}+\frac{1}{q}=a+b \text{ and } \frac{1}{p}\cdot\frac{1}{q}-0=ab-h^2$$

If $O$ be the common origin and any point $P$ be $(X, Y)$ (respectively, $(x, y)$); then

$$OP^2 = X^2 + Y^2 = x^2 + y^2.$$

Thus $$\frac{X^2}{p-\lambda}+\frac{Y^2}{q-\lambda}=\frac{(q-\lambda)X^2+(p-\lambda)Y^2}{(p-\lambda)(q-\lambda)}$$

$$=\frac{(qX^2+pY^2)-\lambda(X^2+Y^2)}{pq-\lambda(p+q)+\lambda^2}$$

$$=\frac{\left(\frac{X^2}{p}+\frac{Y^2}{q}\right)-\frac{\lambda}{pq}(X^2+Y^2)}{1-\lambda\left(\frac{1}{p}+\frac{1}{q}\right)+\frac{\lambda^2}{pq}} \text{ (divide by } pq)$$

which changes to $\frac{(ax^2+2hxy+by^2)-\lambda(ab-h)^2(X^2+Y^2)}{1-\lambda(a+b)+\lambda^2(ab-h^2)}$

## Exercise 3.1

1. If the axes be parallel and through the point (1, –2), transform the equation $y^2 - 4x + 4y + 8 = 0$.
2. Prove that the equations of the lines $x + y - 1 = 0$ and $x - y - 1 = 0$ can be written as $X \pm Y = 0$ by shifting the origin to a suitable point. Find that point.
3. Translate the axes so as to change the equation
$$3y^2 - 2xy + 4y^2 + 8x - 10y + 8 = 0$$
into an equation with linear terms missing.
4. Show that the equation $x^2 + xy + y^2 = 1$ reduces to the form $\frac{x^2}{a^2} + \frac{y^2}{b^2} = 1$ when the axes are rotated to eliminate the cross-product.
5. Find the angle through which the axes may be turned so that the equation $Ax + By + C = 0$ be reduced to $Y = k$ and also determine the value of $k$.
6. What does the equation
$$2x^2 + 4xy - 5y^2 + 20x - 22y - 14 = 0$$
becomes when referred to rectangular axes through the point (–2, –3), the new axes being inclined at an angle 45°, with the old?
7. By suitable transformation and rotation of the co-ordinate axes, transform the equations
   (i) $x^2 - 10xy + y^2 + 2x - 10y - 11 = 0$ to the form $ay^2 - bx^2 = 1$.
   (ii) $8x^2 + 4xy + 5y^2 - 24x - 24y = 0$ to the form $\frac{x^2}{a^2} + \frac{y^2}{b^2} = 1$.
8. If $(x, y)$ and $(X, Y)$ be the co-ordinates of the same point referred to two sets of rectangular axes with the same origin and if $ux + vy$, where $u$ and $v$ are independent of $x$ and $y$, becomes $VX + UY$, show that $u^2 + v^2 = U^2 + V^2$.
9. Prove that $g^2 + f^2$ is an invariant of
$$\phi(x, y) = ax^2 + 2hxy + by^2 + 2gx + 2fy + c = 0$$
under a rotation of axes without changing the origin.
10. Find the point so that the equation $12x^2 - 10\,xy + 2y^2 +$

$11x - 5y + 2 = 0$ referred to parallel axes through it, may transform into the equation from which first degree terms in $x$ and $y$ are missing.

11. Transform the equation
$14x^2 - 4xy + 11y^2 - 36x + 48y + 41 = 0$
to rectangular axes through the point $(1, -2)$ inclined at the angle $\tan^{-1}(-\frac{1}{2})$ to the original axes.

12. Transform the equation
$$9x^2 + 24xy + 2y^2 - 6x + 20y + 41 = 0$$
into a one which is free of $x$, $y$ and $xy$ terms.

13. Transform the equation
$$x^2 - y^2 - 2ax + 2by + c^2 = 0$$
by shifting the origin to $(a, b)$ and rotating the axes through an angle $\pi/4$.

14. If the equation, $ax^2 + 2hxy + by^2 = c$ is changed into one in which there is no term involving $xy$, then show that the transformed equation is
$$(a + b + \lambda)\, x^2 + (a + b - \lambda)\, y^2 = 2c, \text{ where } \lambda = \sqrt{(a-b)^2 + 4h^2}\,.$$

15. If through a translation of axes, the equation
$$x^2 + 2hxy + by^2 + 2gx + 2fy + c = 0$$
transforms into $ax^2 + 2hxy + by^2 = 0$, prove that
$$\Delta = \begin{vmatrix} a & b & g \\ h & b & f \\ g & f & c \end{vmatrix} = 0$$

16. Show that if the equations $ax^2 + 2hxy + by^2 = 1$ and $Ax^2 + 2hxy^2 + By^2 = 1$, represent the same curve, the axes being rectangular, then $(a - b)^2 + 4h^2 = (A - B)^2 + 4H^2$.

17. If the transformation
$x = lX + mY, \qquad y = l'X + m'Y$
transforms $ax^2 + 2hxy + by^2$ into $AX^2 + 2HXY + BY^2$; prove that
$$\begin{vmatrix} A & H \\ H & B \end{vmatrix}^2 = \begin{vmatrix} l & l' \\ m & m' \end{vmatrix}^2 \begin{vmatrix} a & h \\ h & b \end{vmatrix}$$

18. If the expressions $ax^2 + 2hxy + by^2$ and $AX^2 + 2HXY + BY^2$ be transformed by a change from one set of rectangular axes to another with the same origin, prove that $aB - 2hH + bA$ and $aA + 2hH + b\beta$ are both invariant.
19. Transform the equation
$x^2 + 4xy + 4y^2 - 14x - 8y + 5 = 0$
to new axes of $x$ and $y$ whose equations are $x + 2y - 3 = 0$ and $2x - y + 1 = 0$, respectively.
20. Find the equations of the loci represented by $(ax + by + c)^2 = a^2 + b^2$ and $(ax + by + c)(bx - ay + d) = a^2 + b^2$, when the perpendicular lines $ax + by + c = 0$ and $bx - ay + d = 0$ are taken as axes of $x$ and $y$ respectively.
21. If the formula for transformation from one set of axes to another (both rectangular) with the same origin be $x = aX + bY$; $y = cX + dY$, then show that $a^2 + c^2 = b^2 + d^2$. (*Hint*: Use $x^2 + y^2 = X^2 + Y^2$).

# 4. STRAIGHT LINE

## INTRODUCTION

In this chapter, we find the equations of straight lines in various forms. The conditions for two lines to be parallel or perpendicular and for three lines to be concurrent are also obtained.

## 4.1 GENERAL EQUATION OF A LINE

Consider any point on the $x$-axis. Its ordinate is zero i.e. the co-ordinates of every point, on the $x$-axis satisfy the equation, $y = 0$ and conversely any point whose co-ordinates satisfy the equation, $y = 0$, lies on it. Thus, we see that the equation of the $x$-axis is, $y = 0$. Similarly, the equation of the $y$-axis is, $x = 0$.

We see that these equations are of first degree in $x$ and $y$ and are special cases of $ax + by + c = 0$. Now, we prove a theorem:

**Theorem 4.1.** A first degree equation in $x$ and $y$ represents a straight line and conversely.

*Proof.* We know that a line is completely characterised by any two points on it. Thus, with reference to a coordinate system, let $A(x_1, y_1)$ and $B(x_2, y_2)$ be any two points on a given line. If $P(x, y)$ is any point on it, then

$$\text{area } \Delta\, PAB = 0.$$

$$\Rightarrow \quad x(y_1 - y_2) + y(x_2 - x_1) + (x_1 y_2 - x_2 y_1) = 0 \quad (\text{by § } \mathbf{1.3})$$

which is of the form $ax + by + c = 0$.

Conversely, let $ax + by + c = 0$ be any first degree equation in $x$ and $y$. Let $T$ be the locus represented by this equation.

Then $T$ will be a straight line if, whenever $P$ and $Q \in T$, every point, $R$, on the segment $PQ$ is also on $T$.

Taking $P(x_1, y_1)$ and $Q(x_2, y_2) \in T$ and $R$ on $PQ$ with co-ordinates $\left(\frac{x_1 + kx_2}{1+k}, \frac{y_1 + ky_2}{1+k}\right)$ $(k \neq -1)$, we have,

$$ax_1 + by_1 + c_1 = 0 \text{ and } ax_2 + by_2 + c_2 = 0,$$

whence $\quad ax_1 + by_1 + c_1 + k(ax_2 + by_2 + c) = 0.$

$\Rightarrow \quad a\left(\frac{x_1 + kx_2}{k+1}\right) + b\left(\frac{y_1 + ky_2}{k+1}\right) + c = 0.$

$\Rightarrow \quad R \in T$, the locus represented by the equation

$$ax + by + c = 0$$

$\Rightarrow \quad$ the locus, $T$, is a straight line.

## 4.2 PARTICULAR CASES OF THE EQUATION OF A LINE

### 4.2.1. Lines parallel to the axes

First consider any line $AB$ parallel to the $y$-axis at a distance $h$. The abscissa of every point on this line is $h$, and conversely, any point with abscissa $h$ lies on the line $AB$. Hence, we see that equation of the line $AB$ is $x = h$. Thus the equation of any line parallel to the $y$-axis is of the form $x = h$, where $|h|$ is the distance between the $y$-axis and the given line. Similarly, the equation of any line parallel to the $x$-axis at a distance $|k|$ from it, is of the form $y = k$.

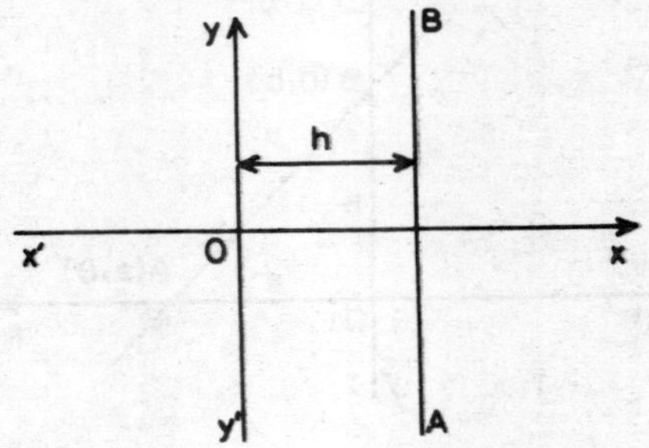

### 4.2.2. Two point form

Let $A(x_1, y_1)$ and $B(x_2, y_2$ be two points on a given line. If $P(x, y)$ is any point on this line, then as in **Theorem 4.1**, we have

$$x(y_1 - y_2) + y(x_2 - x_1) + (x_1y_2 - x_2y_1) = 0$$

$$\Rightarrow x(y_1 - y_2) + y(x_2 - x_1) + (x_1y_2 - x_1y_1) + (x_1y_1 - x_2y_1) = 0$$

(add and subtract $x_1y_1$)

$$\Rightarrow \quad y(x_2 - x_1) - y_1(x_2 - x_1) - x(y_2 - y_1) + x_1(y_2 - y_1) = 0$$

$$\Rightarrow \quad (y - y_1)(x_2 - x_1) = (x - x_1)(y_2 - y_1) \qquad (1)$$

Equation (1) is called the *two point form* of the equation of the line.

**Note.** If the above line through $A$ and $B$ is not parallel to the $y$-axis i.e. if $x_1 \neq x_2$, its equation can also be rewritten as:

$$y - y_1 = \frac{y_2 - y_1}{x_2 - x_1}(x - x_1) \qquad (2)$$

### 4.2.3. Intercept form of a line

Let a line make intercepts $a$ and $b$ on the co-ordinate axes. Then it passes through $(a, 0)$ and $(0, b)$.

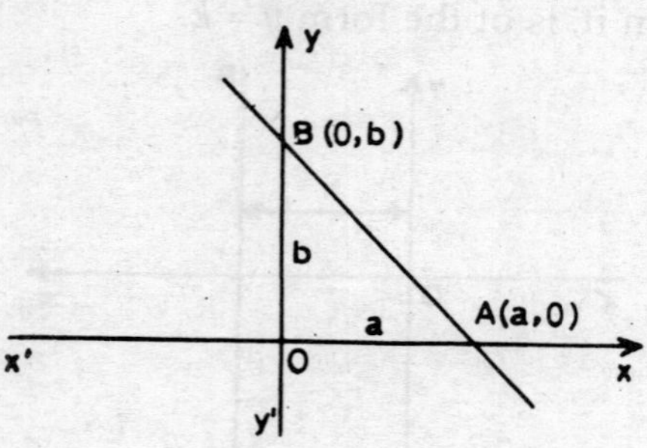


Thus its equation by § **4.2.2**, (1), is

$$(y - 0)(0 - a) = (x - a)(b - 0).$$

$$\Rightarrow \qquad xb + ya = ab.$$

$$\Rightarrow \qquad \frac{x}{a} + \frac{y}{b} = 1.$$

This is called the *intercept form* of the equation of the line.

### 4.2.4. Slope of a line

Let a line make an angle θ (to be measured in the anti-clockwise sense) with the positive direction of the $x$-axis. Then tan θ is called the *slope* or the *gradient* of the line. The slope of a line is generally denoted by the letter '$m$'; thus, $m = \tan \theta$.

Clearly the slope of a horizontal line i.e. a line parallel to the $x$-axis, is zero, and the slope of a vertical line i.e. a line parallel to the $y$-axis (or a line perpendicular to the $x$-axis), is not defined. But in this case, we can find the equation of a line parallel to the $y$-axis, directly.

*Assumption*. We assume in the general discussion hence forth that θ is not an odd multiple of π/2.

**Corollary.** *Slope of a line joining two points* $A(x_1, y_1)$ and $B(x_2, y_2)$.

Let θ be the angle which $AB$ makes with the $x$-axis.

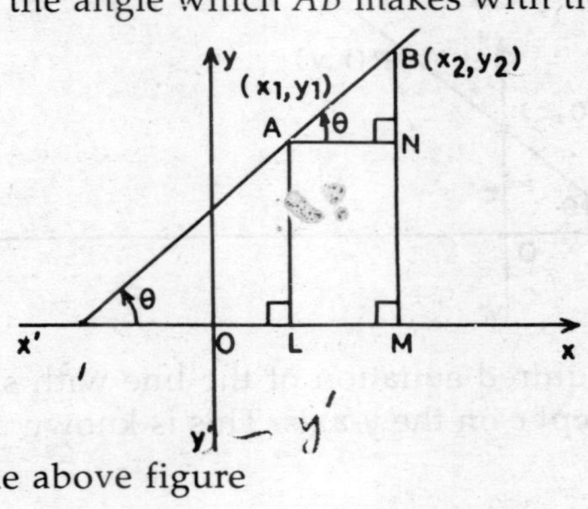


Then by the above figure

$$m = \tan \theta = \frac{NB}{AN} = \frac{MB - MN}{OM - OL} = \frac{y_2 - y_1}{x_2 - x_1}$$

$$\text{i.e. } m = \frac{y_2 - y_1}{x_2 - x_1}$$

### 4.2.5. Point-slope Form

Let a line pass through $(x_1, y_1)$ and make an angle $\theta$ with the $x$-axis. Let $P(x, y)$ be any point on this line. Then by the above Corollary,

$$m = \frac{y - y_1}{x - x_1}$$

when $m = \tan\theta$ is the slope of this line.

$\therefore \quad y = y_1 = m(x - x_1)$.

This is the required equation of a line drawn through $(x_1, y_1)$ whose slope is $m$. This form of the equation a line is known as the *point-slope form*.

### 4.2.6. Slope-intercept form

Let a line make an intercept $c$ on the $y$-axis and have slope $m$. Then the line passes through the point $(0, c)$ and its equation is

$$y - c = m(x - 0) \text{ or } y = mx + c.$$

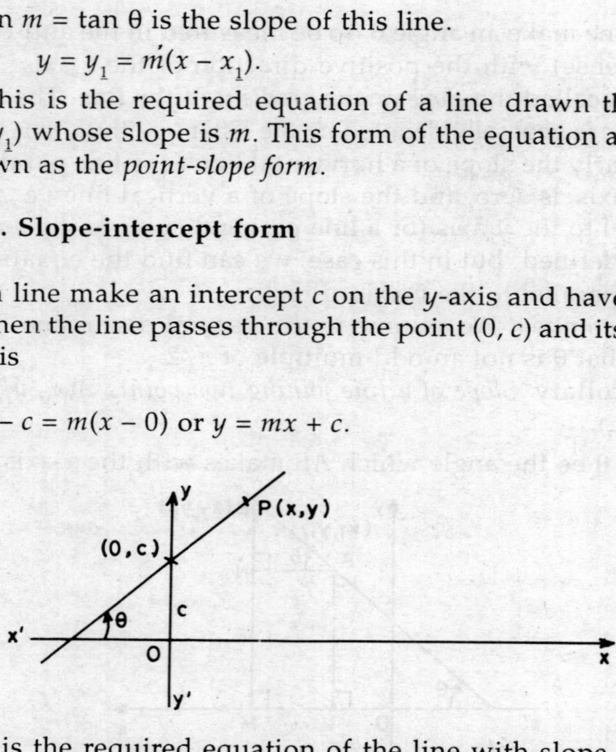


This is the required equation of the line with slope $m$ and having intercept $c$ on the $y$-axis. This is known as the *slope-intercept form*.

### 4.2.7. Parametric Form

Let a line make an angle $\theta$ with the $x$-axis and pass through a point $A(x_1, y_1)$. Let $P(x, y)$ be any point on this line and let $|AP| = r$.

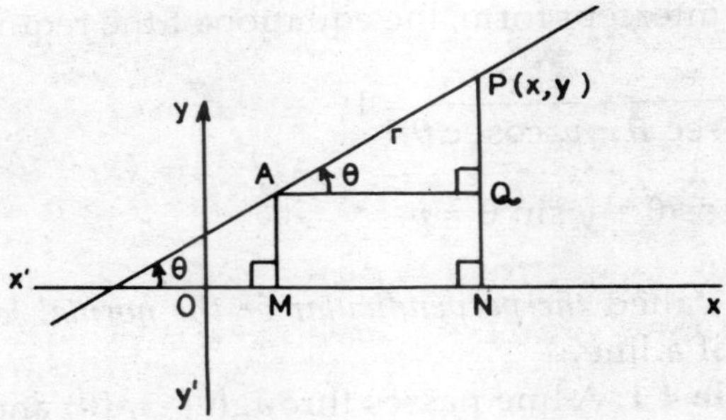


Then, $x - x_1 = AQ = r\cos\theta.$

and $y - y_1 = QP = r\sin\theta.$

$$\Rightarrow \quad \frac{x - x_1}{\cos\theta} = \frac{y - y_1}{\sin\theta} = r$$

These are called *parametric equations* of the given line.

**Corollary.** $x = x_1 + r\cos\theta$, $y = y_1 + r\sin\theta$ are the co-ordinates of any point on this line, in terms of the parameter $r$.

### 4.2.8. Perpendicular or Normal Form

Let $p$ be the length of the perpendicular from the origin on a given line $AB$ and let this perpendicular make an angle $\theta$ with the $x$-axis.

Clearly this line makes intercepts $p\sec\theta$ and $p\cos\theta$ on the co-ordinate axes.

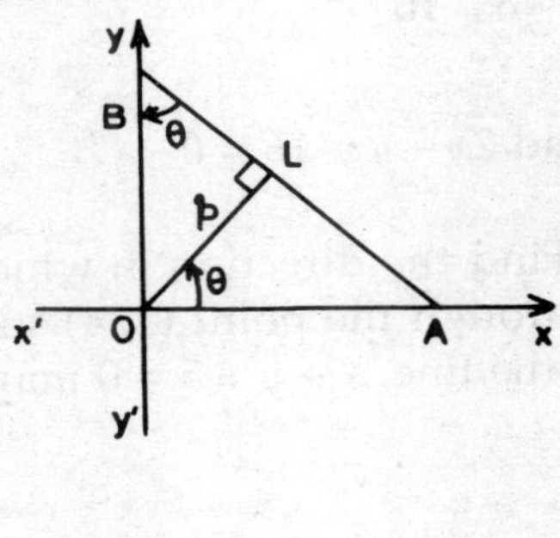

∴ By the intercept form, the equation of the required line is

$$\frac{x}{p\ \sec\theta} + \frac{y}{p\ \operatorname{cosec}\theta} = 1$$

$$\Rightarrow \quad x\cos\theta + y\sin\theta = p.$$

This is called *the perpendicular or the normal form* of the equation of a line.

**Example 4.1.** A line passes through (–3, 10) and the sum of its intercepts on the axes is 8. Find its equation.

*Solution.* Let the equation of the line be $\frac{x}{a} + \frac{y}{b} =$ .

Given : $a + b = 8 \Rightarrow b = 8 - a$.

Also (–3, 10) lies on it.

$$\therefore \quad \frac{-3}{a} + \frac{10}{8-a} = 1$$

$$\Rightarrow \quad -3(8-a) + 10a = a(8-a)$$

$$\Rightarrow \quad a^2 + 5a - 24 = 0$$

$$\Rightarrow \quad a = 3, -8$$

$$\therefore \quad b = 5\ ;\ 16$$

∴ There will be two such lines whose equations are

$$\frac{x}{3} + \frac{y}{5} = 1 \text{ and } \frac{x}{-8} + \frac{y}{16} = 1$$

i.e. $5x + 3y = 15$ and $2x - y + 16 = 0$

**Example 4.2.** Find the direction in which a straight line must be drawn through the point (1, 4) so that its point of intersection with the line, $x + y + 5 = 0$ may be at a distance $5\sqrt{2}$.

*Solution.*

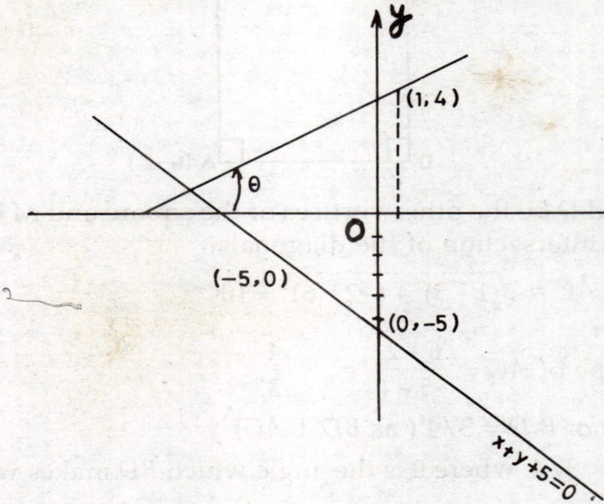


Let the line through (1, 4) make an angle θ with the $x$-axis.
∴ Its equation is

$$\frac{x-1}{\cos\theta} + \frac{y-4}{\sin\theta}.$$

$\Rightarrow \quad x = 1 + r\cos\theta, y = 4 + r\sin\theta.$

Thus $(1 + r\cos\theta, 4 + r\sin\theta)$ lies on $x + y + 5 = 0$, where $r = 5\sqrt{2}$.

$\Rightarrow \quad (1+5\sqrt{2}\cos\theta) + (4 + 5\sqrt{2}\sin\theta) + 5 = 0$

$\Rightarrow \quad \cos\theta + \sin\theta = -\sqrt{2}$ (1)

$\Rightarrow \quad \cos^2\theta + \sin^2\theta + 2\sin\theta\cos\theta = 2$

$\Rightarrow \quad \sin 2\theta = 1$

$\Rightarrow \quad 2\theta = \pi/2, 5\pi/2.$

$\Rightarrow \quad \theta = \pi/4, 5\pi/4.$

( $\theta = \pi/4$ does not satisfy equation (1) ).

**Example 4.3.** The opposite vertices of a square are $A(1, -2)$ and $C(-5, 6)$. Find the other two vertices.

*Solution.*

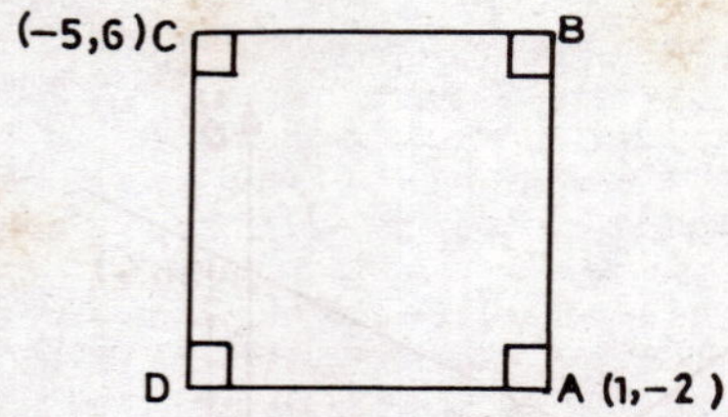


Let $B$ and $D$ be the other vertices of the square and $M$ be the point of intersection of the diagonals.

$\therefore\ BD = AC = \sqrt{(1+5)^2 + (-2-6)^2} = 10$

Also slope of $AC = \dfrac{6+2}{-5-1} = -\dfrac{4}{3}$

$\therefore$ Slope of $B\,D = 3/4$ ( as $BD \perp AC$)

$\Rightarrow \tan\theta = 3/4$, where $\theta$ is the angle which $BD$ makes with $x$-axis.

$\Rightarrow \quad \sin\theta = 3/5;\ \cos\theta = 4/5.$

Since $M$ is the mid point of both $AC$ and $BD$; the coordinates of $M$ are $\left(\dfrac{1-5}{2}, \dfrac{-2+16}{2}\right)$ i.e. (–2, 2)

$\therefore$ The parametric equations of $BD$ are

$$x = -2 + r\cos\theta = -2 + \frac{4}{5}r$$

and $$y = 2 + r\sin\theta = 2 + \frac{3}{5}r$$

Now $B$ and $D$ are at a distance of 5 units on either side of $M$.

$\therefore$ Putting $R = \pm 5$, we get the coordinates of the other vertices of the square as (2, 5) and (– 6, – 1).

**Example 4.4** Find the equations of the lines passing through (– 2, –3) and having an intercept of length 3 units between the lines, $4x + 3y = 12$ and $4x + 3y = 3$.

*Solution.*

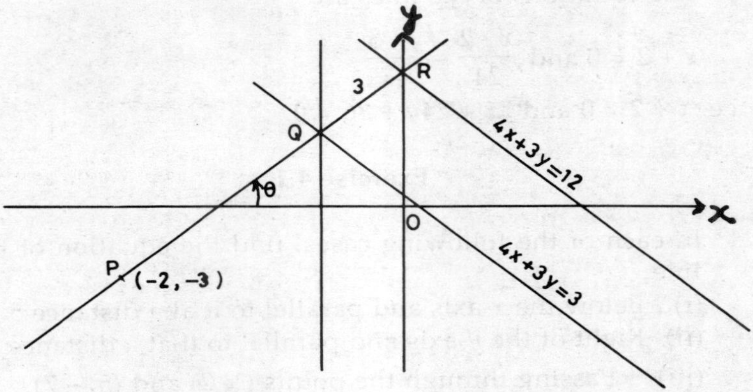


Let $PQR$ be any line through $P$ (−2, −3) making an angle θ with $x$-axis, meeting the given lines at $Q$ and $R$ respectively.

Let $PQ = r$ ; then $PR = r + 3$.

Now the equation of $PQR$ is

$$\frac{x+2}{\cos\theta} = \frac{y+3}{\sin\theta} = r$$

∴ $Q$ and $R$ have co-ordinates respectively as

$$(-2 + r\cos\theta, -3 + r\sin\theta)$$

and $(-2 + (r+3)\cos\theta, -3 + (r+3)\sin\theta)$.

These lie respectively on

$4x + 3y = 3$ and $4x + 3y = 12$.

∴ $4(-2 + r\cos\theta) + 3(-3 + r\sin\theta) = 3$

and $4[-2 + (r+3)\cos\theta] + 3[-3 + (r+3)\sin\theta] = 12$.

Subtracting these, we get

$12\cos\theta + 9\sin\theta = 9$

$\Rightarrow$ $4\cos\theta = 3(1 - \sin\theta)$

$\Rightarrow$ $16\cos^2\theta = 9(1 - \sin\theta)^2$

$\Rightarrow$ $16(1 - \sin^2\theta) = 9(1 + \sin^2\theta - 2\sin\theta)$

$\Rightarrow$ $25\sin^2\theta - 18\sin\theta - 7 = 0$

$\Rightarrow$ $\sin\theta = 1, -7/25$.

$\Rightarrow \qquad \cos\theta = 0\ ,\ 24/25$ (respectively).

$\therefore$ The equations of the lines are

$$x + 2 = 0 \text{ and } \frac{x+2}{24} = \frac{y+3}{-7}$$

i.e. $x + 2 = 0$ and $7x + 24y + 86 = 0$.

## Exercise 4.1

1. In each of the following cases, find the equation of the line
   (i) Below the $x$-axis and parallel to it at a distance 5.
   (ii) Right of the $y$-axis and parallel to it at a distance 2.
   (iii) Passing through the points (1, 2) and (5, – 7).
   (iv) Passing through the points $(a, b)$ and $(c, d)$.
   (v) Passing through (2, 3) and with slope – 5.
   (vi) Having intercepts – 4 and 5.
   (vii) Making equal intercepts on the axes.
   (viii) Making an angle 60° with the $x$-axis and passing through the point (–1, 1).
   (ix) At a perpendicular distance 5 from the origin and this perpendicular making an angle 150° with the $x$-axis.
2. Find the equations of the lines joining the following pairs of points:
   (i) $(at_1^2, 2at_1)$, $(at_2^2, 2at_2)$
   (ii) $(a\cos\theta_1, b\sin\theta_1)$, $(a\cos\theta_2, b\sin\theta_2)$
   (iii) $(a\sec\theta_1, b\tan\theta_1)$, $(a\sec\theta_2, b\tan\theta_2)$
   (iv) $(at_1, a/t_1)$, $(at_2, a/t_2)$.
3. Find the direction in which a straight line must be drawn through the point (1, 2) so that its point of intersection with the line, $x + y = 4$, may be at a distance $\dfrac{\sqrt{6}}{3}$ from this point.
4. Find the equation of the straight line passing through the point (–3, 2) and cutting off intercepts equal in magnitude but opposite in sign from the axes.

5. Prove that the straight line, $ax + by + c = 0$, divides the join of $(x_1, y_1)$ and $(x_2, y_2)$ in the ratio $-(ax_1 + by_1 + c) : (ax_2 + by_2 + c)$.
6. Find the area of the triangle, the co-ordinates of whose vertices are $(x_i, y_i)$; $i = 1, 2, 3$. Hence find the equation of the straight line through $(x_1, y_1)$ and $(x_2, y_2)$.
7. The portion of the line, $\frac{x}{a} + \frac{y}{b} = 1$, lying between the axes is trisected. Find the equations of the lines joining these points to the origin.
8. Find the distance of the point (2, 3) from the line, $2x - 3y + 9 = 0$, measured along a line making an angle 45° with the $x$-axis.
9. Show that the distance of the point $(h, k)$ from the line $ax + by + c = 0$ in a direction making an angle $\theta$ with the axis of $x$ is $-(ah + bk + c) / (a \cos\theta + b \sin\theta)$.
10. A line is such that the portion of it intercepted between the axes is bisected at the point $(\alpha, \beta)$. Prove that its equation is
$$\frac{x}{2\alpha} + \frac{y}{2\beta} = 1$$
11. A variable line passes through a fixed point $(\alpha, \beta)$. Find the equation to the locus of the mid. point of that portion of it, which is intercepted between two given perpendicular lines.
(*Hint:* Take the perpendicular lines as the co-ordinate axes).

## 4.3. REDUCTION OF THE GENERAL EQUATION ax + by + c = 0 OF A LINE TO SPECIAL FORMS

### 4.3.1. Reduction to slope-intercept form

We write $ax + by + c = 0$ as

$$y = -\frac{a}{b}x - \frac{c}{b}.$$

(Clearly this will be possible iff b ≠ 0 i.e. iff the line is not parallel to the $y$-axis).

$$\Rightarrow \text{its slope, } m = -\frac{a}{b} = -\frac{\text{coefficient of } x}{\text{coefficient of } y}$$

$$\text{and} \quad y\text{-intercept} = -\frac{c}{b} = -\frac{\text{constant term}}{\text{coefficient of } y}$$

**4.3.2. Reduction to intercept form**

We write $ax + by + c = 0$ as

$$\frac{x}{(-c/a)} + \frac{y}{(-c/b)} = 1$$

(This will be possible iff line is neither parallel to the $x$-axis nor to the $y$-axis and it does not pass through origin as well).

$$\Rightarrow \quad x\text{-intercept} = -\frac{c}{a} = -\frac{\text{constant term}}{\text{coefficient of } x}$$

$$\text{and} \quad \text{y-intercept} = -\frac{c}{b} = -\frac{\text{constant term}}{\text{coefficient of } y}$$

**4.3.3. Reduction to normal or perpendicular form**

Compare $ax + by + c = 0$ with

$$x \cos \theta + y \sin \theta - p = 0.$$

$$\Rightarrow \quad \frac{\cos\theta}{a} = \frac{\sin\theta}{b} = \frac{-p}{c} = k \text{ (say)}$$

$$\Rightarrow \quad \cos \theta = ak, \sin \theta = bk, p = -ck.$$

$$\Rightarrow \quad 1 = \cos^2 \theta + \sin^2 \theta = (a^2 + b^2)\, k^2$$

$$\Rightarrow \quad k^2 = \frac{1}{a^2 + b^2} \Rightarrow |k| = \frac{1}{\sqrt{a^2 + b^2}}$$

Since $p > 0 \Rightarrow -ck > 0$

$$\Rightarrow \quad \text{we take } k = \frac{-1}{\sqrt{a^2 + b^2}} \text{ or } \frac{1}{\sqrt{a^2 + b^2}}$$

according as $c > 0$ or $c < 0$ ($c = 0$, is not possible – why?).

Thus if $c > 0$, $k = \dfrac{-1}{\sqrt{a^2+b^2}}$

and $\cos\theta = \dfrac{-a}{\sqrt{a^2+b^2}}$ $\sin\theta = \dfrac{-b}{\sqrt{a^2+b^2}}$, $p = \dfrac{c}{\sqrt{a^2+b^2}}$

$\therefore$ the required normal form of the given equation is

$$\frac{-a}{\sqrt{a^2+b^2}}\,x + \frac{-b}{\sqrt{a^2+b^2}}\,y = \frac{c}{\sqrt{a^2+b^2}}$$

Similarly, the normal form, when $c < 0$, is

$$\frac{a}{\sqrt{a^2+b^2}}\,x + \frac{b}{\sqrt{a^2+b^2}}\,y = \frac{-c}{\sqrt{a^2+b^2}}$$

**Corollary.** Perpendicular distance from the origin on the line given by $ax + by + c = 0$ is $\left|\dfrac{c}{\sqrt{a^2+b^2}}\right|$

## 4.4 TO FIND THE PERPENDICULAR DISTANCE FROM THE POINT $(x_1, y_1)$ ON THE LINE: $ax + by + c = 0$.

Shifting the origin to $(x_1, y_1)$, the axes remaining parallel, we have

$$x = X + x_1 \text{ and } y = Y + y_1$$

$\therefore$ $ax + by + c = 0$ transforms to

$$aX + bY + (ax_1 + by_1 + c) = 0.$$

Then by the above corollary,

$$p = \left|\frac{ax_1 + by_1 + c}{\sqrt{a^2+b^2}}\right|$$

**Example 4.5.** Reduce the equation $4x + 3y + 5 = 0$ to its normal form.

*Solution.* We write the equation 4x + 3y + 5 - 0 as

$$-4x - 3y = 5$$

Dividing it by $\sqrt{(-4)^2 + (-3)^2} = 5$, we get the required normal form of the given equation as

$$x\left(\frac{-4}{5}\right) + y\left(\frac{-3}{5}\right) = 1$$

On comparing with $x \cos \theta + y \sin \theta = p$, we have $p = 1$ and $\theta$ is given by

$$\cos \theta = \frac{-4}{5} \text{ and } \sin \theta = \frac{-3}{5}$$

**Example 4.6.** Prove that two lines can be drawn through the point $P(0, a)$ so that their perpendicular distances from the point $Q(2a, 2a)$ will each be equal to $a$. Find also their equations.

*Solution.* Equation of any line through $P(0, a)$ is given by

$$y - a = m(x - 0)$$

$$\Rightarrow \quad mx - y + a = 0 \qquad (1)$$

By hypothesis

$$\frac{m.2a - 2a + a}{\sqrt{m^2 + 1}} = \pm a$$

$$\Rightarrow \quad 2m - 1 = \pm\sqrt{m^2 + 1}$$

$$\Rightarrow \quad 3m^2 - 4m = 0 \Rightarrow m = 0, 4/3.$$

$\therefore$ There are two such lines and their equations are (form (1)).

$$y = a,\ 4x - 3y + 3a = 0.$$

**Example 4.7.** Find the locus of a point which moves such that the square of its distance from the base of an isosceles triangle is equal to the rectangle under its distances from the other sides.

*Solution.*

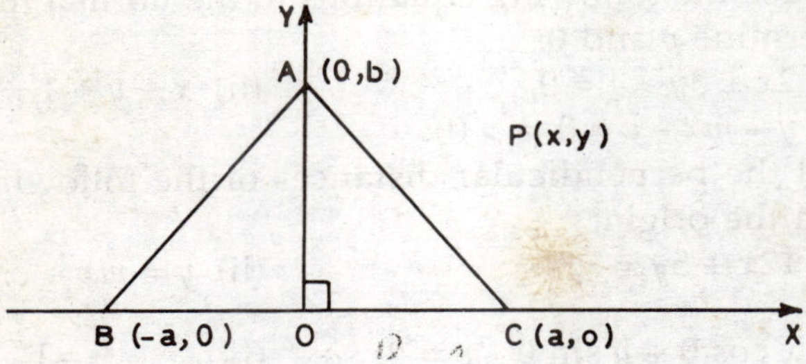


Take the base of the given isosceles triangle as the $x$-axis and the perpendicular from the vertex on the base as $y$-axis; then $O$ is the mid-point of $BC$ (which is origin presently). Let $A$, $B$, $C$ have co-ordinates respectively as $(0, b)$, $(-a, 0)$ and $(a, 0)$. Let $P(x, y)$ be the point whose locus is required. Now the equations of the sides are :

$BC \ : y = 0$

$CA \ : (x/a) + (y/b) = 1$

$AB \ : (x/-a) + (y/b) = 1$

By hypothesis,

$$y^2 = \pm\left[\frac{\frac{x}{a}+\frac{y}{b}-1}{\sqrt{\frac{1}{a^2}+\frac{1}{b^2}}}\cdot\frac{-\frac{x}{a}+\frac{y}{b}-1}{\sqrt{\frac{1}{a^2}+\frac{1}{b^2}}}\right]$$

$$\Rightarrow \quad \left(\frac{1}{a^2}+\frac{1}{b^2}\right)y^2 = \pm\left[\left(\frac{y}{b}-1\right)^2-\frac{x^2}{a^2}\right]$$

as the required locus.

**Exercise 4.2.**

1. Find the slopes of the following lines:
   (i) $2x - 3y + 5 = 0$, (ii) $ax - by + c = 0$.
   (iii) $x \cos\theta + y \sin\theta = p$, (iv) $y - mx - c = 0$.
2. Find the intercepts of the above lines on the co-ordinate axes.

3. Reduce the following equations to the normal form and determine $p$ and $\theta$:
(i) $2x + 3y - 4 = 0$, (ii) $x + y + 1 = 0$,
(iii) $y - mx - c = 0$ $(c > 0)$.
4. Find the perpendicular distances of the following lines from the origin:
(i) $12x + 5y = 7$, (ii) $y = mx + c$,
(iii) $x \cos\theta + y \sin\theta - p = 0$, (iv) $\frac{x}{a} + \frac{y}{b} = 1$
5. Find the perpendicular distances of the above lines from the points:
(a) $(-1, 2)$, (b) $(\alpha, \beta)$.
6. A point moves so that it always remains equi-distant from the lines, $y = m_1x$ and $y = m_2x$. Find its locus.
7. If $p$ be the length of the perpendicular from the origin, upon the line, $\frac{x}{a} + \frac{y}{b} = 1$, prove that

$$\frac{1}{p^2} = \frac{1}{a^2} + \frac{1}{b^2}.$$

8. Find the equations to the straight lines which go through the origin and trisect the portion of the straight line $3x + y = 12$ which is intercepted between the axes of co-ordinates.
9. If $p$ and $p'$ be the perpendiculars from the origin upon straight lines whose equations are $x \sec\theta + y \operatorname{cosec}\theta = a$ and $x \cos\theta - y \sin\theta = a \cos 2\theta$,
prove that $4p^2 + p'^2 = a^2$.
10. Find the points on the line, $y = x$, which are at a distance of 5 units from $4x + 3y = 1$.
11. Prove that the product of the perpendiculars from the two points $\left(\pm\sqrt{a^2 - b^2}, 0\right)$ on the line

$\frac{x}{a} \cos\theta + \frac{y}{b} \sin\theta =$ ; is $b^2$

12. If $x \cos \theta + y \sin \theta = p$, where $p = -\sin \theta \tan \theta$, be a line; prove that the perpendiculars on this line from the points $(m^2, 2m)$; $(mm', m + m')$ and $(m'^2, 2m')$ form a G.P.
13. Prove that the sum of the squares of the perpendiculars from the origin on the lines, $x \sec \alpha + y \operatorname{cosec} \alpha = 2a$ and $x \cos \alpha + y \sin \alpha = a \cos 2\alpha$ is independent of $\alpha$.
14. Show that the perpendicular from the origin, 0, upon the line joining the points $P$ $(a \cos \alpha, a \sin \alpha)$ and $Q$ $(a \cos \beta, a \sin \beta)$ bisects the distance between them. (*Hint*: Prove : $OP = OQ$).
15. If the product of the perpendiculars, from a point $P$ on the lines $x \cos \theta + y \sin \theta = a$ and $x \cos \phi + y \sin \phi = a$, is equal to the square of the perpendicular from $P$ on the line $x\cos\left(\frac{\theta+\phi}{2}\right)+y\sin\left(\frac{\theta+\phi}{2}\right)=a\cos\left(\frac{\theta+\phi}{2}\right)$ then prove that the locus of $P$ is $x^2 + y^2 = a^2$.

## 4.5 TO FIND THE ANGLE BETWEEN THE LINES WHOSE SLOPES ARE GIVEN

Let $l$ and $l'$ be the given lines with their slopes $m_1 = \tan \theta_1$ and $m_2 = \tan \theta_2$ respectively.

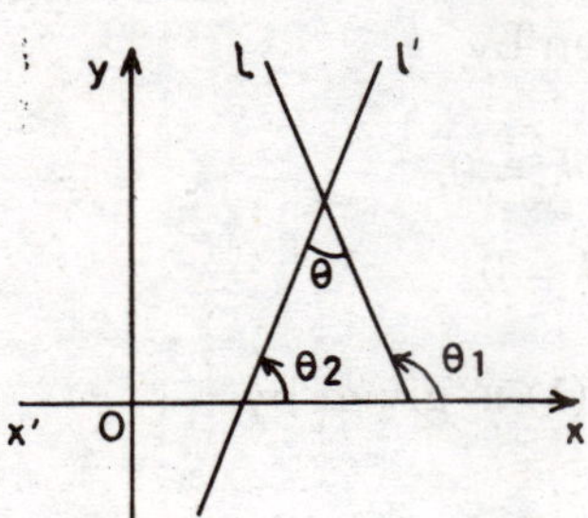


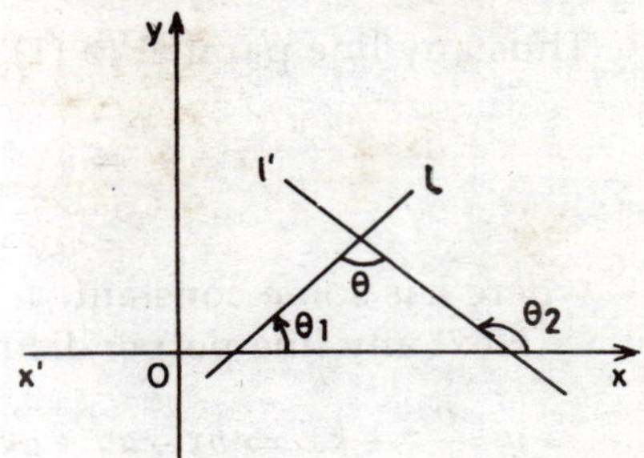


Let $\theta$ be the angle between these lines. Then

$$\theta = \pm (\theta_1 - \theta_2)$$

$$\Rightarrow \quad \tan \theta = \pm \tan (\theta_1 - \theta_2) = \pm \frac{\tan\theta_1 - \tan\theta_2}{1+\tan\theta_1 \tan\theta_2} = \pm \frac{m_1 - m_2}{1+m_1 m_2}$$

**Corollary 1.** Two lines with slopes $m_1$ and $m_2$ are *parallel iff* $\theta = 0$ (or $\pi$) $\Leftrightarrow \tan\theta = 0 \Leftrightarrow m_1 = m_2$.

**Corollary 2.** Two lines with slopes $m_1$ and $m_2$ are perpendicular iff $\theta = \frac{\pi}{2}\left(\text{or } -\frac{\pi}{2}\right) \Leftrightarrow \cot\theta = 0 \Leftrightarrow m_1 m_2 = -1$.

**Remark :** By Corollary 2, if a line has slope $m$ ($\neq 0$), then any line perpendicular to it, has slope $= -\frac{1}{m}$. However, if $m = 0$, i.e. when a line is parallel to the $x$-axis, then any line perpendicular to it is parallel to the $y$-axis and its slope is not defined.

## 4.6 TO FIND THE GENERAL EQUATIONS OF LINES PARALLEL AND PERPENDICULAR TO THE LINE ax + by + c = 0.

The given line is

$$ax + by + c = 0. \tag{1}$$

Its slope $= -a/b$

$\therefore$ The slopes of lines parallel and perpendicular to it are $-a/b$ and $b/a$ respectively.

Thus any line parallel to (1) is given by

$$y = \left(-\frac{a}{b}\right)x + k_1 \Rightarrow ax + by - bk_1 = 0,$$

$$\Rightarrow ax + by + k = 0$$

where $k$ is some constant.

Next any line perpendicular to (1) is given by

$$y = \frac{b}{a}x + k_2 \Rightarrow bx - ay + ak_2 = 0$$

$$\Rightarrow bx - ay + k' = 0$$

where $k'$ is some constant.

**Corollary 1.** The equation of the line parallel to $ax + by + c = 0$, and passing through $(x_1, y_1)$ is

$$a(x - x_1) + b(y - y_1) = 0$$

**Corollary 2.** The equation of the line through $(x_1, y_1)$ and perpendicular to $ax + by + c = 0$ is

$$b(x - x_1) - a(y - y_1) = 0.$$

**4.6.1. To find the distance between two parallel lines.**

If $ax + by + c = 0$, and $ax + by + c' = 0$ be two parallel lines, then the distance between them is the perpendicular distance of any point on one line from the other line.

Let $(\alpha, \beta)$ be any point on $ax + by + c = 0$.

$$\therefore \quad a\,\alpha + b\,\beta + c = 0 \tag{1}$$

Now, perpendicular distance of the point $(\alpha, \beta)$ from the line $ax + by + c' = 0$, is $\left|\dfrac{a\alpha + b\beta + c'}{\sqrt{a^2 + b^2}}\right| = \left|\dfrac{c' - c}{\sqrt{a^2 + b^2}}\right|$

using (1). This is the required distance between the given parallel lines.

**Example 4.8.** Find the general equations of the lines respectively (1) parallel and (2) perpendicular to the line $2x - 5y + 7 = 0$. Also find such lines through the point (4, 6).

*Solution.* (1) Equation of any line parallel to

$$2x - 5y + 7 = 0 \text{ is } \quad 2x - 5y + k = 0$$

(2) Equation of any line perpendicular to

$$2x - 5y + 7 = 0 \text{ is } (-5)\,x - (2)\,y + k_1 = 0$$

or $$5x + 2y - k_1 = 0.$$

We want to determine the equations of such lines which pass through (4, 6).

Hence $$2 \,.\, 4 - 5 \,.\, 6 + k = 0 \Rightarrow k = 22,$$
$$5 \,.\, 4 + 2 \,.\, 6 - k_1 = 0 \Rightarrow k_1 = 32.$$

$\therefore$ The required equations are

$$2x - 5y + 22 = 0$$

and $$5x + 2y - 32 = 0.$$

**Example 4.9.** Show that the equation of the line passing

through the point ($a \cos^3 \theta, a \sin^3 \theta$ ) and perpendicular to the line $x \sec \theta + y \operatorname{cosec} \theta = a$ is

$$x \cos \theta - y \sin \theta = a \cos 2\theta.$$

*Solution.* Equation of any line perpendicular to the line $x \sec \theta + y \operatorname{cosec} \theta = a$

i.e. to $\quad x \sin \theta + y \cos \theta = a \sin \theta \cos \theta$

is $\quad x \cos \theta - y \sin \theta = k.$

Also it is to pass through ($a \cos^3 \theta, a \sin^3 \theta$),

$\therefore\ a \cos^3 \theta . \cos \theta - a \sin^3 \theta . \sin \theta = k$

$$\Rightarrow \quad k = a(\cos^4 \theta - \sin^4 \theta)$$

$$= a (\cos^2 \theta - \sin^2 \theta) (\cos^2 \theta + \sin^2 \theta)$$

$$= a (\cos^2 \theta - \sin^2 \theta ) = a \cos 2 \theta.$$

$\therefore$ The required equation is

$$x \cos \theta - y \sin \theta = a \cos 2 \theta.$$

**Example 4.10.** Find the locus of the foot of the perpendicular from the origin upon the line joining the points ($a \cos \theta, b \sin \theta$) and ($-a \sin \theta, b \cos \theta$); $\theta$ varying in any manner.

*Solution.* Equation of the line joining ($a \cos \theta, b \sin \theta$) and ($-a \sin \theta, b \cos \theta$) is

$$y - b \sin \theta = \frac{b (\sin \theta - \cos \theta)}{a (\cos \theta + \sin \theta)} (x - a \cos \theta) \qquad (1)$$

Equation of the line through (0, 0) and perpendicular to (1) is

$$y - 0 = \frac{-a (\cos \theta + \sin \theta)(x - 0)}{b (\sin \theta - \cos \theta)}$$

$$\Rightarrow \quad y = \frac{-a (\cos \theta + \sin \theta)}{b (\sin \theta - \cos \theta)} x$$

If $by \sin \theta - by \cos \theta = - xa \cos \theta - ax \sin \theta.$

$$\Rightarrow \quad \cos \theta (xa - by) + \sin \theta (by + ax) + 0 = 0 \qquad (2)$$

Also (1) $\Rightarrow \quad ya \cos \theta + ya \sin \theta - ab \sin \theta \cos \theta - ab \sin^2 \theta.$

$$= bx \sin \theta - ab \sin \theta \cos \theta - bx \cos \theta + ab \cos^2 \theta.$$

or $\quad \cos \theta (ay + bx) + \sin \theta (ay - bx) - ab = 0 \qquad (3)$

the required locus will be obtained by eliminating θ from (2) and (3).

Now, these give

$$\frac{\cos\theta}{-ab(by+ax)} = \frac{\sin\theta}{ab(xa-by)} = \frac{1}{(xa-by)(ay-bx)-(by+ax)(ay+bx)}$$

or $\cos\theta = \dfrac{(by+ax)}{2(x^2+y^2)}$, $\sin\theta = \dfrac{(xa-by)}{-2(x^2+y^2)}$

Now $\cos^2\theta + \sin^2\theta = 1$

$\Rightarrow (by + ax)^2 + (ax - by)^2 = 4\,(x^2+y^2)^2$

or $a^2x^2 + b^2y^2 = 2\,(x^2 + y^2)^2$

which is the required locus.

### Exercise 4.3.

1. Find the angles between the following pairs of lines:
   (i) $2x - 3y + 1 = 0,\ 3x + 4y - 1 = 0$.
   (ii) $y + x = 0,\ y - x = 0$.
   (iii) $x = 0,\ y = 0$.
2. Find the general equations of the lines respectively (a) parallel and (b) perpendicular to the following lines:
   (i) $3x - 4y + 1 = 0$.
   (ii) $3x + 4y + 1 = 0$.
   (iii) $lx + my + n = 0$.
   (iv) $lx - my + n = 0$.
   (v) $y = m_1x + c_1$.
3. Find the lines in Question 2, which pass through (a) (1, 2); (b) $(x', y')$.
4. Find the distance between the following parallel lines
   (i) $5x - 12y + 2 = 0$; $5x - 12y - 3 = 0$
   (ii) $y = mx + c$; $y = mx + d$.
5. Find the equations of the lines parallel to $5x - 12y + 26 = 0$ and at a distance of 4 units from it.
6. Find the equations of the lines through (2, – 1) and mak-

ing an angle of 45° with the line, $6x + 5y - 1 = 0$. Show also that these lines are at right angles to one another.

7. Show that the equations of the lines through the origin making an angle $\theta$ with the line, $y = mx + c$, are given by

$$\frac{y}{x} = \frac{m \pm \tan\theta}{1 \mp m\tan\theta}$$

8. If a line joining two points $(a, b)$ and $(c, d)$ subtends an angle $\alpha$ at the origin; then

$$\cos\alpha = (ac + bd) / \sqrt{(a^2 + b^2)(c^2 + d^2)}$$

Also determine the equation of the perpendicular from the origin on this line.

9. The three lines, $x + 2y + 3 = 0$; $x + 2y - 7 = 0$; $2x - y - 4 = 0$ form the three sides of two squares. Find the equations of the fourth side of each square.

10. A diagonal of a square lies along the line, $8x - 15y = 0$ and one vertex of the square is (1, 2). Find the equations of the sides of the square through this vertex.

11. An equilateral triangle has its centroid at the origin and one side on the line, $x + y = 1$. Find the equations of the other two sides.

12. $A$ is (2, 5); $B$ is (4, – 11) and $C$ lies on $9x + 7y + 4 = 0$. Prove that the locus of the centroid of the triangle $ABC$ is a line parallel to the given line.

13. Show that the lines given by $x + y = 0$ ; $(a - b)x + (a + b)y = 2ab$ and $(a + b)x + (a - b)y = 2ab$ form an isosceles triangle whose vertical angle is $2\tan^{-1}(a/b)$. Determine the coordinates of its centroid.

## 4.7 TO FIND THE POINT OF INTERSECTION OF TWO LINES

Let $a_1x + b_1y + c_1 = 0$ and $a_2x + b_2y + c_2 = 0$ be two non-parallel lines. If $(x', y')$ be the co-ordinates of their point of intersection,

then $$a_1x' + b_1y' + c_1 = 0$$

and $\qquad a_2x' + b_2y' + c_2 = 0.$

Solving these equations, we get

$$(x', y') = \left(\frac{b_1c_2 - b_2c_1}{a_1b_2 - a_2b_1}, \frac{c_1a_2 - c_2a_1}{a_1b_2 - a_2b_1}\right)$$

(Note that as the lines are not parallel, they have unequal slopes, whence $a_1b_2 - a_2b_1 \neq 0$).

**Remark.** In solving numerical questions, the reader need not remember the co-ordinates $(x', y')$ given above, but should solve the equations directly.

## 4.8 LINE THROUGH THE INTERSECTION OF TWO LINES

We prove the following:

**Theorem 4.2.** Any line passing through the point of intersection of the lines

$$a_1x + b_1y + c_1 = 0$$

and $\qquad a_2x + b_2y + c_2 = 0$

is precisely of the form

$$\lambda\,(ax_1 + by_1 + c_1) + \mu\,(ax_2 + by_2 + c_2) = 0$$

where $\lambda$, $\mu$ are some constraints, not both zero.

Firstly, we prove that

$$\lambda\,(a_1x + b_1y + c_1) + \mu\,(a_2x + b_2y + c_2) = 0 \qquad (1)$$

where $\lambda$, $\mu$ are any constants (not both zero) is a line passing through the point of intersection of

$$a_1x + b_1y + c_1 = 0 \qquad (2)$$

$$a_2x + b_2y + c_2 = 0 \qquad (3)$$

Since equation (1) is first degree in $x$ and $y$, it represents a straight line. Further the point $(\alpha, \beta)$ lying on both (2) and (3), clearly lies on (1).

Thus the line given by the equation (1) passes through the point of intersection of the lines given by (2) and (3).

Secondly we prove that any line passing through the intersection of (2) and (3) is of the form (1).

Clearly lines (2) and (3) are not parallel.

whence $\dfrac{a_1}{b_1} \neq \dfrac{a_2}{b_2}$ i.e. $a_1b_2 - a_2b_1 \neq 0$

Two cases arise:

*Case (i).* Line through the intersection of (2) and (3) is parallel to the $y$-axis.

Now the point of intersection of (2) and (3) is

$$\left(\frac{b_1c_2 - b_2c_1}{a_1b_2 - a_2b_1}, \frac{c_1a_2 - c_2a_1}{a_1b_2 - a_2b_1}\right)$$

$\therefore$ the equation of the line, in this case, is

$$x = \frac{b_1c_2 - b_2c_1}{a_1b_2 - a_2b_1}$$

$\Leftrightarrow (a_1b_2 - a_2b_1)\, x - (b_1c_2 - b_2c_1) = 0$

$\Leftrightarrow b_2\,(a_1x + b_1y + c_1) + (-\,b_1)\,(a_2x + b_2y + c_2) = 0$

which is of the form (1), with $\lambda = b_2$ and $\mu = -\,b_1$.

Note that $\lambda = 0 = \mu \Rightarrow b_1 = b_2 = 0$; but, then, lines (2) and (3) are both parallel to the $y$-axis i.e. they are non-intersecting lines. Hence at least one of $b_1$ or $b_2$ is non-zero.

*Case (ii).* Line through the intersection of (2) and (3) is not parallel to the $y$-axis. Thus its equation is

$$y - \frac{a_2c_1 - a_1c_2}{a_1b_2 - a_2b_1} = m\left(x - \frac{b_1c_2 - b_2c_1}{a_1b_2 - a_2b_1}\right)$$

where $m$ is the slope of this line.

$$\Leftrightarrow m\,(a_1b_2 - a_2b_1)\,x - (a_1b_2 - a_2b_1)y + (a_2c_1 - a_1c_2) - m\,(b_1c_2 - b_2c_1) = 0$$

$$\Leftrightarrow \quad (a_1a_2 + ma_1b_2 - a_1a_2 - ma_2b_1)x + (a_2b_1 + mb_1b_2 - a_1b_2 - mb_1b_2)\,y + (a_2c_1 + mb_2c_1 - a_1c_2 - mb_1c_2) = 0$$

$$\Leftrightarrow \quad [a_1x(a_2 + mb_2) - a_2x(a_1 + mb_1)] + [b_1y(a_2 + mb_2) - b_2y\,(a_1 + mb_1)] + [c_1(a_2 + mb_2) - c_2\,(a_1 + mb_1)] = 0$$

$$\Leftrightarrow \quad (a_2 + mb_2)(a_1x + b_1y + c_1) - (a_1 + mb_1)(a_2x + b_2y + c_2) = 0$$

Again this is of the form (1), with

$$\lambda = a_2 + mb_2$$

and $$\mu = -(a_1 + mb_1).$$

It may be noted that

$$\lambda = 0 = \mu \Rightarrow m = \frac{-a_1}{b_1} = \frac{-a_2}{b_2}$$

i.e. lines (2) and (3) are parallel, which is against the hypothesis.

Thus any line passing through the intersection of (2) and (3) can always be put in the form (1).

**Remark.** As at least one of $\lambda$ or $\mu$ is non zero, in equation (1) above, we can re-write it as

$$(a_1x + b_1y + c_1) + k(a_2x + b_2y + c_2) = 0$$

where $k = \dfrac{\mu}{\lambda}$, in case $\lambda \neq 0$,

or $k'(a_1x + b_1y + c_1) + (a_2x + b_2y + c_2) = 0$

where $k' = \dfrac{\lambda}{\mu}$, in case $\mu \neq 0$.

We shall, generally, use the first form unless it fails us, in which case we can use the later form.

**Example 4.11.** Find the equation of a line passing through the point of intersection of the lines

$$2x + 3y - 8 = 0$$

and $$3x + 4y - 11 = 0$$

and also through the point (0, 0).

*Solution.* The point of intersection of the lines

$$2x + 3y - 8 = 0 \tag{1}$$
$$3x + 4y - 11 = 0 \tag{2}$$

is $\left(\dfrac{-33+32}{8-9}, \dfrac{-24+22}{8-9}\right) = (1,2)$

The required line, now, passes through (0, 0) and (1, 2), whose equation is

$$y - 2 = \frac{2-0}{1-0}(x-1) \text{ i.e. } 2x - y = 0.$$

**Aliter.** Any line through the intersection of (1) and (2) can be taken as

$$(2x + 3y - 8) + k(3x + 4y - 11) = 0 \qquad (3)$$

This is to pass through (0, 0), whence $k = \dfrac{-8}{11}$. Substitute for $k$ in (3), to get the required equation as $2x - y = 0$.

**Remark.** The above method will fail, if we have to find such a line through (– 3, 5) instead of (0, 0). This is because the point (– 3, 5) lies on (2). In such a case, we start with the second form i.e. with

$$k(2x + 3y - 8) + (3x + 4y - 11) = 0,$$

as the equation of a line passing through their intersection.

**Example 4.12.** Find the equation of the line passing through the point of intersection of

$$x + 5y + 7 = 0, \qquad 3x + 2y - 5 = 0$$

and (i) parallel to the line

$$7x + 2y - 5 = 0$$

(ii) perpendicular to the line

$$7x + 2y - 5 = 0$$

*Solution.* **Method I**

Any line passing through the intersection of the given lines can be taken as

$$(x + 5y + 7) + k(3x + 2y - 5) = 0$$

$$\Rightarrow \quad (1 + 3k)x + (5 + 2k)y + (7 - 5k) = 0 \qquad (1)$$

Its slope

$$-\left(\frac{1+3k}{5+2k}\right)$$

(i) Line given by (1) is to be parallel to

$$7x + 2y - 5 = 0,$$

$$\therefore \quad -\left(\frac{1+3k}{5+2k}\right) = -\frac{7}{2} \Rightarrow k = -\frac{33}{8}$$

substituting this value of $k$ in (1), we get the required equation as

$$7x + 2y - 17 = 0.$$

(ii) For the lines to be perpendicular, we have $m_1 m_2 = -1$

whence $\left[\dfrac{-(1+3k)}{(5+2k)}\right]\left[\dfrac{-7}{2}\right] = -1$

$\therefore$ the required equation is

$25\,(x + 5y + 7) - 17\,(3x + 2y - 5) = 0$

$\Rightarrow 2x - 7y - 20 = 0$

**Method II.** The point of intersection of the given lines is (3 – 2) and slope of the line

$$7x + 2y - 5 = 0 \text{ is } -\frac{7}{2}$$

Hence the equations of the required lines are

$$y + 2 = -\frac{7}{2}\,(x - 3)$$

and $\quad y + 2 = \dfrac{2}{7}\,(x - 3)$ respectively.

i.e. $\quad 7x + 2y - 17 = 0$

and $\quad 2x - 7y - 20 = 0$ respectively.

**Example 4.13.** The sides of a quadrilateral have the equations: $x + 2y = 3$; $x = 1$; $x - 3y = 4$; $5x + y + 12 = 0$. Show that the diagonals of the quadrilateral are at right angles.

*Solution*. **Method I**

Let the quadrilateral be denoted by *ABCD*. Suppose the equations of the sides are:

| | | | | | |
|---|---|---|---|---|---|
| *AB* | : | $x + 2y = 3$ ; | *BC* | : | $x = 1$ |
| *CD* | : | $x - 3y = 4$ ; | *DA* | : | $5x + y + 12 = 0$ |

Solving these, the coordinates of $B$ and $C$ are respectively (1, 1) and (1, – 1).

Any line through $A$ is

$$x + 2y - 3 + \lambda (5x + y + 12) = 0$$

This passes through $C$ (1, –1)

$\therefore \quad \lambda = 1/4$

$\Rightarrow \quad$ The equation of $AC$ is $x + y = 0$.

Similarly the equation of $BD$ will be $x - y = 0$.

We note that product of their slopes = – 1.

$\therefore$ The diagonals; $AC$ and $BD$, are at right angles.

**Method II.**

We solve for the coordinates of $A$, $B$, $C$, and $D$, and find that they are respectively (– 3, 3), (1, 1), (1, –1) and (–2, –2)

Now slope of $AC = \dfrac{-1-3}{1+3} = -1$

Also slope of $BD = \dfrac{-2-1}{-2-1} = 1$

Product of their slopes = – 1

Hence $AC \perp BD$.

**Exercise 4.4.**

1. Find the point of intersection of the lines:
   (i) $3x - 4y + 1 = 0$, $7x + 4y - 1 = 0$,
   (ii) $x - y = 0$, $3x - 2y = 0$,
   (iii) $y = m_1x + c_1$, $y = m_2x + c_2$.
2. Find the equation of the line passing through the intersection of the given lines and satisfying the given condition:
   (i) $x - y - 4 = 0$, $7x + y + 20 = 0$
   and passing through the origin.
   (ii) $4x + 7y - 1 = 0$, $3x - 5y + 8 = 0$
   and passing through the point (– 1, – 1).
   (iii) $3x + 7y + 1 = 0$, $2x - 7y + 1 = 0$

and parallel to the line

$$2x + 3y + 1 = 0.$$

(iv) $3x + 7y + 1 = 0, \quad 2x - 7y + 1 = 0$

and perpendicular to the line

$$2x + 3y + 1 = 0.$$

(v) $x = 0, y = 0$ and passing through $(1, -1)$.

(vi) $x = 0, y = 0$ and equally inclined to the axes (two lines).

3. The three lines

$$3x - 5y + 2 = 0, \quad x + y - 2 = 0 \quad \text{and} \quad 4x - 2y - 3 = 0$$

form a triangle. Find the equations of the altitudes without solving for the vertices.

4. Find the equation of the straight line drawn at right angle to the straight line, $\frac{x}{a} - \frac{y}{b} = 1$ through the point where it meets the $x$-axis.

5. Find the equation to a line passing through (8, 12) and parallel to another line whose intercept from the co-ordinate axes is bisected at (2, 3).

6. (i) Prove that the line given by the equation

$$(3x - 4y - 7) + k\,(5x - 2y - 7) = 0$$

passes through a fixed point for all values of $k$.

(ii) Determine the fixed point through which the line

$$(2k + 3)\,x + (3k + 2)\,y - 3\,(3k + 2) = 0.$$

passes, whatever the value of $k$ may be.

7. Find the equation to the straight line which passes through the point $(-3, 4)$ and is perpendicular to the straight line

$$2x + 3y + 4 = 0.$$

8. Find the equation to the straight line which passes through the intersection of the straight lines

$$2x - 3y + 4 = 0.$$

$$3x + 4y - 5 = 0$$

and is perpendicular to the straight line

$$6x - 7y + 8 = 0.$$

9. Find the equations of the lines through the point of intersection of $3x + 7y - 17 = 0$ and $5x - y - 3 = 0$, which are at a distance of 2 units from the origin.
10. Find the equation of the line joining the origin to the point of intersection of the lines.
$$\frac{x}{a}+\frac{y}{b}=1 \text{ and } \frac{x}{b}+\frac{y}{a}=1.$$
11. A variable line, drawn through the point of intersection of the lines
$$\frac{x}{a}+\frac{y}{b}=1 \text{ and } \frac{x}{b}+\frac{y}{a}=1.$$
meets the co-ordinate axes in $A$ and $B$. Show that the locus of the mid-point of $AB$ is $2xy\,(a + b) - ab(x + y) = 0$.
12. Prove that the diagonals of the parallelogram formed by the four lines: $\sqrt{3}x+y=0$; $\sqrt{3}x+y=1$; $x+\sqrt{3}y=0$ and $x+\sqrt{3}y=1$ are at right angles to one another.
13. Find the lengths of the attitudes of a triangle whose vertices are (0, 0), (1, –1) and (3, 2).

## 4.9. SYSTEM (OR FAMILY) OF LINES

We mention below a few systems (also called the families) of lines, for different values of the constants, appearing in the equation of a line.

1. $y - y_1 = m\,(x - x_1)$.
For $m$ fixed, this equation represents a system of parallel lines, each having slope '$m$' and passing through differ-

ent points $(x_1, y_1)$. Also for $(x_1, y_1)$ fixed and different values of $m$, the above equation represents a family of different lines, all passing through the same point $(x_1, y_1)$.

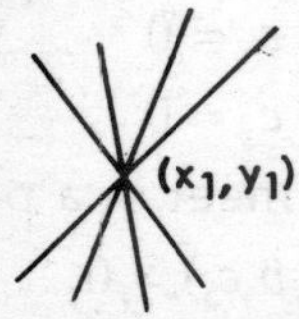


2. $(a_1x + b_1y + c_1) + k\,(a_2x + b_2y + c_2) = 0.$
On changing $k$, we get a family of different lines, passing through the point of intersection of the two given lines

$$a_1x + b_1y + c_1 = 0$$

and
$$a_2x + b_2y + c_2 = 0$$

3. $y = mx + c$.
For different values of $m$, and fixed $c$, this equation represents, a system of lines through the point $(0, c)$.

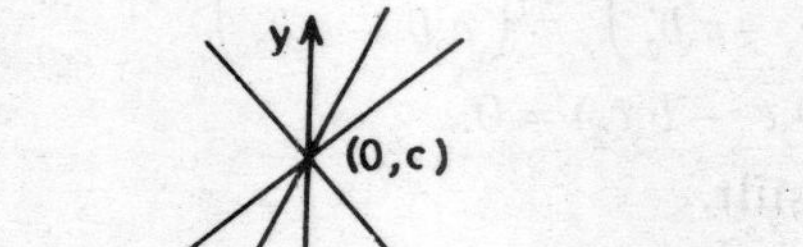


Also for $m$ fixed, and for different values of $c$, the above equation represents a family of parallel lines, making different intercepts on the $y$-axis.

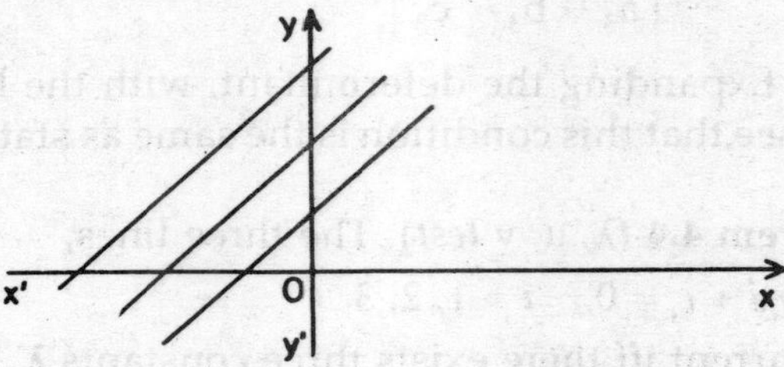

## 4.10. CONCURRENCE OF LINES

**Theorem 4.3.** Three lines given by

$$a_1x + b_1y + c_1 = 0,$$
$$a_2x + b_2y + c_2 = 0$$
and $$a_3x + b_3y + c_3 = 0$$

are concurrent, (i.e. they meet in a point) iff

$$\Sigma a_1 (b_2c_3 - b_3c_2) = 0.$$

(we assume that the lines are not parallel.).

*Proof.* The three given lines are concurrent ⇔ the point of intersection

$$\left(\frac{b_2c_3 - b_3c_2}{a_2b_3 - a_3b_2}, \frac{c_2a_3 - c_3a_2}{a_2b_3 - a_3b_2}\right)$$

of the second and third lines lies on the first i.e. on

$$a_1x + b_1y + c_1 = 0.$$

$$\Leftrightarrow \quad a_1\left(\frac{b_2c_3 - b_3c_2}{a_2b_3 - a_3b_2}\right) + b_1\left(\frac{c_2a_3 - c_3a_2}{a_2b_3 - a_3b_2}\right) + c_1 = 0$$

$$\Leftrightarrow \quad \Sigma a_1 (b_2c_3 - b_3c_2) = 0.$$

Hence the result.

**Corollary.** The three lines,

$$a_ix + b_iy + c_i = 0, i = 1, 2, 3$$

are concurrent iff the determinant

$$\begin{vmatrix} a_1 & b_1 & c_1 \\ a_2 & b_2 & c_2 \\ a_3 & b_3 & c_3 \end{vmatrix} = 0$$

*Proof.* Expanding the determinant, with the help of first row, we see that this condition is the same as stated in **Theorem 4.3.**

**Theorem 4.4** (λ, μ, ν *test*). The three lines,

$$a_ix + b_iy + c_i = 0, \quad i = 1, 2, 3$$

are concurrent iff there exists three constants λ, μ, υ, not all zero, such that

$$\lambda\,(a_1x + b_1y + c_1) + \mu\,(a_2x + b_2y + c_2) + \nu\,(a_3x + b_3y + c_3) \qquad (1)$$

vanishes identically.

*Proof. The condition is sufficient.*

If there exists $\lambda$, $\mu$, $\upsilon$ such that (1) vanishes identically, then co-efficients of $x$, $y$ and the constant term are separately zero,

whence
$$\lambda a_1 + \mu a_2 + \upsilon a_3 = 0,$$
$$\lambda b_1 + \mu b_2 + \upsilon b_3 = 0,$$
$$\lambda c_1 + \mu c_2 + \upsilon c_3 = 0$$

The last two equations yield

$$\frac{\lambda}{b_2c_3 - b_3c_2} = \frac{\mu}{b_3c_1 - b_1c_3} = \frac{\upsilon}{b_1c_2 - b_2c_1} = k \text{ (say)}$$

(Note that $k \neq 0$.)

$\therefore$
$$\lambda = (b_2c_3 - b_3c_2)\,k.$$
$$\mu = (b_3c_1 - b_1c_3)\,k,$$
$$\upsilon = (b_1c_2 - b_2c_1)\,k.$$

Now $\lambda a_1 + \mu a_2 + v a_3 = 0$

$\Rightarrow\ a_1(b_2c_3 - b_3c_2) + a_2(b_3c_1 - b_1c_3) + a_3\,(b_1c_2 - b_2c_1) = 0$

$\Rightarrow\ a_1(b_2c_3 - b_3c_2) + b_1(c_2a_3 - c_3a_2) + c_1(a_2b_3 - a_3b_2) = 0$

$\Rightarrow\ \Sigma\, a_1(b_2c_3 - b_3c_2) = 0.$

Thus the condition of **Theorem 4.3** is satisfied, whence the given lines are con-current.

*The condition is necessary.* Suppose the given lines are con-current. Consider

$$S = \{\lambda(a_1x + b_1y + c_1) + \mu(a_2x + b_2y + c_2) \mid \lambda, \mu \text{ real numbers}\}.$$

The members of this set, equated to zero, give all the lines, through the point of intersection of the lines

$$a_1x + b_1y + c_1 = 0$$

and
$$a_2x + b_2y + c_2 = 0.$$

Since
$$a_3x + b_3y + c_3 = 0,$$

also passes through their point of intersection, we have

$$a_3x + b_3y + c_3 \in S.$$

$\Rightarrow$ there exists real numbers $\lambda$, $\mu$ such that

$$\lambda(a_1x + b_1y + c_1) + \mu(a_2x + b_2y + c_2) = a_3x + b_3y + c_3$$

$$\Rightarrow \quad \lambda(a_1x + b_1y + c_1) + \mu(a_2x + b_2y + c_2) + (-1)(a_3x + b_3y + c_3) = 0.$$

Hence the result.

**Example 4.14.** Show that the lines

$$2x + 3y - 8 = 0,$$

$$x - 5y + 9 = 0,$$

and $\quad 3x + 4y - 11 = 0$

are concurrent.

*Solution.* **Method I.** Solving the first two equations, we see that their point of intersection is (1, 2) which also satisfies the third equation. Hence, the given lines are concurrent.

**Method II.** Substituting the values of $a_1$, $b_1$, $c_1$... etc. in

$$\Sigma\, a_1(b_2c_3 - b_3c_2),$$

we see that

$$\begin{aligned} a_1(b_2c_3 - b_3c_2) &+ b_1(c_2a_3 - c_3a_2) + c_1(a_2b_3 - a_3b_2) \\ &= 2[(-5)\cdot(-11) - 4.9] + 3\,[9.3 - (-11)\ (1)] \\ &\quad + (-8)\,[1.4 - 3\ (-5)] \\ &= 2\,(55 - 36) + 3\,(27 + 11) + (-8)\,(4 + 15) \\ &= 38 + 114 - 152 = 0. \end{aligned}$$

The result then follows by **Theorem 4.3**.

**Method III.** We have

$$\begin{vmatrix} a_1 & b_1 & c_1 \\ a_2 & b_2 & c_2 \\ a_3 & b_3 & c_3 \end{vmatrix} = \begin{vmatrix} 2 & 3 & -8 \\ 1 & -5 & 9 \\ 3 & 4 & -11 \end{vmatrix}$$

$$\begin{vmatrix} 2 & 3 & 0 \\ 1 & -5 & 0 \\ 3 & 4 & 0 \end{vmatrix} = 0 \text{ (operate } C_3 \longrightarrow C_3 + C_1 + 2C_2)$$

whence the lines are concurrent by Corollary to **Theorem 4.3**.

**Method IV.** Suppose.

$\lambda(2x + 3y - 8) + \mu(x - 5y + 9) + \upsilon(3x + 4y - 11) \equiv 0$

$\Rightarrow x(2\lambda + \mu + 3\upsilon) + y(3\lambda - 5\mu + 4\upsilon) + (-8\lambda + 9\mu - 11\upsilon) \equiv 0$

$\Rightarrow 2\lambda + \mu + 3\upsilon = 0,\ 3\lambda - 5\mu + 4\upsilon = 0,\ -8\lambda + 9\mu - 11\upsilon = 0$

$\Rightarrow \lambda = 19,\ \mu = 1,\ \upsilon = -13.$

Hence $19(2x + 3y - 8) + 1(x - 5y + 9) + (-13)(3x + 4y - 11) = 0.$
The concurrence, then, follows by **Theorem 4.4.**

*Note.* The reader is advised to follow **Method I,** in numerical problems.

**Example 4.15.** Show that the altitudes of a triangle are concurrent

*Solution.* Let the vertices *A, B, C* of any $\Delta ABC$ be $(x_i, y_i)$; $i = 1, 2, 3$ respectively and let *AD* be one of its altitudes.

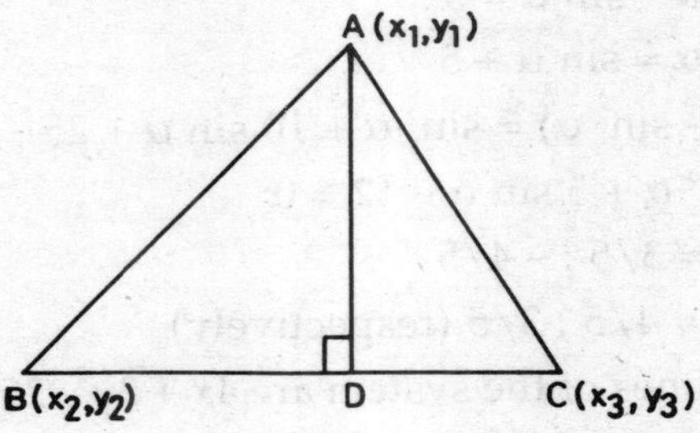


Now the slope of *BC* is $\dfrac{y_3 - y_2}{x_3 - x_2}$.

$\therefore$ the slope of the altitude, *AD* is $-\left(\dfrac{x_3 - x_2}{y_3 - y_2}\right)$

Hence, the equation of the altitude *AD* is

$$y - y_1 = -\left(\frac{x_3 - x_2}{y_3 - y_2}\right)(x - x_1)$$

$\Rightarrow x(x_3 - x_2) + y(y_3 - y_2) + (x_1x_2 - x_1x_3 + y_1y_2 - y_1y_3) = 0$ (1)

By symmetry, the equations of the other two altitudes are

$$x(x_1 - x_3) + y(y_1 - y_3) + (x_2x_3 - x_2x_1 + y_2y_3 - y_2y_1) = 0 \quad (2)$$

and $x(x_2 - x_1) + y(y_2 - y_1) + (x_3x_1 - x_3x_2 + y_3y_1 - y_3y_2) = 0$ (3)

Taking $\lambda = \mu = \upsilon = 1$, we see that

$$\lambda\{\text{L.H.S. of (1)}\} + \mu\,[\text{L.H.S. of (2)}] + \upsilon\,[\text{L.H.S. of (3)}] = 0.$$

Hence, the altitudes of any triangle are concurrent.

**Example 4.16.** Write the equation to the system of lines at a distance 5 units from the origin. Find the equations of the two lines of this system which pass through (7, –1).

*Solution.* The system of lines at a distance 5 units from the origin is

$$x \cos \alpha + y \sin \alpha = 5 \quad (1)$$

Here $\alpha$ is the parameter.

Now (1) passes through (7, – 1).

$\therefore$ $7 \cos \alpha - \sin \alpha = 5$

$\Rightarrow$ $7 \cos \alpha = \sin \alpha + 5$

$\Rightarrow$ $49 (1 - \sin^2 \alpha) = \sin^2 \alpha + 10 \sin \alpha + 25$

$\Rightarrow$ $25 \sin^2 \alpha + 5 \sin \alpha - 12 = 0$

$\Rightarrow$ $\sin \alpha = 3/5 ; - 4/5.$

$\therefore$ $\cos \alpha = 4/5 ; 3/5$ (respectively)

Thus two lines of the system are $4x + 3y - 25 = 0$ and $3x - 4y - 25 = 0$.

### Exercise 4.5

1. Examine whether the following lines are concurrent or not:

(i) $2x - 3y = 7$,
$3x - 4y = 13$,
and $8x - 11y = 33$.

(ii) $x + 2y + 5 = 0$,
$3x - 2y + 1 = 0$,
and $5x - 2y + 7 = 0$.

(iii) $3x - y + 1 = 0$,
$x + y + 1 = 0$,
and $5x - 3y + 1 = 0$.

Also, find the point of concurrence whenever it exists.

2. (i) Show that the medians of a triangle are concurrent.
   (ii) Show that the right bisectors of a triangle are concurrent.
3. Find the condition that the following lines may be concurrent:

   (i) $y = m_1x + c_1$
   $y = m_2x + c_2$
   $y = m_3x + c_3$

   (ii) $ax + by + c = 0$
   $bx + cy + a = 0$
   $cx + ay + b = 0.$
4. Find the value of $k$ for which the lines

   $7x - 8y + 5 = 0$
   $3x - 4y + 5 = 0$
   and $4x + 5y + k = 0$

   may be concurrent.
5. Find the co-ordinates of the orthocentre (where the altitudes meet) of the triangle formed by the lines.

   $3x - 5y + 2 = 0,$
   $x + y - 2 = 0$
   and $4x - 2y - 3 = 0.$

   (See: Question 3, Exercise 4.4).
6. Show that the lines

   $y + t_ix - 2at_i - at_i^3 = 0, i = 1, 2, 3,$

   are concurrent if

   $t_1 + t_2 + t_3 = 0.$
7. The lines

   $ax + 2y + 1 = 0,$
   $bx + 3y + 1 = 0$
   and $cx + 4y + 1 = 0$

   are concurrent. Show that $a$, $b$, $c$ are in A.P.
8. Find the co-ordinates of the orthocentre of the triangle whose vertices are:

   (1, 0); (2, – 4) and (– 5, –2).
9. Two vertices of a triangle are (5, – 1) and (–2, 3). If the orthocentre of the triangle lies at the origin, find the co-ordinates of the third vertex.

10. Prove that the lines given by

$$(b + c)\, x - bcy = a(b^2 + bc + c^2)$$
$$(c + a)\, x - cay = b(c^2 + ca + a^2)$$
and $$(a + b)\, x - aby = c(a^2 + ab + b^2)$$

are concurrent.

## 4.11. COLLINEARITY OF THREE POINTS

We work out an example below to illustrate the collinearity of three points (i.e. three points lying on the same line) in more than one way.

**Example 4.17.** Show that the points $A(1, 2)$, $B(6, 12)$, $C(3, 6)$ are collinear.

*Solution.* **Method I.** (Distance Method) See **Example 1.2.**

**Method II.** (*Slope Method*). The slopes of the line $AB$ and $BC$ are equal (each equal to 2).

$\Rightarrow$ The lines $AB$ and $BC$ are parallel.

But they have the point $B$ in common.

Hence, the lines $AB$ and $BC$ coincide, i.e. $A, B, C$ are collinear.

**Method III.** (*Equation Method*). Equation of the line $AB$ is $y - 2 = 2(x - 1)$ which is satisfied by $(3, 6)$.

Hence $A, B, C$ are collinear.

**Method IV.** (*Area of Triangle Method*). Area of the triangle

$$ABC = \frac{1}{2}[x_1(y_2 - y_3) + x_2(y_3 - y_1) + x_3(y_1 - y_2)]$$

$$= \frac{1}{2}[1(12 - 6) + 6(6 - 2) + 3(2 - 12)] = 0$$

$\Rightarrow$ The points $A, B, C$ are collinear.

### Exercise 4.6.

1. Examine whether the following sets of points are collinear or not:

(i) (0, 0), (2, 3), (4, 6);

(ii) (1, 2), (3, 4), (– 7, 9);
(iii) $(a, b)$, $(c, d)$, (0, 0), $(ad - bc \neq 0)$.

2. Find the condition for collinearity of the following sets of points:
(i) $(a, b)$, $(c, d)$, $(e, f)$;
(ii) $(at_1^2, 2at_1)$, $(at_2^2, 2at_2)$, $(a, 0)$, $(t_1 \neq t_2)$.
3. If the lines $l_1x + m_1y = 1$, $l_2x + m_2y = 1$ and $l_3x + m_3y = 1$ are concurrent, show that the points $(l_1, m_1)$, $(l_2, m_2)$ and $(l_3, m_3)$ are collinear.
4. Prove that the points $(a, b + c)$ ; $(b, c + a)$ ; $(c, a + b)$ are collinear. Find the equation of the line on which they lie.
5. Show that the area of the triangle formed by the lines: $y = mx$, $y = m'x$ and $y = c$ is equal to $c^2 (m \sim m')/2mm'$.
6. Show that the area of the triangle formed by the lines:
$y = m_1 x + c_1$,
$y = m_2x + c_2$ and $y = m_3 x + c_3$ is equal to

$$\frac{1}{2}\left[\frac{(c_2 - c_3)^2}{m_2 - m_3} + \frac{(c_3 - c_1)^2}{m_3 - m_1} + \frac{(c_1 - c_2)^2}{m_1 - m_2}\right]$$

## 4.12. POSITION OF A POINT RELATIVE TO A LINE

The point $(x_1, y_1)$ is on one side or the other of the line, $ax + by + c = 0$ $(b > 0)$ according as $ax_1 + by_1 + c > 0$ or $< 0$.

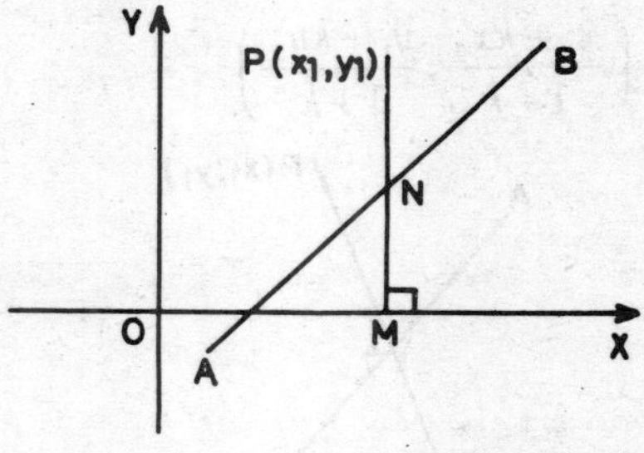


Let $AB$ be the line, $ax + by + c = 0$ and write it in the form so that $b > 0$.

Let $P$ be the point $(x_1, y_1)$. Let the ordinate $MP$, through $P$ meet $AB$ in $N$. Then $N$ will have co-ordinates $(x_1, y')$. As $N$ lies on $AB$,

$\therefore \quad ax_1 + by' + c = 0$

$\Rightarrow \quad y' = -(ax_1 + c)/b$

Now if $P$ lies above $AB$,

$$y_1 > y'$$

$\Rightarrow \quad y_1 > -(ax_1 + c)/b$

$\Rightarrow \quad ax_1 + by_1 + c > 0$, since $b > 0$.

Similarly if $P$ lies below $AB$, then
$y_1 < y' \Rightarrow ax_1 + by_1 + c < 0$.

Thus the point $P\ (x_1, y_1)$ lies on one side or the other of the line, $ax + by + c = 0$ according as the expression $ax_1 + by_1 + c$ is positive or negative .

The point $(x_1, y_1)$ is said to be on the positive (negative) side of the line, $ax + by + c = 0$, if the expression, $ax_1 + by_1 + c$ is positive (respectively negative).

## 4.13. TWO SIDES OF A LINE

Let $ax + by + c = 0$ be any line and $P(x_1, y_1)$, $Q(x_2, y_2)$ be two given points.

It is *required* to find a test by which we can determine whether these points are on the same side or the opposite sides of the given line.

Suppose the line $PQ$ meets the given line $AB$ at $L$. Let $L$ divide $PQ$ in the ratio $k : 1$. Then its co-ordinates are

$$\left(\frac{x_1 + kx_2}{1+k}, \frac{y_1 + ky_2}{1+k}\right).$$

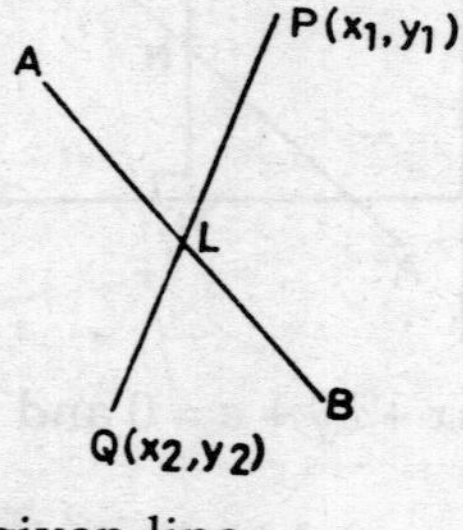


Now $L$ lies on the given line

$$\therefore \qquad a\left(\frac{x_1 + kx_2}{1+k}\right) + b\left(\frac{y_1 + ky_2}{1+k}\right) + c = 0$$

$$\Rightarrow \qquad (ax_1 + by_1 + c) + k(ax_2 + by_2 + c) = 0$$

$$\Rightarrow \qquad k = -\left(\frac{ax_1 + by_1 + c}{ax_2 + by_2 + c}\right)$$

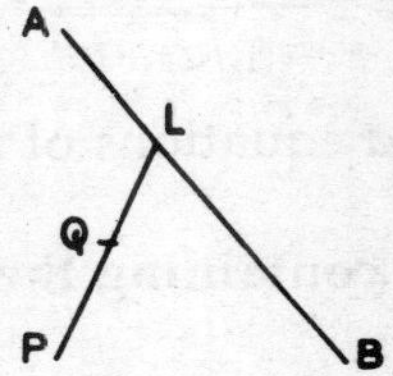


Thus $P$ and $Q$ lie on the same or the opposite sides of the given line according as $L$ divides $PQ$ externally or internally,

i.e. according as $k < 0$ or $k > 0$,

i.e. according as the expressions

$ax_1 + by_1 + c$ and $ax_2 + by_2 + c$

are of the same or the opposite signs.

This is the desired test.

## 4.14 TO FIND THE BISECTORS OF THE ANGLES BETWEEN THE LINES $a_1x + b_1y + c_1 = 0$, $a_2x + b_2y + c_2 = 0$

Let $A'PA$ and $B'PB$ be the given lines and $C'PC$ and $D'PD$ be their bisectors. If $Q(x, y)$ be any point on either of the bisectors, then the perpendicular distances, $QL$ and $QM$, of the point $Q$ on the lines $A'PA$ and $B'PB$, are equal

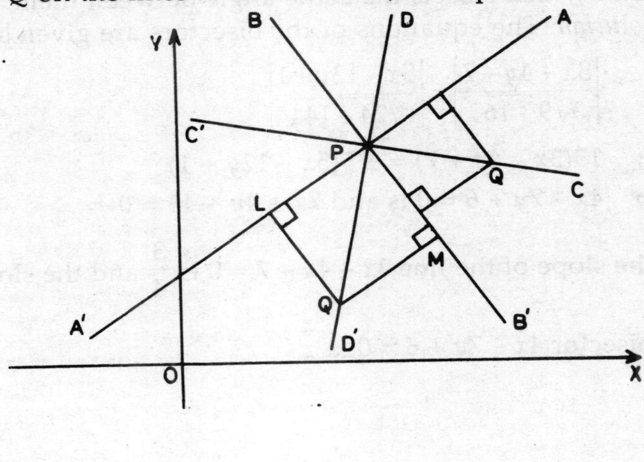

i.e. $$\left|\frac{a_1x+b_1y+c_1}{\sqrt{a_1^2+b_1^2}}\right| = \left|\frac{a_2x+b_2y+c_2}{\sqrt{a_2^2+b_2^2}}\right|$$

$$\Rightarrow \quad \frac{(a_1x+b_1y+c_1)}{\sqrt{a_1^2+b_1^2}} = \pm\frac{(a_2x+b_2y+c_2)}{\sqrt{a_2^2+b_2^2}}$$

which are the required equations of the bisectors.

**Bisector of the angle containing the ORIGIN**

Let $Q$ be any point on the bisector of that angle between the lines in which the origin lies ($\angle A'PB'$ in the above diagram).

Then, since $Q$ lies on the origin side of either line, the perpendiculars, $QL$ and $QM$ from $Q$ to the lines are both positive.

$$\therefore \quad \frac{a_1x+b_1y+c_1}{\sqrt{a_1^2+b_1^2}} = \frac{a_2x+b_2y+c_2}{\sqrt{a_2^2+b_2^2}}$$

(both $c_1$ and $c_2$ are of the same sign) represents the bisector of that angle between the lines which contains the origin.

Clearly then the other bisector does not contain the origin.

**Example 4.18.** Find the bisectors of the angles between the lines $3x - 4y + 7 = 0$ and $5x - 12y + 1 = 0$. Also find the bisector which bisects the acute angle between them.

*Solution.* The equations of the bisectors are given by

$$\left|\frac{3x-4y+7}{\sqrt{9+16}}\right| = \left|\frac{5x-12y+1}{\sqrt{24+144}}\right|$$

$$\Rightarrow \quad 13(3x - 4y + 7) = \pm 5(5x - 12y + 1)$$

$$\Rightarrow \quad 4x - 7y + 6 = 0 \text{ is and } 7x + 4y + 43 = 0$$

The slope of the line $3x - 4y + 7 = 0$ is $\frac{3}{4}$ and the slope of the bisector $4x - 7y + 6 = 0$ is $\frac{4}{7}$

The angle θ between this line and this bisector is given by

$$|\tan\theta| = \left|\frac{\frac{3}{4}-\frac{4}{7}}{1+\left(\frac{3}{4}\right)\left(\frac{4}{7}\right)}\right|$$

$\Rightarrow \tan\theta = 1/8 < 1 \Rightarrow \theta < 45^\circ$

$\therefore$ the angle between the lines = $2\theta < 90^\circ$.

Hence, $4x - 7y + 6 = 0$ is the bisector of the acute angle (and consequently $7x + 4y + 43 = 0$ is the bisector of the obtuse angle) between the given lines.

**Example 4.19.** Find the bisector of the acute angle between the lines

$$3x + 4y = 11 \text{ and } 12x - 5y = 2;$$

which is the bisector of the angle containing the origin?

*Solution*. The equations of the lines are

$$3x + 4y - 11 = 0 \quad (1)$$

$$12x - 5y - 2 = 0 \quad (2)$$

The equations of the two bisectors are

$$\frac{3x+4y-11}{\sqrt{3^2+4^2}} = \pm\frac{12x-5y-2}{\sqrt{12^2+5^2}}$$

i.e. $3x - 11y + 19 = 0$ (3)

and $11x + 3y - 17 = 0$ (4)

As the constant terms in the equations (1) and (2) are of the same sign, the bisector corresponding to positive sign on the right i.e. (3) is the bisector of the angle containing the origin.

Let θ be the angle between the line (1) and the bisector, (3). Then

$$\tan\theta = \frac{\left(\frac{3}{11}\right)-\left(-\frac{3}{4}\right)}{1+\left(\frac{3}{11}\right)\left(-\frac{3}{4}\right)} = \frac{9}{7} > 1$$

As $\tan\theta > 1$, $\theta$ lies between 45° and 90°.

$\therefore$ (3) bisects the obtuse angle between the given line.

Hence (4) i.e. $11x + 3y - 17 = 0$ is the bisector of the acute angle between the lines (1) and (2).

This we can see *alternatively*.

For the given lines (1) and (2)

$a_1a_2 + b_1b_2 = 3.12 + 4.(-5) = 16 > 0$

$\therefore$ By Q. 7 in **Exercise 4.7**, origin lies in the obtuse angle between the lines.

$\therefore$ The bisector of the acute angle corresponds to the negative sign and is, therefore, given by (4).

**Example 4.20.** Find the co-ordinates of the incentre of the triangle formed by the lines: $y + 1 = 0$, $4x - 3y - 7 = 0$ and $8x - 15y + 49 = 0$.

*Solution*. Solving these equations, two at a time, we get their points of intersection as

$A(1, -1)$; $B(-8, -1)$, $C(7, 7)$.

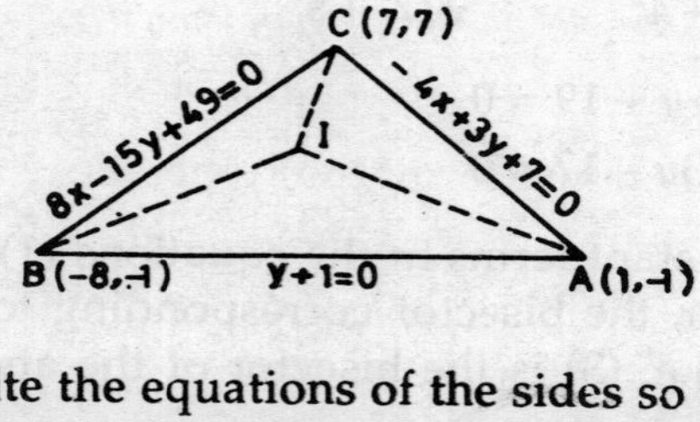


We now rewrite the equations of the sides so that each vertex is on the *positive* side of the opposite line. Hence these are:

$$\begin{aligned} BC &: \quad 8x - 15y + 49 = 0 \\ CA &: \quad -4x + 3y + 7 = 0 \\ AB &: \quad y + 1 = 0 \end{aligned}$$

The bisectors of the $\angle$ A are given by

$$y + 1 = \pm \frac{-4x + 3y + 7}{\sqrt{4^2 + 3^2}}$$

i.e. $2x + y - 1 = 0$ and $x - 2y - 3 = 0$

Now B and C lie on the opposite sides of $2x + y - 1 = 0$.

$\therefore$ $2x + y - 1 = 0$ is the internal bisector of angle $A$.

In the same way, $x - 4y + 4 = 0$ is the internal bisector of angle $B$.

Their point of intersection i.e. the incentre, $I$ is $(0, 1)$.

**Example 4.21.** Show that the line, $lx + my + n = 0$ bisects the angle between, $px + qy + r = 0$ and $(px + qy + r)(l^2 + m^2) = 2(lp + mq)(lx + my + n)$.

*Solution.* The equation of one of the bisectors between the lines

$$px + qy + r = 0$$

and $(px + qy + r)(l^2 + m^2) = 2(lp + mq)(lx + my + n)$, is

$$\frac{px + qy + r}{\sqrt{p^2 + q^2}}$$

$$= \frac{(px + qy + r)(l^2 + m^2) - 2(lp + mq)(lx + my + n)}{[p(l^2 + m^2) - 2l(lp + mq)]^2 + [q(l^2 + m^2) - 2m(lp + mq)]^{1/2}}$$

**Let $D$ be the denominator of the right side.**

$$\begin{aligned} \therefore\ D^2 &= p^2(l^2 + m^2)^2 + 4l^2(lp + mq)^2 - 4lp(l^2 + m^2)(lp + mq) \\ &\quad + q^2(l^2 + m^2)^2 + 4m^2(lp + mq)^2 - 4mq(l^2 + m^2)(lp + mq) \\ &= (p^2 + q^2)(l^2 + m^2)^2 + 4(l^2 + m^2)(lp + mq)^2 - 4(l^2 + m^2)(lp + mq)^2. \\ &= (p^2 + q^2)(l^2 + m^2)^2 \end{aligned}$$

$$\Rightarrow D = (l^2 + m^2)\sqrt{p^2 + q^2}$$

$\therefore$ **(1) simplifies to**

$(px + qy + r)\ (l^2 + m^2)$

$= (px + qy + r)\ (l^2 + m^2) - 2\ (lp + mq)\ (lx + my + n)$

$\Rightarrow lx + my + n = 0$ (as $lp + mq \neq 0$, otherwise the two lines coincide).

## Exercise 4.7

1. Find the bisector of the acute angle between lines
   $4x + 3y + 1 = 0$ and $12x + 5y - 2 = 0$.
2. Find the bisectors of the lines $x + y = 0$ and $x - y = 0$.
3. Find the bisectors of the lines $x = 0$ and $y = 0$.
4. Find the bisectors of the lines $y = m_1x$ and $y = m_2x$.
5. Show that the bisector of the angle between the lines, $x \cos \alpha + y \sin \alpha = p_1$ and $x \cos \beta + y \sin \beta = p_2$ $(p_1, p_2 > 0)$ which contains the origin, is
   $x \cos \alpha + y \sin \alpha - p_1 = x \cos \beta + y \sin \beta - p_2$.
6. The sides $BC$, $CA$, $AB$ of a triangle are given by
   $13x + 16y - 453 = 0$; $19x - 8y - 3 = 0$ ; $x - 4y + 7 = 0$.
   Obtain the equations to the internal bisectors of the angles $A$ and $C$ and hence find the co-ordinates of the incentre of the triangle, $ABC$.
7. If $a_1 a_2 + b_1 b_2$ is positive, prove that the origin lies in the obtuse angle between the lines, $a_1x + b_1y + c_1 = 0$ and $a_2x + b_2y + c_2 = 0$; $c_1$ and $c_2$ being both of the same sign.
8. If $2x - 16y - 1 = 0$, is one of the bisectors of the angles between the two lines one of which is $3x - 4y + 1 = 0$, find the other.
9. Show that the origin is within the triangle whose vertices are $A(2, 1)$, $B(3, -2)$ and $C(-4, -1)$.

**Some miscellaneous solved examples.**

**Example 4.22.** The equations of the sides of a triangle are

$L_i \equiv a_ix + b_iy + c_i\ (i = 1, 2, 3)$.

Show that the co-ordinates of the orthocentre of the triangle satisfy the equations: $\lambda_1 L_1 = \lambda_2 L_2 = \lambda_3 L_3$ where

$\lambda_1 = a_2a_3 + b_2b_3$ ; $\lambda_2 = a_3a_1 + b_3b_1$ ; $\lambda_3 = a_1a_2 + b_1b_2$.

*Solution.* Any line through the point of intersection of $L_1 = 0$ and $L_2 = 0$ is

$$L_1 + \lambda L_2 = 0 \quad (1)$$

$\Rightarrow (a_1x + b_1y + c_1) + \lambda(a_2x + b_2y + c_2) = 0.$

Its slope, $m = -(a_1 + \lambda a_2)/(b_1 + \lambda b_2)$.

This line is to be perpendicular to

$L_3 \equiv a_3x + b_3y + c_3 = 0,$

whose slope, $m' = -a_3/b_3$.

$\therefore \ mm' = -1$

$$\Rightarrow \frac{a_1 + \lambda a_2}{b_1 + \lambda b_2} \cdot \frac{a_3}{b_3} = -1$$

$$\Rightarrow \lambda = -\frac{a_1a_3 + b_1b_3}{a_2a_3 + b_2b_3} = -\frac{\lambda_2}{\lambda_1} \text{ (by hypothesis).}$$

Thus (1) implies $\lambda_1L_1 = \lambda_2 L_2$

Similarly, we get $\lambda_2L_2 = \lambda_3L_3$ and $\lambda_3L_3 = \lambda_1L_1$

whence follows the result.

**Example 4.23.** Prove that the area of the triangle whose sides are

$L_i \equiv a_ix + b_iy + c_i \ (i = 1, 2, 3)$

is $\Delta^2/2C_1C_2C_3$ where $C_i$ is the cofactor of $c_i$ $(i = 1, 2, 3)$ in $\Delta$ given by

$$\Delta = \begin{vmatrix} a_1 & b_1 & c_1 \\ a_2 & b_2 & c_2 \\ a_3 & b_3 & c_3 \end{vmatrix}$$

*Solution.* Let $A_i$, $B_i$, $C_i$ $(i = 1, 2, 3)$ denote the cofactors of $a_i$, $b_i$, $c_i$ respectively in $\Delta$.

The point of intersection of the lines:

$$a_1x + b_1y + c_1 = 0 \text{ and } a_2x + b_2y + c_2 = 0$$

is $\left(\dfrac{b_1c_2 - b_2c_1}{a_1b_2 - a_2b_1}, \dfrac{c_1a_2 - c_2a_1}{a_1b_2 - a_2b_1}\right) = \left(\dfrac{A_3}{C_3}, \dfrac{B_3}{C_3}\right)$

From this it follows that the vertices of the triangle have co-ordinates
$(A_1/C_1, B_1/C_1)$ ; $(A_2/C_2, B_2/C_2)$ ; $(A_3/C_3, B_3/C_3)$.

If $A$ be the area of this triangle, then

$$A = \frac{1}{2}\begin{vmatrix} A_1/C_1 & B_1/C_1 & 1 \\ A_2/C_2 & B_2/C_2 & 1 \\ A_3/C_3 & B_3/C_3 & 1 \end{vmatrix} = \frac{1}{2C_1C_2C_3}\begin{vmatrix} A_1 & B_1 & C_1 \\ A_2 & B_2 & C_2 \\ A_3 & B_3 & C_3 \end{vmatrix}$$

$= \Delta^2/2C_1\, C_2\, C_3$ (by determinants).

**Example 4.24.** Given $n$ lines and a fixed point, $O$. Through $O$ is drawn a line meeting these lines in the points $P_1, P_2, \ldots, P_n$ and on it is taken a point $P$ such that

$$\frac{n}{OP} = \sum_{i=1}^{n} \frac{1}{OP_i}$$

show that the locus of $P$ is a line.

*Solution.* Let $O$ be the origin and the $n$ fixed lines be

$n$ fixed lines be

$a_i x + b_i y = 1$ ($i = 1, 2, 3, \ldots n$).

Let the line through $O$ be $y = mx$.

This meets the given lines at the points whose co-ordinates are given by

$$P_i\left(\frac{1}{a_i + mb_i}, \frac{m}{a_i + mb_i}\right); \; (i = 1,2, \ldots n)$$

Let $P(\alpha, \beta)$ be any point on, $y = mx$.

$\Rightarrow \quad \beta = m\,\alpha \Rightarrow m = \beta / \alpha$ (1)

By hypothesis,

$$\frac{n}{OP} = \sum_{i=1}^{n} \frac{1}{OP_i}$$

$$\frac{n}{\sqrt{\alpha^2+\beta^2}} = \frac{1}{\sqrt{1+m^2}}\sum_{i=1}^{n}(a_i + mb_i)$$

$$\frac{n}{\sqrt{\alpha^2+m^2\alpha^2}} = \frac{1}{\sqrt{1+m^2}}(A+mb)$$

where $A = \sum_{i=1}^{n} a_i$ and $B = \sum_{i=1}^{n} b_i$

$$\frac{n}{\alpha} = A + \frac{\beta}{\alpha}B \qquad \text{[using (1)]}$$

$\therefore$ The locus of $P(\alpha, \beta)$ is
$Ax + By = n$, which is a line.

**Example 4.25.** Find the equations to the diagonals of the parallelogram formed by the lines:

$ax + by + c = 0$ ; $\quad ax + by + d = 0$;
$a'x + b'y + c' = 0$ ; $\quad a'x + b'y + d' = 0$.

Show that the parallelogram will be a rhombus if

$(a^2 + b^2)(c' - d')^2 = (a'^2 + b'^2)(c - d)^2$

and also find its area.

*Solution*. The sides of the parallelogram are

$$ax + by + c = 0 \qquad (1)$$
$$a'x + b'y + c' = 0 \qquad (2)$$
$$ax + by + d = 0 \qquad (3)$$
$$a'x + b'y + d' = 0 \qquad (4)$$

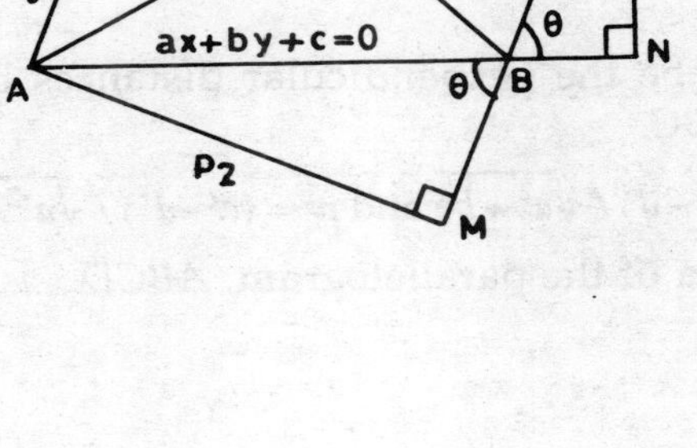

The equation of the line through the intersection of (1) and (4) is

$$ax + by + c + \lambda\,(a'x + b'y + d') = 0$$

and through the intersection of (2) and (3) is $(ax + by + d) + \mu\,(a'x + b'\,y + c') = 0$.

These equations will represent the same line, $AC$, if

$$\lambda = \mu \text{ and } c + \lambda\, d' = d + \lambda c'$$

$$\Rightarrow \quad \lambda = \mu = (c - d)/(c' - d').$$

$\therefore$ The equation of the diagonal, $AC$, is

$$(c' - d')\,(ax + by + c) + (c - d)\,(a'x + b'y + d') = 0$$

In the same way, the equation of the diagonal, $BD$, is

$$(c' - d')\,(ax + by + c) - (c - d)\,(a'x + b'y + c') = 0.$$

The slopes of these diagonals are respectively

$$m_1 - \frac{a(c' - d') + a'(c - d)}{b(c' - d') + b'(c - d)}$$

and $$m_2 - \frac{a(c' - d') + a'(c - d)}{b(c' - d') + b'(c - d)}$$

The parallelogram will be a rhombus, if the diagonals $AC$ and $BD$ are perpendicular i.e. if $m_1 \cdot m_2 = -1$ which yields

$$a^2\,(c' - d')^2 - a'^2\,(c - d)^2 = -\,[b^2\,(c' - d')^2 - b'^2\,(c - d)^2]$$

$$\Rightarrow (a^2 + b^2)\,(c' - d')^2 + (a'^2 + b'^2)\,(c - d)^2.$$

Let $\angle BAD = \theta$.

$$\therefore \quad \tan\theta = \frac{\left(-\dfrac{a'}{b'}\right) - \left(-\dfrac{a}{b}\right)}{1 + \left(-\dfrac{a'}{b'}\right)\left(-\dfrac{a}{b}\right)} = \frac{ab' - a'b}{aa' + bb'}$$

Let $p_1$ and $p_2$ be the perpendicular distances between $AB$, $DC$ and $AD$, $BC$.

$$\therefore \quad p_1 = (c - d)/\sqrt{a^2 + b^2} \text{ and } p_2 = (c' - d')/\sqrt{a'^2 + b'^2}$$

Now $A$ = Area of the parallelogram, $ABCD$

$= AB \cdot p_1 = p_1\, p_2 \operatorname{cosec} \theta.$

$= p_1 p_2 \sqrt{1 + \cot^2 \theta}$

$$\frac{c-d}{\sqrt{a^2+b^2}} \cdot \frac{c'-d'}{\sqrt{a'^2+b'^2}} \left[1 + \left(\frac{aa'+bb'}{ab'-a'b}\right)^2\right]^{1/2}$$

$= (c-d)\,(c'-d')/(ab'-a'b).$

**Miscellaneous Exercise on Chapter 4.**

1. Find the equation of the line through the point $(a, b)$ and parallel to the line.
$$\frac{x}{a} + \frac{y}{b} = 1$$
Find the intercepts made by the above line on the co-ordinate axes and also the length of the perpendicular from the origin on the above line.
2. Obtain the equations of the straight lines passing through the point (3, 2) and making an angle of 45° with the line
$$x - 2y = 3.$$
3. Obtain the equations of the lines passing through the intersection of $4x - 3y - 1 = 0$ and $2x - 5y + 3 = 0$ and equally inclined to the axes.
4. The equation of a given line is $3x - 4y + 1 = 0$. Explain giving reasons, how this line is related to
   (i) the point (1, 1),
   (ii) the line $4x + 3y = 2$,
   (iii) the line $6x - 8y = 10$.
5. Examine if the lines $y = 2x$ and $y = -2x$ are
   (i) parallel,
   (ii) perpendicular,
   (iii) equally inclined to the axes,
   (iv) coincident.
6. Find the correct answer:
The equation of the straight line passing through the point (3, – 1) and perpendicular to the straight line

$2x - 3y = 1$ is
(i) $2x - 3y = 9$,
(ii) $2y + 3x = 7$,
(iii) $2x + 3y = 3$,
(iv) $2y - 3x = -11$,

7. The equations of the sides $BC$, $CA$, $AC$ of a triangle are
$$u_i = a_i x + b_i y + c_i = 0, \; i = 1, 2, 3.$$
Show that the equation of the line through $A$ and parallel to $BC$ is
$$\frac{u_3}{a_3b_1 - a_1b_3} + \frac{u_2}{a_1b_2 - a_2b_1} = 0$$
and the line through $A$ and bisecting $BC$ is
$$\frac{u_3}{a_3b_1 - a_1b_3} - \frac{u_2}{a_1b_2 - a_2b_1} = 0$$

8. Find the equations of the diagonals of the parallelogram formed by the lines
$$ax + by + c = 0, \qquad ax + by + c' = 0$$
$$a'x + b'y + c = 0, \qquad a'x + b'y' + c' = 0$$
and show that the parallelogram will become rhombus if
$$a^2 + b^2 = a'^2 + b'^2$$
(*Hint:* See **Example 4.25**).

9. Show that the lines
$$(b - c)x + (c - a)y + (a - b) = 0$$
$$(c - a)x + (a - b)y + (b - c) = 0$$
$$(a - b)x + (b - c)y + (c - a) = 0$$
are concurrent.

10. Show that the points (0, 0) and (3, 4) are on the opposite sides of the line $y = 2x - 1$.

11. (a) Show that the origin is within the triangle, the equations of whose sides are:
$$x - 7y + 25 = 0,$$
$$5x + 3y + 11 = 0,$$
and $3x - 2y - 1 = 0$.

b) Find that bisector of the angle between the lines given by

$$4x + 3y + 1 = 0,$$

and $$12x + 5y + 13 = 0,$$

which contains the origin.

12. Find the equation of the locus of a point which moves in a plane so that the sum of the squares of its distances from the two lines $7x - 4y - 10 = 0$ and $4x + 7y + 5 = 0$ is always equal to 3.
13. Find the area of the triangle formed by the lines $y = x$, $y = 2x$ and $y = 3x + 4$.
14. Show that the area of the triangle formed by the lines $y = m_1x + c_1$, $y = m_2x + c_2$ and $x = 0$ is

$$\frac{1}{2}\frac{(c_1 - c_2)^2}{|m_1 - m_2|}$$

15. Prove that the area of the triangle formed by the straight lines $y = m_ix + \frac{a}{m_i}$, $i = 1, 2, 3$ is

$$\frac{1}{2}\frac{a^2}{m_1^2m_2^2m_3^2}[(m_1 - m_2)(m_2 - m_3)(m_3 - m_1)]$$

16. Show that the orthocentre of the triangle formed by the lines $y = m_ix + \frac{a}{m_1}$, $i = 1, 2, 3$ is

$$\left[-a, a\left(\frac{1}{m_1} + \frac{1}{m_2} + \frac{1}{m_3} + \frac{1}{m_1m_2m_3}\right)\right]$$

17. A line meets the co-ordinate axes at $A$, $B$ such that the centroid of the triangle $OAB$ is at $(a, b)$. Show that the equation of the line is

$$\frac{x}{a} + \frac{y}{b} = 3$$

18. A variable line is at a constant distance $p$ from the origin, and meets the co-ordinate axes in $A$, $B$. Show that the locus of the centroid of the $\Delta OAB$ is $x^{-2} + y^{-2} = 9p^{-2}$.
19. A variable line passes through a fixed point $(a, b)$ and meets the co-ordinate axes in $A$, $B$. Show that the locus

of intersection of the lines through $A$, $B$ parallel to the co-ordinate axes is

$$\frac{a}{x}+\frac{b}{y}=1$$

20. Sum of the distances of any number of fixed points from a variable line is zero, show that the line passes through a fixed point.
21. A variable line makes intercepts on the co-ordinate axes the sum of whose squares is constant and equal to $k^2$. Show that the locus of the foot of the perpendicular from the origin to this line is $(x^2 + y^2)^2 (x^{-2} + y^{-2}) = k^2$.
22. Prove that no line can be drawn through the point $(4, -5)$ so that its distance from $(-2, 3)$ will be equal to 12.
23. The base $BC$ of a triangle $ABC$ is bisected at the point $(a, b)$ and the equations to the sides $AB$ and $AC$ are $ax + by = 1$ and $bx + ay = 1$ respectively. Show that the equation to the median through $A$ is $(2ab - 1)(ax + by - 1) = (a^2 + b^2 - 1)(bx + ay - 1)$
24. A line is drawn through the point $P(1, 2)$ and intersects the line, $x - 2y + 4 = 0$ in $Q$. If $PQ = \sqrt{2}$, find the direction of the first line.
25. If the vertices of a triangle have integral co-ordinates, prove that the triangle cannot be equilateral.
26. Prove that the area of the parallelogram formed by the lines

$$3x + 4y = 7a_1; \qquad 3x + 4y = 7a_2;$$
$$4x + 3y = 7b_1; \qquad 4x + 3y = 7b_2$$

is $7(a_1 - a_2)(b_1 - b_2)$.

(*Hint:* See **Example 4.25**).

27. Show that the origin and the point $(-1, -2)$ lie in the same angle between the lines

$$7x + 5y - 18 = 0 \text{ and } 9x - 2y + 19 = 0.$$

28. A transversal cuts the sides, $AB$, $BC$ and $CA$ of a triangle $ABC$ in $P$, $Q$ and $R$ respectively. Prove that

$$\frac{AP}{PB} \cdot \frac{BQ}{QC} \cdot \frac{CR}{RA} = -1$$

(*Hint:* Use Q. 5. of Exercise 4.1).

29. A pair of lines pass, one through each of two fixed points $(a, o)$ and $(-a, o)$ and always enclose an angle $(\pi/4)$ between them. Find the locus of their point of intersection.
30. The base of a triangle is fixed and the difference of the base angles is given. Find the locus of the vertex. (*Hint*: Take the base, $BC = 2a$, as the $x$-axis and a line through its mid-point, perpendicular to $BC$ as $y$-axis and $\alpha$ as the given difference of angles .

# 5. PAIR OF STRAIGHT LINES

## INTRODUCTION

We obtain in this chapter, the condition for the general equation of second degree to represent a pair of lines. Also the equation of the bisectors of a pair of lines is obtained.

Consider the equation

$$x^2 - 5xy + 4y^2 = (x - 4y)(x - y) = 0.$$

By § 2.4, the locus consists of two lines given by $x - 4y = 0$ and $x - y = 0$. each passing through the origin. Hence, the equation $x^2 - 5xy + 4y^2 = 0$ represents a pair of straight lines through the origin.

## 5.1. HOMOGENEOUS EQUATIONS

We shall now study such equations in detail. First we deal with second degree homogeneous equations.

**Theorem 5.1.** A second degree homogeneous equation in $x$ and $y$, always represents two straight lines, real or imaginary, through the origin and conversely, the equation of a pair of lines through the origin is a second degree homogeneous in $x$ and $y$.

*Proof.* Let $ax^2 + 2hxy + by^2 = 0$ (1)

be and second degree homogeneous equation. Suppose first $b \neq 0$.

Now $(1) \Rightarrow b\left(\frac{y}{x}\right)^2 + 2h\left(\frac{y}{x}\right) + a = 0$

$\Rightarrow \quad bm^2 + 2hm + a = 0$ (2)

where $\quad m = \frac{y}{x}$

This being a second degree equation in $m$, has two values of $m$, say $m_1$ and $m_2$ (which may be real or imaginary).

$\therefore$ (2) $\Rightarrow bm^2 + 2hm + a = b(m - m_1)(m - m_2)$.

Replacing $m = y/x$, in above, and multiplying by $x^2$, we get

$$ax^2 + 2hxy + by^2 = b(y - m_1x)(y - m_2x).$$

Thus the locus represented by equation $ax^2 + 2hxy + by^2 = 0$ is same as the locus represented by $b(y - m_1x)(y - m_2x) = 0$, but this clearly represents two straight lines $y = m_1x$ and $y = m_2x$, each of them passing through the origin.

Next let $b = 0$, then (1) $\Rightarrow ax^2 + 2hxy = 0$, which represents two lines given by

$$x = 0 \text{ and } ax + 2hy = 0$$

both of which pass through the origin again.

Conversely, let $y = m_1x$ and $y = m_2x$ be two straight lines through the origin. Their joint equation is

$$(y - m_1x)(y - m_2x) = y^2 - (m_1 + m_2)\,xy + m_1m_2x^2 = 0,$$

which is a second degree homogeneous equation in $x$ and $y$.

Hence the theorem.

**Thorem 5.2.** Every $n$th degree homogeneous equation

$$a_0x^n + a_1x^{n-1}y + a_2x^{n-2}y^2 + \ldots\ldots + a_{n-1}xy^{n-1} + a_ny^n = 0$$

in $x$ and $y$, represents $n$ straight lines through the origin and conversely.

*Proof.* Similar to above.

**Remark 1.** From the above discussion, we see that if $ax^2 + 2hxy + by^2 = 0$ represents the lines $y = m_1x$ and $y = m_2x$, then $ax^2 + 2hxy + by^2 = b(y - m_1x)(y - m_2x) = b[y^2 - (m_1 + m_2)\,xy + m_1m_2x^2]$.

Comparing the coefficients of $xy$ and $x^2$ on both sides, we get

$$2h = -b(m_1 + m_2) \text{ and } a = bm_1m_2$$

$$\left.\begin{aligned} \text{i.e.}\quad & m_1 + m_2 = -\frac{2h}{b} \\ \text{and}\quad & m_1m_2 = \frac{a}{b} \end{aligned}\right\} \qquad (3)$$

Thus $ax^2 + 2hxy + by^2 = 0$ represents the lines $y = m_1x$ and $y = m_2x$, where $m_1$, $m_2$ are given by the formulae (3). These formulae will be used throughout this chapter without stating them explicitly.

**Remark 2.** The two lines given by $ax^2 + 2hxy + by^2 = 0$ will be real or imaginary

according as their slopes given by (2) are so,

i.e. according as discriminant of (2) $\geq$ or $< 0$,

i.e. according as discriminant $h^2 \geq$ or $< ab$.

**Example 5.1.** Find the product of the perpendiculars drawn from the point $(x', y')$ on the lines $ax^2 + 2hxy + by^2 = 0$.

*Solution.* Let the lines represented by $ax^2 + 2hxy + by^2 = 0$ be $y = m_1x$ and $y = m_2x$.

Therefore $m_1 + m_2 = -\dfrac{2h}{b}$ and $m_1 m_2 = a/b$.

The lengths of the perpendiculars from $(x', y')$ on these lines are

$$\left|\frac{y'-m_1x'}{\sqrt{1+m_1^2}}\right| \text{ and } \left|\frac{y'-m_2x'}{\sqrt{1+m_2^2}}\right|$$

$$\text{Their product} = \left|\frac{(y'-m_1x')(y'-m_2x')}{\sqrt{(1+m_1^2)(1+m_2^2)}}\right|$$

$$= \left|\frac{(y'^2-(m_1+m_2)x'y'+m_1m_2x'^2)}{\sqrt{1+m_1^2+m_2^2+m_1^2m_2^2)}}\right|$$

$$= \left|\frac{(y'^2-(m_1+m_2)x'y'+m_1m_2x'^2)}{\sqrt{1+(m_1+m_2)^2-2m_1m_2+m_1^2m_2^2)}}\right|$$

$$= \left|\frac{(y'^2+2h/b.x'y'+a/b.x'^2}{\sqrt{1+4h^2/b^2-2a/b+a^2/b^2}}\right|$$

$$= \left| \frac{ax'^2 + 2hx'y' + by'^2}{\sqrt{(a-b)^2 + 4h^2}} \right|$$

**Example 5.2.** Find the condition that one of the lines given by

$$ax^2 + 2hxy + by^2 = 0$$

may be perpendicular to one of the lines given by

$$a'x^2 + 2h'xy + b'y^2 = 0.$$

*Solution.* Let $y = mx$ be one of the lines represented by

$$ax^2 + 2hxy + by^2 = 0$$

then $$bm^2 + 2hm + a = 0 \quad (1)$$

Also, let $y = m'x$ be one of the lines represented by

$$a'x^2 + 2h'xy + b'y^2 = 0$$

which is perpendicular to $y = mx$.

This implies $b'm'^2 + 2h'm' + a' = 0$ (2)

Now $$m' = \frac{-1}{m}$$

Therefore (2) reduces to $\dfrac{b'}{m^2} - \dfrac{2h'}{m} + a' = 0$

$$\Rightarrow \quad a'm^2 - 2h'm + b' = 0 \quad (3)$$

Solving (3) and (1), we get

$$\frac{m^2}{-2h'a - 2hb'} = \frac{m}{bb' - aa'} = \frac{1}{2ha' + 2h'b}$$

$$m^2 = \frac{-(h'a + hb')}{(ha' + h'b)}, m = \frac{bb' - aa'}{2(ha' + h'b)}$$

Eliminating $m$, we obtain,

$$4(ha' + h'b)(h'a + hb') + (bb' - aa')^2 = 0,$$

as the required condition.

**Example 5.3.** If the lines $ax^2 + 2hxy + by^2 = 0$ be two sides of a parallelogram and the line, $lx + my = 1$, be one of its diagonals, show that the equation of the other diagonal, is

$$y(bl - hm) = x(am - hl).$$

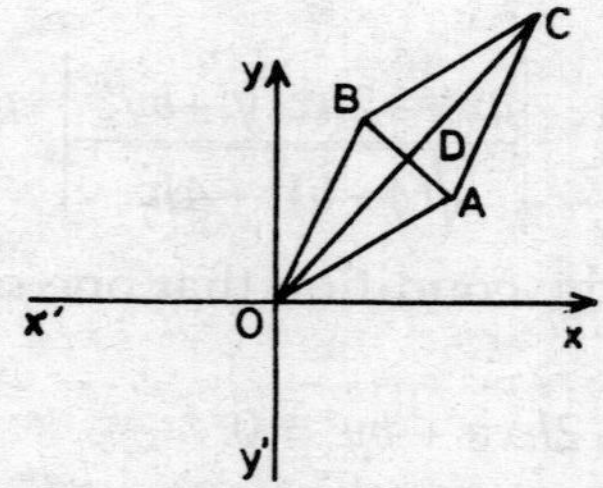


*Solution*. Clearly $lx + my = 1$ does not pass through the origin, so it represents the diagonal $AB$, whereas lines given by

$$ax^2 + 2hxy + by^2 = 0$$

are $OA$ and $OB$ (see the above figure).
Let the lines $OA$ and $OB$ be $y = m_1x$ and $y = m_2x$

$$\Rightarrow \quad m_1 + m_2 = -\frac{2h}{b} \text{ and } m_1m_2 = \frac{a}{b}$$

Solving $y = m_1x$ and $lx + my = 1$, we get

$$x(l + mm_1) = 1 \Rightarrow x = \frac{1}{l + mm_1}$$

$$\text{and so } y = \frac{m_1}{l + mm_1}$$

and hence the point $A$ is

$$\left(\frac{1}{l+mm_1}, \frac{m_1}{l+mm_1}\right)$$

Similarly, the point $B$ is

$$\left(\frac{1}{l+mm_2}, \frac{m_2}{l+mm_2}\right)$$

Let $D$ be the middle point of $AB$,
then, the co-ordinates of $D$ are

$$= \left(\frac{\frac{1}{l+mm_1} + \frac{1}{l+mm_2}}{2}, \frac{\frac{m_1}{l+mm_1} + \frac{m_2}{l+mm_2}}{2}\right)$$

$$= \left( \frac{2l + m(m_1 + m_2)}{2(l + mm_1).(l + mm_2)}, \frac{2mm_1m_2 + l(m_1 + m_2)}{2(l + mm_1).(l + mm_2)} \right)$$

$\therefore$ the equation of the diagonal $OD$, is

$$y = \left( \frac{2mm_1m_2 + l(m_1 + m_2)}{2l + m(m_1 + m_2)} \right) x$$

$$\Rightarrow y = \left[ \frac{2ma/b + l\left(\frac{-2h}{b}\right)}{2l + m\left(\frac{-2h}{b}\right)} \right] x = \left( \frac{am - hl}{bl - hm} \right) x$$

$$\Rightarrow y(bl - hm) = x(am - hl).$$

**Example 5.4.** If one of the lines given by $ax^2 + 2hxy + by^2 = 0$ coincides with one of the lines given by $a'x^2 + 2h'xy + b'y^2 = 0$ and the other lines given by them are perpendicular, then

$$\frac{ha'b'}{b' - a'} = \frac{h'ab}{b - a} = \frac{1}{2}\sqrt{-ab\ a'b'}$$

*Solution.* Let the two lines represented by

$$ax^2 + 2hxy + by^2 = 0 \quad (1)$$

be $y = m_1x$ and $y = m_2x$.

$$\therefore \quad m_1 + m_2 = -(2h/b) \quad (2)$$

and

$$m_1m_2 = a/b \quad (3)$$

By hypothesis, the lines represented by $a'x^2 + 2h'xy + b'y^2 = 0$, are

$$y = -\frac{1}{m_1}x \text{ and } y = m_2x.$$

$$\therefore \quad -\frac{1}{m_1} + m_2 = -(2h'/b') \quad (4)$$

and

$$\left(-\frac{1}{m_1}\right).m_2 = a'/b' \quad (5)$$

From (3) and (5),

$$-m_2^2 = aa'/bb'$$

$$\Rightarrow m_2 = \sqrt{-aa'/bb'} = \sqrt{-aba'b'}\,/\,bb'$$

This along with (3) implies

$$m_1 = \frac{a}{b}\cdot\frac{bb'}{\sqrt{-aba'b'}} = -\frac{\sqrt{-aba'b'}}{a'b}$$

Substituting these values of $m_1$ and $m_2$ in (2) and (4), we get

$$-\frac{\sqrt{-aba'b'}}{a'b} + \frac{\sqrt{-aba'b'}}{bb'} = -\frac{2h}{b}$$

$$\Rightarrow \sqrt{-aba'b'}\left[-\frac{1}{a'} + \frac{1}{b'}\right] = -2h$$

$$\Rightarrow \frac{ha'b'}{b'-a'} = \frac{1}{2}\sqrt{-aba'b'} \qquad (6)$$

and $$\frac{a'b}{\sqrt{-aba'b'}} + \frac{\sqrt{-aba'b'}}{bb'} = -\frac{2h'}{b'}$$

$$\Rightarrow \frac{-\sqrt{-aba'b'}}{ab'} + \frac{\sqrt{-aba'b'}}{bb'} = -\frac{2h'}{b'}$$

$$\Rightarrow \sqrt{-aba'b'}\left[-\frac{1}{a} + \frac{1}{b}\right] = -2h'$$

$$\Rightarrow \frac{h'ab}{b-a} = \frac{1}{2}\sqrt{-aba'b'} \qquad (7)$$

The result follows from (6) and (7).

## Exercise 5.1

1. Find the equations of the lines represented by
   (i) $x^2 - 5xy - 6y^2 = 0$,
   (ii) $xy = 0$,
   (iii) $x^2 - 2xy + y^2 = 0$,
   (iv) $x^2 + 2xy + 4y^2 = 0$,
   (v) $x^3 + 6x^2y + 11\,xy^2 + 6y^3 = 0$.
2. Show that the equation of the pair of lines through the

origin perpendicular to the pair of lines

$$ax^2 + 2hxy + by^2 = 0$$

is $$bx^2 - 2hxy + ay^2 = 0.$$

3. Find the condition that one of the straight lines given by the equation

$$ax^2 + 2hxy + by^2 = 0$$

may coincide with one of those given by

$$a'x^2 + 2h'xy + b'y^2 = 0.$$

4. Show that the two straight lines

$$x^2(\tan^2\theta + \cos^2\theta) - 2xy\tan\theta + y^2\sin^2\theta = 0$$

make with the axis of x, angles, such that the difference of their tangents is 2.

(*Hint*. Show: $|m_1 - m_2| = 2$)

5. Prove that one of the lines

$$ax^2 + 2hxy + by^2 = 0,$$

will bisect an angle between the co-ordinate axes if

$$(a + b)^2 = 4h^2.$$

(*Hint*: If $y = mx$ be one of the lines,
then $a + 2hm + bm^2 = 0$, also $m = \pm 1$).

6. Show that the area of the triangle formed by the lines

$$ax^2 + 2hxy + by^2 = 0$$

and $$lx + my + n = 0 \text{ is}$$

$$\frac{n^2\sqrt{h^2 - ab}}{am^2 - 2hlm + bl^2}$$

7. The equations of pairs of opposite sides of a rectangle are

$$x^2 - 7x + 6 = 0 \text{ and } y^2 - 14y + 40 = 0.$$

Find the equation of its diagonals.

8. Find the equation of the pair of lines through the origin and making an angle $\theta$ with the line $y + x = 0$.

9. Show that the lines $x^2 + 4xy + y^2 = 0$ form with the line $x - y = 6$ an equilateral triangle of area $6\sqrt{3}$.

10. Find the centroid of the triangle formed by the lines

$$ax^2 + 2hxy + by^2 = 0$$

and $$px + qy + r = 0.$$

11. The slope of one of the lines given by

$$ax^2 + 2hxy + by^2 = 0$$

is $k$ times the other. Prove that

$$4kh^2 = ab(1 + k)^2.$$

12. Show that one of the lines,

$$ax^2 + 2hxy + by^2 = 0,$$

makes the same angle with $x$-axis as the other makes with $y$-axis.

13. If $\tan 2\alpha = 2h/(a - b)$, show that the triangle formed by the lines,

$ax^2 + 2hxy + by^2 = 0$ and $x \cos \alpha + y \sin \alpha = p$ is isosceles.

14. Prove that the triangle formed by the lines

$ax^2 + 2hxy + by^2 = 0$ and $lx + my = 1$, is

(i) right angled if $(a + b)(al^2 + 2hlm + bm^2) = 0$

(ii) isosceles if $h(l^2 - m^2) = lm(a - b)$.

15. If $p_1, p_2$ be the perpendiculars from $(x, y)$ on the lines given by

$ax^2 + 2hxy + by^2 = 0$, prove that

$$(p_1^2 + p_2^2)[(a - b)^2 + 4h^2]$$
$$= 2(a - b)(ax^2 - by^2) + 4h(a + b)xy + 4h^2(x^2 + y^2).$$

16. Show that the centroid $(x', y')$ of the triangle with sides,

$ax^2 + 2hxy + by^2 = 0$ and $lx + my = 1$, is given by

$$\frac{x'}{bl - hm} = \frac{y'}{am - hl} = \frac{2}{3(am^2 - 2hlm + bl^2)}$$

## 5.2. TO FIND THE ANGLE BETWEEN THE LINES GIVEN BY $ax^2 + 2hxy + by^2 = 0$.

If $y = m_1x$ and $y = m_2x$ be the lines represented by the given equation, and if $\theta$ be the angle between these lines, we have

$$\tan \theta = \pm \left(\frac{m_1 - m_2}{1 + m_1m_2}\right) \text{ by § 4.5}$$

$$\Rightarrow \theta = \pm \tan^{-1}\left(\frac{\sqrt{(m_1+m_2)^2-4m_1m_2}}{1+m_1m_2}\right)$$

$$= \pm \tan^{-1}\left(\frac{2\sqrt{h^2-ab}}{a+b}\right)$$

(on substituting the values of $m_1 + m_2$ and $m_1m_2$)

**Corollary 1.** The lines given by $ax^2 + 2hxy + by^2 = 0$ are parallel if $\theta = 0$ (or $\pi$) i.e. iff $h^2 = ab$.

**Corollary 2.** The lines given by $ax^2 + 2hxy + by^2 = 0$ are perpendicular iff $\theta = 90°$ i.e. iff $\cot\theta = 0$ i.e. iff $a + b = 0$ i.e. iff coefficient of $x^2$ + coefficient of $y^2 = 0$.

## 5.3. TO FIND THE EQUATION OF THE BISECTORS OF THE ANGLES BETWEEN THE LINES GIVEN BY $ax^2 + 2hxy + by^2 = 0$. (1)

Let the lines given by (1) be $OA$: $y = m_1x$ and $OB$: $x = m_2x$. Then

$$m_1 + m_2 = -\frac{2h}{b} \text{ and } m_1 \cdot m_2 = \frac{a}{b}.$$

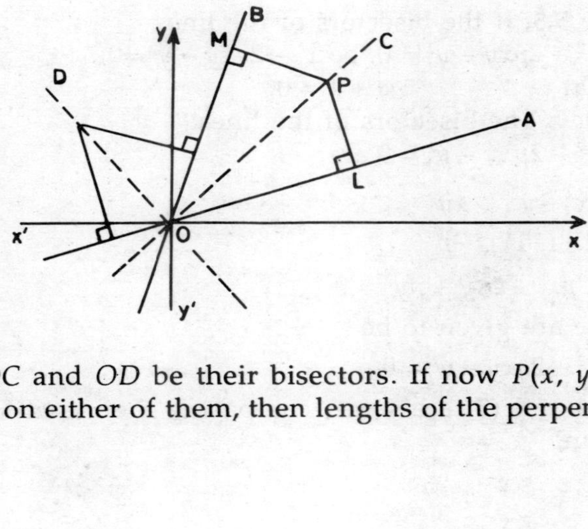


Let $OC$ and $OD$ be their bisectors. If now $P(x, y)$ be any point on either of them, then lengths of the perpendiculars from $P$ on $OA$ and $OB$ are equal.

$$\Rightarrow \left|\frac{y-m_1x}{\sqrt{1+m_1^2}}\right| = \left|\frac{y-m_2x}{\sqrt{1+m_2^2}}\right|$$

$\Rightarrow (y - m_1x)^2 (1 + m_2^2) = (y - m_2x)^2 (1 + m_1^2).$

$\Rightarrow (m_1^2 - m_2^2)x^2 - 2(1 - m_1m_2)(m_1 - m_2)xy - (m_1^2 - m_2^2)y^2 = 0.$

$\Rightarrow (m_1 + m_2)(x^2 - y^2) = 2(1 - m_1m_2)\, xy.$

(Cancel $m_1 - m_2 \neq 0$ )

$$\Rightarrow \frac{-2h}{b}(x^2 - y^2) = 2\left(1 - \frac{a}{b}\right)xy$$

$\Rightarrow h(x^2 - y^2) = (a - b)\, xy$ (2)

$$\Rightarrow \frac{x^2 - y^2}{a-b} = \frac{xy}{h} \qquad (3)$$

(if $a \neq b$ and $h \neq 0$).

Thus (2) or (3) is the required equation of the bisectors of the lines given by (1).

**Corollary.** The bisectors of any two intersecting lines given by $ax^2 + 2hxy + y^2 = 0$ are at right angles.

Their bisectors are given by (3).

Clearly: Coefficient of $x^2$+ Coefficient of $y^2$

$= h + (-h) = 0.$

**Example 5.5.** If the bisectors of the lines

$x^2 - 2pxy - y^2 = 0$, be $x^2 - 2qxy - y^2 = 0$,

show that $pq + 1 = 0.$

*Solution.* The bisectors of the lines

$x^2 - 2pxy - y^2 = 0$ are

$$\frac{x^2 - y^2}{1-(-1)} = \frac{xy}{-p}$$

$\Rightarrow \quad - px^2 - 2xy + py^2 = 0$ (1)

But these are given to be

$x^2 - 2qxy - y^2 = 0$ (2)

As (1) and (2) are same, comparing their coefficients, we obtain,

$$\frac{-p}{1} = \frac{1}{q} = \frac{p}{-1}$$

$$\Rightarrow pq = -1 \;\Rightarrow pq + 1 = 0.$$

**Example 5.6.** Show that the pair of lines given by

$$a^2x^2 + 2h(a + b)\,xy + b^2y^2 = 0$$

is equally inclined to the pair given by

$$ax^2 + 2hxy + by^2 = 0.$$

*Solution.* Equation of the bisectors of first pair is

$$\frac{x^2 - y^2}{a^2 - b^2} = \frac{xy}{h(a+b)} \Rightarrow \frac{x^2 - y^2}{a - b} = \frac{xy}{h}$$

which are also the bisectors of the second pair.

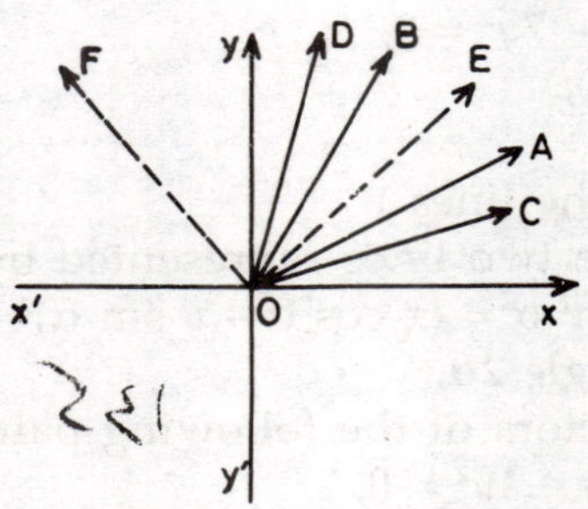


Let $OA$, $OB$; $OC$, $OD$ be the lines of the first and second pairs respectively and $OE$ and $OF$ be their bisectors.

We have

$$\angle COE = \angle DOE$$

and $\angle AOE = \angle BOE$

$\Rightarrow \quad \angle COE - \angle AOE = \angle DOE - \angle BOE$

$\Rightarrow \quad \angle AOC = \angle BOD$

which proves the result.

**Example 5.7.** Show that the line, $y = mx$, bisects the angle between the lines $ax^2 - 2hxy + by^2 = 0$ if $h(1 - m^2) + m(a - b) = 0$.

*Solution.* The equation of the bisectors of the line-pair, $ax^2 + 2hxy + by^2 = 0$ is given by

$$\frac{x^2 - y^2}{a - b} = \frac{xy}{-h} \tag{1}$$

one of these lines is, $y = mx$ (2)

Eliminating $x$ and $y$ between (1) and (2), we get

$$\frac{1 - m^2}{a - b} = \frac{m}{-h}$$

$$\Rightarrow \qquad h(1 - m^2) + m(a - b) = 0,$$

which is the required condition.

**Exercise 5.2.**

1. Find the angles between the following pairs of lines:
   (i) $x^2 - 3xy - 4y^2 = 0$.
   (ii) $2x^2 - 5xy - 7y^2 = 0$,
   (iii) $x^2 - y^2 = 0$,
   (iv) $xy = 0$.
   (*Hint:* Draw the lines.)
2. Prove that the two lines represented by the equation
   $$(x^2 + y^2) \sin^2 \alpha = (x \cos \alpha - y \sin \alpha)^2$$
   include an angle $2\alpha$.
3. Find the bisectors of the following pairs of lines:
   (i) $2x^2 - 3xy - 4y^2 = 0$,
   (ii) $3x^2 + 5xy - 8y^2 = 0$,
   (iii) $xy = 0$,
   (iv) $x^2 - y^2 = 0$.
4. If the pair of lines $x^2 - 2axy - y^2 = 0$ bisects the angles between the lines $x^2 - 2bxy - y^2 = 0$,
   show that the later pair also bisects the angles between the former pair.
5. Prove that the straight lines
   $$ax^2 + 2hxy + by^2 + \lambda(x^2 + y^2) = 0$$
   have the same pair of bisectors for all values of $\lambda$.
6. Find the condition so that $y = mx$ be one of the bisectors of the lines
   $$ax^2 + 2hxy + by^2 = 0.$$

7. Show that the equation

$$\lambda y^2 + (1 - \lambda^2)\, xy - \lambda x^2 = 0$$

where $\lambda$ is any constant, represents a pair of perpendicular lines.

8. Show that the angle between the lines given by

$$(a + 2hm + bm^2)x^2 + 2[(b - a)m - (m^2 - 1)h]xy + (am^2 - 2hm + b)\, y^2 = 0$$

is the same whatever be the value of $m$.

9. Show that the lines

$$(1 - \cos\theta \tan\alpha)y^2 - (2\cos\theta + \sin^2\theta \tan\alpha)xy + \cos\theta\,(\cos\theta + \tan\alpha)x^2 = 0$$

include an angle $\alpha$ between them.

10. Prove that the lines given by

$$(x^2 + y^2)(\cos^2\theta \sin^2\alpha + \sin^2\theta) = (x \tan\alpha - y \sin\theta)^2$$

are inclined at angle, $2\alpha$.

11. Show that the two lines given by

$$x^2 \sin^2\alpha \cos^2\theta + 4xy \sin\alpha \sin\theta + y^2 [\,4\cos\alpha - (1 + \cos\alpha)^2 \cos^2\theta\,] = 0$$

meet at angle, $\alpha$.

## 5.4. GENERAL EQUATION OF SECOND DEGREE

We have seen before that every second degree homogeneous equation represents a pair of straight lines. A natural question arises: Does a general second degree equation

$$ax^2 + 2hxy + by^2 + 2gx + 2fy + c = 0$$

also represent a pair of straight lines? We shall answer this question in this section.

Consider for example, two such equations:

$$2x^2 + 5xy + 3y^2 - 5x - 7y + 2 = 0$$

and $\quad x^2 + y^2 - 2x - 7y - 1 = 0.$

The first of these equations can be written as

$$(x + y - 2)(2x + 3y - 1) = 0$$

and hence, it represents two straight lines

$$x + y - 2 = 0 \quad \text{and} \quad 2x + 3y - 1 = 0,$$

while, the second equation cannot be put into two linear factors (in fact, we shall see in Chapter 6, that it represents a circle).

We prove below a theorem which gives the necessary and sufficient condition for a general second degree equation to represent a pair of lines.

**Theorem 5.3.** The necessary and sufficient (N.A.S.) condition for a general equation of second degree

$$ax^2 + 2hxy + by^2 + 2gx + 2fy + c = 0 \qquad (1)$$

to represent a pair of straight lines is that

$$\mathbf{abc + 2fgh - af^2 - bg^2 - ch^2 = 0}$$

Before proving the theorem, we first prove a lemma.

**Lemma.** The quadratic polynomial $ax^2 + bx + c$ is a perfect square if and only if

$$b^2 - 4ac = 0.$$

*Proof:* Clearly $a \neq 0$.

Now $ax^2 + bx + c = \dfrac{1}{a}(a^2x^2 + abx + ac)$

$$= \frac{1}{a}\left[\left(ax + \frac{b}{2}\right)^2 + \frac{4ac - b^2}{4}\right]$$

$$= \left[\frac{1}{\sqrt{a}}\left(ax + \frac{b}{2}\right)\right]^2 + \frac{4ac - b^2}{4a}$$

which is a perfect square, if $b^2 - 4ac = 0$.

Conversely, let $ax^2 + bx + c \equiv (\alpha x + \beta)^2$

$\Rightarrow \quad a = \alpha^2,\ b = 2\,\alpha\,\beta$

and $\quad c = \beta^2$

$\Rightarrow \quad b^2 - 4ac = 4\alpha^2\beta^2 - 4a^2\beta^2 = 0$

**Proof of the theorem.** As the given equation (1), is of second degree, at least one of $a$, $b$ or $h$ is non-zero.

We study the various cases:

**Case 1.** Let $b \neq 0$.

Rewrite the equation (1) as

$by^2 + 2(hx + f)\,y + (ax^2 + 2gx + c) = 0$

$$\Rightarrow y = \frac{-2(hx + f) \pm \sqrt{4(hx + f)^2 - 4b(ax^2 + 2gx + c)}}{2b}$$

$$= \frac{-(hx+f) \pm \sqrt{(h^2-ab)x^2+2(hf-bg)x+(f^2-bc)}}{b} \quad (2)$$

Now the equation (1) will represent a pair of liness iff (2) gives two first degree equations. The necessary and sufficient condition for that is, that the quadratic polynomial under the radical sign in (2), is a perfect square, which by the above Lemma is so iff,

$$4\,(hf-bg)^2 - 4(h^2-ab)\,(f^2-bc) = 0$$

i.e. iff $b\,(abc + 2fgh - af^2 - bg^2 - ch^2) = 0$

i.e. iff $abc + 2fgh - af^2 - bg^2 - ch^2 = 0$ $(\because b \neq 0)$

**Case 2.** Let $b = 0$, $a \neq 0$.

In this case, we solve equation (1) for $x$ and get the required condition as in Case 1.

**Case 3.** Let $b = 0$, $a = 0$ but $h \neq 0$.

The theorem now reduces to:

the necessary and sufficient condition for

$$2hxy + 2gx + 2fy + c = 0$$

to represent a pair of lines is that

$$2fgh - ch^2 = 0.$$

*Proof:* Clearly $2hxy + 2gx + 2fy + c = 0$ will represent a pair of lines iff we can factorise its left hand side into two linear factors.

Now $2hxy + 2gx + 2fy + c$

$$= \frac{2}{h}\left[h^2xy + ghx + fhy + \frac{ch}{2}\right]$$

$$= \frac{2}{h}\left[(hx+f)(hy+g) + \left(\frac{ch}{2} - fg\right)\right]$$

Thus we can factorise

$$2hxy + 2gx + 2fy + c = 0$$

in two linear factors iff

$$ch - 2fg = 0$$

i.e. iff $2fgh - ch^2 = 0. \quad (\because h \neq 0)$

Hence the result.

We now give two more proofs to show that the condition in **Theorem 5.3** is necessary. These proofs are of some interest in their own way.

*First Proof.* Let equation (1), above, represent a pair of lines. Assuming that these lines are not parallel, we suppose further that their point of intersection is $(x', y')$. Shifting the origin to $(x', y')$, the equation (1) transforms to:

$$a(X + x')^2 + 2h(X + x')(Y + y') + b(Y + y')^2 + 2g(X + x') + 2f(Y + y') + c = 0.$$

Now this equation represents a pair of lines through the new origin, and consequently by **Theorem 5.1,** it is homogeneous in $X$ and $Y$,

whence, coefficient of $X$ = coefficient of $Y$ = constant term=0

$$\Rightarrow \quad ax' + hy' + g = 0 \tag{2}$$

$$hx' + by' + f = 0 \tag{3}$$

and $ax'^2 + 2hx'y' + by'^2 + 2gx' + 2fy' + c = 0,$

which implies $(ax' + hy' + g)x' + (hx' + by' + f)y' + (gx' + fy' + c) = 0$ .

$$\Rightarrow \quad gx' + fy' + c = 0 \tag{4}$$

Eliminating $x', y'$ from (2), (3) and (4), we get the determinant

$$\begin{vmatrix} a & h & g \\ h & b & f \\ g & f & c \end{vmatrix} = 0$$

$\Rightarrow abc + 2fgh - af^2 - bg^2 - ch^2 = 0$ as the required condition.

**Note 1.** The reader not familiar with determinants, can proceed like this:

Solving (2) and (3) we get

$$(x', y') = \left(\frac{hf - bg}{ab - h^2}, \frac{gh - af}{ab - h^2}\right)$$

and then substituting for $x'$ and $y'$ in (4), we get the required condition.

**Corollary.** The point of intersection of the lines given by (1) is $\left(\dfrac{hf-bg}{ab-h^2}, \dfrac{gh-af}{ab-h^2}\right)$ (if these lines are not parallel).

**Note 2.** By making $ax^2 + 2hxy + by^2 + 2gx + 2fy + c = 0$ homogeneous with the help of a new variable $z$, i.e.

$$ax^2 + 2hxy + by^2 + 2\,gxz + 2fyz + cz^2 = 0$$

and then differentiating it partially w.r.t. $x$, $y$, $z$ and finally putting $z = 1$, we obtain the equations:

$$ax + hy + g = 0,\ hx + by + f = 0,\ gx + fy + c = 0,$$

which are same as equations (2), (3) and (4) respectively.

**Note 3. If** $ab - h^2 = 0$, the lines given by (1) are parallel. In this case, the method followed in the above proof fails and we follow the following method.

*Second Proof.* Let the lines represented by equation (1) be

$$lx + my + n = 0 \text{ and } l'x + m'y + n' = 0$$

$$\therefore \quad ax^2 + 2hxy + by^2 + 2gx + 2fy + c$$

$$\equiv \quad (lx + my + n)\,(l'x + m'y + n').$$

Comparing the co-efficients of various terms on both sides, we get,

$$ll' = a,\ mm' = b,\ nn' = c, \tag{1}$$

$$lm' + l'm = 2h,\ ln' + l'n = 2g,\ mn' + m'n = 2f$$

We now eliminate $l, m, n, l'$ $m'$ and $n'$ from these equations.

Clearly, we have

$$\begin{vmatrix} l & l' & 0 \\ m & m' & 0 \\ n & n' & 0 \end{vmatrix} \times \begin{vmatrix} l' & l & 0 \\ m' & m & 0 \\ n' & n & 0 \end{vmatrix} = 0$$

($\because$ each determinant is zero)

Multiplying these determinants (row by row), we have

$$\begin{vmatrix} 2ll' & lm'+l'm & ln'+l'n \\ lm'+l'm & 2mm' & mn'+m'n \\ ln'+l'n & mn'+m'n & 2nn' \end{vmatrix} = 0$$

Using (1), this reduces to

$$\begin{vmatrix} 2a & 2h & 2g \\ 2h & 2b & 2f \\ 2g & 2f & 2c \end{vmatrix} = 0 \Rightarrow \begin{vmatrix} a & h & g \\ h & b & f \\ g & f & c \end{vmatrix} = 0$$

$$\Rightarrow \quad abc + 2fgh - af^2 - bg^2 - ch^2 = 0$$

which is the required condition.

**Note.** The reader not familiar with determinants can proceed as follows:

Now $(lm' + l'm)(ln' + l'n)(mn' + m'n) = 8fgh$

$\Rightarrow 2ll'mm'nn' + ll'(m^2n'^2 + m'^2n^2) + mm'(n^2l'^2 + n'^2l^2)$
$+ nn'(l^2m'^2 + l'^2m^2) = 8fgh$

$\Rightarrow 2ll'mm'nn' + ll'[(mn' + m'n)^2 - 2mm'nn']$
$+ mm'[(nl' + n'l)^2 - 2nn'll'] + nn'[(lm' + l'm)^2 - 2ll'mm']$
$= 8fgh.$

Using (1), this becomes

$2abc + a[4f^2 - 2bc] + b[4g^2 - 2ca] + c[4h^2 - 2ab] - 8fgh = 0$

$\Rightarrow \quad abc + 2fgh - af^2 - bg^2 - ch^2 = 0$ as the required condition.

**Theorem 5.4.** If the equation

$$ax^2 + 2hxy + by^2 + 2gx + 2fy + c = 0$$

represents two straight lines, then the second degree homogeneous part, $ax^2 + 2hxy + by^2$ equated to zero, represents two straight lines through the origin and parallel to the given lines.

*Proof.* We proceed as in the Second Proof above. Let the lines given by the general equation be

$$lx + my + n = 0 \text{ and } l'x + m'y + n' = 0.$$

The equation of the lines through the origin and parallel to these lines are,

$$lx + my = 0 \text{ and } l'x + m'y = 0.$$

Their joint equation is

$$(lx + my)(l'x + m'y) = ll'x^2 + (lm' + l'm)xy + mm'y^2 = 0$$

$\Rightarrow$ $ax^2 + 2hxy + by^2 = 0$, [using (1) of Second Proof].

Hence the result.

**Corollary.** If the general equation

$$ax^2 + 2hxy + by^2 + 2gx + 2fy + c = 0$$

represents two straight lines, the angle $\theta$ between the lines is given by

$$\theta = \pm\tan^{-1}\left(\frac{\sqrt{h^2 - ab}}{a+b}\right)$$

These lines are parallel iff $h^2 = ab$ and perpendicular iff $a + b = 0$.

**Example 5.8.** Find the values of $\lambda$, so that the equation

$$x^2 - \lambda xy + 2y^2 + 3x - 5y + 2 = 0$$

represents a pair of straight lines.

*Solution.* We know that the N.A.S. condition for the general second degree equation to represent a pair of lines is

$$abc + 2fgh - af^2 - bg^2 - ch^2 = 0.$$

Here

$$a = 1, b = 2, c = 2, f = -5/2, g = 3/2 \text{ and } h = -\lambda/2.$$

Thus, we get

$$1.2.2 + 2\left(\frac{-5}{2}\right)\left(\frac{3}{2}\right)\left(\frac{-\lambda}{2}\right) - 1\left(\frac{25}{4}\right) - 2\left(\frac{9}{4}\right) - 2\left(\frac{\lambda^2}{4}\right) = 0$$

which simplifies to $2\lambda^2 - 15\lambda + 27 = 0$

$$\Rightarrow \quad \lambda = 3, 9/2,$$

and the equations of the line-pairs are

$$x^2 - 3xy + 2y^2 + 3x - 5y + 2 = 0$$

and $2x^2 - 9xy + 4y^2 + 6x - 10y + 4 = 0.$

**Example 5.9.** Prove that the equation

$$3y^2 - 8xy - 3x^2 - 29x + 3y - 18 = 0$$

represents two lines. Find their point of intersection and

the angle between them.

*Solution.* **Method I.**
Here

$a = -3,\ b = 3,\ c = -18,\ f = 3/2,\ g = -29/2$ and $h = -4$.

Now $abc + 2fgh - af^2 - bg^2 - ch^2$

$$= (-3).3.(-18) + 2.\left(\frac{-29}{2}\right).\left(-\frac{3}{2}\right)(-4) + 3.\left(\frac{3}{2}\right)^2$$

$$-3.\left(\frac{-29}{2}\right)^2 + 18.(-4)^2 = 0.$$

Thus the given equation represents a pair of lines.

Their point of intersection $(x', y')$ is given by

$$ax' + hy' + g = 0,\ hx' + by' + f = 0$$

[by **Theorem 5.3**]

$\Rightarrow \quad -3x' - 4y' - 29/2 = 0$ and $-4x' + 3y' + 3/2 = 0$

whence $(x', y') = \left(-\frac{3}{2}, \frac{-5}{2}\right)$

Since $a + b = (-3) + 3 = 0$

$\Rightarrow$ the lines are at right angles.

**Method II.** If the given equation represents a pair of lines, then by **Theorem 5.4,** the homogeneous part,

$$3y^2 - 8xy - 3x^2 = 0$$

represents a pair of lines through the origin, and parallel to the given lines.

Clearly $\quad 3y^2 - 8xy - 3x^2 = (3y + x)(y - 3x)$.

Let $\quad 3y^2 - 8xy - 3x^2 - 29x + 3y - 18$

$$\equiv (3y + x + p)(y - 3x + q)$$

$$\equiv (3y^2 - 8xy - 3y^2) + (q - 3p)x + (3q + p) + pq.$$

$\Rightarrow \quad q - 3p = -29,\ 3q + p = 3$ and $pq = -18$.

Solving first two of them,

$$p = 9,\ q = -2$$

which satisfy the third equation, as well.

Thus the given equation represents two lines, viz.

$3y + x + 9 = 0$ and $y - 3x - 2 = 0$.

On solving them, their point of intersection is $\left(\frac{-3}{2}, \frac{-5}{2}\right)$.

Again slopes of these lines are $-1/3$ and 3.

Since their product $= -1$, the lines are perpendicular.

**Method III.** Arrange the given equation as a quadratic in $y$ (or in $x$) as:

$$3y^2 + (-8x + 3)y + (-3x^2 - 29x - 18) = 0$$

$$\Rightarrow \quad y = \frac{-(-8x+3) \pm \sqrt{(-8x+3)^2 + 12(3x^2 + 29x + 18)}}{6}$$

$$= \frac{(8x-3) \pm (10x+15)}{6}$$

$$\Rightarrow y = 3x + 2 \text{ and } y = -\frac{x}{3} - 3$$

as the equations of the required lines.

Now proceed as in Method II.

**Example 5.10.** Show that if the equation

$$ax^2 + 2hxy + by^2\ 2gx + 2fy + c = 0$$

represents two parallel lines, then $h^2 = ab$, $bg^2 = af^2$ and that the distance between these parallel lines is $2\sqrt{\frac{g^2 - ac}{a(a+b)}}$.

*Solution.* We are given that

$$ax^2 + 2hxy + by^2 + 2gx + 2fy + c = 0$$

represents two parallel lines. Let these lines be $lx + my + n = 0$ and $lx + my + n' = 0$.

$\therefore \quad ax^2 + 2hxy + by^2 + 2gx + 2fy + c \equiv (l\,x + my + n)(lx + my + n')$.

Comparing coefficients, we get

$$a = l^2,\ b = m^2,\ c = nn',\ h = lm,\ 2g = l(n + n'),$$
$$2f = m(n + n').$$

Now $\quad h^2 = l^2m^2 = ab$

and $\quad bg^2 = m^2.\left[\dfrac{l(n+n')}{2}\right]^2 = l^2\left[\dfrac{m(n+n')}{2}\right]^2 = af^2$

Also, the distance between these parallel lines is

$$= \left|\frac{n'-n}{\sqrt{l^2+m^2}}\right| \qquad \text{[by § 4.6.1]}$$

$$= \frac{\sqrt{(n+n')^2-4nn'}}{\sqrt{l^2+m^2}} = \frac{\sqrt{\left(\dfrac{2g}{l}\right)^2-4c}}{\sqrt{a+b}}$$

$$= \frac{\sqrt{\dfrac{4g^2}{a}-4c}}{\sqrt{a+b}} = 2.\sqrt{\frac{g^2-ac}{a(a+b)}}$$

**Note.** It is easy to see that

$$\sqrt{\frac{g^2-ac}{a(a+b)}} = \sqrt{\frac{f^2-bc}{b(a+b)}}$$

(use $bg^2 = af^2$)

**Example 5.11.** Show that the general equation of second degree

$$ax^2 + 2hxy + by^2 + 2gx + 2fy + c = 0$$

represents a pair of parallel straight lines if
$h = \sqrt{ab}$ and $g\sqrt{b} = f\sqrt{a}$ or ($h = -\sqrt{ab}$ and $g\sqrt{b} = -f\sqrt{a}$).

Also find the distance between these lines.

(*Compare* with Example 5.10 above and also see *note* below).

*Solution*. We have $abc + 2fgh - af^2 - bg^2 - ch^2$

$$= abc + 2fg\sqrt{ab} - af^2 - bg^2 - c.ab$$

(Substituting the value of $h = \sqrt{ab}$)

$$= -(f\sqrt{a} - g\sqrt{b})^2 = 0.$$

Thus, the given equation represents a pair of lines.

Alternately, we can use $h = -\sqrt{ab}$ and $g\sqrt{b} = -f\sqrt{a}$ to arrive at the same conclusion.

Now $h = \sqrt{ab}$ or $-\sqrt{ab} \Rightarrow h^2 = ab$, whence the given lines are parallel.

For the distance between these lines, proceed as in Example 5.10.

**Note.** In some books, the conditions in Example 5.11 are stated as $h^2 = ab$ and $bg^2 = af^2$. We give an example below to show that these conditions are not sufficient because

$$h^2 = ab \Rightarrow h = \pm\sqrt{ab}$$

and $\quad bg^2 = af^2 \Rightarrow g\sqrt{b} = \pm f\sqrt{a}$,

whereas we actually need either

$g\sqrt{b} = f\sqrt{a}$ and $h = \sqrt{ab}$ (or $g\sqrt{b} = -f\sqrt{a}$ and $h = -\sqrt{ab}$).

Of course, as per a necessary condition both $h = \pm\sqrt{ab}$, $g\sqrt{b} = \pm f\sqrt{a}$ imply that $h^2 = ab$ and $bg^2 = af^2$.

Consider, for example, the second degree equation.

$$x^2 + 2xy + y^2 + 2x - 2y + c = 0 \tag{1}$$

Here $\quad a = 1, b = 1, h = 1, g = 1, f = -1.$

and $\quad h = \sqrt{ab}, g\sqrt{b} = -f\cdot\sqrt{a}.$

whence $\quad h^2 = ab$ and $bg^2 = af^2$.

But $abc + 2fgh - af^2 - bg^2 - ch^2 = c - 2 - 1 - 1 - c = -4 \neq 0$

i.e. the equation (1) does not represent a pair of lines at all.

**Example 5.12.** Find the equations of the two lines, which together with those given by the equation $8x^2 - 14xy + 3y^2 + 10x + 10y - 25 = 0$ form a parallelogram whose diagonals intersect at the point (3, 2). Find the angle between any pain of adjacent sides of the parallelogram.

*Solution.* The given equation can be written as

$$3y^2 + (10 - 14x)y + (8x^2 + 10x - 25) = 0$$

$$\Rightarrow \quad y = \frac{(14x - 10) \pm \sqrt{(10 - 14x)^2 - 12(8x^2 + 10x - 25)}}{6}$$

$$= 4x - 5, (2x + 5)/3$$

$$\Rightarrow \quad 4x - y - 5 = 0 \text{ and } 2x - 3y + 5 = 0 \quad (1)$$

These intersect at (2, 3). Since the point of intersection of the diagonals of the parallelogram is (3, 2); therefore the other end of the diagonal is (4, 1).

Now the equations of the lines through (4, 1) and parallel to the lines in (1) are

$$y - 1 = 4(x - 4) \text{ and } y - 1 = \frac{2}{3}(x - 4)$$

$$\Rightarrow \quad 4x - y - 15 = 0 \text{ and } 2x - 3y - 5 = 0.$$

Their joint equation is

$$(4x - y - 15)(2x - 3y - 5) = 0$$

$$\Rightarrow \quad 8x^2 - 14xy + 3y^2 - 50x + 50y + 75 = 0.$$

Also the angle, $\theta$, between the line pair (1) is given by

$$\tan\theta = \frac{4 - (2/3)}{1 + 4(2/3)} = \frac{10}{11}$$

$$\Rightarrow \quad \theta = \tan^{-1}(10/11).$$

## Exercise 5.3

1. Examine whether or not the following equations represent a pair of lines :
   (i) $x^2 + xy - 6y^2 + 7x + 31y - 18 = 0$,
   (ii) $x^2 + y^2 - 7x - 4y - 2 = 0$
2. Show that the following equation represents a pair of lines. Find also the angle between them :
   $6x^2 + 13xy + 6y^2 + 8x + 7y + 2 = 0$.
3. Find the lines given by the following equation. Also find their point of intersection and the angle between them:
   $2x^2 + 7xy + 3y^2 + 8x + 14y + 8 = 0$.
4. Find the value of $k$ so that the following equations may represent a pair of straight lines:
   (i) $kxy + 5x + 3y + 2 = 0$,
   (ii) $12x^2 + 2kxy + 2y^2 + 11x - 5y + 2 = 0$.

5. For what value of $\lambda$ does the equation
$$12x^2 - 10xy + 2y^2 + 11x - 5y + \lambda = 0$$
represent two straight lines? Show that the angle between them is $\tan^{-1}(1/7)$.

6. Show that the following lines form a rectangle:
$$6x^2 + 5xy - 6y^2 + x + 8y - 2 = 0$$
and $$6x^2 + 5xy - 6y^2 + 13x - 13y - 5 = 0.$$

7. Find the equations of the straight lines parallel to the straight lines
$$2x^2 - xy - 6y^2 + 7x + 21y - 15 = 0$$
and passing through (i) (0, 0), (ii) $(x_1, y_1)$.

8. Show that the equation
$$2(x + 2)^2 + 3(x + 2)(y - 2) - 2(y - 2)^2 = 0$$
represents two straight lines. Where do they intersect?

9. Show that the equation
$$x^2 + 2xy + y^2 - 8ax - 8ay - 9a^2 = 0$$
represents two parallel lines and find the distance between them.

10. Find the condition that the general equation of the second degree
$$ax^2 + 2hxy + by^2 + 2gx + 2fy + c = 0$$
should represent:
(i) two straight lines,
(ii) two parallel lines,
(iii) a pair of coincident lines.

11. A parallelogram is such that two of its adjacent sides are along the lines, $ax^2 + 2hxy + by^2 = 0$ and its centre is at $(\alpha, \beta)$. Find the joint equation of the other two sides.

12. Show that $y^2 - 4y + 3 = 0$ and $x^2 + 4xy + 4y^2 + 5x + 10y + 4 = 0$ represent lines forming a parallalogram and find the length of its sides.

13. If $ax^2 + 2hxy + by^2 + 2gx + 2fy + c = 0$
and $ax^2 + 2hxy + by^2 - 2gx - 2fy + c = 0$
each represents a pair of lines, prove that the area of the parallelogram enclosed by them is $2c/\sqrt{h^2 - ab}$.

14. Find the equations of the diagonals of the parallelogram formed by the lines : $L^2 - cL = 0$ and $L'^2 - cL' = 0$ where $L = x\cos\theta + y\sin\theta - p$ *and*

$L' = x \cos \theta' + y \sin \theta' - p'$.

15. A variable line drawn through $(\alpha, \beta)$ meets the lines, $ax^2 + 2hxy + by^2 = 0$ in $P$ and $Q$. Show that the locus of the mid. point of $PQ$ is
$ax^2 + 2hxy + by^2 = \alpha\,(ax + hy) + \beta\,(hx + by)$.

## 5.5 TO FIND THE EQUATION OF THE LINES JOINING THE ORIGIN TO THE POINTS OF INTERSECTION OF A LINE AND A CURVE

Let $\quad lx + my + n = 0 \qquad (1)$

and $\quad S = ax^2 + 2hxy + by^2 + 2gx + 2fy + c = 0 \qquad (2)$

be the given equations of a line and a curve respectively. By solving (1) and (2) simultaneously, we get that the line (1) meets the curve (2) in two points (which may be real or imaginary).

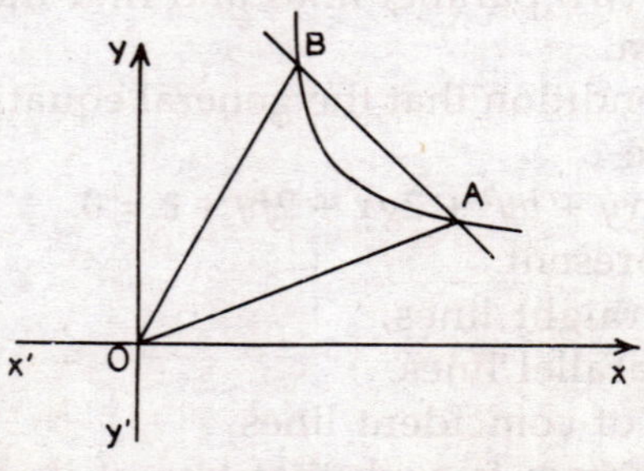


Let these points be $A$ and $B$.

We want the joint equation of the lines $OA$ and $OB$.

One way, is to get the co-ordinates of $A$ and $B$ and, then write the equations of $OA$ and $OB$.

We, however, proceed in a different way.

Write (1) as

$$lx + my = -n$$

$$\Rightarrow \qquad \frac{lx + my}{-n} = 1$$

and makes $S = 0$, homogeneous, with the help of it, to get

$$S' = ax^2 + 2hxy + by^2 + 2gx\left(\frac{lx+my}{-n}\right) + 2fy\left(\frac{lx+my}{-n}\right) + c\left(\frac{lx+my}{-n}\right)^2 = 0$$

We claim that $S' = 0$ is the required equation of the pair of lines $OA$ and $OB$.

As $S' = 0$ is a homogeneous equation of second degree, it represents two lines through the origin.

Let the co-ordinates of $A$ be $(\alpha, \beta)$.

Then $A$ lies on the line $AB$ and also on the curve $S = 0$

$$\therefore \quad l\alpha + m\beta + n = 0 \tag{3}$$

and $$a\alpha^2 + 2h\,\alpha\beta + b\beta^2 + 2g\alpha + 2f\beta + c = 0 \tag{4}$$

Now,

$S'(\alpha, \beta)$

$$a\alpha^2 + 2h\alpha\beta + b\beta^2 + 2g\alpha\left(\frac{l\alpha+m\beta}{-n}\right) + 2f\beta\left(\frac{l\alpha+m\beta}{-n}\right) + c\left(\frac{l\alpha+m\beta}{-n}\right)^2$$

$= a\alpha^2 + 2h\,\alpha\beta + b\beta^2 + 2g\alpha + 2f\beta + c$ (using (3))

$= 0.$ (using (4))

Hence $S' = 0$ passes through the point $A$.

Similarly $S' = 0$ passes through the second point $B$.

$\therefore$ $S' = 0$ represents the lines $OA$ and $OB$.

**Example 5.13.** Prove that the straight lines joining the origin to the points of intersection of the straight line $hx + ky = 2hk$ and the curve $(x - k)^2 + (y - h)^2 = c^2$ are at right angles if $h^2 + k^2 = c^2$.

*Solution.* The equation of the curve is

$$x^2 + y^2 - 2kx - 2hy + (h^2 + k^2 - c^2) = 0$$

We homogenise this equation with the help of the equation of the line,

$$hx + ky = 2hk$$

i.e. $$\frac{hx+ky}{2hk} = 1$$

Thus the required homogeneous equation is

$$x^2 + y^2 - 2kx\left(\frac{hx+ky}{2hk}\right) - 2hy\left(\frac{hx+ky}{2hk}\right)$$
$$+(h^2 + k^2 - c^2)\left(\frac{hx+ky}{2hk}\right)^2 = 0$$

Multiplying throughout by $4h^2k^2$, we get

$$4h^2k^2x^2 + 4h^2k^2y^2 - 4hk^2x(hx + ky) - 4h^2ky(hx + ky)$$
$$+ (h^2 + k^2 - c^2)(h^2x^2 + k^2y^2 + 2hkxy) = 0$$

These lines are at right angles, iff

coefficient of $x^2$ + coefficient of $y^2 = 0$

$\Leftrightarrow \quad (h^2 + k^2)(h^2 + k^2 - c^2) = 0.$

$\Leftrightarrow \quad h^2 + k^2 = c^2 \; ( \quad h^2 + k^2 \neq 0).$

**Example 5.14.** Show that two of the straight lines represented by the equation

$$ax^3 + 3bx^2y + 3cxy^2 + dy^3 = 0 \; (d \neq 0, a \neq 0),$$

are at right angles if $a^2 + 3ac + 3bd + d^2 = 0$.

*Solution.* The given equation is a homogeneous equation of third degree and hence it represents three straight lines through the origin.

Let $y = mx$, be any of these lines.

Substituting, $\dfrac{y}{x} = m$ in the given equation, we get

$$dm^3 + 3cm^2 + 3bm + a = 0 \tag{1}$$

Let $m_1$, $m_2$, $m_3$ be its roots

then $m_1.m_2.m_3 = - \; -\dfrac{a}{d}$

But, two of these lines, say with slopes, $m_1$ and $m_2$, are at right angles,

$\therefore \quad m_1 m_2 = -1.$

Thus $m_3 = \dfrac{a}{d}$, which is a root of (1).

$$\therefore \; d\left(\frac{a}{d}\right)^3 + 3c\left(\frac{a}{d}\right)^2 + 3b\left(\frac{a}{d}\right) + a = 0$$

$\Rightarrow \quad a^2 + 3ac + 3bd + d^2 = 0$ (cancelling $a$)

**Example 5.15.** Prove that the equation

$$m(x^3 - 3xy^2) + y^3 - 3x^2y = 0$$

represents three straight lines equally inclined to each other.

*Solution.* Dividing by $x^3$, the given equation can be written as

$$m\left(1-\frac{3y^2}{x^2}\right)+\frac{y^3}{x^3}-3\frac{y}{x}=0$$

Writing $\frac{y}{x}=\tan\theta$, we have

$$m(1 - 3\tan^2\theta) + (\tan^3\theta - 3\tan\theta) = 0$$

$$\Rightarrow \quad m=\frac{3\tan\theta-\tan^3\theta}{1-3\tan^2\theta}=\tan 3\theta$$

Let $m = \tan\beta$.

Then $\tan 3\theta = \tan\beta$.

$$\Rightarrow \quad 3\theta = \beta, \beta + \pi, \beta + 2\pi.$$

$$\Rightarrow \quad \theta=\frac{\beta}{3},\frac{\beta}{3}+\frac{\pi}{3},\frac{\beta}{3}+\frac{2\pi}{3}$$

Thus the given equation represents three lines through the origin and they are inclined at angles $\frac{\beta}{3},\frac{\beta}{3}+\frac{\pi}{3}$ and $\frac{\beta}{3}+\frac{2\pi}{3}$ with the $x$-axis. Clearly they are equally inclined to each other.

**Example 5.16.** Prove that if two of the lines represented by

$$ax^4 + bx^3y + cx^2y^2 + dxy^3 + ay^4 = 0$$

bisects the angle between the other two, then

$$c + 6a = 0 \text{ and } b + d = 0.$$

*Solution.* Let the equation of one pair be

$$px^2 + 2qxy + ry^2 = 0.$$

The equation of its bisectors is

$$\frac{x^2-y^2}{p-r}=\frac{xy}{q}$$

$$\Rightarrow \qquad x^2 - \frac{(p-r)}{q}xy - y^2 = 0$$

By hypothesis,

$$ax^4 + bx^3y + cx^2y^2 + dxy^3 + ay^4$$

$$\equiv (px^2 + 2qxy + ry^2).\left(x^2 - \frac{(p-r)}{q}xy - y^2\right)$$

$$\equiv px^4 + \left(2q - \frac{p(p-r)}{q}\right)x^3y - 3(p-r)x^2y^2$$

$$-\left(2q - \frac{r(p-r)}{q}\right)xy^3 - ry^4$$

Comparing the respective coefficients, we get

$$p = a,$$

$$2q - \frac{p(p-r)}{q} = b,$$

$$-3(p-r) = c,$$

$$-\left(2q + \frac{r(p-r)}{q}\right) = d$$

and $\qquad -r = a$

Now $c + 6a = -3p + 3r + 6a = -3a - 3a + 6a = 0$

and $\qquad b + d = 2q - \dfrac{p(p-r)}{q} - 2q - \dfrac{r(p-r)}{q}$

$$= -\frac{a(p-r)}{q} + \frac{a(p\quad r)}{p} = 0$$

**Example 5.17.** Prove that if all chords, of $ax^2 + 2hxy + by^2 + 2gx + 2fy + c = 0$, subtend a right angle at the origin, the equation must represent two straight lines at right angles through the origin.

*Solution.* Let $lx + my = 1$ be any chord. Homogenising the given curve with its help, we get

$$ax^2 + 2hxy + by^2 + 2(gx + fy)(lx + my) + c\,(lx + my)^2 = 0.$$

These subtend a right angle at the origin.

$\therefore$ Coeff of $x^2$ + coeff. of $y^2 = 0$

$\Rightarrow (a + 2gl + cl^2) + (b + 2fm + cm^2) = 0$

As $l$ and $m$ are arbitary, so

$$a + b = 0, \quad g = 0, \quad f = 0 \quad \text{and} \quad c = 0$$

$\therefore$ The given curve reduces to $ax^2 + 2hxy - ay^2 = 0$, which represents two lines through the origin at right angles.

## Exercise 5.4

1. Show that the straight lines joining the origin to the points of intersection of $3x^2 + 4xy - 4x + 1 = 0$ and $2x + y = 1$ are at right angles.
2. Find the equations of the straight lines joining the origin to the points of intersection of $x^2 + y^2 - 4x - 2y = 0$ and $x^2 + y^2 - 2x - 4y = 4$.
3. Prove that the lines joining the origin to the points of inter section of $x^2 + y^2 + 2gx + c = 0$ and $x^2 + y^2 + 2fy - c = 0$ are at right angles if $g^2 - f^2 = 2c$.
4. Find the angle between the lines joining the origin to the points common to $7x^2 - 4xy - 8y^2 + 2x - 4y + 8 = 0$ and $3x - y = 2$.
5. Find the equation of the pair of straight lines joining the points of intersection of $y = 3x + 2$ and $x^2 + 2xy + 3y^2 + 4x + 8y - 11 = 0$ with the origin.
6. The curve, $x^2 + y^2 + 2gx + 2fy + c = 0$, intercepts on the line $px + qy = 1$, a length which subtends a right angle at the origin, show that
$$c\,(p^2 + q^2) + 2(gp + fq + 1) = 0$$
7. Find the equation to the pair of lines joining the origin to the points of intersection of the line, $y = mx + c$ and the circle, $x^2 + y^2 = a^2$ and prove that they are at right angles if $2c^2 = a^2\,(1 + m^2)$. Hence find the condition that the line, $y = mx + c$ may touch the circle.

## Some Miscellaneous Solved Examples

**Example 5.18.** A parallelogram is formed by the lines, $ax^2 + 2hxy + by^2 = 0$ and the lines through $(p, q)$ parallel to them. Prove that the equation of the diagonal not passing through the origin is

$$(2x - p)(ap + hq) + (2y - q)(hp + bq) = 0.$$

Also show that the area of the parallelogram is

$$|ap^2 + 2hpq + bq^2| / 2\sqrt{h^2 - ab}.$$

*Solution.* Let $S_1 \equiv ax^2 + 2hxy + by^2 = 0$ (1)

and $S_2 \equiv a(x - p)^2 + 2h(x - p)(y - q) + b(y - q)^2 = 0$

(the other pair of parallel lines). Then the equation of the required diagonal is

$$S_1 - S_2 = 0$$

$$\Rightarrow a(2xp - p^2) - 2h(pq - xq - py) + b(2yq - q^2) = 0$$

$$\Rightarrow (ap + hq)(2x - p) + (hp + bq)(2y - q) = 0.$$

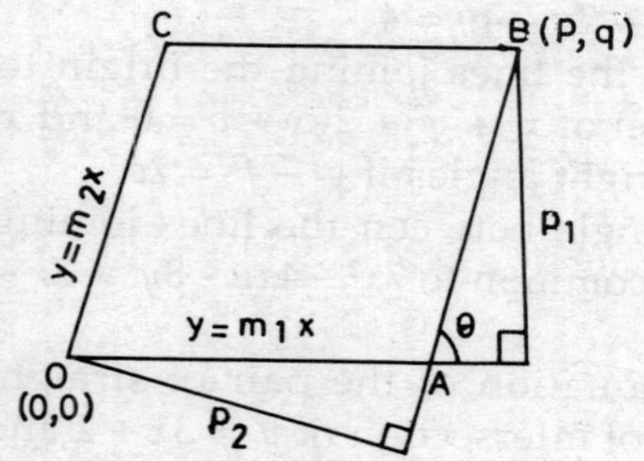


Let the lines given by (1) be

$$y = m_1 x \quad \text{and} \quad y = m_2 x$$

$$\therefore\ m_1 + m_2 = -(2h/b) \text{ and } m_1 m_2 = a/b \qquad (2)$$

Also $p_1 = \dfrac{|q - m_1 p|}{\sqrt{1 + m_1^2}}$; $p_2 = \dfrac{|q - m_2 p|}{\sqrt{1 + m_2}}$

and $\tan\theta = \dfrac{2\sqrt{h^2 - ab}}{a + b}$

$\therefore$ Area of the parallelogram

$$= p_1p_2 \operatorname{cosec}\theta$$

$$= p_1p_2\sqrt{1+\cot^2\theta}$$

$$= \frac{|q-m_1p|}{\sqrt{1+m_1^2}}\ \frac{|q-m_2p|}{\sqrt{1+m_2^2}}\cdot\sqrt{1+\frac{(a+b)^2}{4(h^2-ab)}}$$

$$= \frac{|q^2-(m_1+m_2)pq+m_1m_2p^2|}{\sqrt{1+m_1^2m_2^2+(m_1+m_2)^2-2m_1m_2}}\quad\frac{\sqrt{(a-b)^2+4h^2}}{2\sqrt{h^2-ab}}$$

$$= \frac{|ap^2+2hpq+bq^2|}{2\sqrt{h^2-ab}}\quad \text{using (2) and simplifying.}$$

**Example 5.19.** If one of the bisectors of $ax^2 + 2hxy + by^2 = 0$ passes through the point of intersection of the lines $ax^2 + 2hxy + by^2 + 2gx + 2fy + c = 0$; then show that $h(f^2 - g^2) + (a - b)fg = 0$.

*Solution*. We have

$$ax^2 + 2hxy + by^2 = 0 \quad (1)$$

and $$ax^2 + 2hxy + by^2 + 2gx + 2fy + c = 0 \quad (2)$$

The equation of the bisectors of (1) are given by

$$\frac{x^2-y^2}{a-b} = \frac{xy}{h} \quad (3)$$

Also the point of intersection of the lines given by (2) is

$$\left(\frac{hf-bg}{ab-h^2},\ \frac{gh-af}{ab-h^2}\right)$$

(see § **5.4,** Note 1).

By hypothesis, this lies on (3).

$$\therefore\ \frac{1}{a-b}\left[\left(\frac{hf-bg}{ab-h^2}\right)^2-\left(\frac{gh-af}{ab-h^2}\right)^2\right]$$

$$= \frac{1}{h}\left(\frac{hf-bg}{ab-h^2}\right)\left(\frac{gh-af}{ab-h^2}\right)$$

$\Rightarrow h[(hf - bg)^2 - (gh - af)^2] = (a - b)(hf - bg)(gh - af)$

$\Rightarrow h(f^2 - g^2) + (a - b) fg = 0,$

(on cancelling $h^2 - ab \neq 0$, otherwise the given pairs of lines are parallel).

**Example 5.20.** Show that if two of the lines, $ax^3 + bx^2y + cxy^2 + dy^3 = 0$ $(a \neq 0)$, make complementary angles with $x$-axis is, in anticlock wise sense, then

$$a(a - c) + d(b - d) = 0.$$

*Solution.* The given equation can be written as

$$dm^3 + cm^2 + bm + a = 0, \text{ where } m = y/x.$$

Let its roots be $m_1, m_2\ m_3$

$\therefore \quad m_1 m_2 m_3 = -a/d.$ (1)

By hypothesis, when $m_1 = \tan\theta$

then $m_2 = \tan(90° - \theta) = \cot\theta$.

$\therefore \quad m_1\ m_2 = 1.$

From (1), $m_3 = -a/d$.

Since $m_3$ is a root of the above cubic, we have

$$d(-a/d)^3 + c(-a/d)^2 + b(-a/d) + a = 0$$

$\Rightarrow -(a^3/d^2) + (ca^2/d^2) - (ab/d) + a = 0$

$\Rightarrow -a^2 + ca - bd + d^2 = 0$

$\Rightarrow a(a - c) + d(b - d) = 0.$

**Miscellaneous Exercise on Chapter 5.**

**Section A (For all students)**

1. The slope of one of the lines given by $ax^2 + 2hxy + by^2 = 0$ is square of the other, prove that $ab(a + b) - 6abh + 8h^3 = 0$.
2. Show that the equation of the lines which are at a distance $k$ from the point $(\alpha, \beta)$ and passing through the origin is $(\alpha y - \beta x)^2 = k^2(x^2 + y^2)$.
3. Find the equations of the straight lines through (1, 4) and respectively (i) parallel and (ii) perpendicular to the lines
$x^2 - 3xy + 2y^2 = 0$.

4. Show that $x^2 - y^2 - 2xy \tan\theta + 2ay \sec\theta - a^2 = 0$ represents for all values of $\theta$, two straight lines which pass through two fixed points. Find also the locus of their point of intersection.

5. Show that the bisectors of the angles between the lines
$$(Ax + By)^2 = 3(Bx - Ay)^2$$
are respectively parallel and perpendicular to the line
$$Ax + By + C = 0.$$

6. Show that the line $Ax + By + C = 0$ and the two lines
$$(Ax + By)^2 = 3(Bx - Ay)^2$$
form an equilateral triangle of area
$$\frac{C^2}{\sqrt{3}(A^2 + B^2)}.$$

7. If the pair of straight lines
$$ax^2 + 2hxy - ay^2 = 0$$
and $$bx^2 + 2gxy - by^2 = 0$$
be such that each bisects the angle between the other, then prove that $hg + ab = 0$.
(*Hint:* See Example **5.5**).

8. Show that the angle between one of the lines given by
$$ax^2 + 2hxy + by^2 = 0$$
and one of the lines given by
$$ax^2 + 2hxy + by^2 + \lambda(x^2 + y^2) = 0$$
is equal to the angle between the other two lines of these systems.

9. (i) If the equation,
$$ax^2 + 2hxy + by^2 + 2gx + 2fy + c = 0,$$
represents two straight lines, prove that the square of the distance of their point of intersection from the origin is
$$\frac{c(a+b) - f^2 - g^2}{ab - h^2}$$
(ii) Further, if the lines are perpendicular, prove that
this distance $= \dfrac{f^2 + g^2}{h^2 + b^2}$

10. If $ax^2 + 2hxy + by^2 + 2gx + 2fy + c = 0$

represents a pair of straight lines equidistant from the origin, prove that

$$f^4 - g^4 = c(bf^2 - ag^2).$$

11. Show that if the equation

$$ax^2 + 2hxy + by^2 + 2gx + 2fy + c = 0$$

represents two parallel lines, then

$$\frac{a}{h} = \frac{h}{b} = \frac{g}{f}$$

(*Hint:* See Example **5.10**).

12. Prove that the equation

$$x^3 - 3xy^2 + \sqrt{3}(y^3 - 3x^2y) = 0$$

represents three lines equally inclined to one another.

13. Show that for all values of $\lambda$, the lines joining the origin to the points common to

$$x^2 + 2hxy - y^2 + gx + fy = 0$$

and $$fx - gy = \lambda$$

are at right angles.

14. Find the condition for the triangle formed by the lines

$$ax^2 + 2hxy + by^2 = 0$$

and $$lx + my = 1$$

to be isosceles.

### Section B (For Honours Students)

1. A point moves so that its distance between the feet of the perpendiculars from it on the lines

$$ax^2 + 2hxy + by^2 = 0$$

is a constant $2k$. Show that the equation of its locus is

$$(x^2 + y^2)(h^2 - ab) = k^2 [(a - b)^2 + 4h^2]$$

2. The orthocentre of the triangle formed by the lines

$$ax^2 + 2hxy + by^2 = 0$$

and $$lx + my = 1$$

is $(x', y')$. Show that

$$\frac{x'}{l} = \frac{y'}{m} = \frac{a + b}{am^2 - 2hlm + bl^2}$$

3. A triangle has the lines

$$ax^2 + 2hxy + by^2 = 0$$

for two of its sides and the point $(p, q)$ for its orthocentre.

Prove that the third side is

$(a + b)(px + qy) = aq^2 - 2hpq + bp^2$.

4. Show that the distance from the origin to the orthocentre of the triangle formed by the lines

$$\frac{x}{p} + \frac{y}{q} = 1$$

and $\quad ax^2 + 2hxy + by^2 = 0$

is $\dfrac{(a+b)pq.(p^2+q^2)^{1/2}}{ap^2 - 2hpq + bq^2}$

5. Let

$f_1(x, y) = ax^2 + 2hxy + by^2 = 0$

and let $\quad f_{i+1}(x, y) = 0$

denote the equation of the bisectors of

$$f_i(x, y) = 0$$

for all $\quad i = 1, 2, 3...$

(i) Find $\quad f_2(x, y) = 0$

and $\quad f_3(x, y) = 0$.

(ii) If

$$f_1(x, y) = 0$$

represents the equation of a pair of perpendicular lines, show that

$$f_3(x, y) = 0$$

is same as

$$f_1(x, y) = 0.$$

(iii) Deduce that

$$f_{n+2}(x, y) = 0$$

is same as

$$f_n(x, y) = 0$$

for all $n \geq 2$.

6. Show that equation

$lmx^2 + (l^2 + m^2)xy + lmy^2 + lm(l - m)(x - y) - l^2m^2 = 0$

represents two lines equidistant from the origin.

7. Prove that the equation of the pair of lines passing through the points where the lines

$$S = ax^2 + 2hxy + by^2 + 2gx + 2fy + c = 0,$$

meet the axes is

$$cS + 4(fg - ch)xy = 0.$$

(*Hint:* Find the value of $k$ so that

$$S + kxy = 0$$

represents a pair of lines).

8. If

$$ax^2 + 2hxy + by^2 + 2gx + 2fy + c = 0$$

represents a pair of straight lines, prove that the area of the triangle formed by the lines and the axis of $x$ is

$$\frac{g^2 - ac}{\sqrt{h^2 - ab}}.$$

9. Find the condition that two of the lines represented by

$$ay^4 + bxy^3 + cx^2y^2 + dx^3y + ex^4 = 0$$

may be at right angles.

10. Show that if

$$x^4 + bx^3y + cx^2y^2 + dxy^3 + ey^4 = 0$$

represents two pairs of perpendicular lines, then

$$b + d = 0$$

and $$e = 1.$$

11. Show that the equation

$$a(x^4 + y^4) - 4bxy(x^2 - y^2) + 6cx^2y^2 = 0$$

represents two pairs of lines at right angles and that if

$$2b^2 = a^2 + 3ac.$$

the two pairs will coincide.

12. Show that the straight line joining the origin to the points of intersection of the curves

$$ax^2 + 2hxy + by^2 + 2gx = 0$$

and $$a'x^2 + 2h'xy + b'y^2 + 2g'x = 0$$

are at right angles if

$$g'(a + b) = g(a' + b').$$

13. Show that all chords of the curve

$$3x^2 - y^2 - 2x + 4y = 0,$$

which subtend a right angle at the origin, pass through a fixed point.
(*Hint:* Take $ax + by = 1$ as any such chord).

14. Prove that the equation of the bisectors of the lines given by
$$ax^2 + 2hxy + by^2 + 2gx + 2fy + c = 0$$
intersecting at $(x_1, y_1)$ is
$$\frac{(x-x_1)^2-(y-y_1)^2}{a-b}=\frac{(x-x_1)(y-y_1)}{h}$$
(*Hint:* Shift the origin to $(x_1, y_1)$ ).

15. If
$$ax^2 + 2hxy + by^2 + 2gx + 2fy + c = 0$$
represents two lines, prove that the product of the perpendiculars from the origin on these lines is
$$\frac{|c|}{\sqrt{(a-b)^2+4h^2}}$$

16. A line is drawn through $(p, q)$ so that the portion of the line bounded by $ax^2 + 2hxy + by^2 = 0$ is bisected at this point. Prove that its equation is
$(ap + hq)x + (hp + bq)\, y = ap^2 + 2hpq + bq^2$.

17. If $a$ and $b$ are positive real numbers, for what range of values of $k$, can a real $\lambda$ be found such that the equation
$ax^2 + 2\,\lambda xy + by^2 + 2k\,(x + y) + 1 = 0$ represents a pair of lines.

18. Show that if the portion of the line, $lx + my = 1$, falling inside the circle, $x^2 + y^2 = a^2$, subtends an angle 45° at the origin, then
$4[a^2(l^2+m^2)-1] = [a^2\,(l^2 + m^2) - 2]^2$.

19. A pair of perpendicular lines is drawn through (0, 0) forming with the line, $2x + 3y = 6$, an isosceles triangle which subtends a right angle at (0, 0). Find the equation of the line pair and the area of this triangle.

20. If $ax^2 + 2hxy + by^2 + 2gx + 2fy + c = 0$, represents a pair of lines, then show that the area of the triangle formed by the bisectors of the angles between them and $x$-axis

is given by

$$(g^2 - ac)\sqrt{(a-b)^2 + 4h^2} \,/\, 2h\,(h^2 - ab)$$

21. A variable line is drawn through $A(\alpha, \beta)$ meeting the line-pair, $ax^2 + 2hxy + by^2 = 0$, in $P$ and $Q$. A point $R$ is taken on the line such that $AR$ is the harmonic mean between $AP$ and $AQ$. Prove the locus of $R$ is

$$(a\,\alpha + h\,\beta)\,x + (hx + b\,\beta)\,y = 0$$

22. Show that the locus of a point such that the product of the perpendiculars let fall from it on the three lines represented by,
$ay^3 + by^2x + cyx^2 + dx^3 = 0$ is constant $= k^3$, is
$ay^3 + by^2x + cyx^2 + dx^3 = k^3\sqrt{(a-c)^2 + (b-d)^2}$.

# 6. CIRCLE

## INTRODUCTION

In this chapter, we discuss the various properties and results concerning a circle, including the concepts of orthogonal circles, co-axal system of circles and their limiting points.

**Definition of a Circle.** A circle is the locus of a point which moves in a plane, so that its distance from a *fixed point* in the plane is always *constant*.

The fixed point is called the *centre* of the circle and the constant distance is called its *radius*.

## 6.1 TO FIND THE EQUATION OF A CIRCLE WHOSE CENTRE AND RADIUS ARE GIVEN

Let $(\alpha, \beta)$ be the centre of the circle and $r$ be its radius. If $(x, y)$ be any point on it, then

$$(x - \alpha)^2 + (y - \beta)^2 = r^2$$

which is the required equation of the circle.

This equation can be written as

$$x^2 + y^2 - 2\,\alpha x - 2\,\beta y + (a^2 + \beta^2 - r^2) = 0,$$

which is of the form

$$x^2 + y^2 + 2gx + 2fy + c = 0.$$

We observe that this equation has three characteristics:

(1) It is a second degree equation,

(2) Coefficient of $x^2$ = coefficient of $y^2$, and

(3) Coefficient of $xy = 0$.

*Conversely,* we prove that an equation having these three characteristics, always represents a circle.

Let $ax^2 + ay^2 + 2g'x + 2f'y + c' = 0$ (1)

be any such equation.

Condition (1) $\Rightarrow$ $a \neq 0$.

On dividing (1) by $a$, we get

$$x^2+y^2+2\frac{g'}{a}x+2\frac{f'}{a}y+\frac{c'}{a}=0 \qquad (2)$$

Set

$$\frac{g'}{a}=g,\ \frac{f'}{a}=f \text{ and } \frac{c'}{a}=c.$$

Equation (2) becomes

$$x^2 + y^2 + 2gx + 2fy + c = 0$$

$\Rightarrow (x^2 + 2gx + g^2) + (y^2 + 2fy + f^2) = g^2 + f^2 - c,$

$\Rightarrow [x - (-g)]^2 + [y - (-f)]^2 = (\sqrt{g^2 + f^2 - c})^2$

whence the distance of the moving point $(x, y)$ from a fixed point $(-g, -f)$ is always $\sqrt{g^2+f^2-c}$.

Thus the equation (1) represents a circle with centre at

$$(-g,-f)=\left(-\frac{g'}{a},-\frac{f'}{a}\right)$$

and $\qquad$ radius $=\sqrt{g^2+f^2-c}=\dfrac{1}{a}\sqrt{g'^2+f'^2-c'a}$

**Corollary 1.** The centre and the radius of the circle

$$x^2 + y^2 + 2gx + 2fy + c = 0$$

are $(-g, -f)$ and $\sqrt{g^2+f^2-c}$ respectively.

**Corollary 2.** $x^2 + y^2 + 2gx + 2fy + c = 0$ represents a real circle, a point circle or an imaginary circle according as

$$g^2 + f^2 - c >, = \text{ or } < 0,$$

where we call a circle with zero radius to be a *point circle.*

**Corollary 3.** The equation of the circle with centre at (0, 0) and radius = $a$ is $x^2 + y^2 = a^2$.

**Note.** The equation

$$x^2 + y^2 + 2gx + 2fy + c = 0$$

of a circle contains three unknown quantities $g, f$ and $c$; hence for determining the equation of a circle, three conditions are needed.

### 6.1.1 To find the equation of a circle passing through three non-collinear points.

Let $P(x_1, y_1)$, $Q(x_2, y_2)$ and $R(x_3, y_3)$ be the given non-collinear points.

$\Rightarrow$ area $\Delta PQR \neq 0$

$$\Rightarrow x_1(y_2 - y_3) + x_2(y_3 - y_1) + x_3(y_1 - y_2) \neq 0 \quad \text{(A)}$$

Let $x^2 + y^2 + 2gx + 2fy + c = 0$ (1)

be any circle through these points.

$$\Rightarrow x_1^2 + y_1^2 + 2gx_1 + 2fy_1 + c = 0 \quad (2)$$

$$x_2^2 + y_2^2 + 2gx_2 + 2fy_2 + c = 0 \quad (3)$$

and $x_3^2 + y_3^2 + 2gx_3 + 2fy_3 + c = 0$ (4)

Subtracting (2) from (3) and also from (4) yields

$$2g(x_2 - x_1) + 2f(y_2 - y_1) + (x_2^2 + y_2^2 - x_1^2 - y_1^2) = 0 \quad (5)$$

and $2g(x_3 - x_1) + 2f(y_3 - y_1) + x_3^2 + y_3^2 - x_1^2 - y_1^2) = 0$ (6)

Multiplying (5) by $y_3 - y_1$, (6) by $y_2 - y_1$, and subtracting, we obtain

$$2g[(x_2 - x_1)(y_3 - y_1) - (x_3 - x_1)(y_2 - y_1)]$$
$$+ (x_2^2 + y_2^2 - x_1^2 - y_1^2)(y_3 - y_1)$$
$$- (x_3^2 + y_3^2 - x_1^2 - y_1^2)(y_2 - y_1) = 0 \quad (7)$$

Now coefficient of $2g$

$$= x_2(y_3 - y_1) - x_1(y_3 - y_1) - x_3(y_2 - y_1) + x_1(y_2 - y_1)$$
$$= x_1(y_2 - y_3) + x_2(y_3 - y_1) + x_3(y_1 - y_2) \neq 0. \text{ [using (A)]}$$

Thus we get a unique value of $g$, from (7).

Similarly we get only one value of $f$ from (5) or (6) and only one value of $c$, from (2), (3) or (4).

**Note.** We can also obtain the equation of the required circle through the given points by eliminating $g, f, c$ from (1), (2), (3) and (4), with the help of determinants.

Thus $\begin{vmatrix} x^2+y^2 & x & y & 1 \\ x_1^2+y_1^2 & x_1 & y_1 & 1 \\ x_2^2+y_2^2 & x_2 & y_2 & 1 \\ x_3^2+y_3^2 & x_3 & y_3 & 1 \end{vmatrix} = 0$

is the desired equation.

**Example 6.1.** Find the equation of the circle circumscribing the triangle formed by the lines:

$x + y = 6,\ 2x + y = 4,\ x + 2y = 5.$

*Solution.* **Method I**

Solving the given equations in pairs, we get the vertices of the triangle as

$$(-2, 8),\ (1, 2) \text{ and } (7, -1).$$

We require the circle through these points.

Let $x^2 + y^2 + 2gx + 2fy + c = 0$ (1)

be the equation of the circle through them.

Then $68 - 4g + 16f + c = 0$ (2)

$5 + 2g + 4f + c = 0$ (3)

$50 + 14g - 2f + c = 0$ (4)

Solving (2), (3) and (4), for $g, f$ and $c$, we get

$$g = -\frac{17}{2},\ f = -\frac{19}{2} \text{ and } c = 50$$

Thus the required circle is given by

$$x^2 + y^2 - 17x - 19y + 50 = 0.$$

**Method II.** Consider the equation

$$\lambda(x + y - 6)(2x + y - 4) + \mu(2x + y - 4)(x + 2y - 5) + (x + 2y - 5)(x + y - 6) = 0 \quad (1)$$

It is easily seen that the curve represented by (1) passes through the vertices of the given triangle. We want (1) to represent a circle, for which we must have:

(1) Coefficient of $x^2$ = coefficient of $y^2$, and
(2) Coefficient of $xy = 0$.

Thus, we get $2\lambda + 2\mu + 1 = \lambda + 2\mu + 2$
and $3\lambda + 5\mu + 3 = 0$,

whence $\lambda = 1$ and $\mu = -\frac{6}{5}$

Thus, the equation of the required circle is

$$(x + y - 6)(2x + y - 4) - \frac{6}{5}(2x + y - 4)(x + 2y - 5) + (x + 2y - 5)(x + y - 6) = 0$$

which, on simplifying, reduces to

$$x^2 + y^2 - 17x - 19y + 50 = 0.$$

**6.1.2. To find the equation of a circle, extremities of one of whose diameters are $(x_1, y_1)$ and $(x_2, y_2)$.**

Clearly centre of the desired circle is

$$\left(\frac{x_1 + x_2}{2}, \frac{y_1 + y_2}{2}\right)$$

and its radius

$$= \frac{1}{2}\sqrt{(x_1 - x_2)^2 + (y_1 - y_2)^2}$$

$\therefore$ its equation is

$$\left(x - \frac{x_1 + x_2}{2}\right)^2 + \left(y - \frac{y_1 + y_2}{2}\right)^2 = \frac{1}{4}\left[(x_1 - x_2)^2 + (y_1 - y_2)^2\right] \quad (1)$$

**Aliter.** Call the points $(x_1, y_1)$ and $(x_2, y_2)$ as $A$ and $B$ respectively. Let $P(x, y)$ be any point on the circle, whose diameter is $AB$. Then

$$\angle APB = \frac{\pi}{2} \text{ (angle in a semicircle)}$$

$\Rightarrow$ Product of slopes of $AP$ and $BP$ is equal to $-1$.

$$\Rightarrow \frac{y - y_1}{x - x_1} \cdot \frac{y - y_2}{x - x_2} = -1$$

$$\Rightarrow (x - x_1)(x - x_2) + (y - y_1)(y - y_2) = 0 \quad (2)$$

**Note.** (1) and (2) turn out to be the same equation, after simplification.

**Example 6.2.** Find the equation of the circle passing through (1, 2), (0, –1) and whose centre lies on the line $x + 2y - 5 = 0$.

*Solution.* Let the equation of the circle be

$x^2 + y^2 + 2yx + 2fy + c = 0$

It passes through (1, 2) and (0, –1).

$\therefore \quad 2g + 4f + c + 5 = 0,$

and $\quad -2f + c + 1 = 0.$

Also the centre $(-g, -f)$ lies on $x + 2y - 5 = 0$.

$\therefore \quad -g - 2f - 5 = 0.$

On solving these equations, we obtain

$g = -11, f = 3$, and $c = 5$.

$\therefore$ the equation of the required circle is

$$x^2 + y^2 - 22x + 6y + 5 = 0.$$

**Example 6.3.** A circle intersects a constant length, $2a$, on the axis of $x$ and a constant length, $2b$, on the axis of $y$. If the circle varies, show that the locus of its centre is $x^2 - y^2 = a^2 - b^2$.

*Solution.* Let $x^2 + y^2 + 2gx + 2fy + c = 0 \qquad (1)$

be the circle.

As (1) intercepts a length, $2a$, on $x$ –axis; putting, $y = 0$, in (1), we get

$$x^2 + 2gx + c = 0.$$

Let its roots be $x_1$ and $x_2$

$\therefore \quad x_1 + x_2 = -2g$ and $x_1 x_2 = c$.

By hypothesis,

$$2a = |x_2 - x_1| = \sqrt{(x_1 + x_2)^2 - 4x_1x_2}$$

$\Rightarrow \quad a = \sqrt{g^2 - c}$

$\Rightarrow \quad g^2 = a^2 + c \qquad (2)$

In the same way, we get

$f^2 = b^2 + c \qquad (3)$

Now $\quad (2) - (3) \Rightarrow$

$$g^2 - f^2 = a^2 - b^2.$$

$\therefore$ The locus of $(-g, -f)$ is

$$(-x)^2 - (-y)^2 = a^2 - b^2$$

i.e. $\quad x^2 - y^2 = a^2 - b^2.$

**Example 6.4.** Show that the locus of a point which moves such that the sum of squares of its distances from the vertices of a triangle is constant, is a circle whose centre is the centroid of the triangle.

*Solution.* Let $ABC$ be the given triangle. Let $BC$ be taken as $x$-axis, its mid point, $O$, as the origin and a line through $O$, perpendicular to $BC$ as $y$-axis. Let $BC = 2a$, then the coordinates of B and C are $(-a, 0)$ and $(a, 0)$ respectively. Let the co-ordinates of $A$ be $(p, q)$. Let the co-ordinates of the point $P$ whose locus is required be $(h, k)$.

By hypothesis,

$$PA^2 + PB^2 + PC^2 = c \text{ (a constant)}$$

$$\Rightarrow \quad [(h - p)^2 + (k - q)^2] + [(h + a)^2 + k^2] + [(h - a)^2 + k^2] = c$$

$$\Rightarrow \quad 3\,(h^2 + k^2) - 2ph - 2qk + (p^2 + q^2 + 2a^2) = c$$

$\therefore$ The locus of $(h, k)$ is

$$x^2 + y^2 - \frac{2}{3}px - \frac{2}{3}qy + \frac{1}{3}(p^2 + q^2 + 2a^2) = c/3$$

which represents a circle whose centre is $\left(\frac{1}{3}p, \frac{1}{3}q\right)$ which is the centroid of the triangle $ABC$.

### Exercise 6.1

1. Find, in each case, the equation of a circle whose:
   (i) Centre $(-1, 2)$, radius $= 3$;
   (ii) Centre $(a, b)$, radius $= a + b$;
   (iii) Centre $(2, 3)$ and passing through the point $(3, 5)$;
   (iv) Radius $= a$, and passing through two points on the $x$-axis at a distance $d$ from the origin.
2. Examine which of the following equations, represent real, point or imaginary circles. Whenever possible find the

centre and the radius.

(i) $x^2 + y^2 - 6x + 2y - 3 = 0$,

(ii) $2x^2 + 2y^2 - 7x + 8y - 4 = 0$,

(iii) $x^2 + y^2 + 4x + 6y + 13 = 0$,

(iv) $x^2 + y^2 + 2x + 3y + 7 = 0$.

3. Find the equations of the circles passing through the points

(i) $(-1, 2), (-2, 1), (2, -1)$;

(ii) $(a, 0), (0, b)$ and $(0, 0)$.

4. Find the equations of the circles circumscribing the triangles formed by the lines:

(i) $x - y + 3 = 0, x + 2y = 0, x + y - 1 = 0$;

(ii) $x + y = 2, 2x + 3y - 5 = 0, x + 2y - 7 = 0$.

5. Find the equations of the circles whose diameters have end points :

(i) $(3, 4)$ $(1, -2)$;

(ii) $(a, 0), (0, b)$;

(iii) $(0, 0), (a, b)$.

6. Find, in each of the following cases, the equation of a circle

(i) passing through $(3, 5), (5, 3)$ and centre lying on
$$2x + 3y - 7 = 0.$$

(ii) passing through $(1, -3)$ and centre lying on
$$2x + y - 5 = 0 \text{ and } x + 2y - 7 = 0$$
(or having these lines as diameters).

(iii) passing through the points $(1, 3), (-1, -2)$ and having its centre on the $y$-axis.

7. A circle with its centre on the $y$-axis, passes through the origin and the point $(a, b)$. Find its equation.

8. Find the equation of that diameter of the circle
$$x^2 + y^2 - 4x - 6y + 11 = 0,$$
which passes through $(3, 4)$

9. Prove that the centes of the circles
$$x^2 + y^2 = 1, x^2 + y^2 + 4x + 8y - 1 = 0$$
and $x^2 + y^2 - 6x - 12y + 1 = 0$ are collinear.

10. Show that the points $(1, -6), (5, 2), (7, 0)$ and $(-1, -4)$ are the vertices of a cyclic quadrilateral.

11. *ABCD* is a square whose side is $a$. Taking *AB* and *AD* as the axes of co-ordinates, prove that the equation of the circumcircle of the square is $x^2 + y^2 = a(x + y)$.

12. Find the equation of the locus of a point which moves in a plane so that the sum of squares of its distances from the lines

$$7x - 4y - 10 = 0 \text{ and } 4x + 7y + 5 = 0$$

is always equal to 3. Show that the locus represents a circle. Find the centre and the radius of this circle.

13. Repeat Question 12, taking the lines as

$$ax + by + c = 0 \text{ and } bx - ay + d = 0,$$

and sum of squares as $r^2$.

14. Given a circle $x^2 + y^2 + 2gx + 2fy + c = 0$. Prove that the constant $c$ in this equation is equal to the rectangle under (or the product of) the segments of the chords through the origin.

15. Find the locus of the centre of a circle which passes through the origin and cuts off a length $2a$ from the line, $x = b$.

16. A circle touches a given line and cuts of a constant length, $2l$, from another line, perpendicular to the former; find the locus of its centre.

17. Prove that the equation of the chord of the circle, $x^2 + y^2 = a^2$, joining two points $(x_1, y_2)$ and $(x_2, y_2)$ on it, is $x(x_1 + x_2) + y(y_1, +y_1) = x_1x_2 + y_1y_2 + a^2$.

18. $P$ is a point on the circle, $x^2 + y^2 + 2x + 6y - 15 = 0$; $Q$ is a point on the line, $7x + y + 3 = 0$ and the perpendicular bisector of $PQ$ is the line, $x - y + 1 = 0$. Prove that there are two positions of the point $P$ and find their positions.

19. Find the circle which touches both axes and the line, $3x - 4y = 12$.

20. Prove that the locus of a point, the sum of the squares of whose distances from the vertices of a square is constant, is a circle.

## 6.2 TO FIND THE POSITION OF A POINT P($x_1$, $y_1$) w.r.t. A GIVEN CIRCLE:

$$x^2 + y^2 + 2gx + 2fy + c = 0.$$

Let its centre be $C(-g, -f)$. Join $CP$ and let it meet the circle at $Q$. Then the point $P$ is outside, on or within the circle according as

$$|CP| >, = \text{ or } < |CQ|.$$

i.e. according as $\quad CP^2 >, = \text{ or } < r^2.$

i.e. according as $\quad (x_1 + g)^2 + (y_1 + f)^2 >, = \text{ or } < g^2 + f^2 - c.$

i.e. according as $\quad x_1^2 + y_1^2 + 2gx_1\ 2fy_1 + c >, = \text{ or } < 0.$

### § 6.2.1. Parametric equations of a circle

Let $P(x, y)$ be any point on the circle $x^2 + y^2 = a^2$. Join $OP$ and let it make an angle $\theta$, with the $x$-axis.

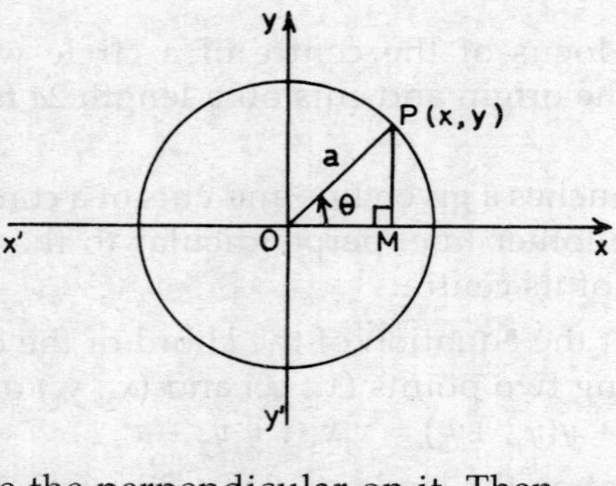


Let $PM$ be the perpendicular on it. Then

$$x = OM = a \cos \theta$$

and $\quad y = MP = a \sin \theta.$

Thus the co-ordinates of any point on the given circle are

$$(a \cos \theta, a \sin \theta)$$

and $\quad x = a \cos \theta, y = a \sin \theta$

are its parametric equations.

**Corollary.** Any point on the circle

$$(x - \alpha)^2 + (y - \beta)^2 = r^2 \text{ is}$$

$(\alpha + r \cos \theta, \beta + r \sin \theta)$ and its parametric equations are

$$x = \alpha + r \cos \theta, y = \beta + r \sin \theta.$$

**Note.** $\theta$ is just a parameter, in above and it takes different values. One can use any other symbol in place of $\theta$, e.g.

$$x = a \cos t,\ y = a \sin t.$$

## 6.3. CIRCLE THROUGH THE POINTS OF INTERSECTION OF TWO GIVEN CIRCLES

Let $S_1 = x^2 + y^2 + 2g_1x + 2f_1y + c_1 = 0$ (1)

and $S_2 = x^2 + y^2 + 2g_2x + 2f_2y + c_2 = 0$ (2)

be two given circles.

Consider the equation

$$\lambda S_1 + \mu S_2 = 0,$$

where $\lambda$, $\mu$ are arbitrary real numbers, not both zero and $\lambda \neq -\mu$.

Using (1) and (2), $\lambda S_1 + \mu S_2 = 0$ becomes

$$(\lambda + \mu)(x^2 + y^2) + 2(\lambda g_1 + \mu g_2)x + 2(\lambda f_1 + \mu f_2) + (\lambda c_1 + \mu c_2) = 0$$

which is a second degree equation in $x$ and $y$, coefficients of $x^2$ and $y^2$ are each equal to $\lambda + \mu$ ($\neq 0$), and there is no term $xy$. Hence it represents a circle. Further any point lying on both (1) and (2) also lies on it.

$\therefore$ $\lambda S_1 + \mu S_2 = 0$, ($\lambda \neq -\mu$), represents a circle passing through the points of intersection of the given circles $S_1 = 0$ and $S_2 = 0$.

**Remark 1.** As in **§ 4.8,** it can be shown that the equation of any circle passing through the points of intersection of the circles can be put in the form

$$\lambda S_1 + \mu S_2 = 0\ (\lambda \neq -\mu).$$

(See also Remark after Theorem 4.2 (**§ 4.8**)).

**Remark 2.** Let $\lambda = -\mu$; then $\lambda S_1 + \mu S_2 = 0$

$\Rightarrow$ $S_1 - S_2 = 0$

$\Rightarrow$ $2(g_1 - g_2)x + 2(f_1 - f_2)y + c_1 - c_2 = 0.$

It is a first degree equation and hence represents a straight line, passing through the points of intersection of the given circles, $S_1 = 0 = S_2$. This line is called their *common chord*.

**Remark 3.** One can show similarly that $S + \lambda L = 0$ is a circle passing through the points of intersection of the circle, $S = 0$ and the line, $L = 0$. Using this, we can write

$$S_1 + \lambda (S_1 - S_2) = 0$$

as the equation of a circle passing through the points of intersection of the circles, $S_1 = 0 = S_2$.

**Example 6.5.** Find the points of intersection of the circle $x^2 + y^2 + 7x - 7y - 8 = 0$ and the line $x - y - 1 = 0$. Also find the length of the chord.

*Solution.* Solving the two equations simultaneously, we have

$$(y + 1)^2 + y^2 + 7(y + 1) - 7y - 8 = 0, \text{ using } x = y + 1.$$

$$\Rightarrow \quad 2y^2 + 2y = 0 \Rightarrow y = 0, -1.$$

$$\therefore \quad x = 1, 0 \text{ respectively.}$$

Thus the points of intersection of the given line and the given circle are (1, 0) and (0, – 1).

Also the length of the chord = $\sqrt{(1-0)^2 + (0+1)^2} = \sqrt{2}$.

**Example 6.6.** Find the points of intersection of the circles

$$x^2 + y^2 + 7x - 7y - 8 = 0$$

and $\quad 2x^2 + 2y^2 + 15x - 15y - 17 = 0.$

*Solution.* The equations of the circles can be written as:

$$S_1 = x^2 + y^2 + 7x - 7y - 8 = 0 \quad (1)$$

$$S_2 = x^2 + y^2 + \frac{15}{2}x - \frac{15}{2}y - \frac{17}{2} = 0 \quad (2)$$

Their common chord is given by

$$S_1 - S_2 = 0 \Rightarrow -\frac{1}{2}x + \frac{1}{2}y + \frac{1}{2} = 0 \Rightarrow x - y - 1 = 0. \quad (3)$$

$\therefore$ The points of intersection of (1) and (2) are the same as those of (1) and (3), which we have found in **Example 6.5.**

**Example 6.7.** Find the equation of the circle passing through the points of intersection of the circles

$$x^2 + y^2 + 8x - 4y + 10 = 0;$$

$$3x^2 + 3y^2 - 2x + 3y - 5 = 0$$

and also passing through the origin.

*Solution.* Let $S_1 = x^2 + y^2 + 8x - 4y + 10 = 0 \quad (1)$

and $\quad S_2 = 3x^2 + 3y^2 - 2x + 3y - 5 = 0 \quad (2)$

The circle passing through the points of intersection of (1) and (2) can be taken as

$$S_1 + \lambda S_2 = 0$$

$$\Rightarrow \quad (x^2 + y^2 + 8x - 4y + 10) + \lambda(3x^2 + 3y^2 - 2x + 3y - 5) = 0 \qquad (3)$$

Now for this to be the required circle, it must pass through (0, 0); whence

$$10 + \lambda(-5) = 0 \Rightarrow \lambda = 2.$$

Substituting this value of $\lambda$ in (3), we get

$$7(x^2 + y^2) + 4x + 2y = 0,$$

which is the equation of the required circle.

**Example 6.8.** Show that the circle on the chord

$$x \cos \alpha + y \sin \alpha = p$$

of the circle

$$x^2 + y^2 = a^2$$

as diameter is

$$x^2 + y^2 - a^2 - 2p(x \cos \alpha + y \sin \alpha - p) = 0.$$

*Solution.* The equation of any circle through their points of intersection is $x^2 + y^2 - a^2 + \lambda(x \cos \alpha + y \sin \alpha - p) = 0$.

Its centre $= \left(\dfrac{-\lambda \cos \alpha}{2}, \dfrac{-\lambda \sin \alpha}{2}\right)$

But the line $x \cos \alpha + y \sin \alpha = p$ is a diameter of the required circle.

$$\therefore \quad \left(\frac{-\lambda \cos \alpha}{2}\right) \cos \alpha + \left(\frac{-\lambda \sin \alpha}{2}\right) \sin \alpha - p = 0$$

$$\Rightarrow \lambda = -2p.$$

Thus the equation of the required circle is

$$x^2 + y^2 - a^2 - 2p(x \cos \alpha + y \sin \alpha - p) = 0.$$

**Example 6.9.** Find the length of the chord $lx + my + n = 0$ of the circle $x^2 + y^2 = a^2$.

*Solution.* Now $lx + my + n = 0 \Rightarrow y = \dfrac{lx + n}{-m}$

$$\therefore \quad x^2 + y^2 = a^2 \Rightarrow x^2 + \left(\frac{lx+n}{-m}\right)^2 = a^2$$

$$\Rightarrow (l^2 + m^2)\, x^2 + 2lnx + (n^2 - a^2m^2) = 0.$$

If $(x_1, y_1)$ and $(x_2, y_2)$ are their points of intersection, then

$$x_1 + x_2 = -\frac{2ln}{l^2 + m^2}$$

and $$x_1 x_2 = \frac{n^2 - a^2m^2}{l^2 + m^2}$$

$$\Rightarrow |x_1 - x_2| = \sqrt{(x_1 + x_2)^2 - 4x_1x_2}$$

$$= \frac{\sqrt{4l^2n^2 - 4(n^2 - a^2m^2)(l^2 + m^2)}}{l^2 + m^2}$$

$$= \frac{2\,|m|\sqrt{a^2(l^2 + m^2) - n^2}}{l^2 + m^2} \quad (1)$$

Now $(x_1, y_1)$, $(x_2, y_2)$ also lie upon $lx + my + n = 0$, whence $lx_1 + my_1 + n = 0$, $lx_2 + my_2 + n = 0$.

Subtracting $y_1 - y_2 = -\dfrac{l}{m}(x_1 - x_2)$

Thus, the length of the chord

$$= \sqrt{(x_1 - x_2)^2 + (y_1 - y_2)^2}$$

$$= \sqrt{(x_1 - x_2)^2 + \frac{l^2}{m^2}(x_1 - x_2)^2}$$

$$= \frac{|x_1 - x_2|\sqrt{l^2 + m^2}}{|m|}$$

$$= 2\sqrt{\frac{a^2(l^2 + m^2) - n^2}{l^2 + m^2}}, \text{ using (1)}$$

**Example 6.10.** A chord of the circle, $x^2 + y^2 - 2ax = 0$, is drawn to pass through the origin. Show that the locus of the centre of the circle described on this chord as diameter is a circle passing through the centre of the given circle.

*Solution*. The given circle is

$$x^2 + y^2 - 2ax = 0 \tag{1}$$

Let any chord of (1) through (0, 0) be

$$mx - y = 0 \tag{2}$$

Equation of the circle passing through the intersection of (1) and (2) is

$$x^2 + y^2 - 2ax + \lambda (mx - y) = 0 \tag{3}$$

As (2) is to be the diameter of (3), centre of (3) lies on (2).

$$\therefore \quad m\left(\frac{2a - \lambda m}{2}\right) - \frac{\lambda}{2} = 0$$

$$\Rightarrow \quad \lambda = 2am/(1 + m^2).$$

$\therefore$ $x$ = $x$-coordinate of the centre of (3)

$$= \frac{2a - \lambda m}{2} = a - \frac{am^2}{1+m^2} = \frac{a}{1+m^2}$$

and $y$ = $y$-coordinate of the centre of (3)

$$= \frac{\lambda}{2} = \frac{am}{1+m^2} \tag{4}$$

whence $y = mx$, i.e. $m = y/x$ and then by using (4),

$$y = \frac{a(y/x)}{1+(y/x)^2}$$

$\Rightarrow x^2 + y^2 - ax = 0$

which is a circle passing through the centre $(a, 0)$ of (1).

**Example 6.11**. Show that the equation of a line meeting the circle, $x^2 + y^2 = a^2$ in two points at equal distance, $d$, from a point $(x_1, y_1)$ on the circumference is, $xx_1 + yy_1 - a^2 + (d^2/2) = 0$.

*Solution*.

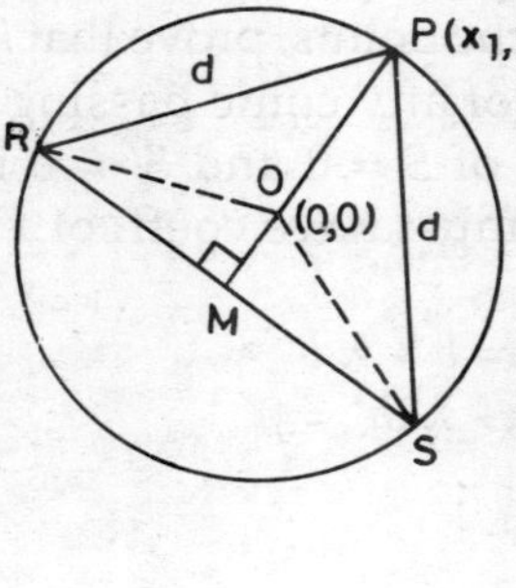

We want the equation of chord, $RS$ where $PR = PS = d$ and $P(x_1, y_1)$ lies on $x^2 + y^2 = a^2$.

$$\therefore \quad x^2_1 + y^2_1 = a^2 \tag{1}$$

From the figure, as $PM \perp RS$, $M$ is the mid. point of $RS$.

$\therefore \quad RM^2 + MO^2 = a^2$

and $\quad RM^2 + (MO + a)^2 = d^2$

$\Rightarrow \quad (MO + a)^2 - MO^2 = d^2 - a^2$

$$\Rightarrow \quad MO = (d^2 - 2a^2)/2a \tag{2}$$

Clearly equation of line, $OP$, is

$$y = \frac{y_1}{x_1} x \text{ i.e. } y_1 x - x_1 y = 0$$

As $RS \perp MP \Rightarrow RS \perp OP$

$\therefore$ Equation of $RS$ is

$$x_1 x + y_1 y + k = 0 \tag{3}$$

Now $a + MO$ = perpendicular from $P$ on (3)

$$= \frac{x_1^2 + y_1^2 + k}{\sqrt{x_1^2 + y_1^2}}$$

$$\Rightarrow a + \frac{d^2 - 2a^2}{2a} = \frac{a^2 + k}{a}; \text{ using (1) and (2)}$$

$$\Rightarrow k = \frac{1}{2} d^2 - a^2$$

$\therefore$ Equation of the chord $RS$ is

$$x_1 x + y_1 y - a^2 + (d^2/2) = 0$$

**Example 6.12.** If the conics whose equations are

$$S \equiv ax^2 + 2hxy + by^2 + 2gx + 2fy + c = 0$$

and $\quad S' \equiv a'x^2 + 2h'xy + b'y^2 + 2g'x + 2f'y + c' = 0$

intersect in four concyclic points, prove that $h'(a - b) = h(a' - b')$.

*Solution*. Equation of any conic passing through the four points of intersection of $S = 0$ and $S' = 0$ is $S + \lambda S' = 0$. For this to be a circle, we must have coeff. of $x^2$ = coeff of $y^2$ and Coeff of $xy = 0$

$$\Rightarrow \quad a + \lambda a' = b + \lambda b'$$

$$\text{and } 2h + \lambda 2h' = 0$$

$$\Rightarrow \qquad \lambda = \frac{a-b}{b'-a'} = -\frac{h}{h'}$$

$$\Rightarrow \qquad h'(a-b) = h(a'-b')$$

**Exercise 6.2**

1. Find the positions of the following points w.r.t. the given circles:
   (i) $(0, 0)$; $x^2 + y^2 + 2x - 3y - 1 = 0$.
   (ii) $(1, -2)$; $2x^2 + 2y^2 - 5x - 2y - 7 =$ -
   (iii) $(0, 2)$; $x^2 + y^2 - 2x - y - 2 = 0$.
2. Find the parametric equations of the following circles:
   (i) $x^2 + y^2 = 9$.
   (ii) $x^2 + y^2 - 2x - 4y - 4 = 0$.
   (iii) $x^2 + y^2 - 2ax - 2by + a^2 + b^2 - r^2 = 0$.
3. Find the co-ordinates of the points on the following circles, for the following parameters:

   (i) $x^2 + y^2 = 4, \theta = \frac{\pi}{2}, 0, \pi, \frac{-\pi}{4}$

   (ii) $(x-1)^2 + (y+2)^2 = 16, \theta = \frac{\pi}{2}, 0, \pi, \frac{-\pi}{4}$
4. Find the points of intersection of:
   (i) $x + y - 1 = 0$ and $x^2 + y^2 - 2x + 2y - 3 = 0$.
   (ii) $x + 3y - 7 = 0$ and $x^2 + y^2 + x - 5y + 4 = 0$.
   Also find the lengths of the chords, in each case.
5. Find the points of intersection of the circles
   $$x^2 + y^2 - 2x + 2y - 3 = 0$$
   and $$x^2 + y^2 + x + 5y - 6 = 0.$$
   Also find the length of their common chord.
6. Write down the equation of a family of circles which pass through the points of intersection of the circles
   $$x^2 + y^2 + 2x + 8y + 9 = 0$$
   and $$x^2 + y^2 + 10x + 2y + 10 = 0$$
   Find the equation of that member of the family which passes through $(1, -1)$.
7. Find the equation and the length of the common chord of the circles:
   $$(x-a)^2 + (y-b)^2 = c^2, (x-b)^2 + (y-a)^2 = c^2$$

8. Find the equation of the circle passing through the intersection of the circle $x^2 + y^2 + 5x - 7y + 2 = 0$, and the line
$$2x - 5y + 2 = 0$$
and having the line $2x + 8y - 9 = 0$ as a diameter.
9. Find the equation of the circle passing through the intersection of the circles:
$$x^2 + y^2 + 5x - 3y + 2 = 0,$$
$$2x^2 + 2y^2 + 3x - 5y + 1 = 0$$
and having
(i) the $y$-axis as a diameter
(ii) the line $3x - y + 1 = 0$ as a diameter.
10. Find the equation of the circle which has the given chord as a diameter:
(i) $3x^2 + 3y^2 + x - 2 = 0$, $x + y - 2 = 0$.
(ii) $x^2 + y^2 - 2ax = 0$, $y = mx$.
(iii) $x^2 + y^2 = a^2$, $lx + my + n = 0$.
(iv) $x^2 + y^2 + 2gx + 2fy + c = 0$, $x$-axis.
11. Find the equation of the circle whose diameter is the common chord of the circles:
(i) $x^2 + y^2 + 2x + 3y + 1 = 0$,
$x^2 + y^2 + 4x + 3y + 2 = 0$.
(ii) $x^2 + y^2 - 4x - 6y - 12 = 0$,
$x^2 + y^2 + 6x + 4y - 12 = 0$.
12. Find the equations of the circles which pass through the origin and cut off equal chords from the lines $y = x$ and $y = -x$.
13. Show that the four points of intersection of the lines $(ax + by + c).(bx + ay + d) = 0$ with the axes are concyclic. Find the centre of the circle through these points.
14. The line, $y = mx + a\sqrt{m^2 + 1 - e^2}$ $(e^2 < 1)$, cuts the lines, $x = \pm a$ in $T$ and $T'$. If $S$ $(ae, 0)$ and $S'$ $(-ae, 0)$ are two other points, show that all these four points are concyclic.
15. Find the length of the common chord of the circles.
$$(x - a)^2 + y^2 = a^2 \text{ and } x^2 + (y - b)^2 = b^2$$
and prove that the equation of the circle whose diameter is this chord is $(a^2 + b^2)(x^2 + y^2) = 2ab(bx + ay)$.

16. Find the locus of the mid points of the chords of a circle which are of constant length = $2b$.
17. If $y = mx$ be the equation of a chord of the circle, $x^2 + y^2 = 2ax$, prove that the equation of the circle of which this chord is the diameter is
$$(1 + m^2)(x^2 + y^2) = 2a(x + my).$$
18. A circle passes through a fixed point $O$ and cuts two perpendicular lines through $O$ in $P$ and $Q$. If the line $PQ$ always passes through a fixed point $(\alpha, \beta)$; find the locus of the centre of this circle.

## 6.4. TANGENTS AND NORMALS TO A CIRCLE (SEE ALSO 8.2, 8.3 )

From Geometry, we know that a *tangent* at a point $P$ to a circle is the line passing through $P$, perpendicular to the line $CP$; $C$ being the centre of the given circle. Also then the line $CP$ (perpendicular to the tangent $PT$, through $P$) is called the *normal* at $P$. The point $P$ is called the point of contact of the tangent $PT$, and it is also called the *foot of the normal*, $CP$.

### 6.4.1. To find the equations of the tangent and the normal at a point P on a given circle.

Let $x^2 + y^2 + 2gx + 2fy + c = 0$ be the equation of a given circle and $P(x', y')$ be a point on it, at which the tangent and the normal are to be drawn.

Now, the centre of the circle is $(-g, -f)$.

Therefore the slope of the normal $CP$ is $\dfrac{y'+f}{x'+g'}$

and the equation of the normal is

$$y - y' = \left(\frac{y'+f}{x+g'}\right)(x - x')$$

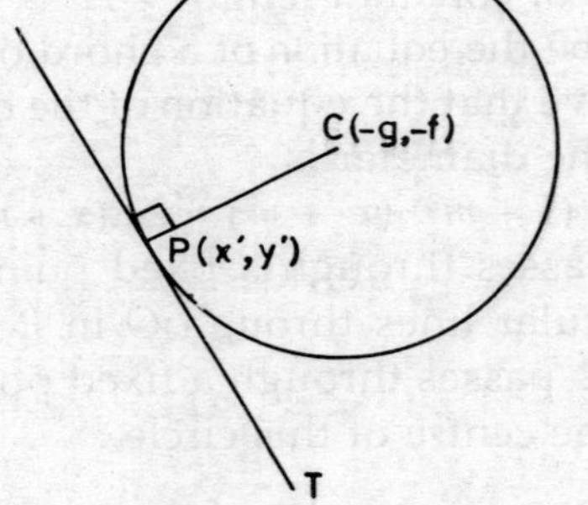


Also the equation of the tangent $PT$ is,

$$y - y' = -\left(\frac{x'+g}{y'+f}\right)(x - x')$$

$\Rightarrow \quad yy' + fy - y'^2 - fy' + xx' - x'^2 + gx - gx' = 0$

$\Rightarrow \quad xx' + yy' + gx + fy - (x'^2 + y'^2 + gx' + fy') = 0 \qquad (1)$

Now $(x', y')$, being a point on the circle, we have

$x'^2 + y'^2 + 2gx' + 2fy' + c = 0$

$\Rightarrow \quad -(x'^2 + y'^2 + gx' + fy') = (gx' + fy' + c) \qquad (2)$

Hence, the required equation of the tangent at $(x', y')$ is

$xx' + yy' + gx + fy + gx' + fy' + c = 0$ (using (1) and (2)

i.e. $xx' + yy' + g(x + x') + f(y + y') + c = 0$.

**Remark.** Method to write the equation of the tangent at a point $(x', y')$:

change $x^2 \to xx', y^2 \to yy'$

$$x \to \frac{x+x'}{2},\ y \to \frac{y+y'}{2}$$

and keep the constant term unchanged.

**Example 6.13.** Find the condition that the line $y = mx + c$ may touch the circle, $x^2 + y^2 = a^2$.

*Solution*. The centre and the radius of the circle, $x^2 + y^2 = a^2$, are (0, 0) and a respectively.

The perpendicular distance from (0, 0) on the line

$$y - mx - c = 0 \text{ is } \left|\frac{0 - m.0 - c}{\sqrt{1+m^2}}\right| = \frac{|c|}{\sqrt{1+m^2}}$$

This must be equal to the radius, $a$.

$$\Leftrightarrow \frac{|c|}{\sqrt{1+m^2}} = a \Rightarrow c^2 = a^2(1+m^2)$$

which is the required condition.

**Example 6.14.** Show that the line $x - y - 3 = 0$ touches the circle $x^2 + y^2 - 4x + 6y + 11 = 0$ and also find the point of contact.

*Solution.* We have the given line is

$$x - y - 3 = 0 \qquad (1)$$

and the given circle is

$$x^2 + y^2 - 4x + 6y + 11 = 0 \qquad (2)$$

Now (1) $\Rightarrow x = y + 3$.

$\therefore$ (2) becomes

$$(y + 3)^2 + y^2 - 4(y + 3) + 6y + 11 = 0$$

$\Rightarrow (y + 2)^2 = 0 \Rightarrow y = -2, -2.$

$\Rightarrow x = 1, 1$; whence the line (1) meets the circle (2) in two coincident points (see § **8.1.1**) and hence (1) is a tangent to (2) at the point $(-1, -2)$ (which is their point of contact).

**Aliter.** The centre and the radius of the circle given by (2), above, are

$$(2, -3) \text{ and } \sqrt{2}, \text{ respectively.}$$

Now perpendicular from $(2, -3)$ on the line (1) is

$$\frac{|2-(-3)-3|}{\sqrt{1+1}} = \sqrt{2} = \text{radius of (2)}$$

$\therefore$ The line (1) touches the circle (2).

Let $(x', y')$ be their point of contact.

Now the equation of the tangent to (2) at $(x', y')$ is

$$xx' + yy' - 2(x + x') + 3(y + y') + 11 = 0$$

$\Rightarrow (x' - 2)x + (y' + 3)y + (-2x' + 3y' + 11) = 0$

which is the same as (1).

$$\therefore \quad \frac{x'-2}{1} = \frac{y'+3}{-1} = \frac{-2x'+3y'+11}{-3} = k \text{ (say)}$$

$\Rightarrow x' = k + 2,\ y' = -k - 3$ and $-2x' + 3y' + 11 = -3k$

$\Rightarrow -2(k + 2) + 3(-k - 3) + 11 = -3k$

$\Rightarrow k = -1.$

Thus the point of contact $(x', y') = (1, -2)$.

**Example 6.15.** Find the equation of the circle through the points of intersection of $x^2 + y^2 = 4$ and $x^2 + y^2 - 2x - 4y + 4 = 0$ and touching the line $x + 2y = 0$.

*Solution.* The given circles are

$$S \equiv x^2 + y^2 - 4 = 0$$

and

$$S' \equiv x^2 + y^2 - 2x - 4y + 4 = 0.$$

Their common chord is given by $S - S' = 0$

i.e. by $\quad x + 2y - 4 = 0$

$\therefore$ the equation of any circle through their intersection is

$$x^2 + y^2 - 4 + \lambda(x + 2y - 4) = 0$$

$$\Rightarrow \quad x^2 + y^2 + \lambda x + 2\lambda y - (4 + 4\lambda) = 0 \quad (1)$$

its centre $= \left(-\dfrac{\lambda}{2}, -\lambda\right)$ and

its radius $= \sqrt{\dfrac{\lambda^2}{4} + \lambda^2 + (4 + 4\lambda)} = \dfrac{1}{2}\sqrt{5\lambda^2 + 16\lambda + 16}$

Since the line $x + 2y = 0$ is to touch the circle (1), we have

$$\left|\frac{-\lambda/2 - 2\lambda}{\sqrt{1^2 + 2^2}}\right| = \frac{1}{2}\sqrt{5\lambda^2 + 16\lambda + 16}$$

$\Rightarrow 5\lambda^2 = 5\lambda^2 + 16\lambda + 16 \Rightarrow \lambda = -1.$

Substituting for $\lambda = -1$, in (1), we get the equation of the required circle as

$$x^2 + y^2 - x - 2y = 0.$$

**Example 6.16.** Find the equation of the circle which touches the $x$-axis at (3, 0) and passes through the point (1, 4).

*Solution.* Let the equation of the required circle be

$$x^2 + y^2 + 2gx + 2fy + c = 0.$$

It passes through (1, 4)

$$\therefore \quad 2g + 8f + c + 17 = 0 \quad (1)$$

Also, the equation of the tangent at (3, 0) is

$$x \cdot 3 + y \cdot 0 + g(x + 3) + f(y + 0) + c = 0 \quad (2)$$

$\Rightarrow (3 + g)x + f.y + (3g + c) = 0$

But this is the same as the $x$-axis, whose equation is $y = 0$ i.e.

$$0.x + 1.y + 0 = 0. \quad (3)$$

Comparing the coefficients in (2) and (3), we get

$$\frac{0}{3+g} = \frac{1}{f} = \frac{0}{3g+c}.$$

These give

$$g = -3,$$
$$c = -3g = 9.$$

Substituting these values in (1), we obtain,

$$f = -5/2.$$

Thus the equation of the required circle is

$$x^2 + y^2 - 6x - 5y + 9 = 0.$$

**Example 6.17.** Find the equations of the tangents to the circle, $x^2 + y^2 + 2gx + 2fy + c = 0$, which are parallel to the line, $x \cos \alpha + y \sin \alpha = p$

*Solution.* Any line parallel to the given line, is given by $x \cos \alpha + y \sin \alpha + k = 0$.

This will be a tangent to the given circle if perpendicular from $(-g, -f)$ on this is equal the radius of the circle.

$$\therefore \quad \frac{k - g\cos\alpha - f\sin\alpha}{\sqrt{\cos^2\alpha + \sin^2\alpha}} = \pm\sqrt{g^2 + f^2 - c}$$

$$\Rightarrow k = g \cos \alpha + f \sin \alpha \pm \sqrt{g^2 + f^2 - c}$$

$\therefore$ Equations of the tangents to the given circle parallel to the given line are

$$x \cos \alpha + y \sin \alpha + g \cos \alpha + f \sin \alpha \pm \sqrt{g^2 + f^2 - c} = 0$$

$$\Rightarrow (x + g) \cos \alpha + (y + f) \sin \alpha = \pm\sqrt{g^2 + f^2 - c}$$

**Example 6.18.** Find the locus of the feet of normals from $(p, q)$ to the circles, $x^2 + y^2 - 2\lambda x = 0$, where $\lambda$ is a parameter.

*Solution.* Let $(x_1, y_1)$ be the foot of the normal from $(p, q)$ to the circles,

$$x^2 + y^2 - 2\lambda x = 0 \quad (1)$$

then $$x_1^2 + y_1^2 = 2\lambda x_1 \quad (2)$$

and the equation of the tangent to (1) at $(x_1, y_1)$ is

$$xx_1 + yy_1 - \lambda(x + x_1) = 0$$

$\therefore$ Its slope $= -\dfrac{x_1 - \lambda}{y_1}$

$\Rightarrow$ Slope of the normal $= y_1/(x_1 - \lambda)$

$\therefore$ Equation of the normal to (1) at $(x_1, y_1)$ is

$$y - y_1 = \frac{y_1}{x_1 - \lambda}(x - x_1)$$

This passes through $(p, q)$

$\therefore$ $(q - y_1)(x_1 - \lambda) = y_1(p - x_1)$ (3)

Eliminating $\lambda$ between (2) and (3), we get

$$(q - y_1)\left(x_1 - \frac{x_1^2 + y_1^2}{2x_1}\right) = y_1(p - x_1)$$

$\Rightarrow$ $(q - y_1)(x_1^2 - y_1^2) = 2x_1y_1(p - x_1)$

$\therefore$ The locus of $(x_1, y_1)$ is

$(q - y)(x^2 - y^2) = 2xy(p - x)$.

**Example 6.19.** $P$ is the point $(a, b)$ and $Q$ is the point $(b, a)$. Find the equation of the circle touching $OP$ and $OQ$ at $P$ and $Q$, where $O$ is the origin.

*Solution.* Let the equation of the required circle be

$x^2 + y^2 + 2gx + 2fy + c = 0$ (1)

Now (1) passes through $P(a, b)$ and $Q(b, a)$.

$\therefore$ $a^2 + b^2 + 2ga + 2fb + c = 0$ (2)

$b^2 + a^2 + 2gb + 2fa + c = 0$.

Subtracting these, we get

$g = f$ (we note that $a \neq b$)

Also equations of $OP$ and $OQ$ are respectively

$bx - ay = 0$ and $ax - by = 0$.

As they touch (1), so

$$\frac{-gb + af}{\sqrt{b^2 + a^2}} = \pm\sqrt{g^2 + f^2 - c}$$

$\Rightarrow f(a - b) = \pm\sqrt{a^2 + b^2}\sqrt{2f^2 - c}$ $(\because g = f)$

$\Rightarrow f^2(a - b)^2 = (a^2 + b^2)(2f^2 - c)$

$\Rightarrow c = f^2(a + b)^2/(a^2 + b^2)$

Using this and $g = f$, (2) implies

$$a^2 + b^2 + 2f(a + b) + \frac{f^2(a+b)^2}{a^2+b^2} = 0$$

$\Rightarrow \quad [a^2 + b^2 + f(a + b)]^2 = 0$

$\Rightarrow \quad f = -(a^2 + b^2)/(a + b) = g.$

$\therefore$ The equation of the desired circle is

$$x^2 + y^2 + 2(x + y)\left[-\frac{a^2+b^2}{a+b}\right] + (a^2+b^2) = 0$$

$\Rightarrow (a + b)(x^2 + y^2) - 2(a^2 + b^2)(x + y) + (a + b)(a^2 + b^2) = 0$

**Exercise 6.3**

1. Find the equations of the tangents and the normals to the following circles, at the given points:
   (i) $x^2 + y^2 = 5$, (2, 1);
   (ii) $x^2 + y^2 + 4x - 8y = 0$, (0, 0);
   (iii) $x^2 + y^2 + 2x - 6y - 7 = 0$, at the points where $y = -1$;
   (iv) $x^2 + y^2 = a^2$, $(a \cos \theta, a \sin \theta)$.
2. Find the equation of the circle whose centre is at the point $(\alpha, \beta)$, and which passes through the origin. Also show that the equation of the tangent to it at the origin is
   $$\alpha x + \beta y = 0.$$
3. Show that the following lines touch the given circles:
   (i) $2x + 3y = 13$, $x^2 + y^2 = 13$.
   (ii) $3x - 4y - 20 = 0$, $x^2 + y^2 - 2x - 4y - 20 = 0$.
   (iii) $x \cos \alpha + y \sin \alpha = a$, $x^2 + y^2 = a^2$ (for all values of $\alpha$).
4. Prove that the straight line, $x + y = 2$, touches the circles $x^2 + y^2 = 2$ and $x^2 + y^2 - 6(x + y) + 10 = 0$ at the same point. Also find this point.
5. Prove that the following lines touch the given circles. Find also the points of contact
   (i) $y = x + c\sqrt{2}$, $x^2 + y^2 = c^2$.
   (ii) $y = mx \pm a\sqrt{1+m^2}$, $x^2 + y^2 = a^2$.
6. Find the condition in each case so that the following lines may touch the given circles:
   (i) $lx + my + n = 0$, $x^2 + y^2 = a^2$.

(ii) $lx + my + n = 0$, $x^2 + y^2 + 2gx + 2fy + c = 0$.

(iii) $px + qy = a^2$, $(x - c)^2 + (y - d)^2 = b^2$.

7. (i) For what value of k, will the line $4x + 2y + k = 0$ touch the circle, $2x^2 + 2y^2 = 5$.

(ii) Find the equations of the tangents to the circle $x^2 + y^2 - 6x + 4y - 12 = 0$ which are parallel to the line

$$4x + 3y + 5 = 0.$$

Also find the points of contact.

8. Find the equation of the circle with centre at (3, 4) and touching the line

$$5x + 12y = 11$$

9. Show that the circles $x^2 + y^2 - 4x + 6y + 8 = 0$ and $x^2 + y^2 - 10x - 6y + 14 = 0$ touch at the point (3, – 1).

10. Find the equation of the circle which touches the axis of $y$ at (0, 2) and cuts off an intercept 3 from the axis of $x$. (Two possible circles)

11. Find the equation of that circle which touches both the coordinate axes and whose centre is (1, 1).

12. Show that the circle $x^2 + y^2 + 4x - 4y + 4 = 0$ touches the co-ordinate axes. Also find the equations of the tangents which make equal intercepts on the co-ordinate axes. (*Hint:* Assume the equations of tangents as $x + ky = a$, where $k = \pm 1$).

13. Find the equation of the circle touching the circle

$$x^2 + y^2 - 15x + 5y = 0$$

at (1, 2) and also passing through the point (0, 2).

14. Find the equations of the tangents to the circles $x^2 + y^2 = a^2$ which are perpendicular to the straight line $y = mx + c$

15. Find the condition that the line, $lx + my + n = 0$ be a normal to the circle, $x^2 + y^2 + 2gx + 2fy + c = 0$.

16. Prove that the length of the chord intercepted by the circle, $x^2 + y^2 = a^2$, on the line, $y = mx + c$ is

$$2\sqrt{\left[a^2(1+m^2)-c^2\right]/\left(1+m^2\right)}.$$

Also prove that the mid point of this chord is

$$\left(-\frac{cm}{1+m^2}, \frac{c}{1+m^2}\right).$$

17. A line moves so that the algebraic sum of the perpendiculars drawn to it from the points $(a, 0)$ and $(-a, 0)$ is constant. Show that the line always touches a fixed circle.
18. Two circles touch the axis of $y$ and intersect in the points $(1, 0)$ and $(2, -1)$. Find their radii and show that they both touch the line, $y + 2 = 0$.
19. The angular points of a triangle are $(0, 0)$, $(48, 20)$ and $(63, 0)$. Prove that its in-circle is
$$x^2 + y^2 - 90x - 18y + 2025 = 0$$
20. Verify that the four lines, $3x + 4y = 5$, $4x - 3y = 5$, $3x + 4y = -5$ and $4x - 3y = -5$ make up a square, circumscribed about the circle, $x^2 + y^2 = 1$.
21. Find the equations of two circles of which the centre is on the line, $4x + 3y = 2$ and which touch the lines, $x + y + 4 = 0$ and $7x - y + 4 = 0$
22. *PQ* and *RS* are parallel tangents to a circle and a tangent at any point *A* cuts them in *Q* and *S* respectively. Show that *AQ*. *AS* is independent of the position of the point *A*.
23. Find the equations of the circles touching the lines,
$$x - 7y + 42 = 0,\ 7x - y - 42 = 0 \text{ and } x + y - 6 = 0.$$
24. Show that the locus of the feet of the perpendiculars drawn from the point $(a, 0)$ on any tangent to the circle, $x^2 + y^2 = a^2$, is $(x^2 + y^2 - ax)^2 = a^2 [(x - a)^2 + y^2]$.

### 6.4.2 To find the length of the tangent from a point to a circle.

Let $\quad x^2 + y^2 + 2gx + 2fy + c = 0$
be the given circle and $P$ $(\alpha, \beta)$ be the given point.

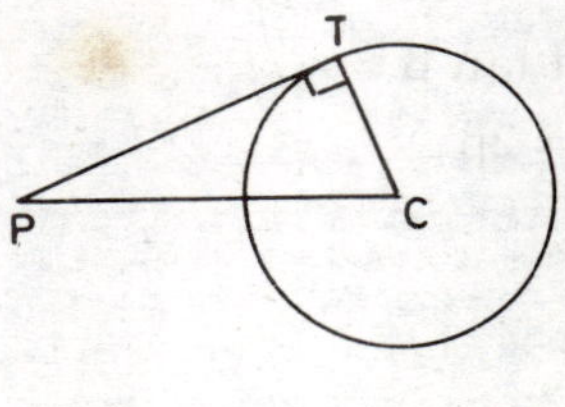

Clearly its centre $C$ is $(-g, -f)$. Draw $PT$ tangent from $P$ to this circle.

Join $CT$. Then from the right angled triangle $PCT$,

$$|PT| = \sqrt{CP^2 - CT^2}$$

$$= \sqrt{\left[(\alpha+g)^2 + (\beta+g)^2\right] - (g^2 + f^2 - c)}$$

$$= \sqrt{\alpha^2 + \beta^2 + 2g\alpha + 2f\beta + c}$$

**Example 6.20.** If $r_1$ and $r_2$ are the radii of the circles $S_1 = 0$, $S_2 = 0$ (the coefficients of $x^2$ and $y^2$ in both are unity), show that the points at which the circles subtend equal angles lie on the circle given by $S_1/r_1^2 = S_2/r_2^2$.

*Solution.*

Let $S_i \equiv x^2 + y^2 + 2g\ x + 2f_i y + c_i = 0$ $(i = 1, 2)$.

Let $S_1$ $(S_2)$ subtend an angle $2\alpha$ $(2\beta)$ at the point, $P(x_1, y_1)$ and $l_1 = PT_1$; $l_2 = PT_4$; $r_1 = C_1T_1$ and $r_2 = C_2T_4$.

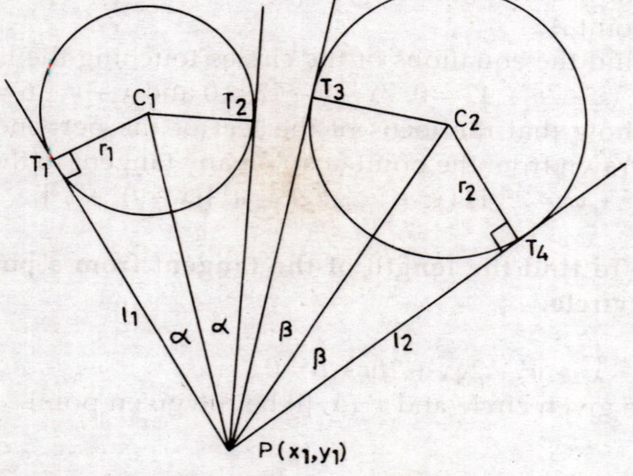


$\therefore$ $\tan\alpha = r_1/l_1$ and $\tan\beta = r_2/l_2$.

By hypothesis, $2\alpha = 2\beta$

$\therefore$ $\quad r_1/l_1 = r_2/l_2$

$\Rightarrow$ $\quad r_1^2\, l_2^2 = r_2^2\, l_1^2$

$\Rightarrow \quad r_1^2 (x_1^2 + y_1^2 + 2g_2x_1 + 2f_2y_1 + c_2)$

$= r_2^2 (x_1^2 + y_1^2 + 2g_1x_1 + 2f_1y_1 + c_1)$

$\therefore$ The locus of $P(x_1, y_1)$ is

$r_1^2 (x^2 + y^2 + 2g_2x + 2f_2y + c_2)$

$= r_2^2 (x^2 + y^2 + 2g_1x + 2f_1y + c_1)$

$\Rightarrow \quad S_1/r_1^2 = S_2/r_2^2.$

**Exercise 6.4.**

1. Find the lengths of the tangents from the following points to the given circles:
   (i) $(0, 0)$, $x^2 + y^2 - 2x - 2y + 1 = 0$.
   (ii) $(2, 3)$, $x^2 + y^2 - 5x + 1 = 0$.
   (iii) $(\alpha, \beta)$, $ax^2 + ay^2 + 2gx + 2fy + c = 0$. $(a > 0)$
2. Show that the locus of a point which moves such that the lengths of the tangents from it, to two given circles is in the ratio $k$, is a
   (i) circle if $k \neq 1$,
   (ii) straight line if $k = 1$.
3. Prove that the length of the tangent drawn from any point on the circle $x^2 + y^2 + 2gx + 2fy + c = 0$ to the circle $x^2 + y^2 + 2gx + 2fy + d = 0$ is $\sqrt{d - c}$.
4. The distances from the origin of the centres of three circles, $x^2 + y^2 - 2\lambda x = a^2$ (where $a$ is a constant and $\lambda$ a parameter) are in G.P. Prove that the lengths of the tangents drawn to them from any point on the circle, $x^2 + y^2 = a^2$, are also in G.P.
5. Find the equation of the circle which is such that the lengths of tangents from the points, $(1, 0)$, $(3, 2)$ and $(6, 0)$ to it are $1$, $\sqrt{2}$ and $\sqrt{3}$ respectively.
6. If the tangent from a point $P$ to the circle, $x^2 + y^2 = a^2$ be equal to the perpendicular from $P$ to the line, $px + qy + r = 0$, find the locus of $P$.
7. Show that if the length of the tangent from a point $P$ to the circle, $x^2 + y^2 = a^2$ be three times the length of the tangent from it to the circle, $(x - a)^2 + y^2 = a^2$, then $P$ lies on the circle, $8(x^2 + y^2) - 18ax + a^2 = 0$.

8. Find the points on the line, $y = x + 1$, the tangents from which to the circle, $x^2 + y^2 = 3x$ are of length 2.

## 6.5. ANGLE OF INTERSECTION OF TWO CIRCLES

Let any two circles intersect at the points $A$ and $B$. Draw tangents at any one of these points, say $A$, to both the circles.

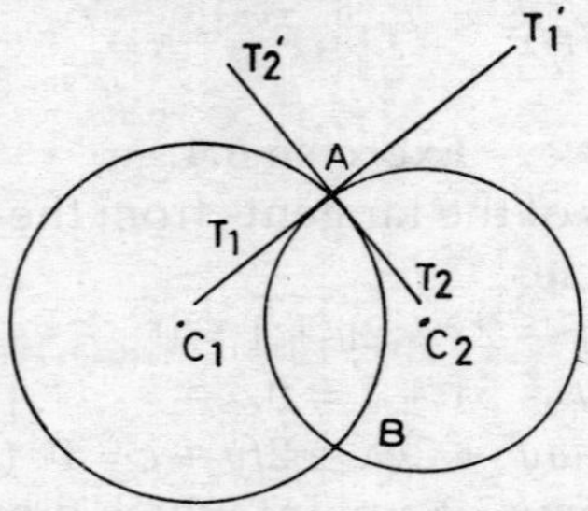


The $\angle T_1AT_2$ between these tangents is defined to be the angle of intersection of these circles.

It is easily seen that the angle between the tangents at $B$, is same as $\angle T_1AT_2$.

### 6.5.1. Orthogonal circles

**Definition.** Two circles which intersect each other at right angles are called *orthogonal*.

**Theorem 6.1.** A necessary and sufficient condition for the circles $x^2 + y^2 + 2g_1x + 2f_1y + c_1 = 0$ and $x^2 + y^2 + 2g_2x + 2f_2y + c_2 = 0$ to be orthogonal is $2g_1g_2 + 2f_1f_2 = c_1 + c_2$.

*Proof. The condition is necessary.*

The circles

$$x^2 + y^2 + 2g_1x + 2f_1y + c_1 = 0$$

and $$x^2 + y^2 + 2g_2x + 2f_2y + c_2 = 0$$

are given to be orthogonal. Their centres are

$$C_1(-g_1, -f_1)$$

and $$C_2(-g_2, -f_2).$$

Also their radii are

$$\sqrt{g_1^2 + f_1^2 - c_1}$$

and $$\sqrt{g_2^2 + f_2^2 - c_2}\,.$$

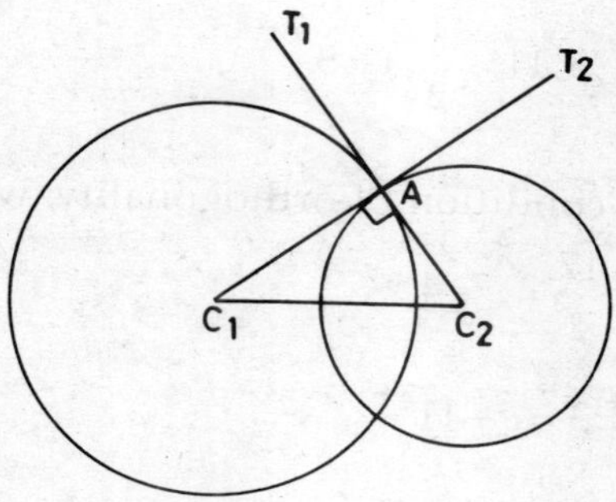


Let $AT_1$, and $AT_2$ be the tangents to them at a point $A$ of their intersection.

By hypothesis

$$\angle T_1AT_2 = 90^\circ.$$

Now, the radii $C_1A$ and $C_2A$ of the two circles are respectively perpendicular to the tangents $AT_1$ and $AT_2$.

$\Rightarrow$ These tangents pass through the centres.

$\Rightarrow \angle C_1AC_2 = 90^\circ$

$\Rightarrow C_1C_2^{\,2} = C_1A^2 + C_2A^2$

$\Rightarrow (g_1 - g_2)^2 + (f_1 - f_2)^2 = (g_1^{\,2} + f_1^{\,2} - c_1) + (g_2^{\,2} + f_2^{\,2} - c_2).$

$\Rightarrow 2g_1g_2 + 2f_1f_2 = c_1 + c_2$ (1)

which is the required necessary condition for orthogonality.

By working the algebra backwards, we can easily see that (1) is also a *sufficient condition* for the given circles to cut orthogonally.

**Example 6.21.** Find the equation of the circle which cuts orthogonally each of the following three circles :

$$x^2 + y^2 + 2x + 17y + 4 = 0,$$
$$x^2 + y^2 + 7x + 6y + 11 = 0,$$

and $$x^2 + y^2 - x + 22y + 3 = 0.$$

*Solution.* Let the equation of the required circle be

$$x^2 + y^2 + 2gx + 2fy + c = 0.$$

Now, $g, f, c$ for the given three circles are

$$1,\frac{17}{2},4;\frac{7}{2},3,11;-\frac{1}{2},11,3$$

respectively.

Applying the condition of orthogonality, we have

$$2g.1+2f.\frac{17}{2}=c+4,$$

$$2g.\frac{7}{2}+2f.3=c+11$$

and $$2g.(-\frac{1}{2})+2f.11=c+3$$

Subtracting first of these equations from second and third (to eliminate $c$), we obtain

$$5g - 11f = 7,$$
$$-3g + 5f = -1.$$

These give

$$g = -3,$$
$$f = -2,$$

and then

$$c = -44.$$

$\therefore$ the equation of the required circle cutting each of the given circles orthogonally, is

$$x^2 + y^2 - 6x - 4y - 44 = 0.$$

**Example 6.22.** Show that the general equation of all circles cutting orthogonally the circles represented by

$$x^2 + y^2 - a_ix - 2b_iy + c_i = 0 \ (i = 1, 2) \text{ is}$$

$$\begin{vmatrix} x^2+y^2 & x & y \\ c_1 & a_1 & b_1 \\ c_2 & a_2 & b_2 \end{vmatrix} + k\begin{vmatrix} x & y & 1 \\ a_1 & b_1 & 1 \\ a_2 & b_2 & 1 \end{vmatrix} = 0$$

*Solution*. The given circles are

$$x^2 + y^2 - 2a_1x - 2b_1y + a = 0 \quad (1)$$
$$x^2 + y^2 - 2a_2x - 2b_2y + c_2 = 0 \quad (2)$$

Let the equation of the circle orthogonal to (1) and (2) be

$$x^2 + y^2 + 2gx + 2fy + k = 0 \quad (3)$$

The condition of orthogonality yields

$$2a_1 g + 2b_1 f + c_1 + k = 0 \quad (4)$$
$$2a_2 g + 2b_2 f + c_2 + k = 0 \quad (5)$$

Eliminating $g$ and $f$ from (3), (4) and (5), we get

$$\begin{vmatrix} x^2+y^2+k & x & y \\ c_1+k & a_1 & b_1 \\ c_2+k & a_2 & b_2 \end{vmatrix} = 0 \begin{vmatrix} x & y & 1 \\ a_1 & b_1 & 1 \\ a_2 & b_2 & 1 \end{vmatrix} = 0$$

$$\begin{vmatrix} x^2+y^2 & x & y \\ c_1 & a_1 & b_1 \\ c_2 & a_2 & b_2 \end{vmatrix} + k \begin{vmatrix} 1 & x & y \\ 1 & a_1 & b_1 \\ 1 & a_2 & b_2 \end{vmatrix} = 0$$

whence follows the result.

### Exercise 6.5.

1. Show that the equation of the circle which passes through (1, 1) and cuts orthogonally each of the circles
$$x^2 + y^2 - 8x - 2y + 16 = 0$$
and $$x^2 + y^2 - 4x - 4y - 1 = 0$$
is $$3x^2 + 3y^2 - 14x + 23y - 15 = 0$$
2. Find the equation of the circle which cuts the circles
$$x^2 + y^2 - 2x + 3y - 7 = 0,$$
$$x^2 + y^2 + 5x - 5y + 9 = 0$$
and $$x^2 + y^2 + 7x - 9y + 29 = 0$$
orthogonally.
3. Find the equation of the circle which cuts orthogonally each of the circles
$$x^2 + y^2 - 2x - 4y + 2 = 0$$
$$x^2 + y^2 - 4x + 2y - 6 = 0$$
and $$x^2 + y^2 + 8x + 4y - 4 = 0$$
4. Find the equation of the circle which cuts each of the circles
$$x^2 + y^2 + 2gx + c = 0,$$
$$x^2 + y^2 + 2g_1x + c = 0$$
and $$x^2 + y^2 + 2hx + 2ky + a = 0$$
orthogonally.

5. Find the equation of the circle which cuts the circle
$$x^2 + y^2 + 5x + 7y - 4 = 0$$
orthogonally, has its centre on the line $x = 2$ and passes through the point $(4, -1)$.
6. Find the equations of the two circles which intersect the circles
$$x^2 + y^2 - 6y + 1 = 0,$$
$$x^2 + y^2 - 4y + 1 = 0$$
orthogonally and touch the line
$$3x + 4y + 5 = 0,$$
7. (a) Find the angles at which the pairs of the following circles intersect :
   (i) $x^2 + y^2 = 25$ and $3(x^2 + y^2) - 5x - 15y = 0$.
   (ii) $x^2 + y^2 + 2gx + c = 0$ and
   $x^2 + y^2 + 2g'x + c = 0$.
   (b) Show that the angle of intersection of the circles
   $x^2 + y^2 = 4$ and $x^2 + y^2 = 2x + 2y$ is $\pi/4$.
8. Prove that the two circles which pass through $(0, a)$ and $(0, -a)$, and touch the line $y = mx + c$ will cut orthogonally if
$$c^2 = a^2 (2 + m^2).$$
9. Find the equation of the circle which cuts the circle,
$x^2 + y^2 - 6x + 4y - 3 = 0$, orthogenerally, touches the axis of $y$ and passes through $(3, 0)$.
10. Find the equation of the circle which cuts the circles given below orthogonally:
$$x^2 + y^2 = a^2 \; ; \; (x - c)^2 + y^2 = a^2$$
and $x^2 + (y - b)^2 = a^2$.
11. If a circle cuts three given circles, $S_i = 0$ $(i = 1, 2, 3)$ orthogonally show that it will cut every circle of the system, $\lambda S_1 + \mu S_2 + \upsilon S_3 = 0$, orthogonally.
12. Show that if a circle, which cuts, $x^2 + y^2 = a^2$, orthogonally and passes through $(p, q)$, it will also pass through
$$\left(\frac{a^2 p}{p^2 + q^2}, \frac{a^2 q}{p^2 + q^2}\right).$$

## 6.6. RADICAL AXIS OF TWO CIRCLES

**Definition.** The locus of a point tangents from which to two circles are equal in length, is called the *radical axis* of these circles.

*To find the radical axis of the circles given by*

$$x^2 + y^2 + 2g_1x + 2f_1y + c_1 = 0$$

and $$x^2 + y^2 + 2g_2x + 2f_2y + c_2 = 0.$$

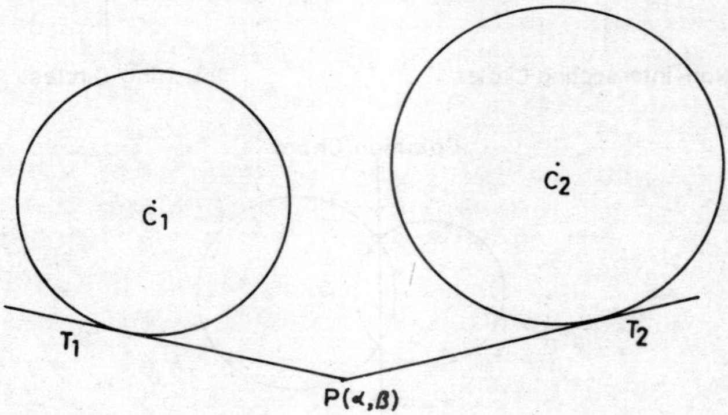


Let $P(\alpha, \beta)$ be a point such that

$|PT_1| = |PT_2|$, where $PT_1$ and $PT_2$

are tangents to the given circles, respectively.

$$\Rightarrow \sqrt{\alpha^2 + \beta^2 + 2g_1\alpha + 2f_1\beta + c_1} = \sqrt{\alpha^2 + \beta^2 + 2g_2\alpha + 2f_2\beta + c_2}$$

(by § **6.4.2.** )

$$\Rightarrow 2(g_1 - g_2)\,\alpha + 2(f_1 - f_2)\,\beta + c_1 - c_2 = 0.$$

$\therefore$ the locus of $(\alpha, \beta)$ is

$$2(g_1 - g_2)\,x + 2(f_1 - f_2)\,y + c_1 - c_2 = 0,$$

which is the required equation of the radical axis of the given circles. Clearly this is a straight line.

**Note 1.** The radical axis of two circles is the same as their common chord.

**Note 2.** If $S_1 = 0$ and $S_2 = 0$ be the equations of the two circles, such that the coefficients of $x^2$ (and also of $y^2$) are equal in both these equations; then their radical axis is given by

$$S_1 - S_2 = 0.$$

**Note 3.** The position of the radical axis of the two circles, geometrically is as shown below:

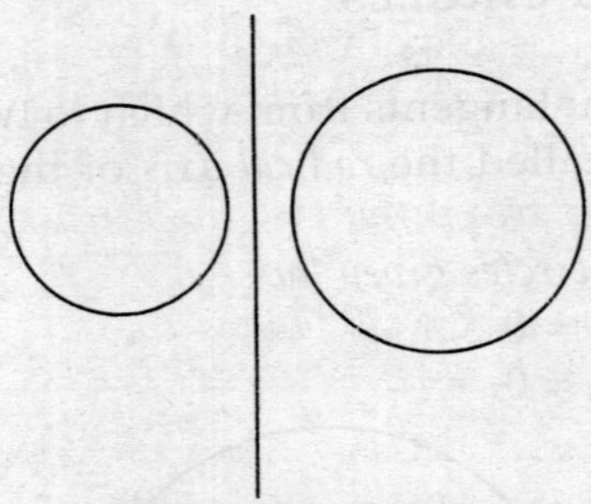

Non-intersecting Circles

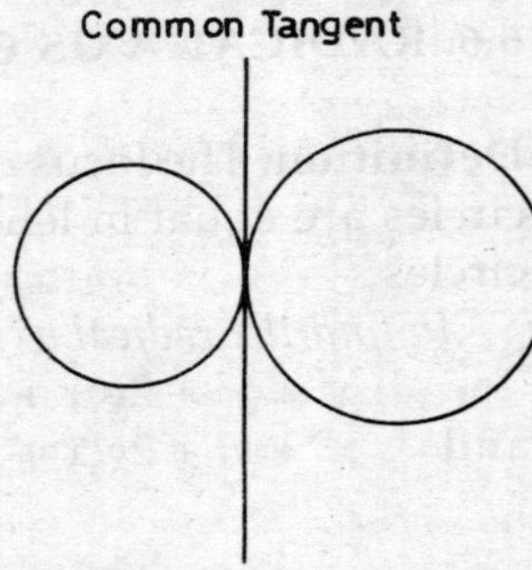


Touching Circles

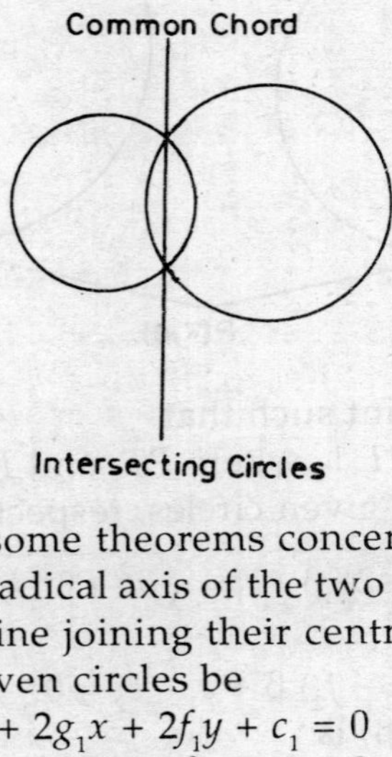


Intersecting Circles

We now prove some theorems concerning radical axis.

**Theorem 6.2**. The radical axis of the two given circles is perpendicular to the line joining their centres.

*Proof.* Let the given circles be

$$S_1 \equiv x^2 + y^2 + 2g_1x + 2f_1y + c_1 = 0$$

and $$S_2 \equiv x^2 + y^2 + 2g_2x + 2f_2y + c_2 = 0$$

Their radical axis is

$$S_1 - S_2 = 2(g_1 - g_2)\,x + 2(f_1 - f_2)y + c_1 - c_2 = 0$$

Its slope $= -\dfrac{g_1 - g_2}{f_1 - f_2}$ (1)

Again centres of the given circles are

$C_1(-g_1, -f_1)$ and $C_2(-g_2, -f_2)$

$\therefore$ the slope of $C_1C_2$ is $\dfrac{f_1 - f_2}{g_1 - g_2}$ (2)

Now (1) and (2), imply that the product of the two slopes is –1.

Thus the radical axis is perpendicular to the line joining the centres of the circles.

**Theorem 6.3**. The radical axes of any three circles taken in pairs are concurrent.

**Proof**. Let the equations of the given circles be

$$S_1 = x^2 + y^2 + 2g_1x + 2f_1y + c_1 = 0$$
$$S_2 = x^2 + y^2 + 2g_2x + 2f_2y + c_2 = 0$$
and $$S_3 = x^2 + y^2 + 2g_3x + 2f_3y + c_3 = 0$$

The radical axes of these circles, taking two at a time are given by

$$S_1 - S_2 = 2(g_1 - g_2)x + 2(f_1 - f_2)y + (c_1 - c_2) = 0$$
$$S_2 - S_3 = 2(g_2 - g_3)x + 2(f_2 - f_3)y + (c_2 - c_3) = 0$$
and $$S_3 - S_1 = 2(g_3 - g_1)x + 2(f_3 - f_1)y + (c_3 - c_1) = 0$$

Clearly $1.(S_1 - S_2) + 1.(S_2 - S_3) + 1.(S_3 - S_1) = 0$
$\Rightarrow$ the three radical axes are concurrent (by § 4.10)

**Definition**. The point of concurrence, where the radical axes of the three circles, taken in pairs, meet is called the *radical centre* of the circles.

**Example 6.23.** Prove that the two circles

$$x^2 + y^2 + 2ax + c = 0$$
and $$x^2 + y^2 + 2by + c = 0$$

touch iff $\frac{1}{a^2} + \frac{1}{b^2} = \frac{1}{c}$

*Solution*.

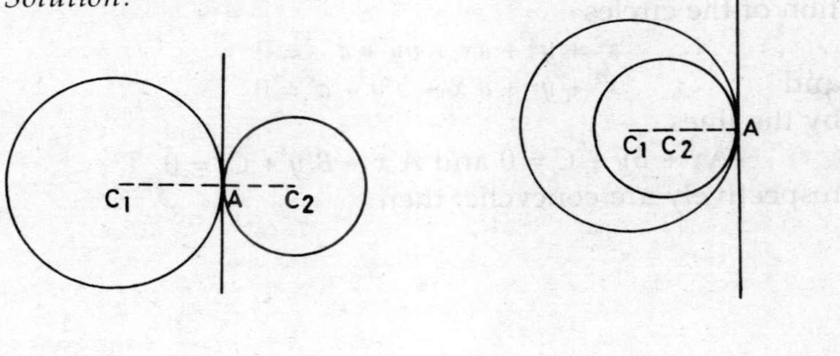

The given circles are

$$x^2 + y^2 + 2ax + c = 0 \quad (1)$$

and $$x^2 + y^2 + 2by + c= 0 \quad (2)$$

They will touch iff their radical axis is a tangent to both of them.

Again this radical axis is given by

$$ax - by = 0 \quad (3)$$

Now (3) will touch both (1) and (2) iff the perpendicular distance of the centre of either circle from (3) is equal to the radius of that circle. In particular, (3) will touch (1) iff

$$\left|\frac{a(-a)-b.0}{\sqrt{a^2+b^2}}\right| = \sqrt{a^2-c}$$

i.e. if $a^4 = (a^2 + b^2)(a^2 - c)$

i.e. if $(a^2 + b^2)\, c = a^2 b^2$

i.e. if $\dfrac{1}{a^2}+\dfrac{1}{b^2}=\dfrac{1}{c}$

**Aliter.** The given circles will touch each other

if $|C_1C_2| = |C_1A| \pm |C_2A|$

i.e. if $\sqrt{a^2+b^2} = \sqrt{a^2-c} \pm \sqrt{b^2-c}$

i.e. if $a^2 + b^2 = a^2 - c + b^2 - c \pm 2\sqrt{a^2-c}\,.\,\sqrt{b^2-c}$

i.e. if $c = \pm\sqrt{a^2-c}\,.\,\sqrt{b^2-c}$

i.e. if $c^2 = (a^2 - c)(b^2 - c)$

i.e. if $\dfrac{1}{a^2}+\dfrac{1}{b^2}=\dfrac{1}{c}$

**Example 6.24.** Prove that if the four points of intersection of the circles

$$x^2 + y^2 + ax + by + c = 0$$

and $$x^2 + y^2 + a'x + b'y + c' = 0$$

by the lines

$$Ax + By + C = 0 \text{ and } A'x + B'y + C' = 0$$

respectively are concyclic; then

$$\begin{vmatrix} a-a' & b-b' & c-c' \\ A & B & C \\ A' & B' & C' \end{vmatrix} = 0$$

*Solution.*

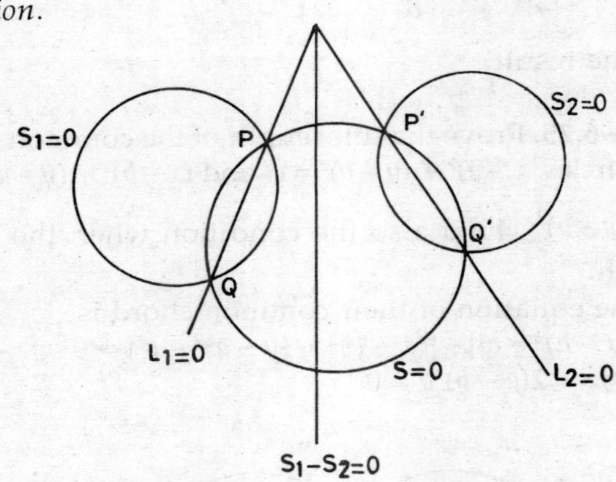


We are given two circles

$S_1 = x^2 + y^2 + ax + by + c = 0$ (1)

$S_2 = x^2 + y^2 + a'x + b'y + c' = 0$ (2)

and two lines

$L_1 = Ax + By + C = 0$ (3)

$L_2 = A'x + B'y + C' = 0$ (4)

We are given (3) meets (I), say, at $P$ and $Q$.

Also (4) meets (2), say, at $P'$ and $Q'$.

Further $P$, $Q$, $P'$ and $Q'$ are given to be concyclic. Let the circle through them be

$S = x^2 + y^2 + 2gx + 2fy + d = 0$ (5)

Now the radical axis of (1) and (2) is

$S_1 - S_2 = (a - a')\,x + (b - b')\,y + (c - c') = 0$ (6)

The radical axis of $S_1 = 0$ and $S = 0$ is

$L_1 = Ax + By + C = 0$ (7)

The radical axis of $S_2 = 0$ and $S = 0$ is

$L_2 = A'x + B'y + C' = 0$ (8)

(note the common chord is their radical axis, in these two cases).

Since the lines (6), (7) and (8) are concurrent (by **Theorem 6.3**)

we have $\begin{vmatrix} a-a' & b-b' & c-c' \\ A & B & C \\ A' & B' & C' \end{vmatrix}$ by (**Theorem 4.3**)

Hence the result.

**Example 6.25.** Prove that the length of the common chord of the two circles: $x-a)^2+(y-b)^2=c^2$ and $(x-b)^2+(y-a)^2=c^2$ is $\sqrt{4c^2-2(a-b)^2}$. Find also the condition when the given circles touch.

*Solution.* The equation of their common chord is

$[(x-a)^2+(y-b)^2-c^2]-[(x-b)^2+(y-a)^2-c^2]=0$

$\Rightarrow -2(a-b)x+2(a-b)y=0$

$\Rightarrow x-y=0$

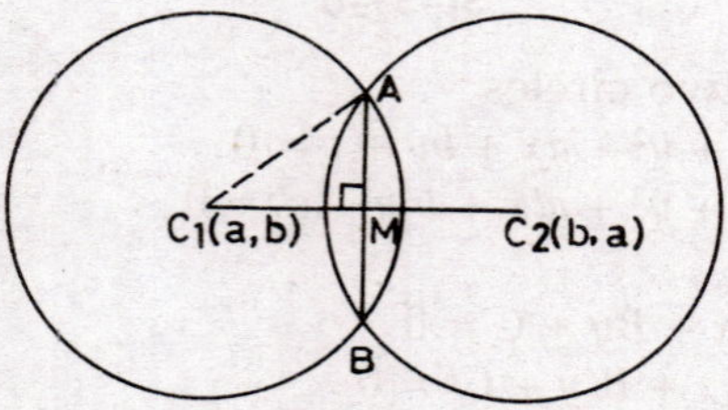


Now $C_1M$ = Perpendicular from $C_1$ on $AB$

$$\frac{|a-b|}{\sqrt{1^2+1^2}}=\frac{|a-b|}{\sqrt{2}}$$

∴ The length of the common chord

$= AB = 2\,AM = 2\sqrt{C_1A^2 - C_1M^2}$

$= 2\sqrt{c^2-\{(a-b)^2/2\}} = \sqrt{4c^2-2(a-b)^2}$

Also when the circles touch, chord $AB$ becomes the common tangent and hence $AB = 0$.

∴ The condition of the tangency is

$$4c^2 - 2(a - b)^2 = 0 \quad \text{i.e.} \quad 2c^2 = (a - b)^2.$$

**Example 6.26.** Two circles intersect at the point $P(x_1, y_1)$ and the line joining the other extremity of the two diameters through $P$ makes an angle $\theta$ with $x$-axis. Prove that the equation of the radical axis of the two circles is

$$(x - x_1) \cos \theta + (y - y_1) \sin \theta = 0.$$

*Solution.* Let the other extremities of the two diameters through $P(x_1, y_1)$ be $Q(x_2, y_2)$ and $R(x_3, y_3)$. Then the circles on $PQ$ and $PR$ as diameters are given by

$$(x - x_1)(x - x_2) + (y - y_1)(y - y_2) = 0$$

and $$(x - x_1)(x - x_3) + (y - y_1)(y - y_3) = 0.$$

Their radical axis is

$$(x - x_1)(x_2 - x_3) + (y - y_1)(y_2 - y_3) = 0 \qquad (1)$$

As $QR$ makes an angle $\theta$ with $x$-axis,

$$\tan \theta = (y_2 - y_3) / (x_2 - x_3) \qquad (2)$$

From (1) and (2),

$$(x - x_1) \cos \theta + (y - y_1) \text{ since } \theta = 0,$$

which is the equation of the radical axis.

## Exercise 6.6

1. Find the radical axes of the following pairs of circles:
   (i) $x^2 + y^2 = 2x$, $x^2 + y^2 - 3y = 5$.
   (ii) $x^2 + y^2 + 2x + 4y = 0$,
   $2x^2 + 2y^2 - 7x - 8y + 1 = 0$.
2. Find the equation of the circle which passes through $(2a, 0)$ and which has with the circle $x^2 + y^2 = a^2$, the radical axis, $x = a/2$.
3. Prove that the circle, with its centre at the radical centre and radius equal to the length of a tangent from its centre to any circle, cuts all the three circles orthogonally.
4. Prove that the radical centre of the three circles described on the sides of a triangle as diameters is the orthocentre of the triangle.
5. Prove that the radical axis of two circles bisects each of their common tangents.
6. Find the radical centre $C$ of the following circles:

$$x^2 + y^2 + 2x + 4y - 2 = 0,$$

$$x^2 + y^2 - x - y + 1 = 0,$$
$$x^2 + y^2 + 2x + 3y + 1 = 0.$$

Find the length of the tangent from $C$ to the three circles and verify that they are equal. Also verify that the circle having its centre at $C$ and radius equal to the length of the tangent cuts each of the given circles orthogonally.

7. Find the radical centre of the circles whose equations are, $x^2 + y^2 + a_1x + b_1 y + c = 0$ ; $x^2 + y^2 + a_2x + b_2y + c = 0$ and $x^2 + y^2 + a_3x + b_3 y + c = 0$. Find also the equation of the circle cutting each of them orthogonally.
8. If two circles cut a third circle orthogonally, prove that the radical axis of the two circles passes through the centre of the third circle.
9. Show that the circles $x^2 + y^2 + 2x - 8y + 8 = 0$ and $x^2 + y^2 + 10x - 2y + 22 = 0$ touch each other externally. Find also the point of contact.
10. Show that the difference of the squares of the lengths of the tangents to two circles from any point in their plane varies as the distance of the point from their radical axis.
11. Show that the two circles $x^2 + y^2 + ax + by = 0$ and $x^2 + y^2 = c^2$ touch each other if $a^2 + b^2 = c^2$.
12. Show that the equation of the common chord of the circles
$$x^2 + y^2 + 2ax + 2by + c = 0$$
and
$$x^2 + y^2 + 2bx + 2ay + c = 0$$
is $x - y = 0$. Hence, show that the circles touch if $(a + b)^2 = 2c$.
13. Find the equations of the lines which touch both the circles $x^2 + y^2 = 4$ and $(x - 4)^2 + y^2 = 1$.
Find also the co-ordinates of the centres of simlitude (i.e. the points of intesection of the transverse (internal) and direct (external) tangents).
14. Find the equation of the circle whose radius is 3 and which touches the circle, $x^2 + y^2 - 4x - 6y - 12 = 0$ internally at the point $(-1, -1)$.
15. Find the equation of the circle whose radius is 5 and which touches the circle, $x^2 + y^2 - 2x - 4y - 20 = 0$, externally at the point $(5, 5)$.

16.(i) Show that the circles,

$$x^2 + y^2 + 2g_i x + 2f_i y + c_i = 0 \quad (i = 1, 2)$$

touch each other if

$$(2g_1g_2 + 2f_1f_2 - c_1 - c_2)^2 = 4(g_1^{\ 2} + f_1^{\ 2} - c_1)(g_2^{\ 2} + f_2^{\ 2} - c_2)$$

(ii) Show that the circles,

$$x^2 + y^2 + 2g_i x + 2f_i\, y = 0 \ (i = 1, 2)$$

touch each other if $f_1g_2 = g_1f_2$.

17.Let 0 be the origin and $A_i\ (x_i, y_i)$ $i = 1, 2$ be two points. If the circles be described on $OA_1$, $OA_2$ as diameters, prove that the length of their common chord is $(x_1\ y_2 - x_2y_1)/A_1A_2$.

18.Show that the circle, $x^2 + y^2 + 2g_1x + 2f_1y + c_1 = 0$. will bisect the circumference of the circle, $x^2 + y^2 + 2g_2x + 2f_2y + c_2 = 0$ if $2g_2\ (g_1 - g_2) + 2f_2\ (f_1 - f_2) = c_1 - c_2$.
(*Hint:* The common chord, $S_1 - S_2 = 0$, will pass through $(-g_2, -f_2)$).

19.Prove that for all constants, $\lambda$ and $\mu$, the circle $(x - a)(x - a + \lambda) + (y - b)(y - b + \mu) = r^2$ bisects the circumference of $(x - a)^2 + (y - b)^2 = r^2$.

## 6.7. CO-AXAL SYSTEM OF CIRCLES

**Definition.** A system (or a family) of circles, every pair of whose members, has the same radical axis, is said to be a *co-axal system (or family)* of circles.

**Theorem 6.4.** If $S_1 = x^2 + y^2 + 2g_1x + 2f_1y + c_1 = 0$ and $S_2 = x^2 + y^2 + 2g_2x + 2f_2y + c_2 = 0$ be two given circles, then the system of circles given by $S_1 + \lambda S_2 = 0$, is co-axal, $\lambda$ being any parameter such that $\lambda \neq -1$.

*Proof.* Let $S_1 + \lambda_1 S_2 = 0$ and $S_i + \lambda_2 S_2 = 0$ be any two distinct members of the given sys tem (i.e. $\lambda_1 \neq \lambda_2$ and $\lambda_2 \neq -1$).

$$\Rightarrow \frac{S_1 + \lambda_1 S_2}{1 + \lambda_1} = 0 \text{ and } \frac{S_1 + \lambda_2 S_2}{1 + \lambda_2} = 0$$

(Note in these equations, the coefficient of $x^2$ is unity). Their radical axis is given by

$$\left(\frac{S_1 + \lambda_1 S_2}{1 + \lambda_1}\right) - \left(\frac{S_1 + \lambda_2 S_2}{1 + \lambda_2}\right) = 0$$

$\Rightarrow \quad (S_1 + \lambda_1 S_2)(1 + \lambda_2) - (S_1 + \lambda_2 S_2)(1 + \lambda_1) = 0$

$\Rightarrow \quad (\lambda_2 - \lambda_1)(S_1 - S_2) = 0$

$\Rightarrow \quad S_1 - S_2 = 0 \qquad (\because \lambda_1 \neq \lambda_2)$

Since this does not contain the parameter, $\lambda$; we see that any two distinct members of the given system of circles have the same radical axis. Thus the system of circles, $S_1 + \lambda S_2 = 0$, is co-axal. This co-axal system of circles is said to be *determined* by the circles $S_1 = 0$ and $S_2 = 0$. Clearly it has the radical axis, $S_1 - S_2 = 0$.

**Theorem 6.5.** If $S = 0$ be a circle and $L = 0$ be a line, then the system of circles given by $S + \lambda L = 0$, ($\lambda$ any parameter) is co-axal and the equation of its radical axis is $L = 0$.

*Proof.* Left to the reader.

**Remark.** Let $S_1 = 0$ and $S_2 = 0$ be two given circles. Then $S_1 + \lambda(S_1 - S_2) = 0$ ($\lambda$ a parameter) is also a co-axal system. In fact, this co-axal system of circles is the same as the system given by $S_1 + \lambda S_2 = 0$.

### 6.7.1. A simplified form of the equation of a co-axal system

We know that the radical axis of any two circles is perpendicular to the line joining their centres (**Theorem 6.2**). Further, every pair of distinct members of a co-axal system of circles, has the same radical axis; hence we deduce that the centres of all the circles in the co-axal system are collinear.

Let, therefore, the line of their centres be the $x$-axis and the radical axis be the $y$-axis.

Let $\qquad x^2 + y^2 + 2g_1 x + 2f_1 y + c_1 = 0$

and $\qquad x^2 + y^2 + 2g_2 x + 2f_2 y + c_2 = 0$

be any two members of the system.

Then $f_1 = f_2 = 0$ ( centres are on the $x$-axis).

Thus the equations reduce to

$$x^2 + y^2 + 2g_1 x + c_1 = 0$$

and $\qquad x^2 + y^2 + 2g_2 x + c_2 = 0.$

Their radical asix is

$$2(g_1 - g_2)\, x + (c_1 - c_2) = 0$$

which is the same as the $y$-axis, i.e. as $x = 0$.

On comparing the respective coefficients, we get

$$\frac{1}{2(g_1 - g_2)} = \frac{0}{c_1 - c_2}$$

whence $c_1 = c_2 = c$ (say).

$\therefore$ the equations of the circles become

$$x^2 + y^2 + 2g_1x + c = 0$$

and $$x^2 + y^2 + 2g_2x + c = 0.$$

consequently we see that the simplest form of a system of co-axal circles, is

$$x^2 + y^2 + 2gx + c = 0,$$

where $g$ is a parameter and $c$ *a* constant. This co-axal system of circles have their centres on the $x$-axis and their radical axis is the $y$-axis.

**Exercise 6.7**

1. Show that the simplest form of a co-axal system of circles can be taken as $x^2 + y^2 + 2fy + c = 0$, where $f$ is a parameter and $c$ a constant. What are their line of centres and the radical axis?

### 6.7.2. Limiting points of a co-axal system of circles

Let $S_1 = x^2 + y^2 + 2g_1x + 2f_1y + c_1 = 0$

and $S_2 = x^2 + y^2 + 2g_2x + 2f_2y + c_2 = 0$

be two given circles. The co-axal system determined by them is

$$S_1 + \lambda S_2 = 0,$$

where $\lambda$ is a parameter (and $\lambda \neq -1$).

$$\Rightarrow \quad \frac{S_1 + \lambda S_2}{1 + \lambda} = 0$$

$$\Rightarrow x^2 + y^2 + 2\left(\frac{g_1 + \lambda g_2}{1 + \lambda}\right)x + 2\left(\frac{f_1 + \lambda f_2}{1 + \lambda}\right)y + \left(\frac{c_1 + \lambda c_2}{1 + \lambda}\right) = 0$$

The radius, $r(\lambda)$, of this circle is given by

$$r(\lambda) = \frac{\sqrt{(g_1 + \lambda g_2)^2 + (f_1 + \lambda f_2)^2 - (c_1 + \lambda c_2)(1 + \lambda)}}{1 + \lambda}$$

Choose $\lambda$, such that $r(\lambda) = 0$.

$\Rightarrow (g_1 + \lambda g_2)^2 + (f_1 + \lambda f_2)^2 - (c_1 + \lambda c_2)(1 + \lambda) = 0.$

$\Rightarrow (g_2^2 + f_2^2 - c_2)\lambda^2 + (2g_1g_2 + 2f_1f_2 - c_1 - c_2)\,\lambda$

$$+ (g_1^2 + f_1^2 - c_1) = 0,$$

which is a quadratic in $\lambda$. Let $\lambda_1, \lambda_2$ be its roots.

Thus we get two circles: $S_1 + \lambda_1 S_2 = 0$ and $S_1 + \lambda_2 S_2 = 0$. which have zero radii.

These two point circles are called the *limiting points* of the co-axal system of circles.

**Note:** We can also proceed by taking the co-axal system as

$$S_1 + \lambda\,(S_1 - S_2) = 0.$$

### 6.7.3 To find the limiting points of the co-axal system $x^2 + y^2 + 2gx + c = 0$, where $g$ is a parameter and c a constant.

The centre and the radius of $x^2 + y^2 + 2gx + c = 0$ are $(-g, 0)$ and $\sqrt{g^2 - c}$.

Let $\sqrt{g^2 - c} = 0 \Rightarrow g = \pm\sqrt{c}$.

Thus we get the two limiting points, of the given co-axal system as

$$(\sqrt{c}, 0) \text{ and } (-\sqrt{c}, 0).$$

**Note 1.** Clearly the above limiting points are real and distinct, real and coincident or imaginary according as c >, = or < 0.

**Note 2.** The radical axis of the co-axal system, given by, $x^2 + y^2 + 2gx + c = 0$ is $x = 0$. It meets the members of this system at the points : $(0, \sqrt{-c})$ and $(0, -\sqrt{-c})$. $\Rightarrow$ All the members of the given co-axal system pass through two fixed points, $(0, \pm\sqrt{-c})$, on the radical axis of the system, which are real and distinct, real and coincident or imaginary according as $c <$ , = or > 0.

Thus the above co-axal system of circles has real and distinct, real and coincident or imaginary limiting points according as the circles of this system do not intersect, touch or intersect (in two real points) the radical axis of this system.

### 6.7.4. Orthogonal co-axal system

To find the equation of a co-axal system of circles cutting each member of a given co-axal system orthogonally.

Let the given co-axal system be

$$x^2 + y^2 + 2gx + c = 0 \tag{1}$$

where $g$ is a parameter and $c$ is a constant.

Let the equation of any member of the orthogonal system be

$$x^2 + y^2 + 2Gx + 2Fy + C = 0. \tag{2}$$

Applying the condition of orthogonality, we get

$$2G\cdot g + 2F\cdot 0 = C + c.$$

$$\Rightarrow \qquad 2G\cdot g = C + c.$$

This equation is true for all real values of $g$; hence $G = 0$ and then $C = -c$.

$\therefore$ the required orthogonal system is given by

$$x^2 + y^2 + 2Fy - c = 0, \tag{3}$$

where $F$ is a parameter (and $c$ is a constant as before).

**Note 1.** The limiting points, $(\pm \sqrt{c}, 0)$ of the system, (1) lie on the orthogonal system, (3), i.e. every circle of the orthogonal co-axal system passes through the limiting points of the given co-axal system.

**Note 2.** The limiting points of the system (3) are $(0, \pm \sqrt{-c})$. Thus the limiting points of the orthogonal system are real and distinct, real and coincident or imaginary according as the limiting points of the system (1) are imaginary, real and coincident or real and distinct.

**Example 6.27.** Find the limiting points of the co-axal system of circles determined by

$$x^2 + y^2 + 4x + 2y + 5 = 0$$

and $$x^2 + y^2 + 2x + 4y + 7 = 0.$$

*Solution*. The equation of their radical axis is $S_1 - S_2 = 0$

$$\Rightarrow \quad 2x - 2y - 2 = 0.$$

$\therefore$ the equation of the co-axal system determined by $S_1 = 0$ and $S_2 = 0$ is

$$x^2 + y^2 + 4x + 2y + 5 + \lambda(2x - 2y - 2) = 0$$

$$\Rightarrow \quad x^2 + y^2 + 2(2 + \lambda)x + 2(1 - \lambda)y + (5 - 2\lambda) = 0.$$

Equating the radius of this circle to zero, we get

$$(2 + \lambda)^2 + (1 - \lambda)^2 - (5 - 2\lambda) = 0$$

$$\Rightarrow \quad \lambda(\lambda + 2) = 0$$

$$\therefore \quad \lambda = 0, -2.$$

Now the limiting points of the coaxal system are the centres of the point-circles i.e. $(-(2 + \lambda), -(1 - \lambda))$.

$\therefore$ For $\lambda = 0$ and $-2$; these points are $(-2, -1)$ and $(0, -3)$

**Example 6.28.** Find the co-axal system, one of whose members is $x^2 + y^2 + 2x + 3y - 7 = 0$ and a limiting point is $(1, -2)$.

*Solution*. The limiting point (i.e. the point circle at $(1, -2)$) is a member of the required co-axal system.

$\therefore$ the equation of this point circle at $(1, -2)$ is

$$(x - 1)^2 + (y + 2)^2 = 0.$$

$$\Rightarrow \quad x^2 + y^2 - 2x + 4y + 5 + 0.$$

Also the given circle is

$$x^2 + y^2 + 2x + 3y - 7 = 0.$$

$\therefore$ The radical axis of these two circles is

$$(x^2 + y^2 + 2x + 3y - 7) - (x^2 + y^2 - 2x + 4y + 5) = 0$$

i.e. $\quad 4x - y - 12 = 0.$

Thus the required co-axal system of circles is given by

$$x^2 + y^2 + 2x + 3y - 7 + \lambda(4x - y - 12) = 0, \ (\lambda \text{ a parameter}).$$

**Example 6.29.** If $A$, $B$, $C$ be the centres of three co-axal circles and $t_1$, $t_2$, $t_3$ be the lengths of the tangents to them from any point, prove that $\overline{BC}\, t_1^2 + \overline{CA} \cdot t_2^2 + \overline{AB} \cdot t_3^2 = 0$.

*Solution*. Let $x^2 + y^2 + 2gx + c = 0$ be the equation of the co-axal system. Let $A(-g_1, 0)$, $B(-g_2, 0)$ and $C(-g_3, 0)$ be the centres of three members of this family and let $P(\alpha, \beta)$ be any point from where the tangents to them are drawn.

We then have $AB = g_1 - g_2$, $BC = g_2 - g_3$, $CA = g_3 - g_1$.

Also $\quad t_i^2 = \alpha^2 + \beta^2 + 2g_i\alpha + c, \ i = 1, 2, 3. \qquad$ (by § **6.4.2**)

Now $\overline{BC}.t_1^2 + \overline{CA}.t_2^2 + \overline{AB}.t_3^2$

$= \Sigma(g_2 - g_3).(\alpha^2 + \beta^2 + 2g_1\alpha + c)$

$= (\alpha^2 + \beta^2 + c).\Sigma(g_2 - g_3) + 2\alpha.\Sigma\, g_1\,(g_2 - g_3)$

= 0, which proves the result.

**Example 6.30**. Prove that the equation of the two given circles can always be put in the form $x^2 + y^2 + ax + b = 0$, $x^2 + y^2 + a'x + b = 0$ and if one of the circles is within the other, then $aa'$ and $b$ are both positive.

*Solution*. The first part follows by § **6.7.1.** For the second part, as one of the circles lies within the other, they have no real points of intersection, whence by § **6.7.3,** $b > 0$.

Also their radical axis is $x = 0$, which does not intersect any of them at real points, which implies both of them lie on the same side of $x = 0$.

$\Rightarrow$ their centres $\left(-\frac{a}{2}, 0\right)$ and $\left(-\frac{a'}{2}, 0\right)$ lie on the same side of $x = 0$.

$\Rightarrow$ $a$ and $a'$ have the same sign.

$\Rightarrow aa' > 0$.

**Example 6.31**. A point $P$ moves so that its distances from two fixed points are in a constant ratio $\lambda$ ($\neq 1$). Prove that the locus of $P$ is a circle. If $\lambda$ varies, show that $P$ generates a system of co-axals of which the fixed points are the limiting points.

*Solution*. Let $A'$ and $A$ be the fixed points. Choose $A'A$ as $x$-axis and the right bisector of $A'A$ as $y$-axis. Let $A'$ $(-a, 0)$ and $A$ $(a, 0)$. Let $P$ $(x, y)$ be any point on the locus. By hypothesis, $AP = \lambda\ \ A'P$

$\Rightarrow (x - a)^2 + y^2 = \lambda^2\,[(x + a)^2 + y^2]$ (1)

which is a circle (as $\lambda \neq 1$).

As $\lambda$ varies, (1) generates a coaxal system of which two members are

$(x - a)^2 + y^2 = 0$ and $(x + a)^2 + y^2 = 0$

These being circles of zero radii, represent the limiting points of this coaxal system. The centres of these point circles are $A(a,0)$ and $A'(-a, 0)$.

$\therefore$ The coaxal system has the fixed points as its limiting points.

**Example 6.32.** A system of circles is drawn through two fixed points; tangents are drawn to these circles parallel to a given line. Find the locus of the points of contact.

*Solution.* As in the previous example, let the fixed points be $(-a, 0)$ and $(a, 0)$.

Let any circle through these be given by the equation

$$x^2 + y^2 + 2gx + 2fy + c = 0$$

$\therefore$ $a^2 + 2ga + c = 0$ and $a^2 - 2ga + c = 0$

$\Rightarrow g = 0$ and $c = -a^2$.

$\therefore$ The equation of the circle through these fixed points is

$$x^2 + y^2 + 2fy - a^2 = 0. \tag{1}$$

Here $f$ is a parameter.

Let $m$ be the slope of the given line. Let any line parallel to it touch (1) at $(x', y')$.

$\therefore$ Equation of the tangent to (1) at $(x', y')$ is $xx' + yy' + f(y + y') - a^2 = 0$

Its slope $-\dfrac{x'}{(y'+f)} = m$ (by hypothesis)

$$\Rightarrow \quad m(y' + f) + x' = 0 \tag{2}$$

But $(x', y')$ lies on (1)

$$\therefore \quad x'^2 + y'^2 + 2fy' - a^2 = 0$$

$$\Rightarrow \quad f = (a^2 - x'^2 - y'^2)/2y'$$

$\therefore$ (2) becomes

$$m\left[y' + \frac{a^2 - x'^2 - y'^2}{2y'}\right] + x' = 0$$

$$\Rightarrow \quad m\,[y'^2 + a^2 - x'^2] + 2x'y' = 0$$

$\therefore$ The locus of $(x', y')$ is

$$m\,(a^2 + y^2 - x^2) + 2xy = 0$$

**Example 6.33.** The circles of the coaxal system, $x^2 + y^2 + 2gx - a^2 = 0$ are taken in pairs, which cut each other orthogonally. Prove that if $p_1, p_2$ are perpendiculars from $(0, a)$; $(0, -a)$ on a common tangent of any such pair of circles, then $p_1p_2 = a^2$.

*Solution.* Let the circles

$$S \equiv x^2 + y^2 + 2g_i\, x - a^2 = 0 \; (i = 1, 2)$$

cut each other orthogonally

$$\therefore \quad g_1g_2 + a^2 = 0 \tag{1}$$

If $lx + my = 1$ is a common tangent of these circles, then

$$\frac{-lg_i - 1}{\sqrt{l^2 + m^2}} = \pm\sqrt{g_i^2 + a^2} \quad (i = 1, 2)$$

$\Rightarrow (lg_i + 1)^2 = (l^2 + m^2)(g_i^2 + a^2)$

$\Rightarrow m^2 g_i^2 - 2lg_i + a^2 (l^2 + m^2) - 1 = 0$

$\Rightarrow m^2g^2 - 2lg + a^2 (l^2 + m^2) - 1 = 0$

has two roots $g_1$ and $g_2$.

$\therefore\; g_1g_2 = [a^2 (l^2 + m^2) - 1]/m^2 = -a^2$ (by (1))

$$\Rightarrow \quad a^2 (l^2 + m^2) = 1 - a^2m^2 \tag{2}$$

$$\text{Now } p_1p_2 = \frac{|ma - 1|}{\sqrt{l^2 + m^2}} \cdot \frac{|-ma - 1|}{\sqrt{l^2 + m^2}}$$

$$= \frac{|1 - m^2a^2|}{l^2 + m^2} = a^2 \qquad \text{(by (2))}$$

## Exercise 6.8

1. Find the co-ordinates of the limiting points of the co-axal system determined by the circles:
   (i) $x^2 + y^2 - 10x - 8y + 5 = 0$ and $x^2 + y^2 + 10x + 2y = 0$.
   (ii) $x^2 + y^2 - 6x - 6y + 4 = 0$ and $x^2 + y^2 - 2x - 4y + 3 = 0$.
   (iii) $x^2 + y^2 + 4x + 8y + 19 = 0$
   and $x^2 + y^2 + 8x + 4y + 19 = 0$.
2. Find the equation of the circle which passes through the origin and belongs to the coaxal system of which the limiting points are (1, 2) and (4, 3).
3. The limiting points of a coaxal system of circles are

(1, 2) and (3, 5). Find the equation of that member of the system which passes through the point (2, 2).

4. Find the equation of the coaxal system of circles, one of whose members is $x^2 + y^2 - 4x - 2y - 5 = 0$ and one limiting point is: (i) (0, 0), (ii) (1, 2).
5. Find the equation of the coaxal system of circles whose limiting points are $(a, b)$ and $(c, d)$.
6. The circle $x^2 + y^2 + 4x - 6y + 3 = 0$ is one of a coaxal system having as radical axis, the line $2x - 4y + 1 = 0$. Find the circles of the system which touch the line $x + 3y - 2 = 0$.
7. (i) Define a coaxal system of circles. Show that the circles
$$x^2 + y^2 + 2ax + 2by + 2\lambda(ax - by) = 0,$$
where $\lambda$ is a parameter, form a coaxal system. Find the equation, of the radical axis of the system.
(ii) Also find the equations of circles which are orthogo nal to this system.
8. If the origin be one of the limiting points of a system of coaxal circles of which $x^2 + y^2 + 2gx + 2fy + c = 0$ is a member, prove that the equation of the system of the circles, cutting this system orthogonally is,
$$(x^2 + y^2)(g + \mu f) + c(x + \mu y) = 0.$$
Find also the other limiting point.
9. Find the limiting points of the system of circles
$$(x^2 + y^2 + 2gx + c) + \lambda(x^2 + y^2 + 2fy + d) = 0$$
and show that the square of the distance between them is
$$\frac{(c-d)^2 - 4f^2g^2 + 4cf^2 + 4dg^2}{f^2 + g^2}$$
10. Find the general equation of the system of circles any pair of which have the same radical axis as the circles:
$$x^2 + y^2 + x - 5y - 3 = 0 \text{ and } x^2 + y^2 + 3x + 4y + 6 = 0.$$
show that the equation to that members of the system which passes through (0, 0) is $3(x^2 + y^2) + 5x - 6y = 0$
11. Prove that the chords of intersection of a given circle and a given coaxal system of circles pass through a fixed point.

12. The point (2,1) is a limiting point of a coaxal system of circles of which, $x^2 + y^2 - 6x - 4y - 3 = 0$ is a member. Find the equation of the radical axis and the other limiting point.
13. Of a given coaxal system of circles, one member can be drawn to pass through a given point and two members can be drawn to touch a given line.
14. Prove that a common tangent to any two circles of a coaxal system subtends a right angle at either limiting point of the system.
15. Prove that the square of the tangent that can be drawn from any point on one circle to another circle is equal to twice the product of the perpendicular distance of the point from the radical axis of the two circles and the distance between their centres.

**Miscellaneous Exercise on Chapter 6.**
**Section A (For all students)**

1. Two circles are called *concentric,* if they have the same centre. (a) Show that for fixed $g$ and $f$, and variable $c$, the equation $x^2 + y^2 + 2gx + 2fy + c = 0$ represents a system of concentric circles.

   (b) Find the equations of the circles concentric with $x^2 + y^2 + 2x + 4y - 1 = 0$ and
   (i) passing through (0, 0)
   (ii) having radius 3.

   (c) Find the equation of the circle concentric with the circle, $x^2 + y^2 - 8x - 10y + 16 = 0$ whose circumference passes through the centre of the circle,
$$x^2 + y^2 + 4x - 6y - 12 = 0$$

2. (i) Show that the equations
$$x = a \cos\theta + b \sin\theta$$
$$y = a \sin\theta - b \cos\theta$$
represent a circle, $\theta$ being a parameter.

(ii) Show that the locus of the point of intersection of the lines,

$$x \cos \theta + y \sin \theta = a,$$
$$x \sin \theta - y \cos \theta = b$$

for different values of θ, is a circle.

(iii) Show that the equations

$$x = a\left(\frac{1-t^2}{1+t^2}\right) \quad \text{and} \quad y = \frac{2at}{1+t^2}$$

represent a circle, t being a parameter.

3. The parameters of two points on the circle $x^2 + y^2 = a^2$ are θ and π/4 and the tangents at these points are perpendicular. Find the values of θ.

4. The tangents at the points $(x_1, y_1)$ and $(x_2, y_2)$ on the circle

$$x^2 + y^2 + 2gx + 2fy + c = 0$$

are perpendicular to each other. Prove that

$$x_1 x_2 + y_1 y_2 + g(x_1 + x_2) + f(y_1 + y_2) + f^2 + g^2 = 0.$$

5. Show that the line

$$(x + g) \cos \theta + (y + f) \sin \theta = k$$

is a tangent to

$$x^2 + y^2 + 2gx + 2fy + c = 0$$

for all values of θ, iff

$$g^2 + f^2 = k^2 + c.$$

6. Lines drawn through the points $(a, 0)$, $(-a, 0)$ make a constant angle α with one another. Show that the locus of their point of intersection are the circles.

$x^2 + y^2 - a^2 \pm 2ay \cot \alpha = 0$

7. Show that the locus of the centre of a circle which cuts the circles,

$$x^2 + y^2 + 2g_1x + 2f_1y + c_1 = 0$$

and $x^2 + y^2 + 2g_2x + 2f_2y + c_2 = 0$ orthogonally, is their radical axis.

(Also see Question 8 Excerise 6.6).

8. Prove that the limiting points of a coaxal system of circles are inverse points with respect to every member of the system.

(**Definition.** Two points $P$ and $Q$, on a line through the centre $C$ of a circle with radius $a$, are called *inverse points*, if $|CP| \cdot |CQ| = a^2$).

9. Show that the equation of the co-axal system whose limiting points are (0, 0) and $(h, k)$ is
$$x^2 + y^2 + \lambda(2hx + 2ky - h^2 - k^2) = 0.$$

10. Prove that the circles
$$x^2 + y^2 - 6x - 2y + 1 = 0$$
$$x^2 + y^2 + 2x - 8y + 13 = 0$$
touch one another and find the equation of the tangent at their point of contact.

11. Prove that the limiting points of the system
$$x^2 + y^2 + 2gx + c + \lambda(x^2 + y^2 + 2fy + k) = 0$$
subtend a right angle at the origin if
$$\frac{c}{g^2} + \frac{k}{f^2} = 2$$

12. If $S_1 = 0$ and $S_2 = 0$ are the equations of two circles, prove that the locus of a point which moves such that the tangents from it to the two circles are in a given ratio $\neq 1$, is a circle, co-axal with the circles $S_1 = 0$ and $S_2 = 0$. (*Hint:* **See Question 2, Exercise 6.4**).

13. Show that the circle
$$x^2 + y^2 + 2fy - d = 0$$
passes through the limiting points of the co-axal system
$$x^2 + y^2 + 2gx + d = 0$$
and cuts every member of the system orthogonally, for all values of $f$. Hence deduce that every circle that passes through the limiting points of a co-axal system cuts every circle of the system orthogonally.
(*Hint:* **See § 6.7.4**).

14. Find the equation to the circles which passes through the point (2, 0) and whose centre in the limit of the point of intersection of the lines : $3x + 5y = 1$ and $(2 + \alpha)\, x + 5\, \alpha^2\, y = 1$ as $\alpha$ tends to 1.

15. Show that the locus of points such that the difference of the squares of the tangents from them to two given circles is constant, is a line parallel to their radical axis.

16. Let $OA$ and $OB$ be tangents from the origin, $O$, to the circle, $x^2 + y^2 + 2gx + 2fy + c = 0$. Prove that the equation of the circumcircle of the triangle, $OAB$ is $x^2 + y^2 + gx + fy = 0$
17. Show that the two circles
$a(x^2 + y^2) + 2gx + 2fy + c = 0$ and
$a'(x^2 + y^2) + 2g'x + 2f'y + c' = 0$, will touch each other, if
$(af' - a'f)(cf' - c'f) + (ag' - a'g)(cg' - c'g) + (ac' - a'c)^2 = (fg' - f'g)^2/4$.

**Section B (For Honours students)**

1. Find the radical centre of the circles
$$x^2 + y^2 + 2g_i x + 2f_i y + c_i = 0 \ (i = 1, 2, 3)$$
Also find the equation of the circle cutting them orthogonally.
2. If $A, B, C$ be the centres and $r_1, r_2, r_3$ be the radii of three co-axal circles, prove that
$r_1^2 \cdot \overline{BC} + r_2^2 \cdot \overline{CA} + r_3^2 \cdot \overline{AB} + \overline{BC} \cdot \overline{CA} \cdot \overline{AB} = 0.$
3. If $A, B, C$ be the centres of the three co-axal circles
$$S_1 = 0, S_2 = 0. S_3 = 0;$$
prove that
$$\overline{BC} \cdot S_1 + \overline{CA} \cdot S_2 + \overline{AB} \cdot S_3 = 0.$$
4. If the equations of two circles whose radii are $a$ and $a'$ respectively, are $S = 0$ and $S' = 0$, then prove that the circles given by
$$\frac{S}{a} \pm \frac{S'}{a'} = 0$$
will cut orthogonally.
5. Show that the circumcircle of the triangle formed by the lines
$$a_i x + b_i y + c_i = 0, \ (i = 1, 2, 3) \text{ is}$$
$$\begin{vmatrix} \dfrac{a_1^2 + b_1^2}{a_1x + b_1y + c_1} & \dfrac{a_2^2 + b_2^2}{a_2x + b_2y + c_2} & \dfrac{a_3^2 + b_3^2}{a_3x + b_3y + c_3} \\ a_1 & a_2 & a_3 \\ b_1 & b_2 & b_3 \end{vmatrix} = 0$$

6. If the circumcircle of the triangle formed by the lines

$$ax + by + c = 0,$$
$$bx + cy + a = 0,$$
and
$$cx + ay + b = 0$$

passes through the origin, show that

$$(b^2 + c^2)(c^2 + a^2)(a^2 + b^2) = abc(b + c)(c + a)(a + b).$$

7. Show that if the points

$$\left(cm_i, \frac{c}{m_i}\right), (i = 1,2,3,4)$$

are concyclic, then $m_1 m_2 m_3 m_4 = 1$.
(*Hint*: they all lie on $xy = c^2$).

8. The line

$$lx + my = 1$$

meets the lines

$$ax^2 + 2hxy + by^2 = 0$$

in $P$ and $Q$. Show that

(i) The circle on $PQ$ as diameter is

$$(am^2 - 2hlm + bl^2)(x^2 + y^2) + 2(hm - bl)x + + 2(hl - am)y + a + b = 0.$$

(ii) The circumcircle of the triangle $OPQ$ is

$$(am^2 - 2hlm + bl^2)(x^2 + y^2) + 2h(mx + ly) + + (a - b)(lx - my) = 0.$$

9. The straight line

$$px + qy = 1$$

makes with

$$ax^2 + 2hxy + by^2 = c,$$

a chord which subtends a right angle at the origin. Show that

$$c(p^2 + q^2) = a + b.$$

Also show that

$$px + qy = 1$$

touches a circle whose radius is

$$\sqrt{\frac{c}{a+b}}.$$

(*Hint:* The line $px + qy = 1$ touches the circle

$$x^2 + y^2 = \frac{c}{a+b}\Big)$$

10. A line passes through a fixed point $(a, b)$. Show that the locus of the foot of the perpendicular drawn to it from the origin is the circle

$$x^2 + y^2 - ax - by = 0.$$

11. A variable line passes through a fixed point $(a, b)$ and cuts the co-ordinate axes in the points $A$, $B$. Show that the locus of the centre of the circle $OAB$ is

$$\frac{a}{x} + \frac{b}{y} = 2.$$

12. A circle of constant radius $r$ passes through the origin $O$, and cuts the axes in $A$, $B$. Show the locus of the foot of the perpendicular from $O$ on the line $AB$ is

$$(x^2 + y^2)^2 (x^{-2} + y^{-2}) = 4r^2.$$

13. If $O$ be the centre of a circle of radius unity and $A$, $B$ be two points on a line through $O$ such that

$$\overline{OA} \,.\, \overline{OB} = 1$$

and if $P$ be a variable point on the circle, show that

$$\overline{PA} \,.\, \overline{PB} = \text{constant}.$$

14. A circle of constant radius $3k$ passes through the origin and meets the axes in $A$, $B$. Show that the locus of the centroid of the triangle $OAB$ is the circle

$$x^2 + y^2 = 4k^2.$$

15. Tangent line at any point of the circle

$$x^2 + y^2 = r^2$$

meets the co-ordinate axes at $A$, $B$. Show that the locus of the point of intersection of lines drawn parallel to the co-ordinate axes through $A$, $B$ is

$$x^{-2} + y^{-2} = r^{-2}.$$

16. Find the common tangents of the circles

$$x^2 + y^2 = 1$$

and

$$x^2 + y^2 - 2x - 6y + 6 = 0.$$

17. Show that the common tangents to the circles

$$x^2 + y^2 - 6x = 0$$

and $$x^2 + y^2 + 2x = 0$$

form an equilateral triangle.

18. If $x \cos \theta + y \sin \theta = 2$ is the equation of a common tangent to $x^2 + y^2 = 4$ and $x^2 + y^2 - 6\sqrt{3}x - 6y + 20 = 0$, find the value of θ. Prove that the angle between the pair of direct common tangents is $\tan^{-1} (4\sqrt{2}/7)$.

19. Find the length of the direct common tangent to the two circles: $(x - a)^2 + (y - b)^2 = r^2$ and $(x - a')^2 + (y - b')^2 = r'^2$. Show that the chord of contact with the first circle of the direct common tangents is $(x - a)(a' - a) + (y - b)(b' - b) = r(r + r')$.

20. Prove that the circum-circle of the triangle formed by the three lines, $x \cos \alpha_i + y \sin \alpha_i = a \sec \alpha_i + b \sin \alpha_i$ $(i = 1, 2, 3)$. passes through the point (o, b). (*Hint:* Use the result in **Q5** of this exercise).

21. Find the equation of the circum circle of the triangle formed by the three lines, $x \cos \alpha_i + y \sin \alpha_i = p_i$ $(i = 1, 2, 3)$, Also show that this circumcircle will pass through the origin, if $p_1 p_2 \sin (\alpha_1 - \alpha_2) + p_2 p_3 \sin (\alpha_2 - \alpha_3) + p_3 p_1 \sin (\alpha_3 - \alpha_1) = 0$.

22. Show that the system of circles given by $x^2 + y^2 + 2\lambda x + 2\mu y + u = 0$ has a common orthogonal circle if λ, μ, υ satisfy an equation of the form, $A\lambda + B\mu + C\upsilon + D = 0$.

23. A circle cuts orthogonally two fixed non-intersecting circles. Prove that it passes through two fixed points on their line of centres.

24. If the circles given by $x^2 + y^2 + 2gx + 2fy + c = 0$ and $x^2 + y^2 + 2g'x + 2f'y + c = 0$ be orthogonal, prove that the circle through their centres and their point of intersection is

$$x^2 + y^2 + (g + g')x + (f + f')y + c = 0.$$

25. Given two circles, $S_i = 0$ $(i = 1, 2)$; show that the locus of the centre of a circle which moves in such a manner as to intersect $S_1 = 0$ orthogonally and to be

met at the ends of a diameter by $S_2 = 0$ is given by $S_1 + S_2 = 0$.
(*Hint:* If $C$ be the centre and $r$ be the radius of the moving circle, then power of $C$ w.r.t. $S_1 = 0$ is $r^2$ and that w.r.t. $S_2 = 0$ is $-r^2$)

# 7. SOME STANDARD CURVES

## INTRODUCTION

In the present chapter, we shall trace the following curves. Actual derivation of these curves will be taken up in the subsequent chapters.

We have already known the following:

(1) $ax + by + c = 0$, which is a straight line.

(2) $ax^2 + 2hxy + by^2 = 0$, which represents a pair of straight lines through the origin.

(3) $ax^2 + 2hxy + by^2 + 2gx + 2fy + c = 0$, which represents a pair of lines iff $abc + 2fgh - af^2 - bg^2 - ch^2 = 0$.

We now take up the tracing of the following curves;

(i) $x^2 + y^2 = a^2$ (Circle),

(ii) $x^2 + y^2 + 2gx + 2fy + c + 0$ (Circle),

(iii) $\frac{x^2}{a^2} + \frac{y^2}{b^2} = 1$ (Ellipse),

(iv) $y^2 = 4ax$ (Parabola).

(v) $\frac{x^2}{a^2} - \frac{y^2}{b^2} = 1$ (Hyperbola),

(vi) $xy = c^2$ (Rectangular Hyperbola).

(vii) $\frac{y^2}{b^2} - \frac{x^2}{a^2} = 1$ (Conjugate Hyperbola).

or $\frac{x^2}{a^2} - \frac{y^2}{b^2} = -1$

### 7.1 TRACING OF CURVES

We mention below some salient points which will be useful

for the tracing of the above curves:

1. If on changing $(x, y)$ to $(-x, y)$ the equation of a curve does not change, then the curve is symmetrical w.r.t. the $y$-axis, as in that case both the points $(x, y)$ and $(-x, y)$ lie on the curve. This is illustrated in the following figure:

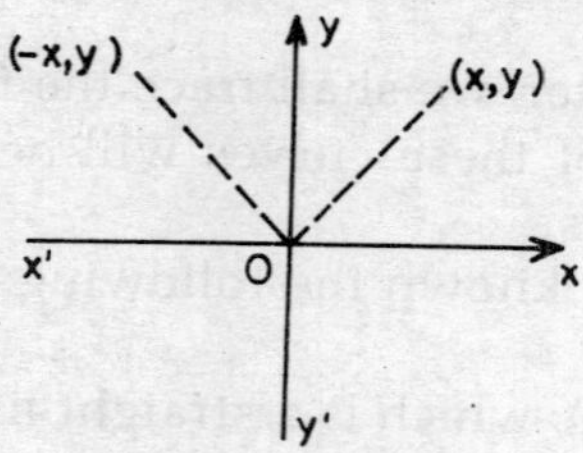


2. Similarly, if on changing $(x, y)$ to $(x, -y)$, the equation of a curve does not change, then the curve is symmetrical w.r.t. the $x$-axis, as is clear from the following figure:

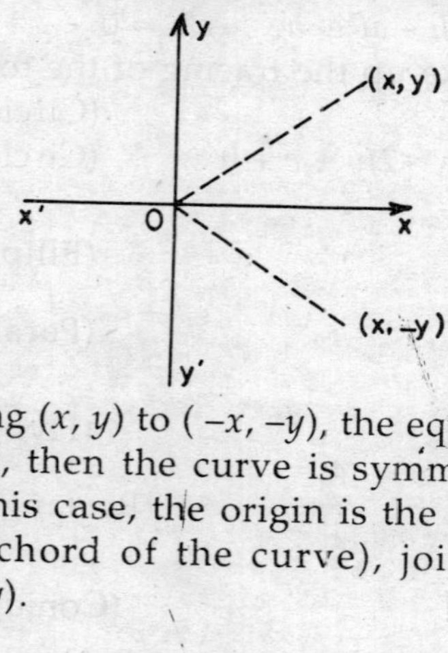


3. If on changing $(x, y)$ to $(-x, -y)$, the equation of a curve does not change, then the curve is symmetrical w.r.t. the origin. Also in this case, the origin is the mid-point of the line (in fact, a chord of the curve), joining the points $(x, y)$ and $(-x, -y)$.

**Definition**. *Central curve.* If each chord of a curve, through a point, is bisected thereat then this point is called the *centre* of the curve and such a curve is called a *central curve.*

4. For the curve to intersect the $y$-axis, we put $x = 0$, in its equation and solve for $y$. Similarly for its intersection with the $x$-axis, we put $y = 0$ and solve for $x$.

5. Next we find the range of values of $x$ (or those of $y$) between which the curve lies; for instance, the curve $x^2 + y^2 = a^2$, lies between the lines, $x = -a$ and $x = a$ (and also between the lines ($y = -a$ and $y = a$). We call such curves as *closed curves*.

We now trace the above mentioned curves:

**(i) $x^2 + y^2 = a^2$ (Circle)**

This curve is symmetrical about both the axes and (0, 0) is its centre. Consequently, it is enough to trace it only in the first quadrant.

Now $\quad x^2 + y^2 = a^2 \;\Rightarrow\; y^2 = a^2 - x^2,$

whence for $y$ to be real,

$$x^2 \leq a^2 \text{ i.e. } 0 \leq x \leq a$$

Similarly $\quad 0 \leq y \leq a$

(These ranges are only for the first quadrant).

Again as $x$ increases from 0 to $a$, $y$ decreases from $a$ to 0. The curve meets the axes at $(a, 0)$ and $(0, a)$. Also equations of the tangents at these points to the curve are $x = a$ and $y = a$ respectively.

Thus the shape of the curve in the first quadrant and by symmetry, the shape of the complete curve (curve is a circle, in the present case) is as shown:

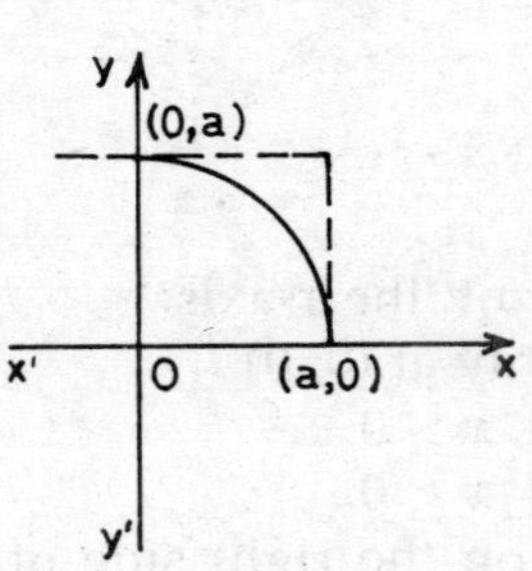


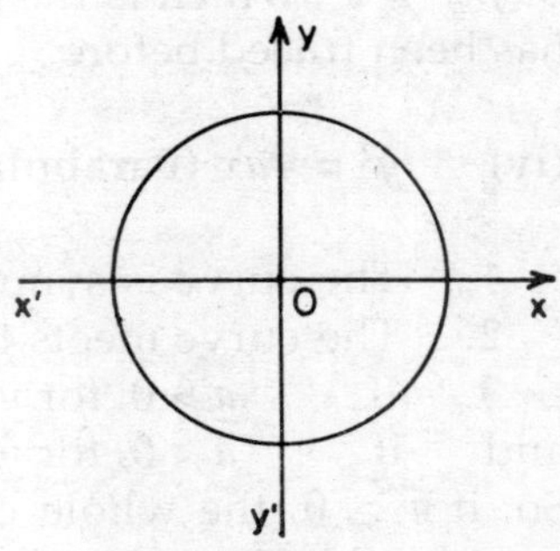


**(ii) $x^2 + y^2 + 2xy + 2fy + c + 0$ (Circle)**

This is similar to (i), with centre at $(-g, -f)$ and radius

$$= \sqrt{g^2 + f^2 - c}.$$

**(iii)** $\dfrac{x^2}{a^2}+\dfrac{y^2}{b^2}=1$ **(Ellipse).**

The discussion for this curve is similar to the curve in (i) ; the only changes are, that while $x$ changes from $-a$ to $a$, $y$ changes from $-b$ to $b$.

The equations of the tangents at $(a, 0)$ and $(0, b)$ are $x = a$ and $y = b$ respectively. It is also a central curve with centre at $(0, 0)$. The curve is as shown below:

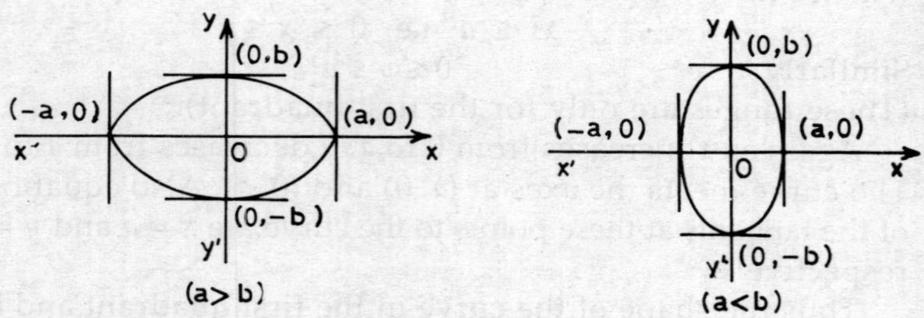


*Remark.* When $a = b$, the equation of the ellipse becomes $x^2 + y^2 = a^2$, which is the equation of a circle, and the same has been traced before.

**(iv)** $y^2 = 4ax$ **(Parabola).**

1. The curve is symmetrical w.r.t. the $x$-axis.
2. The curve meets the axes only at $(0, 0)$.
3. If $a > 0$, for $y$ to be real, $x \geq 0$,

and if $a < 0$, for $y$ to be real $x \leq 0$.

So, if $a > 0$, the whole curve lies on the right side of the $y$-axis and if $a < 0$, the whole curve lies on the left side of the y-axis.

4. As $x$ increases from 0 to $\infty$, $y$ also increases from 0 to $\infty$. Hence he curve is as shown below:

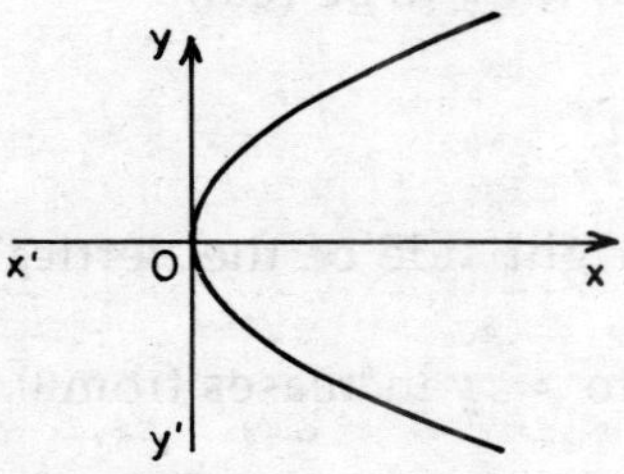


(a>0)

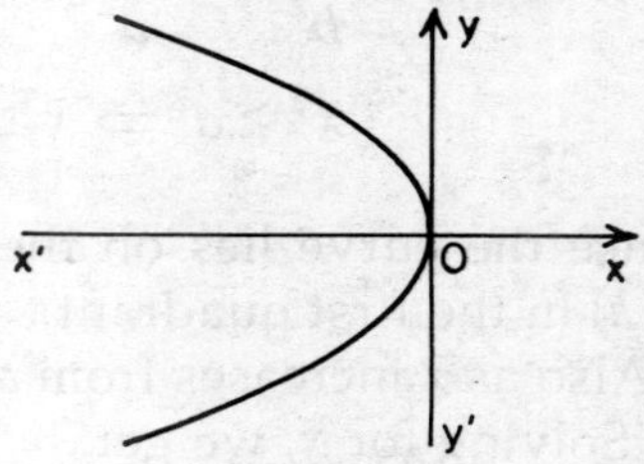


(a<0)

The curve is symmetrical w.r.t. the line $x'Ox$, which is called the *axis of the parabola*. Also the point $O$ is called the *vertex of the parabola*.

*Remark.* $x^2 = 4ay$ is also the equation of a parabola, whose diagram is shown below:

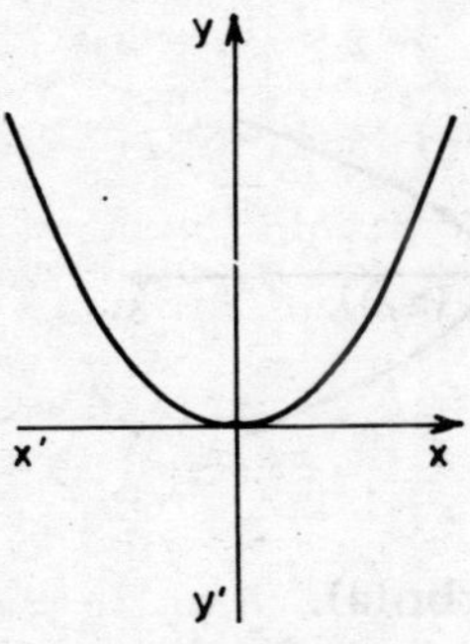


(a>0)

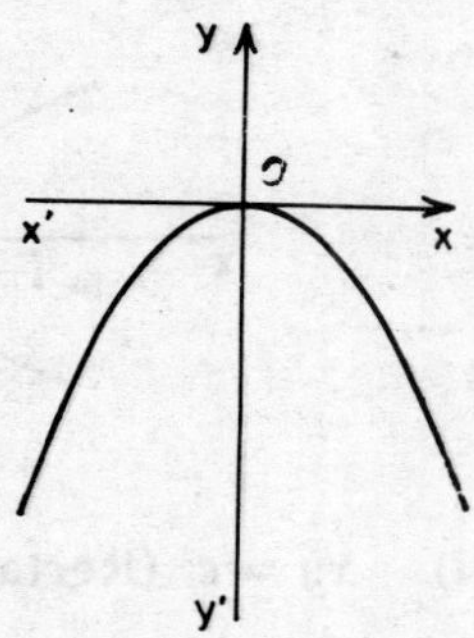


(a<0)

**(v)** $\dfrac{x^2}{a^2}-\dfrac{y^2}{b^2}=1$ **(Hyperbola).**

We have the following facts about this curve:

1. The curve is symmetrical about both the axes and is also a central curve with (0, 0) as its centre. So, we trace the curve only in the first quadrant.
2. Solving for $y$, we get

$$\frac{y^2}{b^2} = \frac{x^2 - a^2}{a^2}, \text{ so for } y \text{ to be real,}$$

$$x^2 \geq a^2 \Rightarrow x \geq a$$

Hence the curve lies on the right side of the vertical line $x = a$, in the first quadrant.

Also as $x$ increases from $a$ to $\infty$, $y$ increases from 0 to $\infty$.

3. Solving for $x$, we get

$$\frac{x^2}{a^2} = \frac{y^2 + b^2}{b^2}$$

whence as $y$ increases from 0 to $\infty$, $x$ increases from $a$ to $\infty$.

4. If we put $y = 0$, we get $x^2 = a^2$, i.e. $x = a$, and the curve meets the $x$-axis at $(a, 0)$.

Also, if we put $x = 0$, we get $y^2 = -b^2$ and hence the curve does not meet the $y$-axis.

Thus the complete curve is as shown:

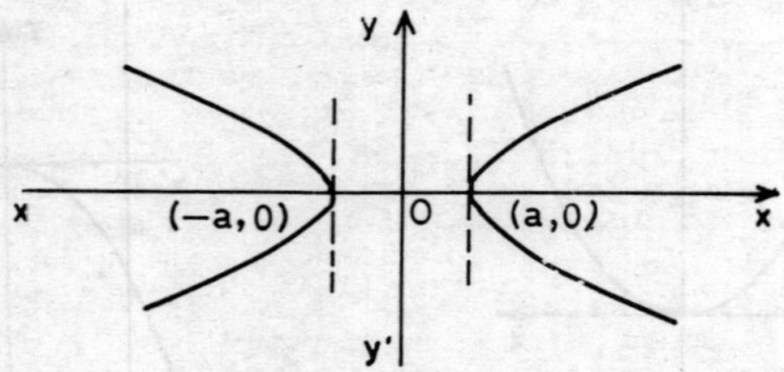


**(vi) $xy = c^2$ (Rectangular Hyperbola).**

We note the following facts about this curve:

1. The curve is not symmetrical about any axis.
2. The curve is a central curve, having (0, 0) as its centre.
3. Only one of $x$ or $y$ cannot be negative, that is, no portion of the curve lies in the second and the fourth quadrant.
4. If we put $y = 0$, we get no solution for $x$, hence the curve does not meet the $x$-axis. Similarly, the curve does not meet the $y$-axis.

5. Solving for $y$, we have, $y = c^2/x$, whence as $x$ increases, $y$ decreases.

Also $\qquad x \to 0 \Rightarrow y \to \infty,$

and $\qquad x \to \infty \Rightarrow y \to 0.$

Hence the curve is as shown:

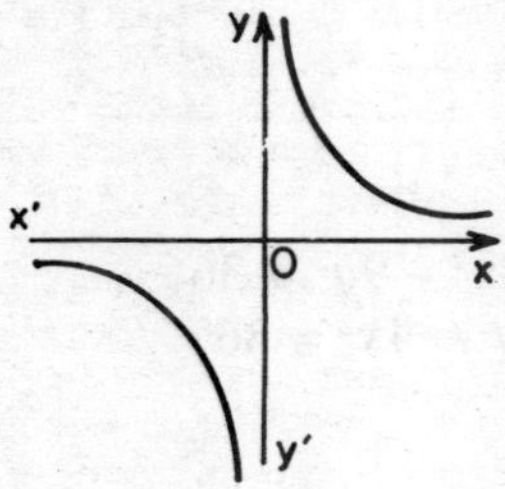


**(vii)** $\dfrac{y^2}{b^2} - \dfrac{x^2}{a^2} = 1$ **(Conjugate Hyperbola)**

or $\qquad \dfrac{x^2}{a^2} - \dfrac{y^2}{b^2} = -1$

The discussion for this curve is similar to the curve in (v). Its shape is as given below:

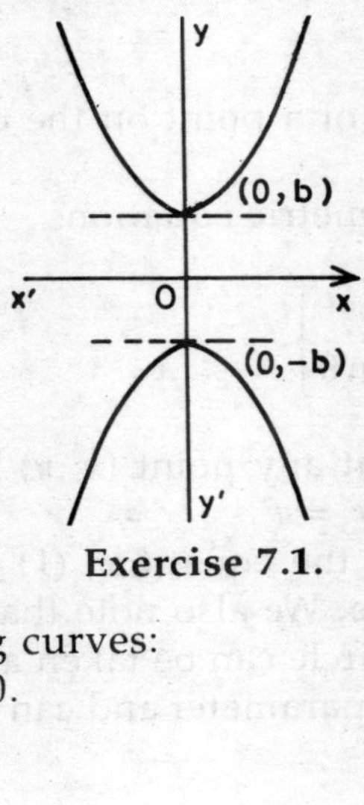


**Exercise 7.1.**

Trace the following curves:

1. $2x + 3y - 4 = 0.$
2. $x^2 + y^2 = 0.$

3. $x^2 - y^2 = 0$.
4. $xy = 0$.
5. $x^2 + y^2 = 2$.
6. $x^2 + y^2 - 6x + 4y - 7 = 0$.
7. $x^2/9 + y^2/4 = 1$ or $4x^2 + 9y^2 = 36$.
8. $x^2/4 + y^2/9 = 1$ or $9x^2 + 4y^2 = 36$.
9. $y^2 = 8x$.
10. $y^2 = -10x$.
11. $x^2 = 4y$.
12. $x^2 = -3y$.
13. $x^2/9 - y^2/4 = 1$ or $4x^2 - 9y^2 = 36$.
14. $y^2/4 + x^2/9 = 1$ or $9y^2 - 4x^2 = 36$.
15. $xy = 2$.
16. $xy = -2$.
17. $\frac{x^2}{4} - \frac{y^2}{9} = 1$ or $9x^2 - 4y^2 = 36$.
18. $\frac{y^2}{9} - \frac{x^2}{4} = 1$ or $4y^2 - 9x^2 = 36$.

## 7.2. PARAMETRIC EQUATIONS OF THE STANDARD CURVES

(Or, the co-ordinates of a point on the curve in terms of a single parameter).

Consider the parametric equations

$$\left.\begin{aligned} x &= a\cos\theta \\ y &= a\sin\theta \end{aligned}\right\} \quad (1)$$

We can easily see that any point $(x, y)$ for all values of $\theta$, always satisfies $x^2 + y^2 = a^2$.

Thus, we say that the equations (1) are the parametric equations of the circle. We also note that the co-ordinatess of any point on this circle can be taken as $(a \cos \theta, a \sin \theta)$.

Here, $\theta$, is called a parameter and can be replaced by any other parameter.

Thus $x = a \cos t$

$y = a \sin t$, $t$ being a parameter.

Thus for $t = 0$, the point on the circle is $(a, 0)$, for $t = \pi/2$, the point is $(0, a)$ and so on. (See also § 6.2.1).

In the same way, we have the following:

| Name | Curve | Parametric Equations | Co-ordinates of any point, on the curve |
|---|---|---|---|
| Parabola | $y^2 = 4ax$ | $\begin{cases} x = at^2, \\ y = 2at \end{cases}$ | $(at^2, 2at)$ |
| | or | $\begin{cases} x = am^2 \\ y = -2am \end{cases}$ | $(am^2, -2am)$ |
| Ellipse | $x^2/a^2 + y^2/b^2 = 1$ | $x = a \cos \theta$, $y = b \sin \theta$. | $(a \cos \theta, b \sin \theta)$ |
| Hyperbola | $x^2/a^2 - y^2/b^2 = 1$ | $\begin{cases} x = a \sec \theta, \\ y = b \tan \theta. \end{cases}$ | $(a \sec \theta, b \tan \theta)$ |
| | or | $\begin{cases} x = a \cosh \theta \\ y = b \sinh \theta \end{cases}$ | $(a \cosh \theta, b \sinh \theta)$ |
| Rectangular hyperbola | $xy = c^2$ | $\begin{cases} x = ct, \\ y = c/t, (t \neq 0) \end{cases}$ | $(ct, c/t), (t \neq 0)$ |
| Conjugate hyperbola | $y^2/b^2 - x^2/a^2 = 1$ | $\begin{cases} x = a \tan \theta \\ y = b \sec \theta \end{cases}$ | $(a \tan \theta, b \sec \theta)$ |
| | or | $\begin{cases} x = a \sinh \theta \\ y = b \cosh \theta \end{cases}$ | $(a \sinh \theta, b \cosh \theta)$. |

## 7.3. POSITION OF A POINT w.r.t. STANDARD CURVES

In § 6.2, we have seen that a point $(x_1, y_1)$ is outside, on or within the circle, $x^2 + y^2 + 2gx + 2fy + c = 0$ according as

$$x_1^2 + y_1^2 + 2gx_1 + 2fy_1 + c >, = \text{ or } < 0.$$

We now find similar tests for the position of a given point w.r.t. other standard curves.

**7.3.1 To find the position of a point $P(x_1, y_1)$ w.r.t. the ellipse $x^2/a^2 + y^2/b^2 = 1$.**

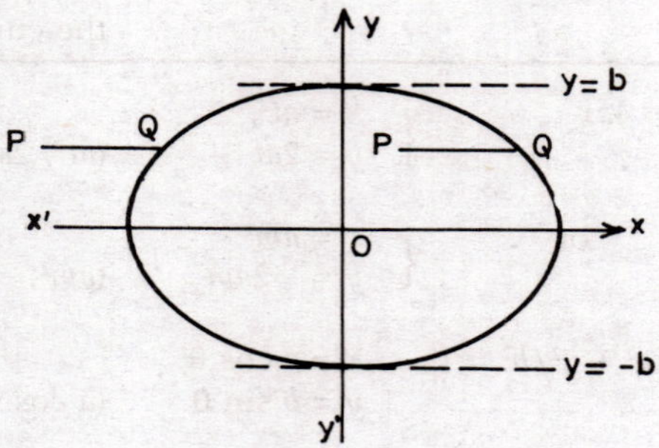


Since the ellipse lies between the lines $y = -b$ and $y = b$, there arise two cases:

**Case I.** $y_1 > b$ or $y_1 < -b \Leftrightarrow y_1^2 > b^2$ ; then the point $P(x_1, y_1)$, clearly lies outside the ellipse.

Again $y_1^2 > b^2 \Rightarrow \dfrac{y_1^2}{b^2} > 1 \Rightarrow \dfrac{x_1^2}{a^2} + \dfrac{y_1^2}{b^2} > 1$

$$\Rightarrow \frac{x_1^2}{a^2} + \frac{y_1^2}{b^2} - 1 > 0 \tag{1}$$

**Case II** $-b \le y_1 \le b \Leftrightarrow y_1^2 \le b^2$.

Draw a line, through $P(x_1, y_1)$, parallel to the $x$-axis meeting the ellipse at $Q(x_2, y_1)$, where $P$ and $Q$ lie in the same quadrant.

$$\Rightarrow \quad \frac{x_2^2}{a^2} + \frac{y_1^2}{b^2} = 1 \qquad (2)$$

Now the point $P(x_1, y_1)$ lies outside, on or within the ellipse according as $x_1^2 >, =$ or $< x_2^2$.

i.e. according as $\dfrac{x_1^2}{a^2} >, =$ or $< \dfrac{x_2^2}{a^2}$

i.e. according as $\dfrac{x_1^2}{a^2} + \dfrac{y_1^2}{b^2} - 1 \quad >, =$ or $< \dfrac{x_2^2}{a^2} + \dfrac{y_1^2}{b^2} - 1$.

i.e. according as $\dfrac{x_1^2}{a^2} + \dfrac{y_1^2}{b^2} - 1 >, =$ or $< 0$ (using (2)).

Thus in either case, we arrive at the conclusion that the point $(x_1, y_1)$ lies outside, on or within the ellipse according as

$$\frac{x_1^2}{a^2} + \frac{y_1^2}{b^2} - 1 >, = \text{ or } < 0.$$

**7.3.2. To find the position of a point $P(x_1, y_1)$ w.r.t. the parabola $y^2 = 4ax$, $(a > 0)$.**

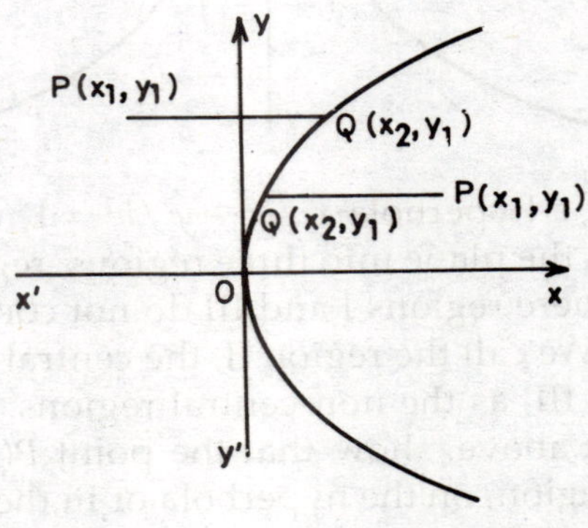


Let the line through $P(x_1, y_1)$ drawn parallel to the $x$-axis meet the parabola at $Q(x_2, y_1)$. Then the point $P(x_1, y_1)$ lies outside, on or within the parabola, $y^2 = 4ax$,

according as $x_1 <, =$ or $> x_2$

i.e.according as $-4ax_1 >, =$ or $< -4ax_2$ (a > 0)

i.e.according as $y_1^2 - 4ax_1 >, =$ or $< y_1^2 - 4ax_2 = 0$

($\because$ $Q(x_2, y_1)$ lies on $y^2 = 4ax$).

Thus the point $P(x_1, y_1)$ is outside, on or inside the parabola, $y^2 = 4ax$ according as $y_1^2 - 4ax_1 >, =$ or $< 0$

*Remark.* We can derive similar results, when the parabolas are in the form $y^2 = -4ax$, $x^2 = 4ay$ and $x^2 = -4ay$.

**7.3.3. To find the position of a point $P(x_1, y_1)$ w.r.t. the hyperbola, $x^2/a^2 - y^2/b^2 = 1$.**

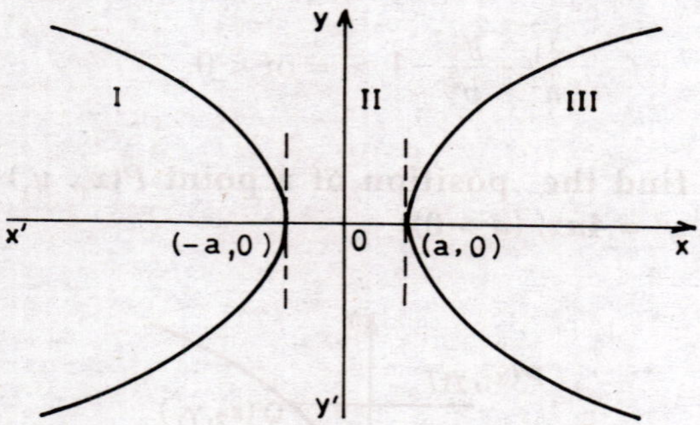


In the case of a hyperbola $x^2/a^2 - y^2/b^2 = 1$, we see that the curve divides the plane into three regions; region II contains the centre, where regions I and III do not contain the centre of the curve. We call the region II, the central region and the regions I and III, as the non-central regions.

We can, as above, show that the point $P(x_1, y_1)$ is in the non-central region, on the hyperbola or in the central region

according as $x_1^2/a^2 - y_1^2/b^2 - 1 >, = < 0$.

**Exercise 7.2.**

1. Find the positions of the following points w.r.t. the given curves:

(i) (0, 0), $x^2/9 + y^2/4 = 1$.
(ii) (0, 0), $y^2 = 4ax$.
(iii) (0, 0), $x^2/16 - y^2/9 = 1$.
(iv) (10, – 2), $x^2/25 + y^2/9 = 1$.
(v) (1, – 2), $y^2 = 16x$.
(vi) (1, – 2), $x^2/16 - y^2/9 = 1$.
(vii) ( – 8, 3), $x^2/16 - y^2/9 = 1$.
(viii) (0, – 3), $x^2/16 - y^2/9 = 1$.
(ix) ( – 4, 0), $x^2/16 - y^2/9 = 1$.
(x) (0, – 4), $x^2/25 + y^2/16 = 1$.
(xi) (1, – 3); $x^2 = - 8y$.
(xii) (3, – 4); $4y^2 - 9x^2 = 36$.

# 8. A LINE AND A CURVE

## INTRODUCTION

In this chapter, we study the intersection of a line and a curve, represented by the general equation of second degree in $x$ and $y$. We also derive the equation of the tangent at a point on this curve, chord of contact and polar of a point w.r.t. it, its chord with a given middle point and pair of tangents from a point to it.

In Chapter 7, we have studied the graphs of the following curves:

(i) Circle: $x^2 + y^2 + 2gx + 2gy + c = 0$,

(ii) Parabola: $y^2 = 4ax$,

(iii) Ellipse: $\frac{x^2}{a^2} + \frac{y^2}{b^2} = 1$,

(iv) Hyperbola: $\frac{x^2}{a^2} - \frac{y^2}{b^2} = 1$,

(v) Rectangular hyperbola: $xy = c^2$.

We now propose to study the relationship of a line and the above curves.

## 8.1. TO FIND THE POINTS OF INTERSECTION OF A LINE AND A CURVE

Let $S = ax^2 + 2hxy + by^2 + 2gx + 2fy + c = 0$ be the given curve and $px + qy + r = 0$ be the given line. For their points of intersection, we solve these equations simultaneously. As at least one of $p$ or $q$ is to be non-zero, we can assume $q \neq 0$ (say), then

$$px + qy + r = 0 \Rightarrow y = -\frac{px + r}{q}.$$

$\therefore$ the equation of the curve becomes

$$ax^2 + 2hx\left(-\frac{px+r}{q}\right) + b\left(-\frac{px+r}{q}\right)^2 + 2gx$$
$$+2f\left(-\frac{px+r}{q}\right) + c = 0$$

$$\Rightarrow (aq^2 - 2hpq + bp^2)x^2 + 2(-hqr - bpr + gq^2 - fpq)x + (br^2 - 2fqr + cq^2) = 0, \quad (1)$$

which being a quadratic in $x$, yields two values of $x$, say, $x_1$ and $x_2$. Let the corresponding values of $y$ be, say, $y_1$ and $y_2$. Thus $(x_1, y_1)$ and $(x_2, y_2)$ are the co-ordinates of the points of intersection of the given curve and the given line i.e. *in general,* a line meets a curve in two points. These points will be real and distinct, real and coincident, or imaginary according as the discriminant, $\Delta$, of the equation (1) >, = or < 0, where

$$\Delta = 4[(-hqr - bpr + gq^2 - fpq)^2 - (aq^2 - 2hpq + bp^2)(br^2 - 2fqr + cd^2)]$$

**Note 1.** If $q = 0$ in $px + qy + r = 0$, we can proceed to eliminate $x$ instead of $y$ as above.

**Note 2.** Throughout this chapter, we take S = 0, as the equation of a curve, where

$$S = ax^2 + 2hxy + by^2 + 2gx + 2fy + c.$$

### 8.1.1. Tangent to a curve

**Definition:** A *tangent* to a curve is a straight line, which meets it in two coincident points.

By the foregoing article, it is clear that the line $px + qy + r = 0$ is a tangent to the curve $S = 0$ iff $\Delta = 0$.

### 8.1.2. To find the number of tangents from a point (x′, y′) to the curve, S = 0.

Let the equation of any line through $(x', y')$ be

$$\frac{x - x'}{\cos\theta} = \frac{y - y'}{\sin\theta} = r \quad (1)$$

Any point on (1) is

$$(x' + r\cos\theta,\ y' + r\sin\theta\ ).$$

If it lies on

$$S = ax^2 + 2hxy + by^2 + 2gx + 2fy + c = 0,$$

then

$$a(x' + r\cos\theta)^2 + 2h(x' + r\cos\theta)\,(y' + r\sin\theta) + b(y' + r\sin\theta)^2$$
$$+ 2g\,(x' + r\cos\theta) + 2f(y' + r\sin\theta) + c = 0$$

$$\Rightarrow (a\cos^2\theta + 2h\sin\theta\cos\theta + b\sin^2\theta)\,r^2$$
$$+ 2[(ax' + hy' + g)\cos\theta + (hx' + by' + f)\sin\theta]r + S' = 0 \qquad (2)$$

where

$$S' = ax'^2 + 2hx'y' + by'^2 + 2gx' + 2fy' + c.$$

Equation (2), being a quadratic in $r$, is has two values of $r$, showing that the line (1) meets the curve, $S = 0$ in two points (a fact, which we have already established in § **8.1** ). Further (1) will be a tangent to $S = 0$, iff the roots of (2) (i.e. two values of $r$ given by (2)) are equal, whence

$$[(ax' + hy' + g)\cos\theta + (hx' + by' + f)\sin\theta]^2$$
$$- (a\cos^2\theta + 2h\sin\theta\cos\theta + b\sin^2\theta\ ).\ S' = 0,$$

$$\Rightarrow [(ax' + hy' + g) + (hx' + by' + f)\tan\theta]^2$$
$$- (a + 2h\tan\theta + b\tan^2\theta)\ .\ S' = 0$$

which being a quadratic in $\tan\theta$, gives two values of $\theta$ and consequently, there are two tangents given by (1).

Thus, in general, two tangents can be drawn from a point to a curve.

We shall now take up the line in the form, $y = mx + c$, to be a tangent to the curves mentioned in the previous chapter.

**Example 8.1.** Find the condition that the line given by $y = mx + c$, may be a tangent to the circle: $x^2 + y^2 = a^2$.

*Solution.* The equations of the line and the circle are

$$y = mx + c \qquad (1)$$
$$x^2 + y^2 = a^2 \qquad (2)$$

Eliminating $y$ from (1) and (2), we get

$$x^2 + (mx + c)^2 = a^2$$

$$\Rightarrow (1 + m^2)x^2 + 2mcx + (c^2 - a^2) = 0 \qquad (3)$$

Now (1) will be a tangent to (2) iff the two roots of (3) are equal,

i.e. iff discriminant of (3) is equal to zero,

i.e. iff $4m^2c^2 - 4(1 + m^2)(c^2 - a^2) = 0$,

i.e. iff $c^2 = a^2(1 + m^2)$.

This is the required condition for (1) to touch (or be a tangent to) (2).

**Corollary.** The equation of any tangent (not parallel to the $y$-axis) to the circle, $x^2 + y^2 = a^2$, can be taken as

$$y = mx + a\sqrt{1+m^2} \text{ or } y = mx - a\sqrt{1+m^2}$$

(This is because $c^2 = a^2(1+m^2) \Leftrightarrow c = \pm a\sqrt{1+m^2}$. Both these equations are the equations of parallel tangents to the circle, $x^2 + y^2 = a^2$.

**Note 1.** It is immediate that the only tangents, parallel to the $y$-axis to the circle, $x^2 + y^2 = a^2$, are $x = \pm a$.

**Note 2.** It can be easily seen that the line, $y = mx + a\sqrt{1+m^2}$ touches the circle, $x^2 + y^2 = a^2$, at the point

$$\left(\frac{-am}{\sqrt{1+m^2}}, \frac{a}{\sqrt{1+m^2}}\right).$$

**8.1.3. Condition for the line, $y = mx + c$, to be a tangent to the other standard curves.**

The line given by, $y = mx + c$, is a tangent to

(i) the ellipse, $\frac{x^2}{a^2} + \frac{y^2}{b^2} = 1$ iff $c^2 = a^2m^2 + b^2$, and that the equation of any tangent to it (not parallel to the $y$-axis) is $y = mx \pm \sqrt{a^2m^2 + b^2}$ (the only tangents parallel to the $y$-axis are $x = \pm a$).

Note that the point of contact of the tangent, $y = mx + \sqrt{a^2m^2 + b^2}$ with the given ellipse is

$$\left(\frac{-a^2m}{\sqrt{a^2m^2+b^2}}, \frac{b^2}{\sqrt{a^2m^2+b^2}}\right).$$

(ii) the hyperbola, $\dfrac{x^2}{a^2} - \dfrac{y^2}{b^2} = 1$ iff $c^2 = a^2m^2 - b^2$, and that the equation of any tangent to it (not parallel to the $y$-axis) is $y = mx \pm \sqrt{a^2m^2 - b^2}$ (the only tangents parallel to the $y$-axis are $x = \pm a$).

Note that the point of contact of the tangent, $y = mx + \sqrt{a^2m^2 - b^2}$ with the given hyperbola, is

$$\left(\frac{-a^2m}{\sqrt{a^2m^2-b^2}}, \frac{b^2}{\sqrt{a^2m^2-b^2}}\right).$$

(iii) the parabola, $y^2 = 4ax$ iff $c = a/m$ and that the equation of any tangent to it (not parallel to the $y$-axis) is $y = mx + \dfrac{a}{m}$ (the only tangent parallel to the $y$-axis is $x = 0$).

Note that the point of contact of the tangent, $y = mx + (a/m)$ with the given parabola is $(a/m^2, 2a/m)$.

**Example 8.2.** (See also **Question 5, Exercise 8.7**).

Find the locus of a point from which two perpendicular tangents can be drawn to the ellipse,

$$\frac{x^2}{a^2} + \frac{y^2}{b^2} = 1.$$

*Solution.* Let $(\alpha, \beta)$ be the point whose locus is required. We know that the equation of any tangent to

$$\frac{x^2}{a^2} + \frac{y^2}{b^2} = 1,$$

is $$y = mx + \sqrt{a^2m^2+b^2} \qquad \text{(§ 8.1.3)}$$

This passes through $(\alpha, \beta)$,

$$\beta - m\alpha = \sqrt{a^2m^2 + b^2}\ .$$

Squaring and simplifying, we get,

$$(a^2 - \alpha^2)m^2 + 2\alpha\beta m + (b^2 - \beta^2) = 0.$$

This, being a quadratic in $m$, has two values of $m$, say $m_1$ and $m_2$.

By hypothesis,

$$m_1m_2 = -1 \Rightarrow \frac{b^2 - \beta^2}{a^2 - \alpha^2} = -1$$

$$\Rightarrow \alpha^2 + \beta^2 = a^2 + b^2.$$

$\therefore$ The locus of $(\alpha, \beta)$ is

$$x^2 + y^2 = a^2 + b^2.$$

**Example 8.3.** A tangent to $y^2 = 8x$ makes an angle of 45° with the line, $y = 3x + 5$. Find its equation and the point of contact.

*Solution*. Here parabola is $y^2 = 8x$ (1)

compare it with $y^2 = 4ax$; $a = 2$.

Now equation of any tangent to (1) is

$$y = mx + \frac{a}{m} = mx + 2/m \tag{2}$$

which touches (1) at

$$\left(\frac{a}{m^2}, \frac{2a}{m}\right) \text{ i.e. at} \left(\frac{2}{m^2}, \frac{4}{m}\right)$$

Also the angle between (2) and the line, $y = 3x + 5$ is 45°.

$$\tan 45° = \pm\frac{m-3}{1+3m}$$

$$\Rightarrow \quad \pm 1 = \frac{m-3}{1+3m}$$

$$\Rightarrow \quad m = -2, \frac{1}{2}.$$

For $m = -2$, the tangent is

$$y = -2x + (2/-2)$$

i.e. $\quad 2x + y + 1 = 0$

whose point of contact is $(\frac{1}{2}, -2)$.

For $m = \frac{1}{2}$, the tangent is

$y = (\frac{1}{2})\, x + 2/(\frac{1}{2})$

i.e. $x - 2y + 8 = 0$

whose point of contact is (8, 8).

**Exercise 8.1**

1. Show that
   (i) $2y - x = 2$, touches $y^2 = 2x$.
   (ii) $2y = x + 4a$, touches $y^2 = 4ax$.
   (iii) $x + y = 1$, touches $y^2 - y + x = 0$.
   Also find the point of contact in each case.
2. (i) Find the equations of those tangents to the hyperbola
   $$4x^2 - 9y^2 = 1,$$
   which are parallel to
   $$4y = 5x + 7.$$
   (ii) Find the equations of the tangents to the parabola,
   $$y^2 = 16x,$$
   which are parallel and perpendicular respectively to the line
   $$2x - y + 5 = 0$$
3. Find the condition that the line
   $$lx + my + n = 0,$$
   may touch the
   (i) circle: $x^2 + y^2 = a^2$,
   (ii) paraboloa: $y^2 = 4ax$,
   (iii) hyperbola: $\frac{x^2}{a^2} - \frac{y^2}{b^2} = 1$,
   (iv) rectangular hyperbola: $xy = c^2$,
   Find also the point of contact in each case.

4. Find the condition that the line,

$$x \cos a + y \sin a = p,$$

may be a tangent to the

(i) circle: $x^2 + y^2 = a^2$,

(ii) parabola: $y^2 = 4ax$,

(iii) hyperbola: $\frac{x^2}{a^2} - \frac{y^2}{b^2} = 1$,

(iv) rectangular hyperbola: $xy = c^2$,

(v) ellipse: $\frac{x^2}{a^2} + \frac{y^2}{b^2} = 1$,

5. Find the locus of a point from where perpendicular tangents can be drawn to the parabola,

$$y^2 = 4ax.$$

6. Find the equations of the tangents having slopes

(a) 1. (b) $-\frac{1}{2}$, (c) n,

for the following curves:

(i) circle: $x^2 + y^2 = a^2$,

(ii) parabola: $y^2 = 4ax$,

(iii) ellipse: $\frac{x^2}{a^2} + \frac{y^2}{b^2} = 1$,

(iv) hyperbola: $\frac{x^2}{a^2} - \frac{y^2}{b^2} = 1$.

7. If the line $lx + my = 1$ touches $x^2 + y^2 = a^2$, then $(l, m)$ lies on a certain circle. Find its equation.

8. Find the equation of each of the tangents from $(-2a, -a)$ to

$$y^2 = 4ax.$$

(*Hint:* Equation of any tangent to the parabola is

$$y = mx + \frac{a}{m})$$

9. Tangents are drawn from the point $(-2, -1)$ to the hyperbola

$$2x^2 - 3y^2 = 6.$$

Find their equations.

(*Hint:* Write the equation of the hyperbola as

$$x^2/3 - y^2/2 = 1).$$

10. A straight line touches the circle $x^2 + y^2 = 2a^2$ and the parabola $y^2 = 8ax$. Show that its equation is $y = \pm (x + 2a)$. Also find the co-ordinates of the point of contact.
11. Find the tangents common to the parabola, $y^2 = 8x$ and the circle,
$$x^2 + y^2 - 12x + 4 = 0.$$
12. Find the equations of the common tangents to the parabolas
$$y^2 = 4ax \text{ and } x^2 = 4by.$$
13. Find the condition that the straight line, $y = mx + c$ may touch the parabola
$$y^2 = 4a(x + a).$$
14. Show that the equation of the tangent to the parabola, $y^2 = 4ax$ which makes an angle $\theta$ with its axis is $y = x \tan \theta + a \cot \theta$.
15. Find all the values of $m$ for which the line, $m^2x - my + a = 0$ touches the circle, $x^2 + y^2 = a^2$ and the parabola, $y^2 = 4ax$.
16. Prove that the parabolas, $y^2 = ax$ and $x^2 = by$ cut one another at an angle
$$\tan^{-1}\left[\frac{3a^{1/3}b^{1/3}}{2(a^{2/3}+b^{2/3})}\right].$$
(*Hint:* $(a^{1/3}\, b^{2/3}, a^{2/3}\, b^{1/3})$ is a point of their intersection. Find tangents to the two parabolas at this point.

Use $$\left.\tan\theta = \frac{m_1 \sim m_2}{1+m_1m_2}\right).$$

## 8.2. TO FIND THE EQUATION OF THE TANGENT AT (x′, y′) TO THE CURVE, S = 0

**Method I.** Let the equation of any line through $(x', y')$ be

$$\frac{x-x'}{\cos\theta} = \frac{y-y'}{\sin\theta} = r. \tag{1}$$

Proceeding as in § **8.1.2,** we see that the value of $r$ are given by

$$(a \cos^2 \theta + 2h \sin \theta \cos \theta + b \sin^2 \theta )r^2 + 2[(ax' + hy' + g) \cos \theta + (hx' + by' + f) \sin \theta].r = 0. \qquad (2)$$

(Here $S' = 0$, as $(x', y')$ is a point on the curves, $S = 0$). Clearly one value of $r$ given by (2) is zero.

Thus for (1) to be a tangent to $S = 0$, at $(x', y')$, we must have the other value of $r$, also zero, whence

$$(ax' + hy' + g) \cos \theta + (hx' + by' + f) \sin \theta = 0$$

$$\Rightarrow (ax' + hy' + g)(x - x') + (hx' + by' + f)(y - y') = 0,$$

using (1).

$$\Rightarrow axx' + h(xy' + x'y) + byy' + gx + fy - (ax'^2 + 2hx'y' + by'^2 + gx' + fy') = 0$$

$$\Rightarrow axx' + h(xy' + x'y) + byy' + gx + fy + (gx' + fy' + c) = 0 \text{ (using } S' = 0)$$

$$\Rightarrow axx' + h(xy' + x'y) + byy' + g(x + x') + f(y + y') + c = 0.$$

Thus, we see that the equation of the tangent at $(x', y')$ to the curve $S = 0$, is

$$axx' + h(xy' + x'y) + byy' + g(x + x') + f(y + y') + c = 0.$$

**Remark**. Method to write the equation of the tangent at $(x', y')$ :

Change $\quad x^2 \to xx', \; y^2 \to yy'$

$$xy \to \frac{xy' + yx'}{2}$$

$$x \to \frac{x + x'}{2}, \; y \to \frac{y + y'}{2}$$

and keep $c$ unchanged.

**Method II.** A *tangent* to a curve at a point $P$ can also be defined as the limiting position of a chord $PQ$, as the point $Q$ approaches $P$. We use this definition and find the equation of the tangent at $P(x', y')$ on the ellipse.

$$\frac{x^2}{a^2} + \frac{y^2}{b^2} = 1.$$

Let $Q(x'', y'')$ be any other point on the ellipse

$$\therefore \qquad \frac{x^2}{a^2}+\frac{y^2}{b^2}=1.$$

$$\therefore \qquad \frac{x'^2}{a^2}+\frac{y'^2}{b^2}=1 \tag{1}$$

and
$$\frac{x''^2}{a^2}+\frac{y''^2}{b^2}=1. \tag{2}$$

Now the equation of the chord $PQ$ is

$$y-y'=\left(\frac{y''-y'}{x''-x'}\right)(x-x'). \tag{3}$$

Subtracting (1) from (2), we get

$$\frac{(x''-x')(x''+x')}{a^2}=-\frac{(y''-y')(y''+y')}{b^2}$$

$$\Rightarrow \frac{y''-y'}{x''-x'}=-\frac{b^2}{a^2}\left(\frac{x'+x'}{y''+y'}\right).$$

Using this, (3) reduces to

$$y-y'=-\frac{b^2}{a^2}\left(\frac{x''+x'}{y''+y'}\right)(x-x') \tag{4}$$

Let $\qquad Q\to P,$

then (4) becomes

$$y-y'=-\frac{b^2}{a^2}\cdot\frac{x'}{y'}(x-x')$$

$$\Rightarrow \frac{(y-y')\,y'}{b^2}=-\frac{(x-x')\,x'}{a^2}$$

$$\Rightarrow \frac{xx'}{a^2}+\frac{yy'}{b^2}=\frac{x'^2}{a^2}+\frac{y'^2}{b^2}=1, \qquad \text{using (1).}$$

Hence, the required equation of the tangent to the ellipse at $(x', y')$ is

$$\frac{xx'}{a^2}+\frac{yy'}{b^2}=1.$$

**Method III.** The reader familiar with calculus can write the equation of the tangent very easily.

Differentiating $S = 0$ w.r.t. $x$, we get

$$2ax + 2h\left(y + x.\frac{dy}{dx}\right) + by\frac{dy}{dx} + 2g + 2f.\frac{dy}{dx} = 0$$

$$\frac{dy}{dx} \text{ at } (x', y') \text{ is } -\frac{ax'+hy'+g}{hx'+by'+f},$$

which is the slope of the tangent to $S = 0$, at $(x', y')$.

Thus the equation of the tangent at $(x', y')$ is

$$y - y' = -\frac{ax' + hy' + g}{hx' + by' + f'}.(x - x')$$

which as in **Method I**, simplifies to

$$axx' + h(xy' + x'y) + byy' + g(x + x') + f(y + y') + c = 0.$$

**Note: Method II,** illustrated above for ellipse, can be easily applied for finding the tangent at a point to any other curve.

### 8.2.1 Equation of the tangent in terms of parametric co-ordinates

By the preceding article, we see that the equation of the tangent at

(i) $(a \cos \theta, a \sin \theta)$, to the circle: $x^2 + y^2 = a^2$
is $x \cos \theta + y \sin \theta = a$.

(ii) $(at^2, 2at)$, to the parabola: $y^2 = 4ax$
is $yt = x + at^2$.

(iii) $(a \cos \theta, b \sin \theta)$, to the ellipse: $\frac{x^2}{a^2} + \frac{y^2}{b^2} = 1$

is $\frac{x \cos \theta}{a} + \frac{y \sin \theta}{b} = 1$

(iv) $(a \sec \theta, b \tan \theta)$, to the hyperbola: $\frac{x^2}{a^2} - \frac{y^2}{b^2} = 1$

is $$\frac{x \sec \theta}{a} - \frac{y \tan \theta}{b} = 1.$$

(v) $\left(ct, \frac{c}{t}\right)$, to the rectangular hyperbola: $xy = c^2$

is $$x + yt^2 = 2ct.$$

**Example 8.4.** Find the condition that the line
$$lx + my + n = 0$$

may touch the ellipse $$\frac{x^2}{a^2} + \frac{y^2}{b^2} = 1$$

Also find the point of contact.

*Solution.* Let the line $lx + my + n = 0$ (1)

touch the ellipse $$\frac{x^2}{a^2} + \frac{y^2}{b^2} = 1 \quad (2)$$

at $(x', y')$.

Now the equation of the tangent at $(x', y')$ to (2) is

$$\frac{xx'}{a^2} + \frac{yy'}{b^2} = 1 \quad (3)$$

As (1) and (3), represent the same line, therefore, on comparing their coefficients, we get

$$\frac{x'}{a^2 l} = \frac{y'}{b^2 m} = \frac{1}{-n}.$$

the point of contact $(x', y') = \left(-\frac{a^2 l}{n}, -\frac{b^2 m}{n}\right)$

To obtain the condition, we eliminate $(x', y')$.

Since $(x', y')$ lies on (2), we have,

$$\frac{x'^2}{a^2} + \frac{y'^2}{b^2} = 1.$$

$$\frac{a^2 l^2}{n^2} + \frac{b^2 m^2}{n^2} = 1 \Rightarrow a^2 l^2 + b^2 m^2 = n^2.$$

**Note 1.** The condition can also be obtained by using the fact that $(x', y')$ lies on the tangent, $lx + my + n = 0$.

**Note 2.** We can also take the point of contact of the line

$lx + my + n = 0$, with the ellipse as $(a \cos \theta, b \sin \theta)$ in place of $(x', y')$ and proceed as above.

**Example 8.5.** Find the equation of the tangent at the point whose parameter is – 1, to the parabola $y^2 = 4ax$.

*Solution.* By § **8.2.1**, the equation of the tangent at $(at^2, 2at)$ to the parabola, $y^2 = 4ax$ is

$$ty = x + at^2.$$

Here $$t = -1.$$

$\therefore$ the required equation of the tangent (to the parabola) at the point, with parameter, – 1 is

$$-y = x + a \quad \Rightarrow \quad x + y + a = 0.$$

## Exercise 8.2

1. With the help of § **8.2,** find the equations of the tangents at $(x', y')$ to the following curves:
   (i) circle: $x^2 + y^2 = a^2$,
   (ii) circle: $x^2 + y^2 + 2gx + 2fy + c = 0$,
   (iii) parabola: $y^2 = 4ax$,
   (iv) ellipse: $x^2/a^2 + y^2/b^2 = 1$,
   (v) hyperbola: $x^2/a^2 - y^2/b^2 = 1$,
   (vi) rectangular hyperbola: $xy = c^2$.
2. Find the equations of the tangents to the given curves at the given points:
   (i) $x^2 + y^2 = 4, (-1, \sqrt{3})$.
   (ii) $2x^2 + y^2 - x - y - 1 = 0$, (1, 1).
   (iii) $x^2 + 2y^2 - 4xy - 5x + 7y - 12 = 0$, (3, – 2).
   (iv) $y^2 = 9x$, (1, –3).
   (v) $4x^2 + 9y^2 = 20$, $(1, \frac{4}{3})$.
3. Find the points of intersection of the line $5y - 4x - 6a = 0$ with the parabola $y^2 = 4ax$ and find the tangents at these points.
4. Find the equations of the tangents to the given curves at the points for which the parameters are given:
   (a) (i) circle: $x^2 + y^2 = a^2$, parameter = $0, -\pi/4, \phi$,
   (ii) ellipse: $x^2/a^2 + y^2/b^2 = 1$, " "
   (iii) hyperbola: $x^2/a^2 - y^2/b^2 = 1$, " "

(b) parabola: $y^2 = 4ax$, parameter $t = 0, -1$ and $p$.

5. Find the condition that the line $lx + my + n = 0$ may touch the following curves, by the method of Example 8.4
(i) $x^2 + y^2 = a^2$,
(ii) $y^2 = 4ax$,
(iii) $x^2/a^2 - y^2/b^2 = 1$.
(iv) $xy = c^2$.
Also find the point of contact, in each case.

6. Prove that the locus of the middle points of the portions of tangents to the ellipse, $x^2/a^2 + y^2/b^2 = 1$, intercepted between the axes is

$$a^2y^2 + b^2x^2 = 4x^2y^2.$$

7. (a) Find the equation of the line joining the points "$\theta_1$" and "$\theta_2$" on
(i) the circle: $x^2 + y^2 = a^2$.
(Here the point "$\theta$" is $(a \cos \theta, a \sin \theta)$).
(ii) the ellipse: $x^2/a^2 + y^2/b^2 = 1$.
(Here the point "$\theta$" is $(a \cos \theta, b \sin \theta)$).
(iii) the hyperbola: $x^2/a^2 - y^2/b^2 = 1$.
(Here the point "$\theta$" is $(a \sec \theta, b \tan \theta)$).
Deduce the equation of the tangent at "$\theta$" in each case.
(b) Find the equation of the line joining the points "$t_1$" and "$t_2$" on
(i) the parabola: $y^2 = 4ax$.
(Here the point "t" is $(at^2, 2at)$).
(ii) the rectangular hyperbola: $xy = c^2$.
(Here the point "$t$" is $(ct, c/t)$).
Deduce the equations of the tangents at "$t$" to the above curves.

8. Show that the tangent lines at the extremities of any diameter of a central conic are parallel.
(*Hint:* Taking the centre as origin, the extremities of a diameter are $(x', y')$ and $(-x', -y')$).

9. Prove that the locus of the foot of the perpendicular from the centre on any tangent to the ellipse

$$\frac{x^2}{a^2}+\frac{y^2}{b^2}=1 \text{ is } (x^2+y^2)^2 = a^2x^2 + b^2y^2.$$

10. A line through the origin meets the circle $x^2 + y^2 = a^2$ at $P$ and the hyperbola $x^2 - y^2 = a^2$ at $Q$. Prove that the locus of the point of intersection of the tangent at $P$ to the circle with the tangent at $Q$ to the hyperbola is the curve.
$$(a^4 + 4y^4)\, x^2 = a^6.$$
11. Prove that the ordinate of the point of intersection of two tangents to a parabola is the arithmetic mean between the ordinates of the points of contact of the tangents.
12. If $P, Q, R$ are the points on the parabola, $y^2 = 4ax$, whose ordinates are in $G.P$, prove that the tangents at $P$ and $R$ intersect on the ordinate through $Q$.

## 8.3 NORMAL

**Definition.** The straight line through a point $P$ of a curve, which is perpendicular to the tangent to it at $P$, is called the *normal* to the curve at $P$.

The point $P$ is called the foot of the normal.

*To find the equation of the normal to the curve,* $S = 0$, *at a point* $P(x', y')$.

Here $\qquad S = ax^2 + 2hyx + by^2 + 2gx + 2fy + c = 0. \qquad (1)$

The equation of the tangent to (1) at $(x', y')$ is

$$axx' + h(xy' + yx') + byy' + g(x + x') + f(y + y') + c = 0$$

Its slope is $-\dfrac{ax'+hy'+g}{hx'+by'+f}$

$\therefore$ the slope of the normal to the curve at $P$ is

$$\frac{hx'+by'+f}{ax'+hy'+g}$$

Thus the equation of the normal to $S = 0$ at $P\,(x', y')$ is

$$y-y'=\frac{hx'+by'+f}{ax'+hy'+g}(x-x')$$

### 8.3.1. Equation of the normal to standard curves

*To find the equation of the normal at P (a cos θ, b sinθ) to the ellipse*

$$\frac{x^2}{a^2}+\frac{y^2}{b^2}=1.$$

The equation of the tangent at $P$ to the given ellipse is

$$\frac{x}{a}\cos\theta+\frac{y}{b}\sin\theta=1$$

Its slope $= -\dfrac{\cos\theta}{a}\Big/\dfrac{\sin\theta}{b}$.

$\therefore$ the slope of the normal is $\dfrac{a\sin\theta}{b\cos\theta}$

$\Rightarrow$ the equation of the normal at $P$ to the given ellipse is

$$y-b\sin\theta=\frac{a\sin\theta}{b\cos\theta}(x-a\cos\theta)$$

$\Rightarrow yb\cos\theta - b^2\sin\theta\cos\theta = xa\sin\theta - a^2\sin\theta\cos\theta$

$\Rightarrow xa\sin\theta - yb\cos\theta = (a^2-b^2)\sin\theta\cos\theta$

$\Rightarrow ax\sec\theta - by\operatorname{cosec}\theta = (a^2-b^2)$

$\Rightarrow \dfrac{ax}{\cos\theta}-\dfrac{by}{\sin\theta}=a^2-b^2$

### 8.3.2. Equations of the normals

We may similarly show that the *equations of the normals* at the given points to the following curves are given as:

(i) Point: $(a\cos\theta, a\sin\theta)$, circle: $x^2+y^2=a^2$.
Normal: $x\sin\theta - y\cos\theta = 0$.

(ii) Point: $(at^2, 2at)$, parabola: $y^2=4ax$.
Normal: $y=-tx+2at+at^3$.
If we write the point as $(am^2-2am)$, the equation of the normal to the parabola, $y^2=4ax$, is

$$y=mx-2am-am^3.$$

(iii) Point: $(a\sec\theta, b\tan\theta)$, hyperbola : $\dfrac{x^2}{a^2}-\dfrac{y^2}{b^2}=1$.
Normal: $ax\cos\theta + by\cot\theta = a^2+b^2$.

$$\Leftrightarrow \frac{ax}{\sec\theta} + \frac{by}{\tan\theta} = a^2 + b^2$$

(iii) Point: $(ct, c/t)$, rectangular hyperbola: $xy = c^2$.

Normal: $ty = t^3x + c\,(1 - t^4)$.

Their proofs are left to the reader as exercises.

**Example 8.6.** Find the condition that the line $lx + my + n = 0$ is a normal to the ellipse $\frac{x^2}{a^2} + \frac{y^2}{b^2} = 1$. Also find the point at which this line is a normal.

*Solution.* We are given the ellipse

$$\frac{x^2}{a^2} + \frac{y^2}{b^2} = 1 \qquad (1)$$

The equation of the normal to (1) at $(a \cos \theta, b \sin \theta)$ is

$$ax \sec \theta - by \operatorname{cosec} \theta = a^2 - b^2 \qquad \textbf{(§ 8.3.1)}$$

This is to be the same as

$$lx + my + n = 0.$$

On comparing their coefficients, we have

$$\frac{a\sec\theta}{l} = \frac{-b\operatorname{cosec}\theta}{m} = \frac{a^2 - b^2}{-n}$$

$$\Rightarrow \cos\theta = \frac{-an}{l(a^2 - b^2)} \qquad (2)$$

and

$$\sin\theta = \frac{bn}{m(a^2 - b^2)} \qquad (3)$$

$\therefore$ the point at which the given line, $lx + my + n = 0$ is normal to (1) is

$$(a \cos \theta, b \sin \theta) = \left(\frac{-a^2 n}{l(a^2 - b^2)}, \frac{b^2 n}{m(a^2 - b^2)}\right)$$

To get the condition, we eliminate $\theta$ from (2) and (3), and get

$$1 = \frac{a^2 n^2}{l^2 (a^2 - b^2)^2} + \frac{b^2 n^2}{m^2 (a^2 - b^2)^2}$$

$$\Rightarrow \quad \frac{a^2}{l^2} + \frac{b^2}{m^2} = \frac{(a^2 - b^2)^2}{n^2}$$

which is the required condition.

We now prove a result which is true only for the case of a circle. (We used this fact in Chapter VI, to deduce the equation of the tangent to the circle at a point).

**Theorem 8.1.** The radius of a circle through the point of contact of a tangent to a circle is perpendicular to the tangent. In other words, the normal at any point to a circle passes through the centre.

*Proof.* Let the equation of the circle be

$$x^2 + y^2 = a^2.$$

The equation of the tangent at $(x', y')$ to it is

$$xx' + yy' = a^2.$$

$$\text{Its slope} = -\frac{x'}{y'}.$$

Also the slope of the line joining the centre (0, 0) and the point $(x', y')$ is $y'/x'$.

Clearly the product of the slopes of these two lines is – 1.

They are perpendicular.

This proves the theorem.

**Example 8.7.** $P$ and $Q$ are ends of a variable chord of the parable, $y^2 = 4ax$ through the point $(-2a, 0)$. Prove that the normals at $P$ and $Q$ to the curve intersect on the parabola.

*Solution.* Let the parabola be

$$y^2 = 4ax \tag{1}$$

Suppose: $P((at_1^2, 2at_1)$ and $Q((at_2^2, 2at_2)$.

$\therefore$ Equation of chord $PQ$ is

$y(t_1 + t_2) = 2x + 2at_1t_2$

By hypothesis, $(-2a, 0)$ lies on it.

$$\therefore \quad t_1t_2 = 2 \tag{2}$$

Also equations of the normals at $P$ and $Q$ are given by

$$y = -t_1 x + 2at_1 + at_1^3$$

and $$y = t_2 x + 2at_2 + at_2^3$$

Then point of intersection is $x = a(2 + t_1^2 + t_2^2 + t_1 t_2)$ ; $y = -at_1 t_2 (t_1 + t_2)$

$\Rightarrow$ $x = a(t_1 t_2 + t_1^2 + t_2^2 + t_1 t_2)$;

$y = -2a(t_1 + t_2)$ (using (2)

$\Rightarrow$ $x = a(t_1 + t_2)^2$; $y = -2a(t_1 + t_2)$

which clearly lies on the parabola (1).

### Exercise 8.3

1. Find the equations of the normals at the points given in Question 2 of Exercise 8.2.
2. Prove § 8.3.2.
3. Find the condition that the line $lx + my + n = 0$ is a normal to the curve
   (i) $y^2 = 4ax$.
   (ii) $\dfrac{x^2}{a^2} - \dfrac{y^2}{b^2} = 1$.
   (iii) $xy = c^2$.
   (iv) $x^2 + y^2 + 2gx + 2fy + c = 0$.
4. Find the condition that the normals to the parabola, $y^2 = 4ax$, at the points, $(at_i^2, \; 2at_i)$, $i = 1, 2, 3$; may be concurrent.
5. Show that the condition for the normal at the points $(x_i, y_i)$, $i = 1, 2, 3$ on the ellipse, $\dfrac{x^2}{a^2} + \dfrac{y^2}{b^2} = 1$ to be concurrent is

$$\begin{vmatrix} x_1 & y_1 & x_1 y_1 \\ x_2 & y_2 & x_2 y_2 \\ x_3 & y_3 & x_3 y_3 \end{vmatrix} = 0$$

6. Find the point of the parabola, $y^2 = 8x$, at which the normal is inclined at an angle of $60°$ to the axis of the parabola.

7. Find the equation of the normal to $y^2 = 8x$, perpendicular to the line, $2x + 6y = 5$. Also find the foot of the normal.
8. Prove that the chord of the parabola, $y^2 = 4ax$ whose equation is, $y - x\sqrt{2} + 4a\sqrt{2} = 0$ is a normal to the parabola and that its length is $6a\sqrt{3}$.; also find the foot of the normal.

## 8.4. CHORD OF CONTACT

**Definition.** The line, joining the points of contact of the two tangents drawn from a point to a curve (see **8.1.3.**) is called the *chord of contact* of this point w.r.t. the given curve.

**8.4.1. To find the equation of the chord of contact of the point P(x′, y′) w.r.t. the curve, S = 0.**

The given curve is

$$S = ax^2 + 2hxy + by^2 + 2gx + 2fy + c = 0$$

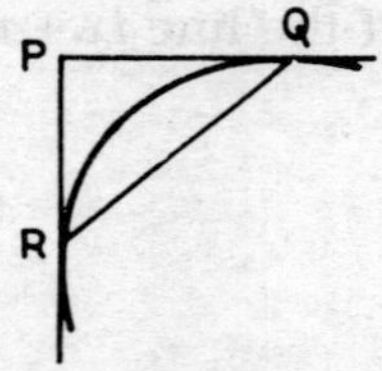


Let $PQ$ and $PR$ be the tangents to it from $P\ (x', y')$. Let $Q$ and $R$ be $(x_1, y_1)$ and $(x_2, y_2)$ respectively. The $QR$ is the chord of contact of $P$ w.r.t. the given curve.

Now the equations of the tangents to $S = 0$, at $Q$ and $R$, are, respectively.

$$axx_1 + h(xy_1 + x_1y) + byy_1 + g(x + x_1) + f(y + y_1) + c = 0$$

and $$axx_2 + h(xy_2 + x_2y) + byy_2 + g(x + x_2) + f(y + y_2) + c = 0.$$

Both these tangents pass through $P(x', y')$.

$$\therefore\ ax'x_1 + h(x'y_1 + x_1y') + by'y_1 + g(x' + x_1) + f(y' + y_1) + c = 0 \qquad (1)$$

and $$ax'x_2 + h(x'y_2 + x_2y') + by'y_2 + g(x' + x_2) + f(y' + y_2) + c = 0 \qquad (2)$$

Now, consider the equation.

$$axx' + h(xy' + x'y) + byy' + g(x + x') + f(y + y') + c = 0, \qquad (3)$$

This, being a first degree equation in $x$ and $y$, represents a straight line and conditions (1) and (2) above, show that this line passes through the points $Q(x_1, y_1)$ and $R(x_2, y_2)$. Thus (3), is the required equation of the chord of contact, $QR$, of the point $P$, w.r.t. the given curve.

**Note.** From the equation of the chord of contact, it is clear that the method of writing this equation is same, as if we are writing the tangent at $(x', y')$. We may also note that if the point $P$ is on the curve, the chord of contact becomes the tangent at the point $P$. Further, it is real iff $P$ lies on or outside the curve.

**Example 8.8.** Tangents are drawn from the Point $(x_1, y_1)$ to the circle, $x^2 + y^2 = a^2$. Prove that the area of the triangle formed by them and the straight line joining their points of contact is

$$\frac{a(x_1^2 + y_1^2 - a^2)^{3/2}}{x_1^2 + y_1^2}.$$

Also find the length of the chord of contact.

*Solution.* Let $P$ be the point $(x_1, y_1)$. In the following figure, clearly $QR$ is the chord of contact of the point $P$, w.r.t. the given circle. Draw $OL$ perpendicular to $QR$, then $OL$ bisects $QR$.

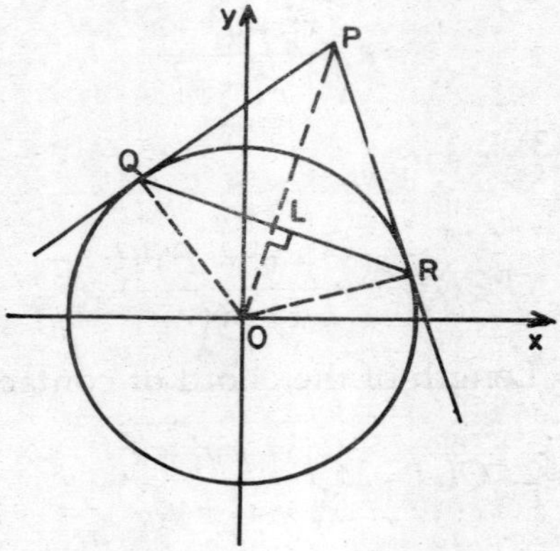

Now, the equation of the chord of contact, $QR$, is

$$xx_1 + yy_1 = a^2.$$

Its slope $= -\dfrac{x_1}{y_1}$.

$\therefore$ the equation of $OL$ is

$$y - 0 = \frac{y_1}{x_1}(x - 0)$$

$$\Rightarrow \quad xy_1 - x_1 y = 0.$$

Clearly, $OL$ passes through $P\ (x_1, y_1)$.

Now area $\Delta\, PQR = 2.$ area $\Delta\, PQL = |PL| \cdot |QL|$ (1)

Now $|PL|$ = perpendicular distance from $P$ on $QR$

$$= \frac{|x_1^2 + y_1^2 - a^2|}{\sqrt{x_1^2 + y_1^2}}. \tag{2}$$

Similarly $$|OL| = \frac{a^2}{\sqrt{x_1^2 + y_1^2}}$$

$$|QL| = \sqrt{OQ^2 - OL^2} = \sqrt{a^2 - \frac{a^4}{x_1^2 + y_1^2}}$$

$$= a\sqrt{\frac{x_1^2 + y_1^2 - a^2}{x_1^2 + y_1^2}} \tag{3}$$

From (1), (2), (3),

$$\text{area } \Delta\, PQR = \frac{a(x_1^2 + y_1^2 - a^2)^{3/2}}{x_1^2 + y_1^2}.$$

Now | QR | = Length of the chord of contact

$$= 2.|QL| = 2a\sqrt{\frac{x_1^2 + y_1^2 - a^2}{x_1^2 + y_1^2}}.$$

**Example 8.9.** Find the locus of points which are such that the chords of contact of tangents drawn from them to the conic, $ax^2 + by^2 = 1$, form a triangle of constant area with the co-ordinate axes.

*Solution*. Let $(\alpha, \beta)$ be the point whose locus is required. Now the chord of contact of $(\alpha, \beta)$ w.r.t. the conic,

$$ax^2 + by^2 = 1$$

is

$$ax\,\alpha + by\,\beta = 1 \qquad (1)$$

This meets the co–ordinate axes at $P$ $(1/a\ x, 0)$ and Q $(0, 1/b\ \beta)$.

By hypothesis,

$$\text{Area of } \Delta\, OPQ = \text{Constant, say } k$$

$$\Rightarrow \quad \frac{1}{2}.\ \frac{1}{a\alpha}.\ \frac{1}{b\beta} = k$$

$$\Rightarrow \quad \alpha\beta = \frac{1}{2abk}$$

$\therefore$ Locus of $(\alpha, \beta)$ is $xy = c$ (a constant).

## Exercise 8.4

1. Find the chord of contact of point $(x', y')$ w.r.t. the following curves:
   (i) circle: $x^2 + y^2 + 2gx + 2fy + c = 0$.
   (ii) circle: $x^2 + y^2 = a^2$.
   (iii) parabola: $y^2 = 4ax$.
   (iv) ellipse: $\dfrac{x^2}{a^2} + \dfrac{y^2}{b^2} = 1$.
   (v) hyperbola: $\dfrac{x^2}{a^2} - \dfrac{y^2}{b^2} = 1$.
   (vi) rectangular hyperbola: $xy = c^2$.
2. Find the chord of contact of origin w.r.t.
   $$ax^2 + 2hxy + by^2 + 2gx + 2fy + c = 0.$$
3. Find the equation of the chord of contact of tangents from the point (3, 4) to the circle $x^2 + y^2 + 4x + 10y + 4 = 0$
4. Tangents are drawn to the circle $x^2 + y^2 = a^2$ at the points

where the line $lx + my + n = 0$ meets it. Find the point of intersection of the tangents.
(*Hint:* The line is the chord of contact of this point.)

5. Show that the locus of the points such that the chords of contact of the tangents drawn from them to the curve $ax^2 + by^2 = 1$ subtend a right angle at its centre is the conic $a^2x^2 + b^2y^2 = a + b$.
6. The chord of contact of tangents drawn from any point on the circle, $x^2 + y^2 = a^2$ to the circle, $x^2 + y^2 = b^2$ touches the circle, $x^2 + y^2 = c^2$. Show that $a$, $b$, $c$ are in G.P.
7. Prove that the area of the triangle formed by tangents from the point $(x_1, y_1)$ to the parabola $y^2 = 4ax$ and its chord of contact is

$$\frac{\left(y_1^2 - 4ax_1\right)^{3/2}}{2a}$$

(*Hint* Compare the equation of the chord of contact with the line joining the points "$t_1$" and "$t_2$" on the parabola).
8. If tangents be drawn to the parabola $y^2 = 4ax$ from a point on the line $x + 4a = 0$, then show that its chord of contact w.r.t. the parabola, will subtend a right angle at the vertex (0, 0).
9. Tangents are drawn to the circle $x^2 + y^2 = a^2$ at the points where a variable chord always at a distance $p$ $(p < a)$ from the centre meets it. Show that the locus of the points of intersection of the tangents is a circle.
10. Tangents to the parabola, $y^2 = 4ax$, at two points $P$ and $Q$ meet in $(x_1, y_1)$. Show that $|PQ|$ is equal to $\frac{1}{a}\sqrt{\left(y_1^2 + 4a^2\right)\left(y_1^2 - 4ax_1\right)}$ and the normals at $P$ and $Q$ intersect at the point $\left(2a - x_1 + \frac{y_1^2}{a}, -\frac{x_1y_1}{a}\right)$
11. Tangents are drawn to the circle, $x^2 + y^2 = a^2$ at the points where $x \cos \alpha + y \sin \alpha = p$ meets it. Show that their point of intersection is

$$\left(\frac{a^2 \cos\alpha}{p}, \frac{a^2 \sin\alpha}{p}\right)$$

12. From a given point $(x_1, y_1)$, tangents $PQ$ and $PR$ are drawn to the circle, $x^2 + y^2 = a^2$. Find the equation of the circum–circle of the triangle, $PQR$.
13. Show that the locus of the points such that chord of contact of tangents from which to the circle, $x^2 + y^2 = a^2$ subtends a right angle at the centre, is a concentric circle whose radius is $\sqrt{2}$ times that of the given circle. Hence show that this is also the locus of the point of intersection of two perpendicular tangents.

### 8.5. POLE AND POLAR

**Definition.** Let $P$ be a given point and $S = 0$ be a given curve. Draw any line $PAB$ from $P$ meeting the curve at $A$ and $B$. Let the tangents to the curve at $A$ and $B$ meet at the point $C$. The locus of the point $C$, as the line $PAB$ varies, is called the *polar* of the point $P$ w.r.t. the given curve and $P$ is called the pole of this polar.

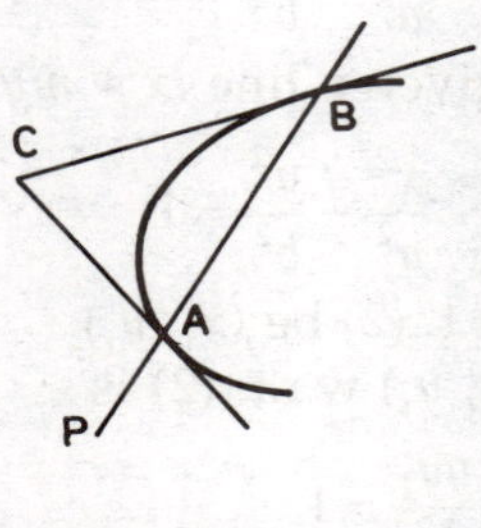


#### 8.5.1. To find the equation of the polar of the point P (x′, y′) w.r.t. to the curve S = 0.

Here $S = ax^2 + 2hxy + by^2 + 2gx + 2fy + c.$

As in the definition of the polar of the point $P$, w.r.t. the curve, $S = 0$, let the co–ordinates of $C$ be $(\alpha, \beta)$.

We have to find the locus of $C$.

Clearly $PAB$ is the chord of contact of $C$ w.r.t. $S = 0$ and its equation is

$$ax\,\alpha + h(x\,\beta + y\,\alpha) + by\,\beta + g(x + \alpha) + f(y + \beta) + c = 0.$$

Since it passes through $P(x', y')$, we have

$$ax'\alpha + h(x'\beta + y'\alpha) + by'\,\beta + g(x' + \alpha) + f(y' + \beta) + c = 0.$$

$\therefore$ the locus of $(\alpha, \beta)$ is

$$axx' + h(xy' + yx') + byy' + g(x + x') + f(y + y') + c = 0.$$

This being a first degree equation, represents a straight line. Hence the polar of a point w.r.t. a given curve is a straight line.

**Note.** As in the case of chord or contact, we note that the method of writing the equation of the polar of a point is same as if we are writing the tangent at the point $(x', y')$. As a matter of fact, we have the following:

**Case 1.** If the point $P$ is outside a curve both chord of contact and polar of the point w.r.t. the curve exist and coincide.

**Case 2.** If the point is on the curve, then its polar exists and is the same as the tangent at the point.

**Case 3.** If the point is inside a curve, only its polar exists.

**Example 8.10.** Find the pole of the line $lx + my + n = 0$ with respect to the hyperbola

$$\frac{x^2}{a^2} - \frac{y^2}{b^2} = 1$$

*Solution*. We are given a line $lx + my + n = 0$ (1)

and a hyperbola $\frac{x^2}{a^2} - \frac{y^2}{b^2} = 1$ (2)

Let pole of (1) w.r.t. (2) be $(x_1, y_1)$.

Then the polar of $(x_1, y_1)$ w.r.t. (2) is

$$\frac{xx_1}{a^2} - \frac{yy_1}{b^2} = 1. \quad (3)$$

Now (1) and (3) represent the same line.

$$\therefore \quad \frac{x_1}{a^2 l} = \frac{-y_1}{b^2 m} = \frac{1}{-n},$$

whence, the pole is $(x_1, y_1) = \left(\frac{-a^2 l}{n}, \frac{b^2 m}{n}\right)$

We now prove some propositions on poles and polars.

**Proposition 8.1.** If the polar of a point $P$ w.r.t, a curve passes through a point $Q$, then the polar of $Q$ w.r.t. the same curve, passes through the point $P$.

*Proof.* Let $S = ax^2 + 2hxy + by^2 + 2gx + 2fy + c = 0$ be the given curve and let $P(x_1, y_1)$ and $Q(x_2, y_2)$ be two points such that the polar of $P$ w.r.t. $S = 0$ passes through the point $Q$. The polar of $P$ w.r.t. $S = 0$, is

$$axx_1 + h(x_1y + xy_1) + byy_1 + g(x + x_1) + f(y + y_1) + c = 0.$$

It passes through $Q\ (x_2, y_2)$.

$$\therefore\ ax_2x_1 + h(x_1y_2 + x_2y_1) + by_2y_1 + g(x_2 + x_1) + f(y_2 + y_1) + c = 0.$$

The symmetry of this result shows that the polar of $Q$ w.r.t. the given curve also passes through the point $P$. Hence the result.

**Corollary.** If the polars of $P$ and $Q$ w.r.t. a curve intersect in $R$, then $R$ is the pole of $PQ$ w.r.t. the curve.

**Proposition 8.2.** If the pole of a line $l$ w.r.t. a curve lies on another line $l'$, then the pole of $l'$ w.r.t. it lies on the line $l$.

*Proof.* Let $S = ax^2 + 2hxy + by^2 + 2gx + 2fy + c = 0$ (1)

be the given curve and $(x_1, y_1)$, $(x_2, y_2)$ be the poles of the lines $l$ and $l'$ respectively w.r.t. $S = 0$.

Then the line $l$ is given by (notice: $l$ is the polar of $(x_1, y_1)$ w.r.t. $S = 0$),

$$axx_1 + h(xy_1 + yx_1) + byy_1 + g(x + x_1) + f(y + y_1) + c = 0 \quad (2)$$

In the same way, $l'$ is given by

$$axx_2 + h(xy_2 + yx_2) + byy_2 + g(x + x_2) + f(y + y_2) + c = 0 \quad (3)$$

Let the pole, $(x_1, y_1)$, of (2) w.r.t. (1) lie on (3), then

$$ax_1x_2 + h(x_1y_2 + y_1x_2) + by_1y_2 + g(x_1 + x_2) + f(y_1 + y_2) + c = 0 \quad (4)$$

Clearly (4) is also the condition that the pole, $(x_2, y_2)$ of (3) w.r.t. (1) lies on (2). Hence the result.

**Definition 1.** *Conjugate points.* Two *points* are said to be *conjugate* if polar of each w.r.t. a curve passes through the other.

**Definition 2.** *Conjugate lines.* Two *lines* are said to be *conjugate* if pole of each line w.r.t. a curve lies on the other line.

**Example 8.11** Show that the locus of the poles of normal

chords of the parabola $y^2 = 4ax$ is $(x + 2a)y^2 + 4a^3 = 0$.

*Solution*. We have to find the locus of a point such that its polar w.r.t. the given curve is also a normal to the curve. So let $(\alpha, \beta)$ be the co-ordinates of a point whose locus is required.

Its polar w.r.t. $y^2 = 4ax$ is

$$y\beta = 2a(x + \alpha).$$

$$\Rightarrow \quad 2ax - \beta y + 2a\alpha = 0 \tag{1}$$

This line is also to be a normal to the parabola, and we know that the equation of the normal to the parabola at a point $(at^2, 2at)$, is

$$tx + y - (2at + at^3) = 0 \qquad \text{(by § 8.3.2)} \quad (2)$$

As (1) and (2) are same,

$$\therefore \quad \frac{t}{2a} = \frac{1}{-\beta} = \frac{-(2at + at^3)}{2a\alpha} = \frac{-(2t - t^3)}{2\alpha}$$

$$\Rightarrow \quad t = \frac{-2a}{\beta}$$

and $\quad 2\alpha = \beta(2t + t^3)$

Eliminating $t$ from these two equations, we get

$$2\alpha = \beta\left(-\frac{4a}{\beta} - \frac{8a^3}{\beta^3}\right) = \frac{-4a\left(\beta^2 + 2a^2\right)}{\beta^2}$$

$$\Rightarrow \quad (\alpha + 2a)\beta^2 + 4a^3 = 0$$

$\therefore$ the locus of $(\alpha, \beta)$ is

$$(x + 2a)y^2 + 4a^3 = 0.$$

**Example 8.12.** Show that the polar of any point on the circle

$$x^2 + y^2 - 2ax - 3a^2 = 0$$

w.r.t. the circle

$$x^2 + y^2 + 2ax - 3a^2 = 0$$

will touch the parabola

$$y^2 + 4ax = 0.$$

*Solution*. Let $(\alpha, \beta)$ be any point on the circle

$$x^2 + y^2 - 2ax - 3a^2 = 0.$$

$$\therefore \quad \alpha^2 + \beta^2 - 2a\,\alpha - 3a^2 = 0. \tag{1}$$

The polar of $(\alpha, \beta)$ w.r.t. the circle

$$x^2 + y^2 + 2ax - 3a^2 = 0$$

is $\quad x\alpha + y\beta + a(x + \alpha) - 3a^2 = 0$

$$\Rightarrow \quad (\alpha + a)x + \beta y + (a\,\alpha - 3a^2) = 0 \qquad (2)$$

We solve the equation (2) and the equation of the parabola

$$y^2 + 4ax = 0$$

simultaneously for their points of intersection.

Now (2) $\Rightarrow \quad x = -\dfrac{\left[\beta y + \left(a\alpha - 3a^2\right)\right]}{\alpha + a}$

Substituting in $y^2 + 4ax = 0$, we get

$$y^2 - \frac{4a\left[\beta y + a\alpha - 3a^2\right]}{\alpha + a} = 0$$

$$\Rightarrow (\alpha + a)y^2 - 4a\beta y - 4a^2(\alpha - 3a) = 0 \qquad (3)$$

Its discriminant

$$\begin{aligned} &= 16a^2\beta^2 + 16a^2(\alpha + a)(\alpha - 3a) \\ &= 16a^2[\beta^2 + \alpha^2 - 2a\,\alpha - 3a^2] \\ &= 0. \qquad \text{(using (1))} \end{aligned}$$

$\Rightarrow$ (3) has equal roots.

Hence, the polar (2) touches the parabola

$$y^2 + 4ax = 0.$$

**Example 8.13.** Show that the locus of the poles with respect to the parabola, $y^2 = 4ax$, of the tangents to the rectangular hyperbola, $x^2 - y^2 = a^2$, is the ellipse $4x^2 + y^2 = 4a^2$.

*Solution.* Let $(\alpha, \beta)$ be the pole of a tangent to the rectangular hyperbola,

$$x^2 - y^2 = a^2 \qquad (1)$$

Its polar w.r.t. the parabola, $y^2 = 4ax$, is

$$y\beta = 2a(x + \alpha) \qquad (2)$$

By hypothesis, this line touches (1).

On solving (1) and (2), we get

$$x^2 - \frac{4a^2(x+\alpha)^2}{\beta^2} = a^2$$

$$\Rightarrow \quad (\beta^2 - 4a^2)x^2 - 8a^2\alpha x - (4a^2\alpha^2 + a^2\beta^2) = 0. \qquad (3)$$

As the line (2) is a tangent, the discriminant of (3) should be zero, whence

$$64a^4\alpha^2 + 4(\beta^2 - 4a^2)\cdot a^2(4\alpha^2 + \beta^2) = 0.$$

Cancelling $4a^2$ and simplifying, we get

$$16a^2\alpha^2 + 4\alpha^2\beta^2 + \beta^4 - 16a^2\alpha^2 - 4a^2\beta^2 = 0.$$

Cancelling $\beta^2$, we get,

$$4\alpha^2 + \beta^2 - 4a^2.$$

$\therefore$ the locus of $(\alpha, \beta)$ is $4x^2 + y^2 = 4a^2$.

**Example 8.14.** Prove that the polar of a given point w.r.t. any one of the circles $x^2 + y^2 - 2kx + c^2 = 0$, where $k$ is a variable, always passes through a fixed point whatever may be the value of $k$.

*Solution.* Let $P(\alpha, \beta)$ be a given point. Its polar w.r.t.

$$x^2 + y^2 - 2kx + c^2 = 0 \text{ is}$$

$$x\alpha + y\beta - k(x + \alpha) + c^2 = 0.$$

$$\Rightarrow \quad (x\alpha + y\beta + c^2) - k(x + \alpha) = 0.$$

For all values of $k$, this line passes through the intersection of the lines

$$x + \alpha = 0, \text{ and } x\alpha + y\beta + c^2 = 0$$

i.e. through the fixed point

$$\left(-\alpha, \frac{\alpha^2 - c^2}{\beta}\right).$$

**Example 8.15.** $PQR$ is any triangle and $P'Q'R'$ is the triangle formed by the polars of $P$, $Q$, $R$ w.r.t. a given circle, so that $Q'R'$, $R'P'$, $P'Q'$ are the polars of $P$, $Q$, $R$ respectively; Prove that the lines $PP'$, $QQ'$, $RR'$ are concurrent.

*Solution.* Let the given circle be $x^2 + y^2 = a^2$ (1)

Let the points be $P(x_1, y_1)$, $Q(x_2, y_2)$ and $R(x_3, y_3)$.

By hypothesis, polars of $P$, $Q$, $R$ w.r.t. (1) are

$Q'R' : xx_1 + yy_1 - a^2 = 0$ (2)

$R'P' : xx_2 + yy_2 - a^2 = 0$ (3)

$P'Q' : xx_3 + yy_3 - a^2 = 0$ (4)

Now the equation of any line through $P'$, the intersection of (3) and (4) is $(xx_2 + yy_2 - a^2) + \lambda\,(xx_3 + yy_3 - a^2) = 0$ (5)

If it passes through $P(x_1, y_1)$ we have

$(x_1x_2 + y_1y_2 - a^2) + \lambda\ (x_1x_3 + y_1y_3 - a^2) = 0$ (6)

Eliminating $\lambda$ between (5) and (6), we get the equation of the line $PP'$, as

$$(xx_2 + yy_2 - a^2)\,(x_1x_3 + y_1y_3 - a^2) - (xx_3 + yy_3 - a^2)$$

$(x_1x_2 + y_1y_2 - a^2) = 0$ (7)

By symmetry, we get the equations of the lines $QQ'$ and $RR'$ as

$$(xx_3 + yy_3 - a^2)(x_2x_1 + y_2y_1 - a^2) - (xx_1 + yy_1 - a^2)(x_2x_3 + y_2y_3 - a^2) = 0 \quad (8)$$

$$(xx_1 + yy_1 - a^2)(x_3x_2 + y_3y_2 - a^2) - (xx_2 + yy_2 - a^2)(x_3x_1 + y_3y_1 - a^2) = 0 \quad (9)$$

Equations (7), (8) and (9) are of the form

$X - Y = 0, \quad Y - Z = 0, \quad Z - X = 0$

Clearly sum of their left sides = 0

the lines $PP'$, $QQ'$, $RR'$ are concurrent.

**Example 8.16.** Prove that polars of a given point w.r.t. a co-axial system of circles always pass through a fixed point and the line joining these points is bisected by the radical axis of the system.

*Solution*. Let the given system of coaxial circles be

$$x^2 + y^2 + 2gx + c = 0 \quad (1)$$

Here $g$ is the parameter.

Let $P(x_1, y_1)$ be the given point. The radical axis of the above system is $\quad x = 0$ (2)

Now polar of $P(x_1, y_1)$ w.r.t. (1) is

$xx_1 + yy_1 + g(x + x_1) + c = 0$

$\Rightarrow (xx_1 + yy_1 + c) + g(x + x_1) = 0$

which for different values of $g$ passes through the point of intersection of the fixed lines:

$xx_1 + yy_1 + c = 0$ and $x + x_1 = 0$

$\therefore$ The polars of $P$ w.r.t. (1) always pass through the fixed point,

$$Q\left(-x_1, \frac{x_1^2 - c}{y_1}\right)$$

The mid–point of $PQ$ is

$$M\left(0, \frac{1}{2}\left(y_1 + \frac{x_1^2 - c}{y_1}\right)\right)$$

which lies on (2).

$\therefore$ The radical axis of the system bisects $PQ$.

**Exercise 8.5.**

1. Proceeding as in § 8.5.1, find the polar of the point $(x_1, y_1)$ w.r.t. the following curves:
   (i) circle: $x^2 + y^2 = a^2$.
   (ii) circle: $x^2 + y^2 + 2gx + 2fy + c = 0$.
   (iii) parabola: $y^2 = 4ax$.
   (iv) ellipse: $\frac{x^2}{a^2} + \frac{y^2}{b^2} = 1$.
   (v) hyperbola: $\frac{x^2}{a^2} - \frac{y^2}{b^2} = 1$.
   (vi) rectangular hyperbola: $xy = c^2$.
2. Write down the polars of the following points w.r.t. the given curves:
   (i) $(1, -1)$, $x^2 + y^2 - 2x + 5y - 1 = 0$.
   (ii) $(0, 0)$, $2x^2 - xy + 3y^2 + 5x - 8y + 1 = 0$.
   (iii) $(\alpha, \beta)$, $2x^2 - 3xy + 5y^2 + 4x + 7y - 1 = 0$.
3. Find the poles of the following lines;
   (i) $2x + 3y = 6$, w.r.t. the circle, $x^2 + y^2 = 5$.
   (ii) $y - 2x = 2$, w.r.t. the parabola, $y^2 = 8x$.
4. Find the pole of the line $lx + my + n = 0$ w.r.t. the following curves:
   (i) circle: $x^2 + y^2 = a^2$.
   (ii) parabola: $y^2 = 4ax$.
   (iii) ellipse: $x^2/a^2 + y^2/b^2 = 1$.
   (iv) rectangular hyperbola: $xy = c^2$.
5. Find the locus of the poles of the normal chords of the following curves:
   (i) ellipse: $x^2/a^2 + y^2/b^2 = 1$.
   (ii) hyperbola: $x^2/a^2 - y^2/b^2 = 1$.
   (ii) rectangular hyperbola: $xy = c^2$.
6. Show that the polar of $(-a, 2a)$ w.r.t. the circle $x^2 + y^2 - 2ax - 3a^2 = 0$ touches the parabola, $y^2 = 4ax$.
7. Prove that the polar of the point $(p, q)$ w.r.t. the circle $x^2 + y^2 = a^2$ touches the circle
   $(x - c)^2 + (y - d)^2 = b^2$ if $b^2 (p^2 + q^2) = (cp + dq - a^2)^2$.

(*Hint:* See Question 6 (iii), Exercise 6.3).

8. If the polar of the point $(x', y')$ w.r.t. the circle, $x^2 + y^2 = a^2$ touches the circle, $(x - a)^2 + y^2 = a^2$, show that $(x', y')$ lies on the curve $y^2 + 2ax = a^2$.
9. Find the locus of the pole w.r.t. the ellipse, $x^2/a^2 + y^2/b^2 = 1$ of any tangent to $x^2 + y^2 = a^2$.
10. Prove that the locus of a point whose polar w.r.t. the parabola $y^2 = 4ax$ touches the circle $x^2 + y^2 = a^2$ is the curve $4x^2 = y^2 + 4a^2$.
11. Find the locus of the poles of the tangents to the parabola $y^2 = 4ax$ w.r.t. the circle $x^2 + y^2 - 2ax = 0$.
12. Show that the locus of the poles of the tangents to the ellipse $x^2/a^2 + y^2/b^2 = 1$ w.r.t. the ellipse $x^2/\alpha^2 + y^2/\beta^2 = 1$ is $\frac{a^2x^2}{\alpha^4} + \frac{b^2y^2}{\beta^4} = 1$.
13. (i) Find the locus of the point whose polars w.r.t. the circles $x^2 + y^2 - 2x + 6y - 2 = 0$ and $x^2 + y^2 - 4x - 4y - 8 = 0$ are mutually perpendicular.

    (ii) The polars of $(x_1, y_1)$, $(x_2, y_2)$ w.r.t. the hyperbola, $x^2/a^2 - y^2/b^2 = 1$ are at right angles. Prove that $\frac{x_1x_2}{a^4} + \frac{y_1y_2}{b^4} = 0$.
14. Chords of the ellipse, $x^2/a^2 + y^2/b^2 = 1$ touch the parabola, $ay^2 = 2b^2x$. Prove that the locus of their poles is the parabola, $ay^2 + 2b^2x = 0$.
15. A variable circle is drawn to touch the $x$–axis at the origin. Find the locus of the pole of the straight line $lx + my + n = 0$ w.r.t. this circle.
16. Prove that the polar of any point on $x^2/a^2 + y^2/b^2 = 1$ w.r.t. $x^2/a^2 - y^2/b^2 = 1$ touches $x^2/a^2 + y^2/b^2 = 1$.
17. Find the locus of the poles w.r.t. the hyperbola $x^2/a^2 - y^2/b^2 = 1$ of any tangent to the circle $x^2 + y^2 = a^2 + b^2$.
18. Find the locus of the poles of the chords of the following curves which subtend a right angle at the centre (0, 0);

    (i) circle: $x^2 + y^2 = a^2$.

(ii) ellipse: $x^2/a^2 + y^2/b^2 = 1$.
(iii) hyperbola: $x^2/a^2 - y^2/b^2 = 1$.

19. Show that the locus of the poles of chords of the parabola, $y^2 = 4ax$ which subtend a constant angle $\alpha$ at its vertex, (0, 0), is the curve
$$4(y^2 - 4ax) = \tan^2 \alpha (x + 4a)^2.$$
Deduce the locus of the poles of the chords of the parabola $y^2 = 4ax$ which subtend a right angle at the origin.

20. The perpendicular from the centre of an ellipse $x^2/a^2 + y^2/b^2 = 1$ on the polar of a point w.r.t. it is constant and is equal to $c$. Prove that the locus of the point is the ellipse
$$x^2/a^4 + y^2/b^4 = 1/c^2.$$

21. Show that the distances of two points from the centre of a circle are proportional to the distances of each from the polar of the other.

22. If $O$ is centre of a circle with radius $a$, and $OP$ meets the polar of $P$, w.r.t. the circle in $Q$, then
(i) $OP$ is perpendicular to the polar of $P$.
(ii) $[OP].[OQ] = a^2$.

23. Show that for a given curve, there exists infinite number of triangles, such that each vertex is pole of the opposite side. (Such a triangle is called *self–polar* or *self conjugate*).

24. The lengths of tangents from two points $A$ and $B$ to a circle are $t$ and $t'$ respectively. If the two points are conjugate w.r.t. the given circle, prove that $AB^2 = t^2 + t'^2$.

25. Show that the limiting points of a co–axial system are conjugate w.r.t. any circle of the system.

26. (i) Prove that the extremities of a diameter of a circle, $S = 0$ which cuts a circle, $S' = 0$ orthogonally are conjugate w.r.t. $S' = 0$.
(ii) (Converse) If $P$, $Q$ is a pair of conjugate points w.r.t. a circle, $S = 0$, prove that the circle on $PQ$ as diameter cuts the circle, $S = 0$, orthogonally.

27 Find the condition that the lines $lx + my + n = 0$ and $l'x + m'y + n' = 0$ are conjugate w.r.t.

(i) circle: $x^2 + y^2 = a^2$,
(ii) parabola: $y^2 = 4ax$,
(iii) ellipse: $x^2/a^2 + y^2/b^2 = 1$,
(iv) hyperbola: $x^2/a^2 - y^2/b^2 = 1$,
(v) rectangular hyperbola: $xy = c^2$,
(vi) $S = ax^2 + 2hxy + by^2 + 2gx + 2fy + c = 0$.

28. The lines $x - \alpha = 0$, $y - \beta = 0$ are conjugate w.r.t. the hyperbola, $xy = c^2$. Prove that $(\alpha, \beta)$ lies on the hyperbola, $xy = 2c^2$.
29. Show that the locus of the poles of tangents to the circle, $x^2 + y^2 = a^2$ w.r.t. the circle, $x^2 + y^2 - 2bx = 0$, is the conic, $(a^2 - b^2)x^2 + a^2y^2 - 2a^2bx + a^2b^2 = 0$.
30. If the pole of a line w.r.t. the circle, $x^2 + y^2 = a^2$ lies on the cirle, $x^2 + y^2 = k^2a^2$, prove that the line will touch the circle, $k^2(x^2 + y^2) = a^2$.
31. Prove that the locus of the poles of the chords of the circle, $x^2 + y^2 = a^2$, which subtend a right angle at $(p, q)$ is $(p^2 + q^2 - a^2)(x^2 + y^2) = 2a^2(px + qy - a^2)$.
32. Find the locus of the poles of the line, $(x/a) + (y/b) = 1$, w.r.t. the circles which touch the co–ordinate axes.
33. Circles are drawn passing through the point $(0, b)$ and touching the circle, $x^2 + y^2 = a^2$. Find the locus of the pole of the axis of $y$ w.r.t. these circles.
34. $P$ is a variable point on the line, $y = b$. Prove that
    (i) Polar of $P$ w.r.t. $y^2 = 4ax$ is fixed in direction.
    (ii) The locus of the foot of the perpendicular from $P$ on its polar is the line, $(4a^2 - b^2)y = 4ab(a - x)$.
35. Show that the locus of the poles of the tangents of $y^2 = 4ax$ w.r.t. $y^2 = 4bx$ is the parabola, $y^2 = 4(b^2/a)x$.

## 8.6. TO FIND THE EQUATION OF THE CHORD OF A CURVE, $S = 0$, IN TERMS OF ITS MIDDLE POINT

Let $S = ax^2 + 2hxy + by^2 + 2gx + 2fy + c = 0$ be the given curve and $(x', y')$ be the co–ordinates of the middle point $M$, of a chord $PQ$ of it. We want to find the equation of $PQ$.

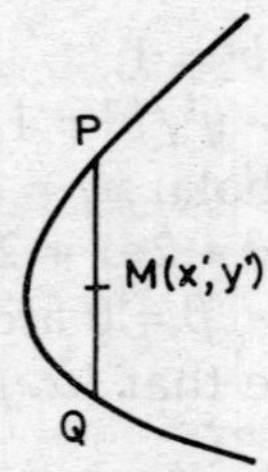


Let the line through $M(x', y')$ be given by

$$\frac{x-x'}{\cos\theta}=\frac{y-y'}{\sin\theta}=r \tag{1}$$

Any point on (1) is

$(x' + r\cos\theta,\ y' + r\sin\theta)$,

which lies on $S = 0$, whence

$$a(x'+r\cos\theta)^2 + 2h(x'+r\cos\theta)(y'+r\sin\theta) + b(y'+r\sin\theta)^2 + 2g(x'+r\cos\theta) + 2f(y'+r\sin\theta) + c = 0.$$

$$\Rightarrow (a\cos^2\theta + 2h\sin\theta\cos\theta + b\sin^2\theta)r^2 + 2[(ax'+hy'+g)\cos\theta + (hx'+by'+f)\sin\theta]r + S' = 0. \quad (2)$$

where

$S' = ax'^2 + 2hx'y' + by'^2 + 2gx' + 2fy' + c.$

As $M$ is the middle point of the chord $PQ$, the values of $r$ for the points $P$ and $Q$, are equal and opposite.

Hence the sum of roots of (2) is zero.

$$\Rightarrow (ax'+hy'+g)\cos\theta + (hx'+by'+f)\sin\theta = 0. \tag{3}$$

Eliminating $\theta$ from (1) and (3), we get

$$(ax'+hy'+g).\ \frac{x-x'}{r} + (hx'+by'+f)\ \frac{y-y'}{r} = 0$$

$$\Rightarrow axx' + hxy' + gx + hx'y + byy' + fy = ax'^2 + hx'y' + gx' + hx'y' + by'^2 + fy'.$$

Adding $gx' + fy + c$ to both sides, we get

$$axx' + h(xy'+x'y) + byy' + g(x+x') + f(y+y') + c = ax'^2 + 2hx'y' + by'^2 + 2gx' + 2fy' + c$$

$$\Rightarrow \qquad T = S'$$

where $T = axx' + h(xy'+x'y) + byy' + g(x+x') + f(y+y') + c$

and $S' = ax'^2 + 2hx'y' + by'^2 + 2gx' + 2fy' + c$.

Thus $T = S'$ is the required equation of the chord $PQ$, whose middle point is $(x', y')$.

**Example 8.17**. Find the middle point of the chord, $9x - 4y - 14 = 0$ of the curve, $2x^2 + xy - 3y^2 - 1 = 0$.

*Solution*. **Method I**. Let $(\alpha, \beta)$ be the middle point of the given chord. Then the equation of the chord is $T = S'$.

$$\Rightarrow 2x\alpha + \frac{1}{2}(x\beta + y\alpha) - 3y\beta - 1 = 2\alpha^2 + \alpha\beta - 3\beta^2 - 1$$

$$\Rightarrow (4\alpha + \beta)x + (\alpha - 6\beta)y - (4\alpha^2 + 2\alpha\beta - 6\beta^2) = 0$$

But this is same as $\quad 9x - 4y - 14 = 0. \qquad (1)$

$$\frac{4\alpha+\beta}{9} = \frac{\alpha-6\beta}{-4} = \frac{4\alpha^2+2\alpha\beta-6\beta^2}{14} = \frac{(2\alpha^2+\alpha\beta-3\beta^2)}{7}$$

$$\Rightarrow \quad -16\alpha - 4\beta = 9\alpha - 54\beta \Rightarrow \alpha = 2\beta. \qquad (2)$$

and $\quad 18\alpha^2 + 9\alpha\beta - 27\beta^2 = 28\alpha + 7\beta$

$$\Rightarrow \quad 72\beta^2 + 18\beta^2 - 27\beta^2 = 56\beta + 7\beta, \qquad \text{using (2).}$$

$$\Rightarrow \quad \beta^2 = \beta$$

$$\therefore \quad \beta = 0, 1;$$

whence from (2),

$$\alpha = 0, 2.$$

$\therefore$ the points are (0, 0) and (2,1).

Clearly (0, 0) does not lie on (1), and so, the middle point of the given chord is (2, 1).

**Method II**. We solve the equations

$$9x - 4y - 14 = 0 \qquad (1)$$

and $\quad 2x^2 + xy - 3y^2 - 1 = 0 \qquad (2)$

simultaneously, for their points of intersection.

Now (1) $\Rightarrow$

$$y = \frac{9x-14}{4}.$$

and then (2) $\Rightarrow$

$$2x^2 + x\left(\frac{9x-14}{4}\right) - 3\left(\frac{9x-14}{4}\right)^2 - 1 = 0$$

$$\Rightarrow \quad 175x^2 - 700x + 604 = 0 \qquad (3)$$

If the given chord (1), meets the curve (2) at the points $(x_1, y_1)$ and $(x_2, y_2)$, then (3) implies

$$x_1 + x_2 = 4 \Rightarrow \frac{x_1 + x_2}{2} = 2 . \tag{4}$$

Now these points lie on (1),

$$\therefore \quad 9x_1 - 4y_1 - 14 = 0$$

and

$$9x_2 - 4y_2 - 14 = 0$$

$$\Rightarrow \quad 9(x_1 + x_2) - 4(y_1 + y_2) - 28 = 0$$

$$\Rightarrow \quad 9.4 - 28 = 4(y_1 + y_2), \qquad \text{using (4)}$$

$$\Rightarrow \quad \frac{y_1 + y_2}{2} = 1 . \tag{5}$$

By (4) and (5), the middle point of the given chord is (2, 1).

**Example 8.18**. Show that the locus of the middle points of the normal chords of the ellipse, $\frac{x^2}{a^2} + \frac{y^2}{b^2} = 1$ is

$$\left(\frac{x^2}{a^2} + \frac{y^2}{b^2}\right)^2 \left(\frac{a^6}{x^2} + \frac{b^6}{y^2}\right) = (a^2 - b^2)^2 .$$

*Solution*. We have to find the locus of the middle points of those chords which are also normal to the ellipse.

Let $(\alpha, \beta)$ be the point, whose locus is required. As it is the middle point of the chord, the equation of the chord is

$$T = S'$$

$$\Rightarrow \quad \frac{x\alpha}{a^2} + \frac{y\beta}{b^2} - 1 = \frac{\alpha^2}{a^2} + \frac{\beta^2}{b^2} - 1.$$

$$\Rightarrow \quad \frac{x\alpha}{a^2} + \frac{y\beta}{b^2} = A, \tag{1}$$

where

$$A = \frac{\alpha}{a^2} + \frac{\beta^2}{b^2}$$

But this chord is also a normal.

Now, the equation of any normal to the ellipse

$$\frac{x^2}{a^2} + \frac{y^2}{b^2} = 1$$

at ($a \cos \theta$, $b \sin \theta$) is

$$a \sec \theta . x - b \operatorname{cosec} \theta . y = a^2 - b^2 \tag{2}$$

(see § 8.3.1)

As (1) and (2) being the same, on comparing their eoefficients, we get

$$\frac{\alpha}{a^3 \sec \theta} = \frac{-\beta}{b^3 \operatorname{cosec} \theta} = \frac{A}{a^2 - b^2}$$

$$\Rightarrow \qquad \cos \theta = \frac{a^3 A}{\alpha(a^2 - b^2)}$$

and $$\sin \theta = \frac{-b^3 A}{\beta(a^2 - b^2)}$$

$$\Rightarrow \qquad 1 = \cos^2 \theta + \sin^2 \theta$$

$$= \frac{a^6 A^2}{\alpha^2 (a^2 - b^2)^2} + \frac{b^6 A^2}{\beta^2 (a^2 - b^2)^2}$$

$$\Rightarrow \qquad A^2 \left( \frac{a^6}{\alpha^2} + \frac{b^6}{\beta^2} \right) = (a^2 - b^2)^2$$

$$\Rightarrow \left( \frac{\alpha^2}{a^2} + \frac{\beta^2}{b^2} \right)^2 . \left( \frac{a^6}{\alpha^2} + \frac{b^6}{\beta^2} \right) = (a^2 - b^2)^2.$$

$\therefore$ the locus of $(\alpha, \beta)$ is

$$\left( \frac{x^2}{a^2} + \frac{y^2}{b^2} \right)^2 \left( \frac{a^6}{x^2} + \frac{b^6}{y^2} \right) = (a^2 - b^2)^2.$$

**Example 8.19.** Show that the locus of the middle points of the chords of the circle,

$$x^2 + y^2 = a^2,$$

which subtend a right angle at $(c, 0)$ is the circle

$$2(x^2 + y^2) - 2cx + (c^2 - a^2) = 0.$$

*Solution.*

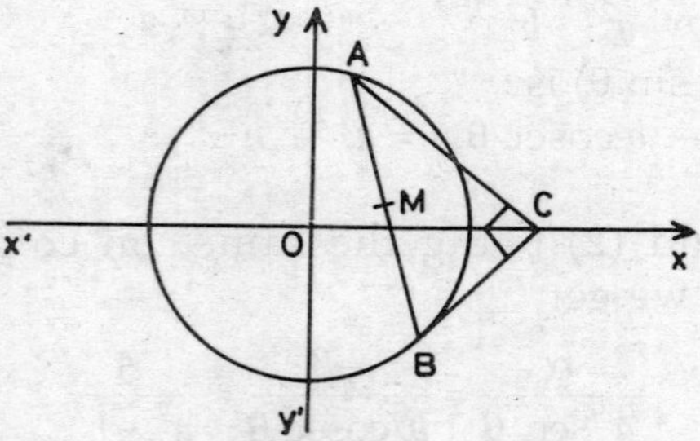


Let $M(\alpha, \beta)$ be the middle point of any chord $AB$ of the circle,

$$x^2 + y^2 = a^2 \qquad (1)$$

By hypothesis, $AB$ subtends a right angle at $C(c, 0)$ i.e.

$$\angle ACB = 90^\circ.$$

We want to find the locus of $(\alpha, \beta)$.

Now the equation of the chord $AB$, with middle point at $(\alpha, \beta)$ is

$$T = S'$$

$$\Rightarrow \quad x\alpha + y\beta = \alpha^2 + \beta^2 \qquad (2)$$

Also the equation of any circle through $A$ and $B$ (the points of intersection of (1) and (2)) is given by

$$x^2 + y^2 - a^2 + \lambda(x\alpha + y\beta - \alpha^2 - \beta^2) = 0. \qquad (3)$$

Its centre is

$$\left(-\frac{\lambda\alpha}{2}, -\frac{\lambda\beta}{2}\right)$$

As $\angle ACB = 90^\circ$, $AB$ is the diameter of the circle (3), $C$ lies on it and the middle point of $AB$ is the centre of (3),

whence $$(\alpha, \beta) = \left(-\frac{\lambda\alpha}{2}, -\frac{\lambda\beta}{2}\right)$$

$$\Rightarrow \lambda = -2.$$

$\therefore$ (3) reduces to

$$x^2 + y^2 - 2\alpha x - 2\beta y + (2\alpha^2 + 2\beta^2 - a^2) = 0.$$

and $C(c, 0)$ lies on it

$$\therefore c^2 - 2\alpha c + (2\alpha^2 + 2\beta^2 - a^2) = 0$$

Thus the focus of $(\alpha, \beta)$ is

$$2(x^2 + y^2) - 2cx + (c^2 - a^2) = 0.$$

**Example 8.20.** Show that the locus of the mid points of the chords of contact of tangents to a given circle drawn from points on another circle is a third circle.

*Solution.* Let the given circle be

$$x^2 + y^2 = a^2 \tag{1}$$

Suppose another given circle be

$$x^2 + y^2 + 2gx + 2fy + c = 0 \tag{2}$$

Let $(x_1, y_1)$ be any point on (2).

$$\therefore x_1^2 + y_1^2 + 2gx_1 + 2fy_1 + c = 0 \tag{3}$$

Now the equation of the chord of contact of tangents from $(x_1, y_1)$ to (1) is

$$xx_1 + yy_1 = a^2 \tag{4}$$

Let $(\alpha, \beta)$ be the mid point of this chord of contact.

$\therefore$ Its equation is

$$T = S_1$$

$$\Rightarrow \quad x\alpha + y\beta - a^2 = \alpha^2 + \beta^2 - a^2$$

$$\Rightarrow \quad x\alpha + y\beta = \alpha^2 + \beta^2 \tag{5}$$

As (4) and (5) represent the same line; comparing their co-efficients, we get

$$\frac{x_1}{\alpha} = \frac{y_1}{\beta} = \frac{a^2}{\alpha^2 + \beta^2}$$

$$\Rightarrow x_1 = \frac{a^2\alpha}{\alpha^2 + \beta^2}; \; y_1 = \frac{a^2\beta}{\alpha^2 + \beta^2}$$

Substituting these values of $x_1, y_1$ in (3), we get

$$\frac{a^4\alpha^2}{(\alpha^2 + \beta^2)^2} + \frac{a^4\beta^2}{(\alpha^2 + \beta^2)^2} + \frac{2ga^2\alpha}{\alpha^2 + \beta^2} + \frac{2fa^2\beta}{\alpha^2 + \beta^2} + c = 0$$

$$\Rightarrow \quad \frac{a^4}{\alpha^2 + \beta^2} + \frac{2(g\alpha + f\beta)a^2}{\alpha^2 + \beta^2} + c = 0$$

$$\Rightarrow \quad c(\alpha^2 + \beta^2) + 2(g\alpha + f\beta)a^2 + a^4 = 0$$

$\therefore$ Locus of $(\alpha, \beta)$ is

$$c(x^2 + y^2) + 2a^2(gx + fy) + a^4 = 0$$

which is clearly a circle.

**Exercise 8.6.**

1. Repeat § 8.6 for the following curves:
   (i) circle: $x^2 + y^2 = a^2$
   (ii) parabola: $y^2 = 4ax$.
   (iii) hyperbola: $\dfrac{x^2}{a^2} - \dfrac{y^2}{b^2} = 1$.
   (iv) rectangular hyperbola: $xy = c^2$.
   (v) ellipse: $\dfrac{x^2}{a^2} + \dfrac{y^2}{b^2} = 1$.
2. Find the equation of the chord of the following curves with given middle point:
   (i) $x^2 + y^2 = 9$, $(1, -2)$.
   (ii) $y^2 = 8x$, $(2, -3)$.
   (iii) $y^2 + 4x = 0$, $(-2, 2)$.
3. Find the middle points of the following chords of the given curves:
   (i) $4x - 3y + 1 = 0$, $y^2 = 8x$.
   (ii) $y + 1 = 0$, $x^2 + 2y^2 - 4x - 8 = 0$.
   (iii) $2x - y - 6 = 0$, $x^2 - y^2 = 5$.
   (iv) $lx + my = n$, $x^2 + y^2 = a^2$.
4. Find the locus of the middle points of the chords of the following curves which pass through a given point $(x', y')$:
   (i) $y^2 = 4ax$,
   (ii) $\dfrac{x^2}{a^2} + \dfrac{y^2}{b^2} = 1$,
   (iii) $\dfrac{x^2}{a^2} - \dfrac{y^2}{b^2} = 1$,
   (iv) $xy = c^2$,
   (v) $x^2 + y^2 = a^2$,

(vi) $ax^2 + 2hxy + by^2 + 2gx + 2fy + c = 0$.

5. Find the locus of the middle points of the normal chords of the following curves:
(i) $y^2 = 4ax$,
(ii) $\frac{x^2}{a^2} - \frac{y^2}{b^2} = 1$,
(iii) $x^2 - y^2 = a^2$,
(iv) $xy = c^2$.

6. Find the locus of the middle points of the chords of the
(i) circle: $x^2 + y^2 = a^2$,
(ii) ellipse: $\frac{x^2}{a^2} + \frac{y^2}{b^2} = 1$,
(iii) hyperbola: $\frac{x^2}{a^2} - \frac{y^2}{b^2} = 1$,
(iv) rectangular hyperbola: $xy = c^2$
which subtend a right angle at the centre.

7. Find the locus of the middle points of the chords of a parabola, $y^2 = 4ax$, which subtend a right angle at its vertex (0, 0). Show that all such chords, $PQ$, pass through a fixed point. Also show that the normals at $P$ and $Q$ meet on the parabola, $y^2 = 16a(x - 6a)$.

8. Prove that the chord of a circle is at right angle to the diameter through its middle point.

9. Tangents are drawn to the circle, $x^2 + y^2 = a^2$, from a variable point $P$ on the line $x = b$. Prove that the locus of the middle points of the chords of contact, is the circle,

$$b(x^2 + y^2) = a^2x.$$

10. From points on the circle, $x^2 + y^2 = a^2$, tangents are drawn to the hyperbola, $x^2 - y^2 = a^2$; prove that the locus of middle points of the chords of contact is the curve

$$(x^2 - y^2)^2 = a^2(x^2 + y^2).$$

11. A tangent to the parabola, $y^2 + 4bx = 0$, meets the parabola $y^2 = 4ax$ at $P$ and $Q$. Prove that the locus of the middle point of $PQ$ is $y^2(2a + b) = 4a^2x$.

12. Find the locus of the middle points of the chords of $y^2 = 4ax$, which touch the circle, $x^2 + y^2 = a^2$.

13. Find the locus of the middle points of a system of parallel chords of the following curves:
    (i) $x^2 + y^2 = a^2$,
    (ii) $ax^2 + 2hxy + by^2 + 2gx + 2fy + c = 0$,
    (iii) $y^2 = 4ax$,
    (iv) $\frac{x^2}{a^2} + \frac{y^2}{b^2} = 1$,
    (v) $\frac{x^2}{a^2} - \frac{y^2}{b^2} = 1$,
    (vi) $xy = c^2$.
    (The locus of the mid points of the system of parallel chords of a curve is called the *diameter* of the curve)
14. Prove that the locus of the middle points of the chords of constant length, $2d$, of the rectangular hyperbola, $xy = c^2$, is
    $$(x^2 + y^2)(xy - c^2) = d^2 xy.$$
15. Find the locus of the middle points of the chords of a circle which are of constant length.
16. Find the mid point of the chord of the circle, $x^2 + y^2 = a^2$ lying along the line, $px + qy = r$.
17. If $y = m(x - h)$ meets the parabola, $y^2 = 4ax$ in $P$ and $Q$, prove that the mid point of $PQ$ is $\left(h + \frac{2a}{m^2}, \frac{2a}{m}\right)$.
18. Find the locus of mid points of the chords of a circle $x^2 + y^2 = a^2$ which subtend a right angle at the point $(\alpha, \beta)$.
19. Show that the locus of the mid points of the chord of constant length, $2l$, of the ellipse, $\frac{x^2}{a^2} + \frac{y^2}{b^2} = 1$ is
    $$\left(\frac{x^2}{a^2} + \frac{y^2}{b^2} - 1\right)\left(\frac{x^2}{a^4} + \frac{y^2}{b^4}\right) + \frac{l^2}{a^2b^2}\left(\frac{x^2}{a^2} + \frac{y^2}{b^2}\right) = 0.$$

## 8.7. PAIR OF TANGENTS FROM A POINT TO A CURVE

We have seen in § 8.1.2. that there are two tangents from a point to a given curve. Now we propose to find their joint equation.

We illustrate the method of finding the joint equation of the pair of tangents from the point $(x', y')$ to the ellipse,

$$\frac{x^2}{a^2}+\frac{y^2}{b^2}=1,$$

$$\Leftrightarrow \qquad b^2x^2 + a^2y^2 - a^2b^2 = 0. \tag{1}$$

Let any line through $(x', y')$ be

$$\frac{x-x'}{\cos\,\theta}=\frac{y-y'}{\sin\,\theta}=r \tag{2}$$

Any point on it is $(x' + r\cos\theta,\ y' + r\sin\theta)$.

Suppose, it lies on (1),

$\therefore\ b^2(x' + r\cos\theta)^2 + a^2(y' + r\sin\theta)^2 - a^2b^2 = 0$

$\Rightarrow (b^2\cos^2\theta + a^2\sin^2\theta)r^2 + 2(b^2x'\cos\theta + a^2y'\sin\theta)r$

$$+ (b^2x'^2 + a^2y'^2 - a^2b^2) = 0 \tag{3}$$

Now the line given by (2) will be a tangent to (1), iff the roots of (3) (i.e. the two values of $r$ given by (3)) are equal, whence the discriminant of (3) is zero,

i.e. $4(b^2x'\cos\theta + a^2y'\sin\theta)^2$

$$- 4(b^2\cos^2\theta + a^2\sin^2\theta)(b^2x'^2 + a^2y'^2 - a^2b^2) = 0 \tag{4}$$

Eliminating $\theta$ from (2) and (4), we get

$$\left[\frac{b^2x'(x-x')}{r}+\frac{a^2y'(y-y')}{r}\right]^2$$

$$=\left[\frac{b^2(x-x')^2}{r^2}+\frac{a^2(y-y')^2}{r^2}\right](b^2x'^2+a^2y'^2-a^2b^2)$$

$\Rightarrow (b^2xx' + a^2yy' - b^2x'^2 - a^2y'^2)^2 = (b^2x^2 + a^2y^2 + b^2x'^2$

$+ a^2y'^2 - 2b^2xx' - 2a^2yy')(b^2x'^2 + a^2y'^2 - a^2b^2).$

Dividing both sides by $a^4b^4$, we get

$$\left[\frac{xx'}{a^2}+\frac{yy'}{b^2}-\left(\frac{x'^2}{a^2}+\frac{y'^2}{b^2}\right)\right]^2$$

$$=\left[\frac{x^2}{a^2}+\frac{y^2}{b^2}+\frac{x'^2}{a^2}+\frac{y'^2}{b^2}-2\left(\frac{xx'}{a^2}+\frac{yy'}{b^2}\right)\right]\left[\frac{x'^2}{a^2}+\frac{y'^2}{b^2}-1\right] \quad (5)$$

Writing $\qquad S=\frac{x^2}{a^2}+\frac{y^2}{b^2}-1,$

$$S'=\frac{x'^2}{a^2}+\frac{y'^2}{b^2}-1,$$

and $\qquad T=\frac{xx'}{a^2}+\frac{yy'}{b^2}-1,$

then (5) reduces to

$[(T + 1) - (S' + 1)]^2 = [(S + 1) + (S' + 1) - 2(T + 1)]S'$

$\Rightarrow (T - S')^2 = (S + S' - 2T)S'$

$\Rightarrow T^2 + S'^2 - 2TS' = SS' + S'^2 - 2TS'$

$\Rightarrow T^2 = SS'$, which is the required equation.

**Note**. We have taken the curve to be an ellipse. In fact, the same method can be applied for any other curve $S = 0$, to get the pair of tangents in the form $T^2 = SS'$, which is the required joint equation of the tangents from the point $(x', y')$ to the given curve.

**Example 8.21.** The angle between the two tangents to the parabola, $y^2 = 4ax$ is constant, and is equal to $\phi$. Prove that the locus of their point of intersection is given by

$$y^2 - 4ax = (a + x)^2.\ \tan^2\phi.$$

*Solution*. Let $(\alpha, \beta)$ be the point of intersection whose locus is required.

Then the equation of pair of tangents from it to the curve is given by

$T^2 = SS' \Rightarrow [y\beta - 2a(x + \alpha)]^2 = (y^2 - 4ax)(\beta^2 - 4a\alpha)$

$\Rightarrow \quad (\beta^2y^2 + 4a^2x^2 + 4a^2\alpha^2 - 4a\beta xy - 4a\alpha\ \beta y + 8a^2\alpha x)$

$$-(\beta^2y^2 - 4a\alpha y^2 - 4a\beta^2 x + 16a^2\alpha x) = 0$$

$$\Rightarrow \quad ax^2 - \beta xy + \alpha y^2 + (\beta^2 - 2a\alpha)x - \alpha\beta y + a\alpha^2 = 0 \qquad (1)$$

$\therefore$ the angle $\phi$, between these lines is given by

$$\tan\phi = \pm\frac{2\sqrt{h^2 - ab}}{a+b} = \pm\frac{2\sqrt{\beta^2/4 - a\alpha}}{a+\alpha}.$$

Squaring and simplifying, we get

$$\tan^2\phi\,(a+\alpha)^2 = (\beta^2 - 4\alpha a).$$

$\therefore$ the locus of $(\alpha, \beta)$ is

$$(a+x)^2\tan^2\phi = y^2 - 4ax.$$

**Corollary**. Find the locus of a point, from where perpendicular tangents can be drawn to the parabola, $y^2 = 4ax$.

*Solution*. By (1) of the above Example, for perpendicular tangents,

'Coefficient of $x^2$ + Coefficient of $y^2$' $= 0$

whence $\quad a + \alpha = 0.$

$\therefore$ the locus of $(\alpha, \beta)$ is

$$x + a = 0.$$

**Example 8.22.** Find the angle between the pair of tangents drawn from $(x_1, y_1)$ to the ellipse, $\frac{x^2}{a^2} + \frac{y^2}{b^2} = 1$. Hence find the locus of the vertices of equilateral triangles circumscribing the ellipse.

*Solution*. The equation of pair of tangents from $(x_1, y_1)$ to the ellipse $\frac{x^2}{a^2} + \frac{y^2}{b^2} = 1$ is given by

$$T^2 = SS'$$

$$\Rightarrow \quad \left(\frac{xx_1}{a^2} + \frac{yy_1}{b^2} - 1\right)^2 = \left(\frac{x^2}{a^2} + \frac{y^2}{b^2} - 1\right)\left(\frac{x_1^2}{a^2} + \frac{y_1^2}{b^2} - 1\right) \qquad (1)$$

The coefficient of $x^2 = \frac{x_1^2}{a^4} - \frac{1}{a^2}\left(\frac{x_1^2}{a^2} + \frac{y_1^2}{b^2} - 1\right)$

$$= \frac{1}{a^2}\left(1 - \frac{y_1^2}{b^2}\right)$$

The coefficient of $y^2 = \frac{1}{b^2}\left(1 - \frac{x_1^2}{a^2}\right)$

The coefficientof $2xy = \frac{x_1 y_1}{a^2 b^2}$.

If θ be the angle between the pair of tangents (1); then

$$\tan\theta = \frac{2\sqrt{h^2 - ab}}{a+b}$$

$$= \frac{2\sqrt{\frac{x_1^2 y_1^2}{a^4 b^4} - \frac{1}{a^2 b^2}\left(1 - \frac{y_1^2}{b^2}\right)\left(1 - \frac{x_1^2}{a^2}\right)}}{\frac{1}{a^2}\left(1 - \frac{y_1^2}{b^2}\right) + \frac{1}{b^2}\left(1 - \frac{x_1^2}{a^2}\right)}$$

$$= \frac{\sqrt{b^2 x_1^2 + a^2 y_1^2 - a^2 b^2}}{a^2 + b^2 - (x_1^2 + y_1^2)}$$

If the tangents form sides of an equilateral triangle, the angle between them is 60°.

$\therefore \quad \sqrt{3}\ (a^2 + b^2 - x_1^2 - y_1^2) = 2\sqrt{b^2 x_1^2 + a^2 y_1^2 - a^2 b^2}$

⇒ the locus of $(x_1, y_1)$ is
$4(b^2x^2 + a^2y^2 - a^2b^2) = 3(x^2 + y^2 - a^2 - b^2)^2$.

## Exercise 8.7.

1. Find the joint equation of the pair of tangents from the point $(x', y')$ to the following curves:
   (i) $x^2 + y^2 = a^2$
   (ii) $y^2 = 4ax$.
   (iii) $\frac{x^2}{a^2} - \frac{y^2}{b^2} = 1$.

(iv) $xy = c^2$.
(v) $x^2 + y^2 + 2gx + 2fy + c = 0$.
(vi) $ax^2 + 2hxy + by^2 + 2gx + 2fy + c = 0$.

2. Show that the equation of the pair of tangents drawn from the origin to the circle
$$x^2 + y^2 + 2gx + 2fy + c = 0 \text{ is}$$
$$(gx + fy)^2 = c(x^2 + y^2).$$
Hence find the locus of the centre of the circle if these tangents are perpendicular.

3. Write down the equation ofthe pair of tangents from
(i) $(4, -5)$ to the circle $x^2 + y^2 = 1$.
(ii) $(-1, -2)$ to the parabola $y^2 = 16x$.

4. Find the locus of a point from where a pair of tangents intersecting at an angle $\theta$ can be drawn to the following curves:
(i) $x^2 + y^2 = a^2$.
(ii) $x^2/a^2 + y^2/b^2 = 1$.
(iii) $x^2/a^2 - y^2/b^2 = 1$.
(iv) $xy = c^2$.

5 Find the locus of a point from where a pair of perpendicular tangents can be drawn to the following curves:
(i) $x^2 + y^2 = a^2$.
(ii) $x^2/a^2 + y^2/b^2 = 1$.
(iii) $x^2/a^2 - y^2/b^2 = 1$.
(**Definition**. *Director Circle*. If the locus of a point tangents from which to a curve, are perpendicular, is a circle, it is called the *director circle* of the curve).

6. Find the locus of the middle points of the chords of contact of the tangents to the ellipse, $x^2/a^2 + y^2/b^2 = 1$, from points on its director circle.
(***Hint***: By Question 5(ii), the equation of the director circle of the given ellipse is $x^2 + y^2 = a^2 + b^2$).

7. A tangent to the ellipse $x^2/a^2 + y^2/b^2 = 1$ meets the ellipse $x^2/a + y^2/b = a + b$ in the points $P$ and $Q$. Prove that the tangents at $P$ and $Q$ are at right angles.
(*Hint*: Let the tangents at $P$ and $Q$ to the given ellipse, meet at $R$. Then $PQ$ is the chord of contact of $R$ w.r.t. it).

8. Show that from a point $(x_1, y_1)$, two tangents are drawn to a circle, $x^2 + y^2 = a^2$, (parabola, ellipse) which are real and distinct, real and coincident or imaginary, according as the point is outside, on or inside the circle, (parabola, ellipse).
9. If 60° be the angle between the tangents drawn from any point $P$ to the circle, $x^2 + y^2 + 2gx + 2fy + c = 0$; find the locus of $P$
10. From any point on the circle, $x^2 + y^2 + 2gx + 2fy + c = 0$, tangents are drawn to the circle, $x^2 + y^2 + 2gx + 2fy + c \sin^2 \alpha + (g^2 + f^2) \cos^2 \alpha = 0$. Prove that the angle between them is $2\alpha$.
11. If $2\alpha$ be the angle between the tangents drawn from the origin to the circle, $x^2 + y^2 + 2gx + 2fy + c = 0$, show that $\cos^2 \alpha = |C|/(g^2 + f^2)$.
12. Show that the angle between the tangents from any point $(x_1, y_1)$ to the circle, $x^2 + y^2 = a^2$ is $\tan^{-1}\left[\dfrac{2a\sqrt{x_1^2 + y_1^2 - a^2}}{x_1^2 + y_1^2 - 2a^2}\right]$. Hence prove that the locus of the point of intersection of perpendicular tangents is $x^2 + y^2 = 2a^2$.

## Miscellaneous Exercise on Chapter 8

1. $A$ and $B$ are two points such that the polar of $A$ w.r.t. a circle passes through $B$; $M$ is the mid point of $AB$; show that the tangents from $M$ to the circle are of length $MA$.
2. If $PQR$ is a variable chord of a circle through a given point $P$, $S$ is taken on the chord such that $PS$ is the harmonic mean between $PQ$ and $PR$. Prove that the locus of $S$ is the polar of $P$ w.r.t the circle.
3. Find the locus of a point the polars of which w.r.t. two given circles, make a given angle with one another.
4. Three circles cut each other orthogonally. Prove that the centres of any two are conjugate w.r.t. the third circle.

5. A certain point has the same polar w.r.t. each of the two given circles; prove that a common tangent subtends a right angle at this point.
6. Prove that the polar of a given point $P$ w.r.t. any member of a coaxial system passes through a fixed point $Q$ and $PQ$ subtends a right angle at either limiting point.
7. Show that there are only two points each of which has the same polar w.r.t. every circle of a coaxial system and that they are the limiting points of the system.
8. The polars of a point $P$ w.r.t. two fixed circles meet in the point $Q$. Prove that the circle on $PQ$ as diameter passes through two fixed points and cuts both the given circles at right angles. Show also that the radical axis of the circles bisects $PQ$.
9. If a circle cuts each of the two circles: $x^2 + y^2 + 2g_ix + c = 0$ (i = 1, 2) orthogonally, show that it belongs to a coaxial system.
10. Show that the equation to the pair of tangents drawn from the origin to the circle, $x^2 + y^2 + 2gx + 2fy + k^2 = 0$ is
$$(gx + fy)^2 = k^2(x^2 + y^2).$$
Hence find the locus of the centre of the circle if these tangents are perpendicular. Also deduce that the tangents intercept on the line $y = h$ is $2hk/(k^2 - g^2)$ times the radius of the circle.
11. If the tangents drawn from an external point to the conic, $ax^2 + 2hxy + by^2 + 2gx + 2fy + c = 0$, be at right angles, show that the locus of their point of intersection is the circle,
$$(ab - h^2)(x^2 + y^2) - 2(hf - bg)x - 2(gh - af)y + c(a^2 + b^2) - f^2 - g^2 = 0.$$
Show also that this circle is concentric with the conic.
12. $P$ is the pole of a chord, $AB$ of a parabola, $y^2 = 4ax$; prove that the perpendiculars from $A$, $P$, $B$ on any tangent to the parabola are in geometric progression.
13. A variable secant is drawn through a fixed point $P(x_1, y_1)$ to meet the parabola, $y^2 = x$ in $R$ and $S$. A point $T$ is taken on this secant such that

$$\frac{1}{PT} = \frac{1}{PR} + \frac{1}{PS}.$$

Prove that $T$ moves on the line,

$$x - 2yy_1 + y_1^2 = 0.$$

14. Show that if a chord of the parabola, $y^2 = 4ax$ touches the parabola, $y^2 = 4bx$, tangents at its extremities meet on the parabola, $by^2 = 4a^2x$ and the normals on the curve $(4a - b)^3 y^2 = 4b^2 (x - 2a)^3$.
15. Show that the locus of the mid points of chords of length $2l$ of the parabola, $y^2 = 4ax$ is the curve $(y^2 + 4a^2)(y^2 - 4ax) + 4a^2l^2 = 0$.
16. From a point $P$ $(c \cos \theta, c \sin \theta)$, the tangents $PA$, $PB$ are drawn to the ellipse, $\frac{x^2}{a^2} + \frac{y^2}{b^2} = 1$. Show that as $\theta$ varies, the mid point of $AB$ describes the curve

$$\left(\frac{x^2}{a^2} + \frac{y^2}{b^2}\right)^2 = \frac{x^2 + y^2}{d^2}.$$

17. Show that the tangents at the extremities of all chords of the ellipse, $\frac{x^2}{a^2} + \frac{y^2}{b^2} = 1$, which subtend a right angle at the centre, intersect on the ellipse, $\frac{x^2}{a^4} + \frac{y^2}{b^4} = \frac{1}{a^2} + \frac{1}{b^2}$.
18. The polar of a variable point w.r.t. the ellipse, $\frac{x^2}{a^2} + \frac{y^2}{b^2} = 1$, always touches a concentric circle of radius $ab$. Prove that its locus is the ellipse, $\frac{b^2x^2}{a^2} + \frac{a^2y^2}{b^2} = 1$.
19. Show that the locus of the pole of any tangent of the circle, $x^2 + y^2 = a^2$ w.r.t. the hyperbola, $x^2 - y^2 = a^2$ is the circle itself.
20. The locus of the mid points of chords of

$\frac{x^2}{a^2} - \frac{y^2}{b^2} = 1$ which touch $\frac{x^2}{a^2} + \frac{y^2}{b^2} = 1$ is the curve -

$$\frac{x^2}{a^4} + \frac{y^2}{b^4} = \left(\frac{x^2}{a^2} - \frac{y^2}{b^2}\right)^2.$$

# 9. CONIC SECTIONS

## INTRODUCTION

So far, we have studied the shapes of the curves, viz., circle, a parabola, an ellipse and a hyperbola, and their common properties. We shall now be concentrating on their individual properties in the sequal. Geometrically, these curves can be obtained as sections of a right circular cone, by a plane in various positions and that's why these are called conic sections. We explain this in § 9.2.

A conic (or a conic section) is given by a second degree equation. Thus a conic means any one of a circle, a parabola, an ellipse or a hyperbola. The last three can be obtained as the locus of a point moving in a special way. We take up this study in this and the following chapters.

## 9.1. CONIC (OR A CONIC SECTION)

**Definition**. A *conic* (or a *conic section*) is the locus of a point, which moves in a plane such that its distance from a fixed point bears a constant ratio to its distance from a fixed line.

The fixed point is called the *focus* and the fixed line is called the *directrix* of the conic. Further the constant ratio is known as the *eccentricity* of the conic and it is denoted by the letter, $e$.

The conic is an *ellipse*, a *parabola* or a *hyperbola* according as $e <, =$ or $> 1$.

Thus, a conic is the locus of a point, which moves such that its distance from the focus is $e$ times its distance from the directrix.

**Example 9.1.** Find the locus of a point, which moves such that its distance from the point (2, 3) is twice its distance from the line $3x - 4y + 1 = 0$. Also name the conic.

*Solution*. Let $(\alpha, \beta)$ be the point, whose locus is required.

Its distance from (2, 3) is $\sqrt{(\alpha-2)^2+(\beta-3)^2}$ and its distance from the line, $3x - 4y + 1 = 0$, is

$$\left|\frac{3\alpha-4\beta+1}{\sqrt{9+16}}\right| = \left|\frac{3\alpha-4\beta+1}{5}\right|.$$

By hypothesis,

$$\sqrt{(\alpha-2)^2+(\beta-3)^2} = 2\left|\frac{3\alpha-4\beta+1}{5}\right|.$$

Squaring and simplifying, we have

$$11\alpha^2 - 96\alpha\beta + 39\beta^2 + 124\alpha + 118\beta - 321 = 0.$$

Thus the locus of $(\alpha, \beta)$ is

$$11x^2 - 96xy + 39y^2 + 124x + 118y - 321 = 0.$$

As $e = 2 > 1$, this conic is a hyperbola.

**Example 9.2.** Find the locus of a point which moves so that its distance from the point $(ae, 0)$ is $e$ times its distance from the line, $x = a/e$, where $e < 1$.

*Solution*. If $(x, y)$ be any such point, whose locus is to be found out, then we have, by hypothesis.

$$\sqrt{(x-ae)^2+y^2} = e.\left|x-\frac{a}{e}\right| = |ex-a|$$

$$\Rightarrow \quad x^2 - 2aex + a^2e^2 + y^2 = e^2x^2 - 2aex + a^2$$

$$\Rightarrow \quad (1 - e^2)x^2 + y^2 = a^2(1 - e^2)$$

$$\Rightarrow \quad \frac{x^2}{a^2}+\frac{y^2}{a^2(1-e^2)} = 1$$

which is the required locus.

Write: $a^2(1 - e^2) = b^2$ (note: $e < 1$).

$\therefore$ (1) becomes

$$\frac{x^2}{a^2}+\frac{y^2}{b^2} = 1.$$

The conic, in this case, is an ellipse.

**Example 9.3.** What conic does the equation

$$25[(x-2)^2+(y-1)^2]=(4x-3y-1)^2$$

represent?

*Solution.* We write the right side of the given equation, so that it appears in perpendicular distance form, i.e.

$$25[(x-2)^2+(y-1)^2]=25\left(\frac{4x-3y-1}{\sqrt{4^2+3^2}}\right)^2$$

$$\Rightarrow \quad \sqrt{(x-2)^2+(y-1)^2} = \left|\frac{4x-3y-1}{\sqrt{4^2+3^2}}\right|.$$

Here $e = 1$.

Thus the given equation represents a parabola. It may further be noted that (2, 1) is the focus and the line, $4x - 3y - 1 = 0$ is the directrix of the paraboia.

## Exercise 9.1.

1. Find the equations of the conics, whose foci, directrices and eccentricities are:

   (i) $(5, -4)$, $2x + 3y - 1 = 0$, $e = 1$;

   (ii) $(0, 0)$, $4x - 5y + 2 = 0$, $e = \frac{1}{2}$;

   (iii) $(0, -1)$, $x - 5 = 0$, $e = 2$;

   (iv) $(2, 0)$, $y - 2 = 0$, $e = 1$;

   (v) $(1, 2)$, $2x + y = 1$, $e = \sqrt{3}$.

   Name the conic in each case.

2. Find the equations of the conics, whose foci, directrices and eccentricities are:

   (i) $(-ae, 0)$, $x = -a/e$, $e < 1$;

   (ii) $(a, 0)$, $x = -a$, $e = 1$;

   (iii) $(0, -a)$, $y = a$, $e = 1$;

   (iv) $(ae, 0)$, $x = a/e$, $e > 1$;

   (v) $(-ae, 0)$, $x = -a/e$, $e > 1$.

   Also name the conic in each case.

3. Determine the conic represented by the following equations. Also find the eccentricity, coordinates of the focus and the equation of the directrix, in each case.
   (i) $13[(x - 1)^2 + (y - 2)^2] = 3(2x + 3y - 2)^2$,
   (ii) $25[(x^2 + y^2 - 2x + 1) = (4x - 3y + 1)^2$,
   (iii) $136(x^2 + y^2) = (5x + 3y + 7)^2$
4. Find the equation of the parabola, whose focus is $(-1, 1)$ and directrix is $x - y + 3 = 0$.
5. Find the equation to the ellipse whose focus is $(-1, 1)$, directrix is $x - y + 3 = 0$ and whose eccentricity is $\dfrac{1}{2}$.
6. Find the equation of the hyperbola whose focus is $(-1, 1)$, directrix is $x - y + 3 = 0$ and whose eccentricity is 2.
7. Find the equation of the parabola, whose focus is $(a, b)$, and the directrix, $(x/a) + (y/b) = 1$.
8. Find the equation of the ellipse with its focus $(4, -3)$, directrix, $x = -1$ and eccentricity, 2/3.
9. Find the equation of the hyperbola whose focus is $(0, 0)$, eccentricity 5/4 and directrix is $x \cos \alpha + y \sin \alpha = p$.

## 9.2 CONIC SECTIONS

As mentioned in the introduction of this chapter, a conic section (or simply a conic) can be obtained geometrically by taking the intersection of a right circular cone with a plane in various positions. In about 200 B.C., the *Greek mathematician Apollonious* studied the curves of intersection of a plane and a cone. Although, the methods at his disposal were elementary, he discovered many properties of these conic sections and was long known as "the great geometer".

The cone was thought of having two nappes, which extended indefinitely far in both directions. A line lying entirely on the cone is called a *generator* of the cone, and all generators of a cone pass through its *vertex*.

As studied by Apollonius, a *conic section* is the curve in which a plane cuts a right circular cone. There are *three types* of conic sections according as the cutting plane is parallel to no generator, to exactly one generator or two generators.

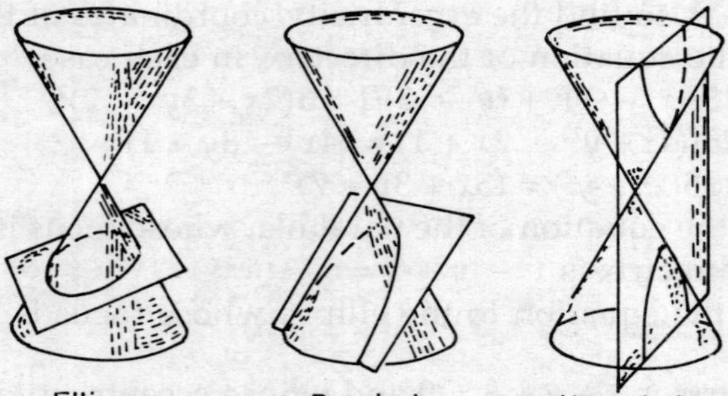


**Type I.** *The cutting plane is parallel to no generator.*
This plane, in fact, meets all the generators of the cone. The curve of intersection, in this case, is called an ellipse. (A circle is a limiting case of the ellipse, when the cutting plane is perpendicular to the axis of the cone). The ellipse reduces to a single point when the intersecting plane passes through the vertex of the cone and contains no generator.

**Type II**. *The cutting plane is parallel to exactly one generator.*
The curve of intersection, in this case, is called a *parabola*. If the intersecting plane passes through the vertex of the cone and it contains exactly one generator, the parabola, degenerates into a straight line. Further, if the vertex of the cone moves far away, the cone approaches the form of a right circular cylinder and the parabola has as a limiting case, two parallel lines.

**Type III**. *The cutting plane is parallel to two generators.*
The curve of intersection, in this case, is called a *hyperbola*. It degenerates into a pair of intersecting lines, when the intersecting plane passes through the vertex of the cone and contains the two generators.

These curves of intersection of a plane and a right circular cone are called conic sections, or simply *conics* (see the above figure).

The purpose of the foregoing historical introduction is to show why these curves came to be known as conic sections. We do not intend to take these intersections as the definition of the various conics. In fact, it involves the three-dimensional concept of a cone and we are confining our attention to plane analytic geometry. In 9.1, we have already defined conics, using only two-dimensional ideas. These two definitions are equivalent (see: Analytic Geometry: R.W. Brink, Appleton Century Crofts, New York).

# 10. PARABOLA

## INTRODUCTION

From Chapter 9, we know that if the eccentricity of a conic is equal to one, then the conic is called a parabola, i.e. it is the locus of points equidistant from the focus and the directrix. We, now, find its equation in the simplest possible form, called the *Standard Form* and study some of its properties in this chapter.

### 10.1. STANDARD FORM OF THE PARABOLA

Let $S$ be the focus and $AB$ be the directrix of a given parabola. Draw $ST \perp AB$ and let $O$ be the middle point of $ST$. Take the line $OS$ as the $x$-axis and the line through $O$, perpendicular to it as the $y$-axis. By definition of the parabola, $O$ lies on it.

Let $[OS] = a$.

Thus the co-ordinates of $S$ are $(a, 0)$ and the equation of the directrix, $AB$, is $x = -a$.

Let $P(x, y)$ be any point on the parabola. Applying the definition of the parabola, we have

$$\sqrt{(x-a)^2 + y^2} = |x + a|$$

$$\Rightarrow \quad x^2 - 2ax + a^2 + y^2 = x^2 + 2ax + a^2$$

$$\Rightarrow \quad y^2 = 4\,ax$$

which is the required equation of the parabola, in the simplest form. This is called the Standard Form of the parabola.

As in Chapter 7, the shape of the parabola is:

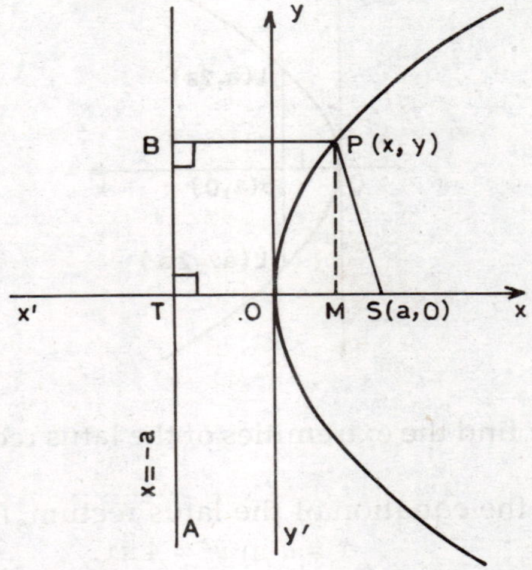


**Definitions.**

1. The point $O$, is called the *vertex* of the parabola. Also the tangent to the parabola, at the vertex is $x = 0$.
2. The line $OS$ i.e. the line joining the vertex, $O$ and the focus, $S$, of the parabola, is called the *axis of the parabola*. Clearly its equation is $y = 0$.
3. Any chord of the parabola, perpendicular to its axis, is called a *double ordinate*.
4. Any chord of the parabola, through its focus, is called a *focal chord*.
5. The double ordinate of the parabola, through its focus is called its *latus rectum*.

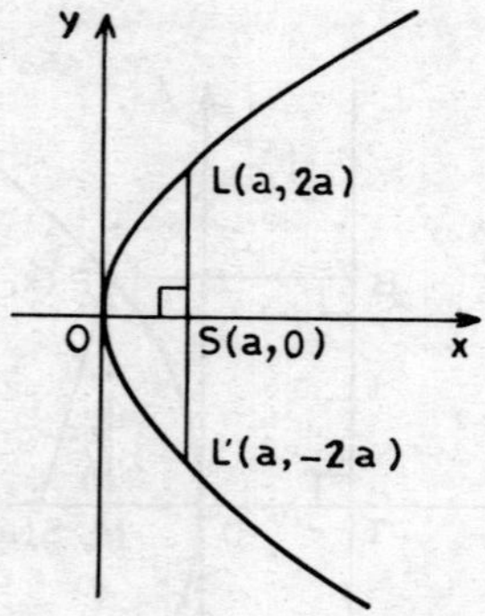


We now find the extremities of the latus rectum. Let these be $L'$ and $L$.

Clearly the equation of the latus rectum, $L'L$ is $x = a$.

Substitute $x = a$ in $y^2 = 4\,ax$.

This means, $y^2 = 4a^2 \Rightarrow y = \pm 2a$

$\therefore$ $L'(a, -2a)$ and $L(a, 2a)$.

Also $L'L$ = length of the latus rectum = $4a$.

*Remark.* The equation of the parabola in its simplest form is $y^2 = 4ax$

i.e. $MP^2 = 4a.OM$,

where $MP$ is the length of the perpendicular from $P$ on the axis of the parabola, $OM$ is the length of the perpendicular from $P$ on the tangent at the vertex $O$, of the parabola and $4a$ is the length of the latus rectum of the parabola.

Thus in general, if $lx + m\,y + n = 0$ $(n > 0)$ be the axis and $mx - ly + n' = 0$ $(n' > 0)$ be the tangent at the vertex and $l$ be the length of the latus rectum of parabola; then its equation is

$$\left(\frac{lx + my + n}{\sqrt{l^2 + m^2}}\right)^2 = l.\frac{|\,mx - ly + n'\,|}{\sqrt{l^2 + m^2}}$$

**Example 10.1.** Find the equation of the parabola whose focus is $S(1, 2)$ and directrix, $x - y + 2 = 0$. Also obtain the co-ordinates of the vertex, length and equation of the latus rectum and the axis.

*Solution*. Let $P(x, y)$ be any point on the parabola. If $PM$ be the perpendicular from $P$ on the directrix, $x - y + 2 = 0$; then by definition

$SP = PM$ (Here $S$ is $(1, 2)$)

$$\Rightarrow \quad SP^2 = PM^2$$

$$\Rightarrow \quad (x-1)^2 + (y-2)^2 = \left(\frac{x-y+2}{\sqrt{2}}\right)^2$$

$$\Rightarrow \quad x^2 + 2xy + y^2 - 8x - 12y + 6 = 0.$$

The axis is the line through $S(1, 2)$, perpendicular to the directrix, $x - y + 2 = 0$.

$\therefore$ Its equation is $x + y = 3$.

Their point of intersection is $T\left(\frac{1}{2}, \frac{5}{2}\right)$.

Now the vertex, being the mid-point of $ST$, the co-ordinates of the vertex are $A\left(\frac{3}{4}, \frac{9}{4}\right)$.

Length of the *latus rectum*

$$= 4.\, AS = 4\sqrt{\left(1-\frac{3}{4}\right)^2 + \left(2-\frac{9}{4}\right)^2} = \sqrt{2}$$

Also latus rectum is a line through the focus $S(1, 2)$ and is parallel to the directrix, $x - y + 2 = 0$.

$\therefore$ Its equation is $x - y + 1 = 0$.

**Example 10.2.** Find the latus rectum, co–ordinates of the vertex and the focus, equations of the axis and the tangent at the vertex of the parabola, $y^2 - 2x = 8y - 9$. Also trace it.

*Solution*. The given equation of the parabola is

$$y^2 - 2x = 8y - 9$$

$$\Rightarrow \quad y^2 - 8y = 2x - 9$$

$$\Rightarrow \quad (y-4)^2 = 2\left(x + \frac{7}{2}\right) \qquad (1)$$

Shifting the origin to $\left(-\frac{7}{2}, 4\right)$ by taking

$$x = X - \frac{7}{2} \text{ and } y = Y + 4,$$

equation (1) transforms to

$Y^2 = 4aX,$ where $4a = 2.$

$\therefore$ the equation of its axis is

$Y = 0$ i.e. $y - 4 = 0$

Also the equation of the tangent to the parabola at the vertex is

$$X = 0 \text{ i.e. } x + \frac{7}{2} = 0.$$

The vertex, $A$, of the parabola is the point of intersection of

$X = 0$ and $Y = 0$ i.e. $\left(-\frac{7}{2}, 4\right)$

Its latus rectum = $4a = 2$.

Its focus is given by

$$X = a, Y = 0 \text{ i.e. } x + \frac{7}{2} = \frac{1}{2}, y - 4 = 0 \text{ i.e. } (-3, 4).$$

Lastly the equation of the directrix of the parabola is

$$X = -a \text{ i.e. } x + \frac{7}{2} = -\frac{1}{2} \text{ i.e. } x = -4.$$

The parabola is shown below:

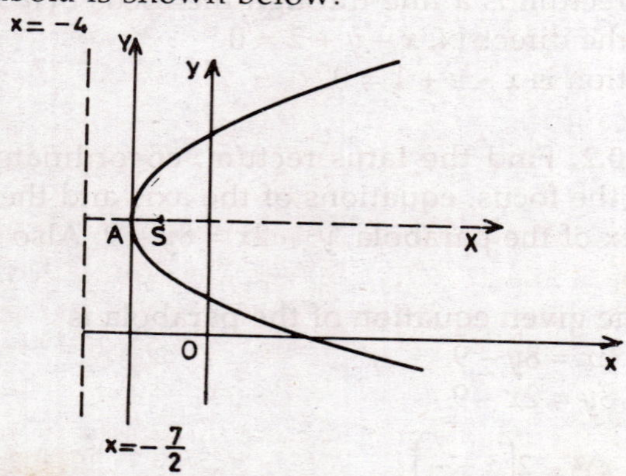


**Example 10.3.** Find the equation of the parabola whose focus is (4, –3) and vertex is (4, 1).

*Solution.* Let $A$ and $S$ denote the vertex and the focus of the parabola respectively.

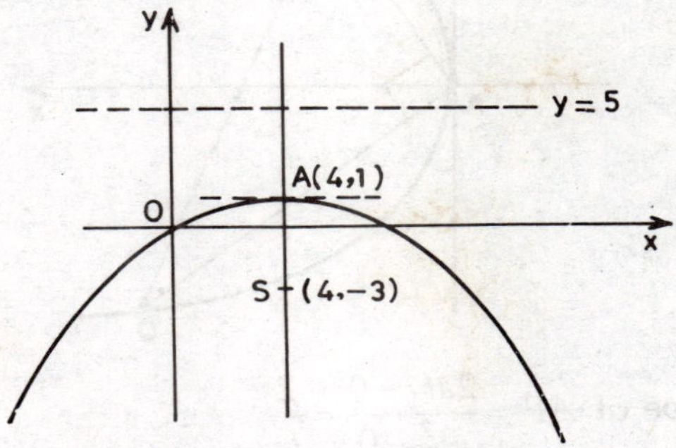


Then clearly

$[AS] = 4 = 'a'$

As the directrix is perpendicular to the axis, $AS$, of the parabola, its equation is

$y = 1 + 4$ i.e. $y = 5$.

Let $P(x, y)$ be any point on the parabola. By definition of the parabola, distance of $P$ from the directrix, $y = 5$ is $SP$.

i.e. $|y-5| = \sqrt{(x-4)^2 + (y+3)^2}$

$\Rightarrow (y - 5)^2 = (x - 4)^2 + (y + 3)^2$

$\Rightarrow (x - 4)^2 = -16(y - 1)$

$\Rightarrow x^2 - 8x + 16y = 0,$

which is the required equation of the parabola.

**Example 10.4.** Two lines $AP$, $AQ$ are drawn through the vertex $A$ of a parabola at right angles to one another, meeting it in $P$ and $Q$. Show that the line $PQ$ meets the axis in a fixed point.

*Solution.* Let the equation of the parabola be $y^2 = 4ax$

Let $P(at_1^2, 2at_1)$ and $Q(at_2^2, 2at_2)$ be the extremities of the chord $PQ$.

Also $A(0, 0)$.

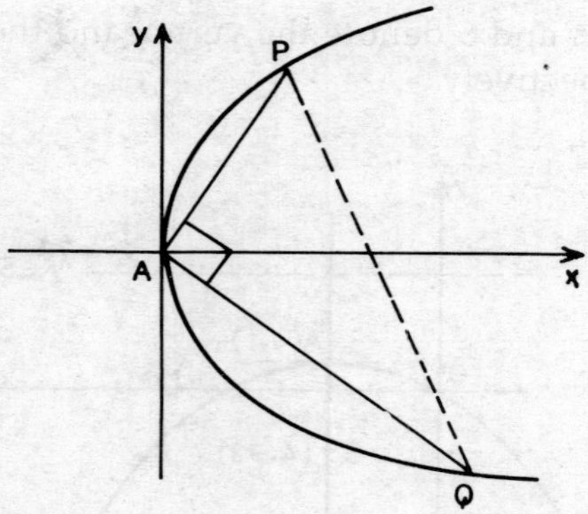


$\therefore$ Slope of $AP = \dfrac{2at_1 - 0}{at_1^2 - 0} = \dfrac{2}{t_1}$

and slope of $AQ = 2/t_2$

By hypothesis, $AP \perp AQ$.

$\therefore$ $\dfrac{2}{t_1} \cdot \dfrac{2}{t_2} = -1 \Rightarrow t_1 t_2 = -4.$

Now the equation of $PQ$ is

$$y - 2at_1 = \frac{2at_1 - 2at_2}{at_1^2 - at_2^2}(x - at_1^2)$$

$\Rightarrow$ $y(t_1 + t_2) = 2x + 2at_1t_2$

which meets the axis ($y = 0$) of the parabola, where

$x = -at_1t_2 = 4a$ ($\because t_1t_2 = -4$)

$\therefore$ $PQ$ meets the axis at $(4a, 0)$

which is a fixed point.

**Example 10.5.** If on a given base, a triangle be described such that the sum of the tangents of the base angles is constant, prove that the locus of the vertices is a parabola.

*Solution.*

Let $BC = 2a$ be the given base of the triangles.

Take $x$-axis along $BC$ and let $y$-axis be the right bisector of $BC$. Then the co-ordinates of $B$ and $C$ are $(-a, 0)$ and $(a, 0)$. Let the vertex $A$ be $(x, y)$.

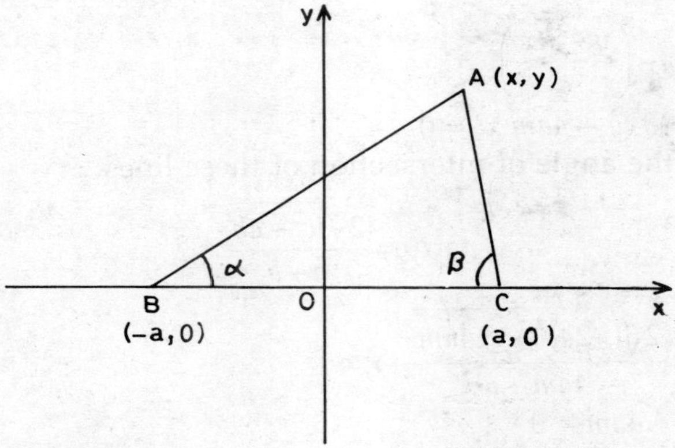


By hypothesis,

$$\tan\alpha + \tan\beta = k \text{ (a constant)}$$

$$\Rightarrow \quad \frac{y-0}{x+a} - \frac{y-0}{x-a} = k$$

$$\Rightarrow \quad y(x-a) - y(x+a) = k(x^2 - a^2)$$

$$\Rightarrow \quad k(x^2 - a^2) - 2ay = 0$$

which represents a parabola.

(See Q. 4 of Exercise 10.1).

**Example 10.6.** Chords of a parabola intersect the axis in a common point *P* which is on the same side of the vertex as focus. If the distance of *P* from the vertex be equal to the length of the latus rectum of the parabola, show that the chords subtend the same angle at the vertex.

*Solution.* Let $y = mx + c$ be the equation of a chord of the parabola, $y^2 = 4ax$.

Since the chords intersect the axis ($y = 0$) of the parabola in a common point *P*, therefore *P* is $(-c/m, 0)$

By hypothesis,

$OP = 4a \Rightarrow -c/m = 4a$

$\Rightarrow c + 4am = 0$ (Here *O* is the vertex).

Now the joint equation of the lines joining the vertex *O* to the points of intersection of $y^2 = 4ax$ and $y = mx + c$ is.

$$y^2 = 4ax\left(\frac{y-mx}{c}\right)$$

$\Rightarrow c\,y^2 - 4axy + 4am\,x^2 = 0.$

Let $\theta$ be the angle of intersection of these lines.

$$\tan\theta = \frac{2\sqrt{h^2-ab}}{a+b}$$

$$= \frac{2\sqrt{(-2a)^2 - c.4am}}{4am+c} \to \infty$$

as $\quad c + 4am = 0$

$\therefore \quad \theta = \pi/2$

$\Rightarrow$ The angle which any chord subtends at the vertex is same $(= \pi/2)$.

## Exercise 10.1

1. Find the latus rectum, co-ordinates of the vertex, focus, equations of the axis, tangent at the vertex, and directrix of the following parabolas. Also trace the curves.
   (i) $y^2 = 8x$. (ii) $x^2 = -4y$.
   (iii) $y^2 - 6y + 2x = 10$. (iv) $x^2 + 2y = 8x - 7$.
2. Find the equation of the parabola having the vertex at (0, 1) and the focus at (0, 0).
3. Find the equation of parabola, given
   (a) focus (5, 3); vertex (4, 3).
   (b) vertex (2, 2); directrix, $x = 6$.
   (c) directrix, $x = -2$; the lower extremity of the latus rectum (6, 2).
4. Prove that the equation of a parabola with vertex $(h, k)$ and axis parallel to the $x$-axis is
   $$(y-k)^2 = 4a(x-h).$$
   Find its focus and the directrix.
5. Find the co-ordinates of a point on the parabola $y^2 = 18x$, where the ordinate is equal to three times its abscissa.
6. Find the equation of the parabola in the standard form $(y^2 = 4ax)$, given that

(a) its focus is (2, 0).
(b) its directrix is $x = -4$.
(c) it passes through (3, –5).
(d) its semi latus rectum is 9.

7. Prove that the equation to the parabola whose vertex and the focus on the axis of $x$ are at distances $a$ and $b$ from the origin respectively, is $y^2 = 4(b - a)(x - a)$.
8. A double ordinate of the parabola, $y^2 = 4ax$, is of length $8a$. Show that the lines joining the vertex to its two ends are at right angles.
9. Find the vertex, the length of the latus rectum and the equations of the axis and tangent at the vertex of the parabola
$$(x - y + 2)^2 = 8\sqrt{2}\ (x + y - 6).$$
10. Find the equation of the parabola of latus rectum 10, having $3x - 4y + 2 = 0$ for its axis and $4x + 3y = 14$ as tangent at the vertex.
11. The focal distance of a point on the parabola, $y^2 = 8x$ is 4. Find its ordinates.
12. Prove that the locus of the centre of a circle which intercepts a chord of given length $2a$ on $x$-axis and passes through a given point on $y$-axis distant $b$ from the origin is the parabola, $x^2 - 2by + b^2 - a^2 = 0$.
13. Show that the parameters of the ends of the latus rectum of a parabola are $\pm$ 1.
14. A variable circle is described to pass through $(a, 0)$ and to touch the line, $y = x$. Show that the locus of the centres of the circles is a parabola and find the equation of its axis.
(*Hint*: See Chapter 13).
15. If through a fixed point $P$ on the axis of the parabola, any chord $Q'PQ$ be drawn, show that the product of ordinates (also abscissae) of $Q'$ and $Q$ will be constant.
16. Show that the length of the chord of the parabola, $y^2 = 4ax$ passing through the vertex and making an angle $\theta$ with the axis is $4\,a \cos\theta \operatorname{cosec}^2\theta$.
17. A line $PQ$ meets the parabola $y^2 = 4ax$ in $R$ such that $PQ$

is bisected at $R$. If the coordinates of $P$ be $(x_1, y_1)$, show that the locus of $Q$ is the parabola,

$$(y + y_1)^2 = 8a(x + x_1).$$

18. If $p_1$, $p_2$ be the lengths of two perpendicular chords of the parabola, $y^2 = 4ax$, drawn through the vertex, prove that

$$(p_1p_2)^{4/3} = 16\, a^2\, (p_1^{\,2/3} + p_2^{\,2/3}).$$

(*Hint*: Let the chord of length $p_1$ make an angle $\theta$ with $x$-axis, then the other chord will make an angle $90^\circ - \theta$ with it).

## 10.2 SUMMARY OF THE RESULTS FROM CHAPTER 8

We summarise here the various results concerning the parabola, $y^2 = 4ax$, for the benefit of the reader.

1. The parameteric equations of the parabola or the co-ordinates of any point on it are
$$x = at^2,\ y = 2at.$$
2. The tangent at $(x', y')$ to it is
$$yy' = 2a(x + x').$$
3. The tangent at $(at^2, 2at)$ to it is
$$ty = x + at^2.$$
4. The normal at $(at^2, 2at)$ to it is
$$y = -\,tx + 2at + at^3.$$
5. The condition that the line, $y = mx + c$, touches it is $c = \dfrac{a}{m}$, and therefore, the equation of any tangent to it (not parallel to the $y$-axis) is
$$y = mx + \frac{a}{m},$$
6. The chord of contact of $(x', y')$ w.r.t. it is
$$yy' = 2a(x + x').$$
7. The polar of $(x', y')$ w.r.t. it is
$$yy' = 2a(x + x').$$
8. The equation of the chord of $y^2 = 4ax$ with middle point is $(x', y')$

$$T = S',$$

where $S = y^2 - 4ax$, $S' = y'^2 - 4ax'$ and $T = yy' - 2a(x + x')$.

9. The equation of the pair of tangents from $(x', y')$ to $y^2 = 4ax$ is $(x', y')$ is $T^2 = SS'$.

**Example 10.7.** If the chord joining the points $(at_1^2, 2at_1)$ and $(at_2^2, 2at_2)$ on the parabola $y^2 = 4ax$, passes through the focus, show that $t_1t_2 = -1$.

*Solution.* Let $P$ be $(at_1^2, 2at_1)$ and $Q$ be $(at_2^2, 2at_2)$.

$\therefore$ the equation of the chord, $PQ$ is

$$y - 2at_1 = \left(\frac{2}{t_1 + t_2}\right)\left(x - at_1^2\right)$$

$$\Rightarrow \quad y(t_1 + t_2) - 2at_1t_2 = 2x.$$

It passes through the focus $(a, 0)$,

$$\therefore \quad -2at_1t_2 = 2a.$$

$$\Rightarrow \quad t_1t_2 = -1.$$

**Example 10.8.** A circle on any focal chord of a parabola as diameter cuts the curve again in $P$ and $Q$. Show that $PQ$ passes through a fixed point.

*Solution.* Let '$t_1$', '$t_2$' be the extremities of a focal chord of the parabola, $y^2 = 4ax$

Then $t_1t_2 = -1$ (by Example 10.7).

The equation of the circle described on the focal chord as diameter is

$$(x - at_1^2)(x - at_2^2) + (y - 2at_1)(y - 2at_2) = 0$$

$$\Rightarrow x^2 + y^2 - a(t_1^2 + t_2^2)x - 2ay(t_1 + t_2) - 3a^2 = 0$$

(using $t_1t_2 = -1$).

Let $(at_2, 2at)$ be a point on it.

$$\therefore \ a^2t^4 + 4a^2t^2 - a^2(t_1^2 + t_2^2)t^2 - 4a^2(t_1 + t_2)t - 3a^2 = 0$$

$$\Rightarrow t^4 + [4 - (t_1^2 + t_2^2)]t^2 - 4(t_1 + t_2)t - 3 = 0 \qquad (1)$$

This equation has four roots, $t_i$ ($i = 1, 2, 3, 4$) where '$t_3$', '$t_4$' correspond to $P$ and $Q$.

From (1), $t_1t_2t_3t_4 = -3$.

$$\Rightarrow \quad t_3t_4 = 3 \ (\because \ t_1t_2 = -1).$$

The equation of the chord $PQ$ is

$$(t_3 + t_4)y = 2x + 2a\, t_3 t_4$$

$$\Rightarrow \quad (t_3 + t_4)y = 2x + 6a$$

which always passes through $(-3a, 0)$.

## 10.3 SOME PROPOSITIONS ON THE PARABOLA

*Proposition* 1. The tangents at the extremities of a focal chord of a parabola intersect at right angles on the directrix.

*Proof.* Let "$t_1$" and "$t_2$" be the extremities of a focal chord. Then by Example 10.7, $t_1 t_2 = -1$.
Now, the equations of the tangents at "$t_1$" and "$t_2$" are

$$t_1 y = x + a t_1^2,$$
$$t_2 y = x + a t_2^2,$$

which have slopes $\dfrac{1}{t_1}$ and $\dfrac{1}{t_2}$ respectively.

Product of these slopes $= \dfrac{1}{t_1} \cdot \dfrac{1}{t_2} = -1$.

Also, the point of intersection of these tangents is

$$(x, y) = (at_1 t_2, a(t_1 + t_2)) = (-a, a(t_1 + t_2))$$

(See Question 8, Ex. 10.2)

which clearly lies on the directrix $x = -a$. Hence the proposition.

*Proposition* 2. The tangent at any point of a parabola bisects the angle between the focal distance of the point and the perpendicular on the directrix from the point.

*Proof.* Let $P(at^2, 2at)$ be any point on the parabola. The equation of the tangent at it is

$$ty = x + at^2$$

which has slope

$$= \frac{1}{t} = \tan \gamma$$

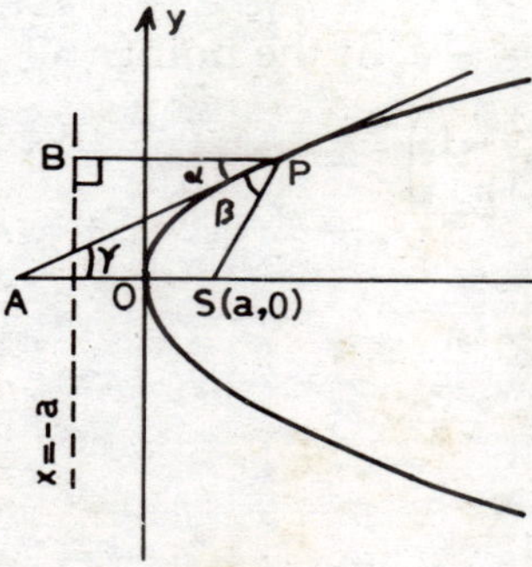


Also, the slope of the chord

$$SP = \frac{2t}{t^2 - 1}$$

$$\tan\beta = \frac{\frac{2t}{t^2-1} - \frac{1}{t}}{1 + \frac{2t}{t^2-1}\cdot\frac{1}{t}} = \frac{1}{t} = \tan\gamma$$

$\Rightarrow \quad \angle\beta = \angle\gamma$

But $\quad \angle\alpha = \angle\gamma$,

($\because PB$ is parallel to the x–axis).

$\therefore \quad \angle\alpha = \angle\beta$

Hence the result.

*Proposition* 3. The portion of a tangent, to a parabola, cut off between the directrix and the curve, subtends a right angle at the focus.

*Proof*. Let $P(at^2, 2at)$ be any point on the parabola. The equation of the tangent at $P$ is

$$ty = x + at^2$$

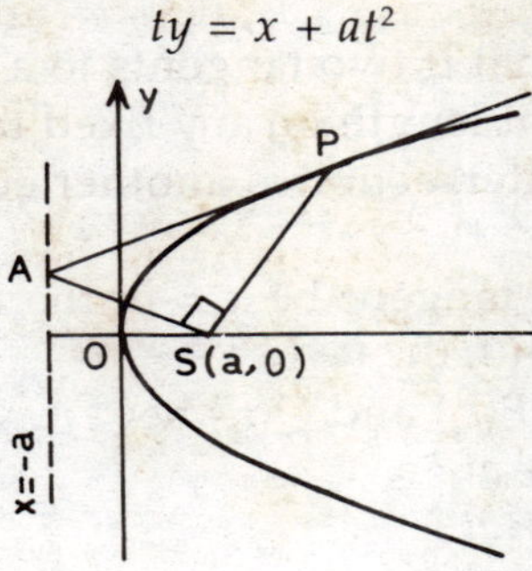

It meets the directrix, $x = -a$, at the point

$$A\left(-a, a.\frac{t^2-1}{t}\right)$$

Now the slope of $PS$ is

$$= \frac{2t}{t^2-1}$$

and the slope of $AS$ is

$$= -\frac{t^2-1}{2t}$$

Clearly their product $= -1$.
Hence the result.

*Proposition* 4. The foot of the perpendicular from the focus on any tangent to a parabola lies on the tangent at the vertex.

*Proof*. Equation of the tangent at any point '$t$' on the parabola, $y^2 = 4ax$ is

$$ty = x + at^2 \qquad (1)$$

Equation of the line through the focus $S(a, 0)$ perpendicular to (1) is

$$y = -tx + at \qquad (2)$$

Solving (1) and (2), we get

$$x = 0, y = at.$$

$\therefore$ The foot of the perpendicular $(0, at)$ lies on the $y$-axis which is the tangent at the vertex.

**Example 10.9.** Show that if two tangents to a parabola, $y^2 = 4ax$ intercept a constant length on any fixed tangent, the locus of their point of intersection is another equal parabola.

*Solution*. Let the fixed tangent be $x = 0$, the tangent to the parabola at the vertex $(0, 0)$.

Let the tangents at $P(t_1)$ and $Q(t_2)$ be

$$yt_1 = x + at_1^2 \qquad (1)$$

$$yt_2 = x + at_2^2 \qquad (2)$$

They meet at $T$ whose co-ordinates are

$$x' = at_1t_2,\ y' = a(t_1 + t_2) \qquad (3)$$

Let (1) and (2) meet $x = 0$ at $R$ and $S$ respectively, then

$$R(0, at_1) \text{ and } S\ (0, at_2)$$

By hypothesis,

$$RS = c \text{ (given)}$$

$$\Rightarrow \quad a^2 (t_1 - t_2)^2 = c^2$$

$$\Rightarrow \quad a^2 [(t_1 + t_2)^2 - 4\, t_1t_2] = c^2$$

$$\Rightarrow \quad a^2 [(y'/a)^2 - 4(x'/a)] = c^2$$

$\therefore$ Locus of $T(x', y')$ is

$$y^2 - 4ax = c^2$$

$$\Rightarrow \quad y^2 = 4a\left(x + \frac{c^2}{4a}\right)$$

which is an equal parabola (i.e. the latera recta of the given parabola and this parabola are of the same length).

## Exercise 10.2

1. If one end of a focal chord of the parabola, $y^2 = 4ax$, is $(at^2, 2at)$, find the co-ordinates of the other end. Also show that the length of this chord is, $a\left(t + \frac{1}{t}\right)^2$ and its equation is $(t^2 - 1)y = 2t(x - a)$.
2. Show that in a parabola, the length of the focal chord, varies inversely as the square of the distance of the vertex of the parabola from the focal chord.
3. If $l$, $l'$ be the lengths of the segments of any focal chord of the parabola, $y^2 = 4ax$, prove that

$$\frac{1}{l} + \frac{1}{l'} = \frac{1}{a}$$

4. Prove that the locus of the middle points of focal chords of the parabola, $y^2 = 4ax$, is $y^2 = 2a(x - a)$.
5. Prove that the circle described on any focal chord of a parabola as diameter touches the directrix of the parabola.
6. If the line, $y = 3x + 1$, touches the parabola, $y^2 = 4ax$, find the length of the latus rectum.

7. Find the equation of the tangents to the parabola, $y^2 = 4ax$ at the ends of its latus rectum and show that they intersect orthogonally on the directrix.
(*Hint*: See Proposition 1).
8. Find the point of intersection of the tangents at the points "$t_1$" and "$t_2$" on the parabola $y^2 = 4ax$. Also show that the ordinate of the point of their intersection is the arithmetic means between the ordinates of the points.
9. Show that the locus of the point of intersection of two perpendicular tangents to a parabola is the directrix.
10. Find the length of the focal chord of a parabola which makes an angle $\theta$ with the axis, given that $l$ is its semi–latus rectum.
11. Show that the equation of the circle described on the chord intercepted by the parabola, $y^2 = 4ax$ on the line, $y = mx + c$ as diameter is
$$m^2(x^2 + y^2) + 2(mc - 2a)x - 4amy + c(4am + c) = 0.$$
12. Prove that equal focal chords are equally inclined to the axis of the parabola.
13. If $PSP'$ is a focal chord of a parabola, $M$ is its mid-point and $MN$ is a line perpendicular to it, meeting the axis in $N$; prove that $SN$ and $MN$ are respectively $A.M.$ and $G.M.$ between the focal distances $SP$ and $S'P$.
14. $P$ and $Q$ are points on the parabola, $y^2 = 4ax$. The circle on $PQ$ as diameter cuts the parabola in $R$ and $S$. If the equation to $PQ$ be $lx + my + a = 0$; prove that $RS$ is the line, $lx - my + (1 + 4l)a = 0$.
15. Circles are described on any two focal chords of a parabola as diameters. Prove that their common chord passes through the vertex of the parabola.
16. The inclinations $\theta$ and $\phi$ of two tangents to the parabola, $y^2 = 4ax$ and given by $\tan\theta = 1/m$ and $\tan\phi = m/2$. Show that as m varies, their point of intersection traces a line parallel to the directrix.
17. If two tangents to a parabola intersect on its latus rectum, prove that they make complementary angles with its axis.
18. Two tangents to the parabola, $y^2 = 4ax$ make angles $\alpha$

and $\beta$ with the axis. Find the locus of their point of intersection if

(i) $\tan\alpha + \tan\beta = c$ (ii) $\tan\alpha . \tan\beta = c$

(iii) $\tan^2\alpha + \tan^2\beta = c$ (iv) $\alpha - \beta = c$

where c is a constant.

19. Prove that the subtangent at any point of a parabola is bisected at the vertex and the subnormal, at that point is always of constant length.

20. Show that the locus of the point of intersection of tangents to the parabola, $y^2 = 4ax$ at points whose

(i) abscissae are in the ratio $\mu : 1$ is

$$y^2 = (\mu^{1/4} + \mu^{-1/4})^2 ax$$

(ii) ordinates are in the ratio $p^2 : q^2$ is

$$y^2 = \left(\frac{p}{q} + \frac{q}{p}\right)^2 ax.$$

21. $P(h, k)$ is a fixed point in the plane of the parabola, $y^2 = 4ax$. Through $P$ a variable secant is drawn to cut the parabola in $Q$ and $R$. $T$ is a point on $QR$ such that

(i) $PQ.PR = OT^2$

(ii) $PQ + PR = OT$.

Find the locus of $T$ in each case.

(*Hint*. Take the variable secant in the form:

$$\frac{x-h}{\cos\theta} = \frac{y-k}{\sin\theta} = r)$$

22. Show that the locus of the point of intersection of the tangents to a parabola, $y^2 = 4ax$,

(i) which intercept a constant length $d$ on the directrix is

$$(y^2 - 4ax)(x + a)^2 = d^2x^2.$$

(ii) which form a triangle of constant area $c^2$ with the tangent at the vertex is

$$x^2(y^2 - 4ax) = 4c^4.$$

23. Two unequal parabolas have a common axis and concavities in opposite direction. If any line parallel to the common axis meets them in $P$ and $Q$, prove that the locus of the mid-point of $PQ$ is another parabola.

24. (a) Show that two parabolas having the same focus and

their axes in opposite directions intersect at right angles.
(b) Show that the two parabolas, $x^2 + 4a(y - a - 2b) = 0$ and $y^2 = 4b(x + b - 2a)$ intersect at right angles at the common end of their latera recta.

25. Two equal parabolas have the same vertex and their axes are at right angles. Show that they cut again at angle $\tan^{-1}(3/4)$.
26. Prove that the circle circumscribing the triangle formed by three tangents to a parabola, passes through its focus.

### 10.3.1 Some Problems on pole, polars and chords with a given mid-point.

**Example 10.10.** *A point P is such that the line through it perpendicular to its polar w.r.t. the parabola, $y^2 = 4ax$ touches the parabola, $x^2 = 4by$. find the locus of P.*

*Solution.* Let $P$ be $(h, k)$.

Its polar w.r.t. to $y^2 = 4ax$ is

$$y.k = 2a(x + h) \tag{1}$$

Now the equation of a line perpendicular to (1) through $P$ is

$$kx + 2ay = hk + 2ak \tag{2}$$

By hypothesis, (2) touches $x^2 = 4\,by$

$$\therefore \quad x^2 = 4b\left(\frac{hk + 2ak - kx}{2a}\right)$$

(on eliminating $y$ between them)

$$\Rightarrow \quad ax^2 + 2bkx - 2bk(h + 2a) = 0$$

This must have equal roots.

$$4b^2k^2 - 4a\{-2bk(h+2a)\} = 0$$

$$\Rightarrow \quad bk + 2a(h+2a) = 0$$

$\therefore$ Locus of $(h, k)$ is

$$2ax + by + 4a^2 = 0.$$

**Example 10.11.** Find the locus of the mid-points of chords of the parabola, $y^2 = 4ax$ which subtend a constant angle $\alpha$ at the vertex.

*Solution*. Here vertex is (0, 0).

Let $(x', y')$ be the mid-point of a chord of the parabola,

$$y^2 = 4ax \qquad (1)$$

$\therefore$ The equation of the chord with mid-point $(x', y')$ is

$$T = S'$$

$$\Rightarrow \quad yy' - 2a(x + x') = y'^2 - 4ax'$$

$$\Rightarrow \quad yy' - 2ax = y'^2 - 2ax' \qquad (2)$$

Homogenising (1) with the help of (2), we get the equation of the lines joining the vertex (0, 0) to the points of intersection of (1) and (2) as

$$y^2 = 4ax.\frac{yy' - 2ax}{y'^2 - 2ax'}$$

$$\Rightarrow \quad 8a^2x^2 - 4ay'xy + (y'^2 - 2ax')y^2 = 0.$$

These include an angle $\alpha$.

$$\therefore \quad \tan\alpha = \frac{2\sqrt{h^2 - ab}}{a + b}$$

$$= \frac{2\sqrt{4a^2y'^2 - 8a^2(y'^2 - 2ax')}}{8a^2 + y'^2 - 2ax'}$$

$$\Rightarrow \quad \tan^2\alpha(y'^2 - 2ax' + 8a^2)^2 = 16a^2(4ax' - y'^2)$$

$\therefore$ The locus of $(x', y')$ is

$$(y^2 - 2ax + 8a^2)^2 \tan^2\alpha = 16a^2(4ax - y^2).$$

## Exercise 10.3.

1. If a perpendicular be let fall from any point $P$ upon its polar w.r.t. a given parabola; prove that the distance of the foot of this perpendicular from the focus is equal to the distance of the point $P$ from the directrix.
2. If a chord of a parabola, $y^2 = 4ax$ subtends a right angle at its focus, show that the locus of the pole of this chord w.r.t. the given parabola is the rectangular hyperbola, $x^2 - y^2 + 6ax + a^2 = 0$.
3. Two tangents $OP$ and $OQ$ are drawn to the parabola, $y^2 = 4x$, from the point $O(4, 5)$. Prove that the line joining the mid-points of $OP$ and $OQ$ touches the parabola.

4. $P$ is a variable point on the tangent at the vertex of the parabola, $y^2 = 4ax$. Prove that the locus of the foot of the perpendicular from $P$ on its polar w.r.t. the parabola is the circle, $x^2 + y^2 - 2ax = 0$.
5. Show that the locus of the poles of the chords of the parabola, $y^2 = 4ax$ which are at a constant distance $l$ from the vertex, is
$$l^2y^2 + 4a^2(l^2 - x^2) = 0.$$
6. Any chord of the parabola and the perpendicular to it through its pole w.r.t. a parabola, meet the axis of the parabola at $P$ and $Q$. Prove that $P$ and $Q$ are equidistant from the focus of the parabola.
7. The perpendicular from a point $O$ on its polar w.r.t. a parabola meets the polar in the point $M$ and cuts the axis in $G$; the polar meets the axis in $T$ and the ordinate through $O$ intersects the curve in $P$ and $P'$; show that the points $G$, $M$, $P$, $P'$ and $T$ lie on a circle whose centre is $S$.
8. If two tangents to a parabola make equal angles with a fixed line, show that their chord of contact must pass through a fixed point.
9. Show that the locus of the mid-points of chords of the parabola, $y^2 = 4ax$ such that the focal distances of its extremities are in the ratio 2 : 1 is
$$9(y^2 - 2ax)^2 = 4a^2(2x - a)(4x + a).$$
10. Prove that the chord of the parabola, $y^2 = 4ax$ which is bisected at the point $(h, k)$ is parallel to the polar of $(h, k)$ w.r.t. the parabola.
11. Prove that the mid-point of the intercept made on a tangent to a parabola by the tangents at two points $P$ and $Q$ lies on the tangent which is parallel to $PQ$.

## 10.4 TO FIND THE EQUATION OF THE NORMAL TO THE PARABOLA, IN TERMS OF ITS SLOPE, m.

By § 10.2, we know that the equation of the normal to the parabola, $y^2 = 4ax$, at $(at^2, 2at)$ is
$$y = - tx + 2at + at^3. \qquad (1)$$

If $m$ be the slope, then

$$m = -t.$$

Now (1) becomes

$$y = mx - 2am - am^3,$$

which is the required equation of the normal to the parabola, at $(am^2, - 2am)$,
where $m$ is its slope.

**Definition**. *Co–normal Points.*

The points, on a parabola, normals at which to it are concurrent are called *co–normal points*.

We now prove a theorem, concerning them.

**Theorem 10.1.** In general, three normals can be drawn to a given parabola from a given point, one of which is always real. Also sum of the ordinates of the feet of these co-normal points is zero.

*Proof*. Let $y^2 = 4ax$ be the given parabola and $(\alpha , \beta)$ be the given point.

The equation of any normal to this parabola at $(am^2, - 2am)$ is

$$y = mx - 2am - am^3.$$

Let it pass through $(\alpha , \beta)$,

$$\therefore \quad \beta = mx - 2am - am^3$$

$$\Rightarrow \quad am^3 + (2a - \alpha)m + \beta = 0. \tag{1}$$

which being a cubic in $m$, has three values of $m$. Consequently, three normals, in general, can be drawn from a point to the parabola. Since complex roots of a real equation, always occur in conjugate pairs and (1) being an odd degree equation, it has at least one real root. This proves the first part of the theorem.

For the second part, let $m_i$ $(i = 1, 2, 3)$, be the three roots of (1).

$$\therefore \quad m_1 + m_2 + m_3 = 0 \tag{2}$$

Suppose $y_1, y_2, y_3$ to be the ordinates of the feet of these three co-normal points, then

$$y_1 = - 2am_i \qquad (i = 1, 2, 3)$$

$$\Rightarrow \quad y_1 + y_2 + y_3 = - 2a(m_1 + m_2 + m_3) = 0. \tag{3}$$

Hence the result.

**Corollary 1**. The sum of the slopes of three concurrent normals is zero.

This follows from equation (2).

**Corollary 2.** The centroid of the triangle formed by the feet of three concurrent normals (from a point) always lies on the axis of the parabola.

This follows from equation (3) above $\left(y = \dfrac{y_1 + y_2 + y_3}{3} = 0\right.$
$\Rightarrow$ Centroid lies on the $x$-axis i.e. the axis of the parabola).

### 10.4.1. Circle through co–normal points

*To find the equation of the circle passing through the three (co-normal) points on the parabola, normals at which pass through a given point, $(\alpha, \beta)$.*

Let $P_i(am_1^2, -2am_i)$; $i = 1, 2, 3$ be the three points on the parabola, $y^2$ = wax, normas at which pass through $(\alpha, \beta)$. Then as in Theorem 10.1, $m_i$ $(i = 1, 2, 3)$ are the roots of

$$am^3 + (2a - \alpha)m + \beta = 0 \tag{1}$$

Let the equation of the circle through $P_1, P_2, P_3$ be

$$x^2 + y^2 + 2gx + 2fy + c = 0 \tag{2}$$

If the point $(am^2, -2am)$ lies on it, then

$$a^2m^4 + 4a^2m^2 + 2gam^2 - 4fam + c = 0 \tag{3}$$

This is a fourth degree equation in $m$, and hence there are four values of $m$; say $m_1$, $m_2$, $m_3$ and $m_4$ such that the circle passes through the points $(am_1^2, -2am_i)$, $i = 1, 2, 3, 4$.

Also, $\quad m_1 + m_2 + m_3 + m_4 = 0.$

But $\quad m_1 + m_2 + m_3 = 0,$ by (1)

$\therefore \quad m_4 = 0$

$\Rightarrow \quad (am_4^2, -2am_4) = (0, 0).$

Thus, this circle passes through the vertex (0, 0).

Using (2), we get, $\quad c = 0$.

Putting $c = 0$, in (3) and cancelling $am$, we get

$$am^3 + 2(2a + g)m - 4f = 0 \tag{4}$$

Now, equations (4) and (1) are same, whence comparing

their coefficients, we get

$$1 = \frac{4a+2g}{2a-\alpha} = -\frac{4f}{\beta}$$

$$\Rightarrow \qquad 2g = -(2a+\alpha), 2f = -\frac{\beta}{2}$$

$\therefore$ The equation of the required circle is

$$x^2 + y^2 - (2a+\alpha)x - \frac{\beta}{2}y = 0$$

**Corollary 1.** The algebraic sum of the ordinates of the four points of intersection of a circle and a parabola is zero.

Sum of the ordinates

$= \quad -2a(m_1+m_2+m_3+m_4)$

$= \quad -2a \times 0 = 0$ (from above).

**Corollary 2.** The common chords of a circle and a parabola are in pairs equally inclined to the axis of the parabola.

If $P, Q, R, S$ be the points of intersection of the circle and the parabola with their co-ordinates $(am_i^2, -2am_i)(i = 1, 2, 3, 4)$; then the equation of $PQ$ is

$$y(m_1 + m_2) = -2x - 2am_1m_2.$$

Also the equation of $RS$ is

$$y(m_3 + m_4) = -2x - 2am_3m_4.$$

$\therefore$ Slopes of the chords $PQ$ and $RS$ are respectively

$\frac{-2}{m_1+m_2}$ and $\frac{-2}{m_3+m_4}$.

But $m_1 + m_2 + m_3 + m_4 = 0$.

$\Rightarrow \quad m_3 + m_4 = -(m_1 + m_2)$.

$\therefore$ Their slopes are equal in magnitude and opposite in sign.

$\therefore$ The chords $PQ$ and $RS$ are equally inclined with the axis. This is likewise true for the two pairs $PR$, $QS$ and $PS$, $QR$.

**Corollary 3**. The circle through co-normal points passes through the vertex (0, 0) of the parabola.

**Corollary 4.** The centroid of the four points, in which a circle intersects a parabola, lies on the axis of the parabola.

This follows from Corollary 1.

**Note:** The centroid of $n$ points $(x_i, y_i)$; $i = 1, 2, .., n$ is defined as $\left(\frac{1}{n}\sum_{i=1}^{n} x_i, \frac{1}{n}\sum_{i=1}^{n} y_i\right)$.

**Example 10.12**. The normal at the point $(at_1^2, 2at_1)$ on the parabola $y^2 = 4ax$ meets it again at $(at_2^2, 2at_2)$. Prove that

$$t_2 = -\left(t_1 + \frac{2}{t_1}\right) \Leftrightarrow t_1^2 + t_1t_2 + 2 = 0$$

*Solution*. The equation of the normal at $(at_1^2, 2at_1)$ is

$$y = -t_1x + 2at_1 + at_1^3. \tag{1}$$

But, this is also the line joining the points

$$(at_1^2, 2at_1) \text{ and } (at_2^2, 2at_2)$$

whose equation is

$$y - 2at_1 = \left(\frac{2at_2 - 2at_1}{at_2^2 - at_1^2}\right)(x - at_1^2)$$

$$\Rightarrow \qquad y - 2at_1 = \left(\frac{2}{t_2 + t_1}\right)(x - at_1^2)$$

which can be written as

$$y = \frac{2}{t_2 + t_1} . x + \frac{2at_2t_1}{t_2 + t_1} \tag{2}$$

Comparing the respective coefficients in (1) and (2), we have

$$-t_1 = \frac{2}{t_2 + t_1}$$

$$\Rightarrow \qquad t_2 + t_1 = -\frac{2}{t_1}$$

which gives

$$t_2 = -\left(t_1 + \frac{2}{t_1}\right)$$

Hence the result.

**Example 10.13.** Find the locus of the point of intersection of two normals to the parabola $y^2 = 4ax$, which are at right angles to each other.

*Solution*. Let $(\alpha, \beta)$ be such a point.

Now the equation of the normal at $(am^2, -2am)$ to the given parabola,

$$y^2 = 4ax,$$

is $\quad y = mx - 2am - am^3.$

It passes through $(\alpha, \beta)$,

$$\therefore\ am^3 + (2a - \alpha)m + \beta = 0 \qquad (1)$$

Let $m_1$, $m_2$ and $m_3$ be its roots, which are the slopes of the three normals, from $(\alpha, \beta)$, to the given parabola. As two of these normals are at right angles, we have,

$$m_1 . m_2 = -1 \quad \text{(say)}.$$

Also from (1)

$$m_1 m_2 m_3 = -\frac{\beta}{a},$$

$$\therefore \qquad m_3 = \frac{\beta}{a}$$

But $m_3$ is a root of (1).

$$\therefore \qquad a.\frac{\beta^3}{a^3} + (2a - \alpha)\frac{\beta}{a} + \beta = 0$$

$$\Rightarrow \qquad \beta^2 + a(2a - \alpha) + a^2 = 0$$

$$\Rightarrow \qquad \beta^2 = a(\alpha - 3a)$$

Thus the locus of $(\alpha, \beta)$ is,

$$y^2 = a(x - 3a).$$

**Example 10.14.** If the normal at $P$ to the parabola, $y^2 = 4ax$, makes an angle $\theta$ with the axis; show that:

(i) it will cut the parabola again at $Q$ at an angle $\tan^{-1}$

($\frac{1}{2}$ tan θ) and $PQ = 4a \sec\theta \operatorname{cosec}^2\theta$.

(ii) the area of the triangle formed by it and the tangents at its extremities is $4a^2(\sec\theta \operatorname{cosec}\theta)^3$.

*Solution.* Let $P$ on $y^2 = 4ax$ be $(at^2, 2at)$. The equation of the normal at $P$ is

$$y = -tx + 2at + at^3. \tag{1}$$

Its slope $= -t = \tan\theta$ (given)

Also it meets the parabola again at $Q(at'^2, 2at')$ where

$$t' = -\left(\frac{2}{t} + t\right) \tag{2}$$

(see Example 10.12).

(i) The angle between (1) and the parabola at $Q$ is the angle between the normal and tangent to the parabola at $Q(t')$.

The tangent at $Q$ is

$$t'y = x + at'^2$$

$$\therefore \text{ Its slope} = \frac{1}{t'} = -\frac{t}{2+t^2}$$

Let $\phi$ be the angle of intersection,

$$\therefore \tan\phi = \frac{m_1 - m_2}{1 + m_1 m_2} = \frac{-t + \left(\frac{t}{2+t^2}\right)}{1 + \frac{t^2}{2+t^2}}$$

$$= -t/2.$$

$$\therefore \phi = \tan^{-1}(-t/2) = \tan^{-1}\left(\frac{1}{2}\tan\theta\right).$$

Now $PQ^2 = (at^2 - at'^2)^2 + (2at - 2at')^2$

$$= a^2(t + t')^2 (t - t')^2 + 4a^2 (t - t')^2$$

$$= a^2(t - t')^2 [(t + t')^2 + 4]$$

$$= 16a^2 (1 + t^2)^3/t^4 \quad [\because t + t' = -2/t$$

$$t - t' = 2\left(t + \frac{1}{t}\right). \text{ (using (2))]}$$

$$= 16a^2 \sec^6\theta/\tan^4\theta \quad (\therefore t = -\tan\theta)$$

$= \quad 16a^2 \sec^2\theta \operatorname{cosec}^4\theta$

$\Rightarrow PQ = 4a \sec\theta \operatorname{cosec}^2\theta.$

(ii) Let the tangents to the parabola at $P(t)$ and $Q(t')$ meet at $R$.

$\therefore$ The co-ordinates of $R$ are

$(att', a(t + t')).$

$\therefore$ $p$ = The perpendicular from $R$ on (1)

$$= \left|\frac{at^2t' + a(t+t') - 2at - at^3}{\sqrt{t^2+1}}\right|$$

$$= \left|a(t'-t)\sqrt{t^2+1}\right| = \left|2a\left(t^2+1\right)^{3/2}/t\right|$$

$= 2a \sec^3\theta.\cot\theta = 2a \sec^2\theta.\ \operatorname{cosec}\theta$

$\therefore$ Area of the required triangle

$$= \frac{1}{2}.PQ.p = 4\,a^2 \sec^3\theta \operatorname{cosec}^3\theta.$$

**Example 10.15**. $P$, $Q$ are the points '$t_1$', '$t_2$' on the parabola, $y^2 = 4ax$. If the normals at $P$ and $Q$ meet on the parabola; prove that

(i) the product of their slopes = 2.

(ii) the product of the ordinates of the two points = $8a^2$.

(iii) the line $PQ$ passes through a fixed point on the axis.

(iv) the mid-point of $PQ$ lies on the parabola, $y^2 = 2a(x + 2a)$.

(v) the locus of the pole of $PQ$ is $x = 2a$.

(vi) the ordinates of $P$ and $Q$ are the roots of the quadratic, $y^2 + ky + 8a^2 = 0$ where $k$ is the ordinate of the point of intersection of the normals.

*Solution*. The given parabola is

$$y^2 = 4ax \tag{1}$$

The normals at $P(t_1)$ and $Q(t_2)$ are given by

$$y = -t_1x + 2at_1 + at_1^3 \tag{2}$$

and

$$y = -t_2x + 2at_2 + at_2^3 \tag{3}$$

Now (2) and (3) intersect at

$$R[a(2 + t_1^2 + t_2^2 + t_1t_2), -at_1t_2(t_1 + t_2)] \tag{4}$$

This lies on (1) (by hypothesis)

$\therefore \quad t_1^2 t_2^2 (t_1 + t_2)^2 = 4(2 + t_1^2 + t_2^2 + t_1 t_2)$

$= 4(2 + (t_1 + t_2)^2 - t_1 t_2).$

$\Rightarrow \quad (t_1 + t_2)^2 (t_1^2 t_2^2 - 4) + 4(t_1 t_2 - 2) = 0$

$\Rightarrow \quad (t_1 t_2 - 2)[(t_1 t_2 + 2)(t_1 + t_2)^2 + 4] = 0$

$\Rightarrow \quad t_1 t_2 = 2$

$\Rightarrow \quad (-t_1)(-t_2) = 2$ and $(2at_1).(2at_2) = 8a^2$

This proves (i) and (ii).

Now the equation of $PQ$ is

$$y(t_1 + t_2) = 2x + 2at_1 t_2$$

$$= 2x + 4a \tag{5}$$

which passes through $(-2a, 0)$ a fixed point on the axis of the parabola, for all positions of $P$ and $Q$. This is (iii).

Let $M(x, y)$ be the mid-point of $PQ$.

$$\therefore \quad x = \frac{at_1^2 + at_2^2}{2} = \frac{a}{2}\left[(t_1 + t_2)^2 - 2t_1 t_2\right]$$

$$= \frac{a}{2}\left[(t_1 + t_2)^2 - 4\right] \qquad (\because t_1 t_2 = 2)$$

and $\quad y = \dfrac{2at_1 + 2at_2}{2} = a(t_1 + t_2)$

Eliminating $t_1$ and $t_2$ from these, we get

$$x = \frac{a}{2}\left[\frac{y^2}{a^2} - 4\right] \Rightarrow y^2 = 2a(x + 2a)$$

This is (iv).

Let the pole of $PQ$ be $(x', y')$.

$\therefore$ Polar of $(x', y')$ w.r.t. (1) is

$$yy' = 2ax + 2ax' \tag{6}$$

Comparing the coefficients of (5) and (6), we get

$$\frac{y'}{t_1 + t_2} = \frac{2a}{2} = \frac{2ax'}{4a}$$

$$\Rightarrow \quad x' = 2a.$$

The locus of $(x', y')$ is $x = 2a$.

This is (v).

Now the equation whose roots are $2at_1$ and $2at_2$ (ordinates of $P$ and $Q$) is

$$y^2 - 2a(t_1 + t_2)y + 4a^2 t_1 t_2 = 0$$

$\Rightarrow \quad y^2 - 2a(t_1 + t_2)y + 8a^2 = 0 \qquad (\because t_1 t_2 = 2)$

$\Rightarrow \quad y^2 + ky + 8a^2 = 0$

where $k = -2a(t_1 + t_2)$

$= -at_1 t_2 (t_1 + t_2)$

= ordinate of the point of intersection of the normals (by (4)). This proves (vi).

**Example 10.16.** Show that the shortest normal chord of the parabola, $y^2 = 4ax$, is $6a\sqrt{3}$ and its inclination to the axis is $\tan^{-1}(\sqrt{2})$.

*Solution.* Let the normal to the given parabola at $P(at^2, 2at)$ meet it again at $Q(at'^2, 2at')$,

then $t' = -\left(t + \dfrac{2}{t}\right)$ (see Example 10.12).

Let $\quad l = PQ.$

$\therefore \; l^2 = 16a^2(1 + t^2)^3/t^4 \qquad (1)$

$$\left[\therefore \; t + t' = -2/t \text{ and } t - t' = 2\left(t + \frac{1}{t}\right)\right]$$

(as in Example 10.14)

For maxima or minima,

$$\frac{dl^2}{dt} = 0$$

$$\Rightarrow \quad 16a^2 . \frac{t^4 . 3(1+t^2)^2 . 2t - (1+t^2)^3 . 4t^3}{t^8} = 0$$

$$\Rightarrow \quad \frac{32a^2}{t^5} . (1+t^2)^2 (t^2 - 2) = 0$$

$$\Rightarrow \quad t = \sqrt{2}$$

When $t^2 < 2$, $\dfrac{dl^2}{dt}$ is negative and when $t^2 > 2$, $\dfrac{dl^2}{dt}$ is positive

$\therefore \; t^2 = 2$ yields the minimum value of $l^2$ and therefore of $l$.

From (1), $l^2 = 16\, a^2\, (1 + 2)^3/4 = 108\, a^2$

$\therefore$ The shortest normal, has length $= l = 6\sqrt{3}\, a$.

Also the slope of the normal at $P$ is $-t = \tan\theta$

$\Rightarrow \theta = \tan^{-1}(-t) = \tan^{-1}\left(\sqrt{2}\right)$ (acute angle).

### Exercise 10.4

1. Find the normal to the parabola, $y^2 = 4ax$, at the point $(at^2, 2at)$ and show that from any general point of the plane, there are three normals to the parabola, and that the centroid of the triangle formed by the feet of these normals lies on the axis of parabola.
2. Find the point of intersection of normals at $(am_1^2, -2am_1)$ and $am_2^2, -2am_2)$.
3. Show that the locus of the point of intersection of the normals at the ends of a focal chord of the parabola, $y^2 = 4ax$ is
$$y^2 = a(x - 3a).$$
4. Find the equations of the normals to the parabola, $y^2 = 4x$ which pass through the point (3, 0). Find also the co-ordinates of the feet of the normals.
5. If three normals from a point to the parabola, $y^2 = 4ax$ cut its axis in the points, whose distances from the vertex are in A.P., show that the point lies on the curve,
$$27ay^2 = 2(x - 2a)^3.$$
6. If the normals at three points $P, Q, R$ of the parabola
$$y^2 = 4ax$$
meets in a point $O$ and $S$ be its focus, prove that
(i) $[SP].[SQ].[SR] = a.OS^2$.
(ii) $|OP|.|OQ|.|OR| = a.|OL|.|OM|$ where $|OL|$ and $|OM|$ are the lengths of the tangents from $0$ to the parabola.
7. Show that the tangent at one extremity of a focal chord of a parabola is parallel to the normal at the other extremity.
8. If the normals at two points of a parabola be inclined to the axis of the parabola at angles $\theta$ and $\phi$, such that
$$\tan\theta \,.\, \tan\phi = 2,$$

show that they intersect on the parabola.

9. Show that the locus of points such that two of its normls drawn from them to the parabola $y^2 = 4ax$ coincide is
$$27ay^2 = 4(x - 2a)^2.$$

10. Find the point on the parabola, $y^2 = 4ax$, the normal at which — terminated by the axis — is equal in length to the latus rectum.

11. Prove that the length of the intercept on the normal at the point $P(t)$ of the parabola, $y^2 = 4ax$, made by the circle described on the line joining the focus and $P$ as diameter is $a\sqrt{1+t^2}$.

12. The normals at $P$, $Q$ the ends of a focal chord of a parabola meet the parabola again at $P'$, $Q'$ respectively. Show that $PQ$ is parallel to $P'Q'$ and $P'Q' = 3.PQ$.

13. The normal at a point $P$ of the parabola $y^2 = 4ax$ whose focus is $S$ meets it again at $Q$. If $PQ$ subtends a right angle at its vertex, show that
$$SQ = \frac{1}{3}SQ = a.$$

14. $PQ$ is a chord of a parabola normal at $P$; $AQ$ is drawn from the vertex $A$ and through $P$, a line is drawn parallel to $AQ$ meeting the axis in $R$. Show that $AR$ is double the focal distance of $P$.

15. The normal at $P$ to a parabola meets the axis in $G$, and the normals at $Q$ and $R$ pass through the mid-point of $PG$. Prove that $QR$ passes through the foot of the directrix.

16. Show that the distance between a tangent to the parabola, $y^2 = 4ax$ and the parallel normal is $a \sec^2 \theta \operatorname{cosec} \theta$, where $\theta$ is the inclination of either to the axis of the parabola.

17. From the point where any normal to the parabola, $y^2 = 4ax$ meets the axis, is drawn a line perpendicular to this normal. Show that this normal always touches an equal parabola.

18. Tangents are drawn at the ends of a normal chord of the parabola, $y^2 = 4ax$. Show that the locus of their point

of intersection is the curve, $(x + 2a)y^2 + 4a^3 = 0$.

19. Show that the locus of a point from which two normals can be drawn making complementary angles with the axis is a parabola.
20. Find the locus of a point $P$ when the three normals drawn from it are such that one bisects the angle between the other two.
21. Prove that the normals at the points $P$, $Q$ on the parabola, $y^2 = 4ax$ meet on the normal at $(4am^2/l^2, 4am/l)$ where $lx + my = 1$ is the equation of the chord $PQ$.
22. The normals at three points $P$, $Q$ and $R$ of parabola, $y^2 = 4ax$ meet in a point $O(h, k)$ and $S$ is the focus. Prove that if $OP$ and $OQ$ make complementary angles with the axis, then the tangent at $R$ is parallel to $SO$.
23. The normals at $P$, $Q$, $R$ on the parabola, $y^2 = 4ax$ meet in a point on the line, $y = k$. Prove that the sides of the triangle $PQR$ touch the parabola, $x^2 = 2ky$.
24. Show that the locus of points such that the sum of the angles which the three normals drawn from them to a parabola make with the axis is constant, is a line.
25. If the three normals drawn from any point to the parabola cut the line, $x = 2a$ in points whose ordinates are in A.P., prove that the tangent of the angles which the normals make with the axis are in G.P.

**Example 10.17.** Prove that the area of the triangle inscribed in the parabola $y^2 = 4ax$ is

$$\frac{1}{8a}\left|(y_1 - y_2)(y_2 - y_3)(y_3 - y_1)\right|,$$

where $y_1$, $y_2$ and $y_3$ are the ordinates of its vertices.

*Solution*. Let $(at_i^2, 2at_i)$, $i = 1, 2, 3$, be the vertices of a triangle inscribed in the parabola $y^2 = 4ax$. Now area of the triangle

$$= \frac{1}{2}\,|\,[x_1(y_2 - y_3) + x_2(y_3 - y_1) + x_3(y_1 - y_2)]\,|$$

$$= \frac{1}{2}\,|\,[at_1^2(2at_2 - 2at_3) + at_2^2(2at_3 - 2at_1) + at_3^2(2at_1 - 2at_2)]\,|$$

$= |a^2(t_1^2t_2 - t_1^2t_3 + t_2^2t_3 - t_2^2t_1 + t_3^2t_1 - t_3^2t_2]|$ (1)

Also

$$\frac{1}{8a}\left|(y_1 - y_2)(y_2 - y_3)(y_3 - y_1)\right|$$

$$= \left| \frac{1}{8a} . 8a^3(t_1 - t_2)(t_2 - t_3)(t_3 - t_1) \right|$$

$$= a^2 \; |(t_1 - t_2)(t_2 - t_3)(t_3 - t_1)|$$

which is the same as (1).

**Example 10.18**. Prove that two lines, one a tangent to the parabola, $y^2 = 4a(x + a)$ and the other to the parabola, $y^2 = 4b(x + b)$ which are at right angles to one another, meet on the line, $x + a + b = 0$. Show also that this line is the common chord of the two given parabolas.

*Solution*. The given parabolas are

$$y^2 = 4a(x + a) \qquad (1)$$

$$y^2 = 4b(x + b) \qquad (2)$$

$\therefore$ The equations of any tangents to (1) and (2) respectively are

$$y = m(x + a) + a/m \qquad (3)$$

and

$$y = m'(x + b) + b/m' \qquad (4)$$

By hypothesis, (3) and (4) are at right angles.

$$\therefore \qquad mm' = -1$$

$\therefore$ (4) reduces to

$$y = -\frac{1}{m}(x + b) - bm \qquad (5)$$

Now (3) – (5) $\Rightarrow$

$$0 = x\left(m + \frac{1}{m}\right) + a\left(m + \frac{1}{m}\right) + b\left(m + \frac{1}{m}\right)$$

$$\Rightarrow x + a + b = 0 \text{ as } m + \frac{1}{m} \neq 0.$$

which is the required result.

Also the common chord of (1) and (2) is obtained by substracting (2) from (1) and it is this very equation.

**Example 10.19.** The normals at the points $Q$, $R$ on the parabola, $y^2 = 4ax$ meet on the parabola at the point $P$. Show that the locus of the

(i) orthocentre of the triangle $PQR$ is the parabola, $y^2 = a(x + 6a)$.

(ii) circumcentre of the triangle $PQR$ is the parabola, $y^2 = \frac{1}{2}a(x-a)$.

*Solution.* (i) Let the co-ordinates of $P, Q, R$ be $(at_i^2, 2at_i)$; $i = 1, 2, 3$.

Now the equation of the side $QR$ is

$$y(t_2 + t_3) = 2x + 2a\,t_2t_3$$

The equation of the altitude through $P$ on $QR$ is

$$y - 2at_1 = -\left(\frac{t_2+t_3}{2}\right)(x - at_1^2) \quad (1)$$

Similarly the equation of the altitude through $Q$ on $RP$ is

$$y - 2at_2 = -\left(\frac{t_3+t_1}{2}\right)(x - at_2^2) \quad (2)$$

If $(\alpha, \beta)$ be the point of intersection of (1) and (2) (i.e. $(\alpha, \beta)$ is the orthocentre of the triangle, $PQR$; then

$$\alpha = -4a - a(t_1t_2 + t_2t_3 + t_3t_1)$$

and $$\beta = 2a(t_1 + t_2 + t_3) + \frac{a}{2}(t_1 + t_2)(t_2 + t_3)(t_3 + t_1).$$

(on solving (1) and (2))

Also $t_1, t_2, t_3$ are the roots of

$$at^3 + (2a - h)t - k = 0 \quad (3)$$

where $(h, k)$ is the point where the three normals at $P, Q, R$ meet.

(by Equation (1) of Theorem 10.1, taking $m = -t$)

$t_1 + t_2 + t_3 = 0$,

$t_1t_2 + t_2t_3 + t_3t_1 = (2a - h)/a$ and $t_1t_2t_3 = k/a$.

$\therefore \quad \alpha = -4a - a(2a - h)/a = h - 6a$

and $\quad \beta = -k/2$.

But $(h, k)$ is the same point as $P$ which lies on the parabola.

$\therefore \quad k^2 = 4ah$

$\Rightarrow \quad (-2\beta)^2 = 4a(\alpha + 6a)t_1t_2 + t_2t_3 + t_3t_1 =.$

$\therefore \quad$ Locus of $(\alpha, \beta)$ is $y^2 = a(x + 6a)$.

(ii) The circle through three co-normal points, normals through which meet at $(h, k)$ is given by

$$x^2 + y^2 - (h + 2a)x - \frac{1}{2} ky = 0 \qquad (4)$$

(See § 10.4.1).

Here normals at $Q, R, P$ concur at $(h, k)$ which coincides with $P$.

$$\therefore \quad k^2 = 4ah \qquad (5)$$

Let $(x_1, y_1)$ be the circumcentre of the triangle $PQR$ (i.e. the centre of the circle, given by (4)).

$$\therefore \; x_1 = \frac{1}{2}(h + 2a) \text{ and } y_1 = k/4 \qquad (6)$$

Now (5) and (6) imply

$$(4y_1)^2 = 4a(2x_1 - 2a)$$

$\therefore$ Locus of $(x_1, y_1)$ is

$$y^2 = \frac{1}{2} a(x - a).$$

**Example 10.20**. Show that the locus of the middle points of a system of parallel chords of a parabola is a line which is parallel to the axis of the parabola.

*Solution*. Let the equation ofthe parabola be $y^2 = 4ax$ and $(\alpha , \beta)$ be the middle-point of one such chord. Then equation of the chord is

$$T = S'$$

$$\Rightarrow \quad y\beta - 2ax = \beta^2 - 2a\alpha,$$

which has slope $\dfrac{2a}{\beta}$.

If $m$ is the slope of the parallel chords, then

$$\frac{2a}{\beta} = m$$

$$\Rightarrow \qquad \beta = \frac{2a}{m}$$

Hence the locus of $(\alpha, \beta)$ is

$$y = \frac{2a}{m},$$

which is a line parallel to the axis of the parabola.

**Definition**. A line which bisects a system of parallel chords of a parabola, is called a *diameter* of the parabola.

*Note*. This diameter, $y = 2a/m$ meets the parabola, $y^2 = 4ax$ at $V(a/m^2, 2a/m)$ which is called the vertex of the diameter.

**Example 10.21.** Prove that the length of any focal chord of the parabola, $y^2 = 4ax$ is 4 times the distance of the focus from the point where the diameter bisecting the chord meets the parabola.

*Solution*. Let $PQ$ be any focal chord where $P$ and $Q$ have co-ordinates $(at_i^2, 2at_i)$ $(i = 1, 2)$.

$\therefore\ t_1t_2 = -1$ (See Example 10.7).

Also $m$ = slope of the chord $PQ$

$= 2/(t_1 + t_2)$.

Now the diameter bisecting $PQ$ meets the parabola at $V\left(\dfrac{a}{m^2}, \dfrac{2a}{m}\right)$.

Now $PQ^2 = (at_1^2 - at_2^2)^2 + (2at_1 - 2at_2)^2$

$= a^2(t_1 - t_2)^2[(t_1 + t_2)^2 + 4]$ (1)

Also $(4SV)^2 = 16\left[\left(a - \dfrac{a}{m^2}\right)^2 + \left(0 - \dfrac{2a}{m}\right)^2\right]$

$$= 16a^2\left[\left\{1 - \left(\frac{t_1 + t_2}{2}\right)^2\right\}^2 + (t_1 + t_2)^2\right]$$

$= a^2[\{(t_1 + t_2)^2 - 4\}^2 + 16(t_1 + t_2)^2]$

$= a^2[t_1 + t_2)^2 + 4]^2$

$= a^2\{(t_1 + t_2)^2 - 4t_1t_2][(t_1 + t_2)^2 + 4]$

$(\therefore\ t_1t_2 = -1)$

$= a^2(t_1 - t_2)^2[(t_1 + t_2)^2 + 4]$ (2)

$\therefore\ PQ^2 = (4SV)^2$ (using (1) and (2))

$\Rightarrow |PQ| = 4\ |SV|$.

**Example 10.22.** Prove that the locus of the point of intersection of the normals at the ends of a system of parallel chords of a parabola is a line which is normal to the given parabola.

*Solution.* Let the given parabola be

$$y^2 = 4ax \tag{1}$$

Any normal to (1) at the point with parameter '$t$' is

$$y = -tx + 2at + at^3 \tag{2}$$

Consider any chord $PQ$ of the system where $P(t_1)$ and $Q(t_2)$, having slope = $m$.

$$\therefore \quad m = 2/(t_1 + t_2) \tag{3}$$

Now the normals at $P$ and $Q$ intersect at $(x_1, y_1)$, where

$$x_1 = 2a + a(t_1^2 + t_2^2 + t_1 t_2); \quad y_1 = -at_1 t_2(t_1 + t_2).$$

(See Example 10.15)

$$\Rightarrow \quad x_1 - 2a = a[(t_1 + t_2)^2 - t_1 t_2]$$

$$= a[(2/m)^2 + (my_1/2a)] \quad \text{(using (3))}$$

$$\Rightarrow \quad my_1/2 = x_1 - 2a - 4a/m^2$$

$$\Rightarrow y_1 = (2/m)\,x_1 - (4a/m) - (8a/m^3).$$

$\therefore$ The locus of $(x_1, y_1)$ is

$$y = (2/m)x + 2a(-2/m) + a(-2/m)^3$$

which is normal to (1) at the point whose parameter is $t = -2/m$ (See: Equation (2) above).

## Miscellaneous Exercise on Chapter 10.

1. (i) Prove that the polar of the focus of a parabola is the directrix of the parabola (or *alternately*).
   Prove that directrix is the locus of the poles of the focal chords of a parabola).
   (ii) Deduce or prove directly that the polar (or chord of contact) of any point on the directrix of a parabola passes through the focus.
   (**Note**. The question (ii) can be restated as: Prove that the polars of all points on the directrix of a parabola are concurrent.)
2. If $ST$ be the perpendicular to the tangent at a point $P$ of a parabola, then $T$ lies on the tangent at the vertex $A$ and

$$ST^2 = |AS|\,|SP|.$$

3. (i) Prove that the area of the triangle inscribed in a parabola is twice the area of the triangle formed by the tangents at the vertices.

   (ii) Prove that the area of the triangle formed by the normals to the parabola, $y^2 = 4ax$ at the points '$t_1$', '$t_2$', '$t_3$' is

$$\frac{1}{2}a^2(t_1 - t_2)(t_2 - t_3)(t_3 - t_1)(t_1 + t_2 + t_3)^2.$$

4. (i) Prove that the orthocentre of the triangle formed by any three tangents to a parabola (i.e. the triangle circumscribing a parabola), lies on its directrix.

   (ii) Prove that the orthocentres of the triangles formed by three tangents and the corresponding three normals to a parabola are equidistant from the axis.

5. $PQ$ is a double ordinate of the parabola, $y^2 = 4ax$. Show that the locus of its points of trisection is $9y^2 = 4ax$.

6. Show that the locus of the middle points of all tangents drawn from points on the directrix to the parabola, $y^2 = 4ax$ is

$$y^2(2x + a) = a(3x + a)^2.$$

7. A line is drawn so that the two circles, $x^2 + y^2 + 2ax + c = 0$ and $x^2 + y^2 + 2bx + c = 0$ intercept equal lengths upon it. Prove that the line touches the parabola.

   (*Hint*. Take the line as $y = mx + \lambda/m$ and determine $\lambda$ (using the hypothesis).

8. A chord of the parabola $y^2 = 4ax$ subtends a right angle at the vertex. Prove that the tangents at the extremities of the chord intersect on the line,

$$x + 4a = 0.$$

9. If a chord which is normal to a parabola at one end, subtends a right angle at its vertex, then prove that it is inclined at an angle $\tan^{-1}\sqrt{2}$ with its axis.

10. $PN$ is an ordinate of a parabola. A straight line is drawn parallel to the axis to bisect $NP$ and meet the curve in $Q$. Prove that $NQ$ meets the tangent at the vertex $A$ in a point $T$ such that

$$|AT| = \frac{2}{3}|NP|.$$

11. (i) Show that the locus of the point of intersection of the polars of a variable point on the line, $y = c$, w.r.t. the parabolas $y^2 = 4ax$ and $x^2 = 4ay$ is

$$2ax^2 - cxy + 4a^2(y + c) = 0.$$

(ii) The polar of a point $P$ w.r.t. $y^2 = 4ax$ meets the curve in $Q$ and $R$. Show that if $P$ lies on $lx + my + n = 0$, then the mid-point of $QR$ lies on the parabola

$$l(y^2 - 4ax) + 2a(lx + my + n) = 0.$$

12 (i) The normal at any point $P$ of the parabola, $y^2 = 4ax$ meets the axis in $G$ and the tangent at its vertex in $H$. If $A$ be the vertex and the rectangle $AGBH$ is completed, prove that the locus of $B$ is $x^3 = a(2x^2 + y^2)$.

(ii) A variable tangent to a parabola, $y^2 = 4ax$, meets its axis at $T$ and the tangent at its vertex $A$ at $Y$. If the rectangle $TAYG$ is completed, show that the locus of $G$ is

$$y^2 + ax = 0.$$

13. If the tangent and normal at any point $P$ of a parabola meet its axis in $T$ and $G$ respectively, then show that

$$|ST| = |SG| = |SP|,$$

where $S$ is its focus.

14. Find the length of the side of an equilateral triangle inscribed in the parabola, $y^2 = 4ax$ so that one of the angular points is at its vertex.

15. A variable chord of a given parabola passes through a fixed point. The circle on this chord as diameter cuts the parabola again at two other points. Prove that the lines joining these two points passes through another fixed point whose ordinate is equal in magnitude to the ordinate of the first point.

16. A parabola is drawn to pass through $A$ and $B$, the ends of a diameter of a given circle of radius = $a$ and to have as directrix, a tangent to a concentric circle of radius = $b$; the axes being $AB$ and a perpendicular diameter. Show that the locus of the focus of the parabola is

$$\frac{x^2}{b^2} + \frac{y^2}{b^2 - a^2} = 1$$

17. If the normals at the points $A$, $B$, $C$ on the parabola , $y^2 = 4ax$ meet at $(h, k)$; then
(a) the centroid of the triangle $ABC$ is $(2(h - 2a)/3, 0)$(i.e. it lies on the axis).
(b) the orthocentre of the triangle $ABC$ is $(h - 6a, k/2)$.

18 (i) Prove that the tangent to a parabola at its point of intersection with a diameter is parallel to the system of chords bisected by the diameter.
(ii) Prove that the tangents at the ends of a system of parallel chords of a parabola intersect on the diameter bisecting the chords.

19. If the diameter through any point $O$ of a parabola meets any chord in $P$ and the tangents at the ends of this chord meet the diameter in $Q$ and $R$, show that $OP^2 = OQ.PR$.

20. $P$, $Q$ and $R$ are three points on a parabola and the chord $PQ$ meets the diameter through $R$ in $T$. Ordinates $PM$ and $QN$ are drawn to this diameter. Show that $RM.RN = RT^2$.

# 11. ELLIPSE

## INTRODUCTION

As already defined in Chapter 9, an ellipse is the locus of a point which moves such that its distance from a fixed point, called *focus*, is always $e$ times its distance from a fixed line, called *directrix*, where $e < 1$, $e$ being the *eccentricity* of the ellipse. We now find its equation in the *standard form* and study some of its properties in this chapter.

## 11.1 TO FIND THE EQUATION OF THE ELLIPSE IN THE STANDARD FORM.

Let $S$ be a focus and $ZM$ be a directrix of the given ellipse. Draw $ST \perp ZM$. Divide $ST$ internally and externally in the ratio $e : 1$ at $A$ and $A'$, respectively.

$$\therefore \quad \overline{SA} = e.\overline{AT} \text{ and } \overline{A'S} = e.\overline{A'T} \tag{1}$$

whence the points, $A$ and $A'$, lie on the ellipse.

Set $\quad \overline{A'A} = 2a$, $(a > 0)$.

Let C be the middle point of $A'A$. Take C to be the origin, the line $CS$ as the axis of $x$ and a line through C, perpendicular to $CS$ as the axis of $y$.

Now equations (1) imply

$$\overline{A'S} + \overline{SA} = e\left(\overline{A'T} + \overline{AT}\right) \tag{2}$$

and

$$\overline{A'S} - \overline{SA} = e\left(\overline{A'T} - \overline{AT}\right) \tag{3}$$

We express each side of (2) and (3), in terms of the origin C.

Thus (2) $\Rightarrow \overline{A'C} + \overline{CS} + \overline{CA} - \overline{CS}$

$$= e.\left(\overline{A'C} + \overline{CT} + \overline{CT} - \overline{CA}\right)$$

$$\Rightarrow \quad \overline{CT} = \frac{a}{e} \qquad \left(\because \overline{A'C} = \overline{CA} = a\right)$$

$\Rightarrow$ the equation of the directrix , $ZM$, of the ellipse is

$x = \frac{a}{e}$.

Again (3) $\Rightarrow \left(\overline{A'C} + \overline{CS}\right) - \left(\overline{CA} - \overline{CS}\right)$

$$= e.\left[\left(\overline{A'C} + \overline{CT}\right) - \left(\overline{CT} - \overline{CA}\right)\right]$$

$\Rightarrow \quad \overline{CS} = ae$

$\Rightarrow$ the co-ordinates of the focus, $S$, are $(ae, 0)$.

Let, now, $P(x, y)$ be any point on the ellipse, then by definition

$$\overline{PS} = e.\overline{PQ}$$

where $\quad PQ \perp ZM.$

$$\Rightarrow \quad \sqrt{(x - ae)^2 + y^2} = e.\left|x - \frac{a}{e}\right|$$

$$\Rightarrow \quad \frac{x^2}{a^2} + \frac{y^2}{a^2(1-e^2)} = 1 \text{ (as in Example 9.2)}$$

Set $\quad b^2 = a^2(1 - e)^2$

(Notice here: $e < 1 \Rightarrow b$ is real and $b < a$)

$$\therefore \quad \frac{x^2}{a^2} + \frac{y^2}{b^2} = 1$$

This is the equation of the ellipse in the simplest possible form and as such it is called the Standard Form of the ellipse.

By the method of tracing a curve (see Chapter 7) the shape of the above ellipse is:

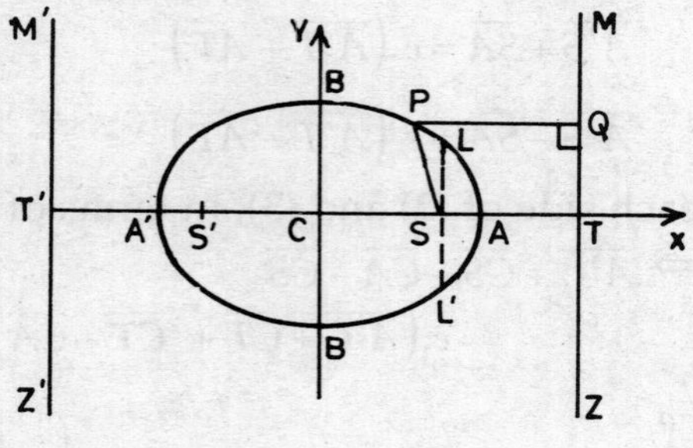

## Definitions

1. *Centre.* The point C is called the *centre* of the ellipse.
2. *Axes*. The lines $A'A$ and $B'B$ are called the *principal* axes of the ellipse. $A'A$ is called its *major axis* and its length is $2a$. Also $B'B$ is called its *minor axis* and its length is $2b$.

Clearly $A(a, 0)$, $A'(-a, 0)$, $B(0, b)$ and $B'(0, -b)$. These four points are called the *vertices* of the ellipse.
3. *Latus Rectum.* A line through the focus, $S$, of the ellipse, perpendicular to its major axis, is called its *latus rectum*.

If we denote it by $L'L$ (see the previous figure) then its equation is $$x = ae.$$

This meets the ellipse

$$\frac{x^2}{a^2} + \frac{y^2}{b^2} = 1$$

where $$e^2 + \frac{y^2}{b^2} = 1$$

$$\Rightarrow \qquad y^2 = b^2(1 - e^2) = b^2 . \frac{b^2}{a^2}$$

$$\Rightarrow \qquad y = \pm\frac{b^2}{a} \cdot$$

$\therefore$ $L$ and $L'$ have co–ordinates $\left(ae, \; \pm\dfrac{b^2}{a}\right)$

and $[L'L]$ i.e. the length of the latus rectum $= \dfrac{2b^2}{a}$.

4. The line, $PN$, perpendicular to the major axis from a point $P$ of the ellipse is called the *ordinate* of $P$. If $PN$ produced meets the ellipse again in $P'$, then $P'NP$ is called the double *ordinate* of $P$. Note that the latus rectum is a double ordinate drawn through the focus of the ellipse.
5. *Length the radius vector from the centre.* Let $P$ be any point on the ellipse at a distance $r$ from the centre C of the ellipse. Let $CP$ be inclined at angle $\theta$ with the major axis. If the co–ordinates of $P$ be $(x, y)$, then

$$x = r\cos\theta,\ y = r\sin\theta$$

and $\frac{x^2}{a^2}+\frac{y^2}{b^2}=1$ becomes

$$\frac{(r\cos\theta)^2}{a^2}+\frac{(r\sin\theta)^2}{b^2}=1$$

$\Rightarrow r^2 = a^2b^2/(b^2\cos^2\theta + a^2\sin^2\theta)$

$\quad = a^2b^2/[b^2 + (a^2 - b^2)\sin^2\theta]$

$\Rightarrow$ The greatest value of $r^2$ is $a^2b^2/[b^2 + (a^2 - b^2).0]$ i.e. $a^2$ (this corresponds to $\theta = 0$) and the least value of $r^2$ is $a^2b^2/[b^2+(a^2- b^2).1]$ i.e. $b^2$ (this corresponds to $\theta = \pi/2$)

When $\theta = \alpha$, $r = \pm ab/(b^2\cos^2\alpha + a^2\sin^2\alpha)$, so that any line through $C$, the centre intersects the ellipse in two points equidistant from the centre.

**Note 1.** By the symmetry of the curve, or directly, one can see that if the point $S'(-ae, 0)$ is taken as the focus and the line $Z'M'$ with the equation, $x = -a/e$ as the directrix, one gets the same equation of the ellipse.

Thus we see that an ellipse has *two foci*, $S(ae, 0)$ and $S'(-ae, 0)$ and also *two directrices*, $x = a/e$ and $x = -a/e$ respectively. Consequently it has *two latera recta*, $x = \pm ae$.

**Note 2**. The equation $\frac{x^2}{a^2}+\frac{y^2}{b^2}=1$, when $a < b$, has the following figure. Its foci are $(0, be)$ and $(0, -be)$, the directrices are

$$y=\frac{b}{e} \text{ and } y=-\frac{b}{e}$$

and major axis $= 2b$,
minor axis $= 2a$.

Also the eccentricity is given by

$$a^2 = b^2(1 - e^2),$$

and latus rectum $=\frac{2a^2}{b}$

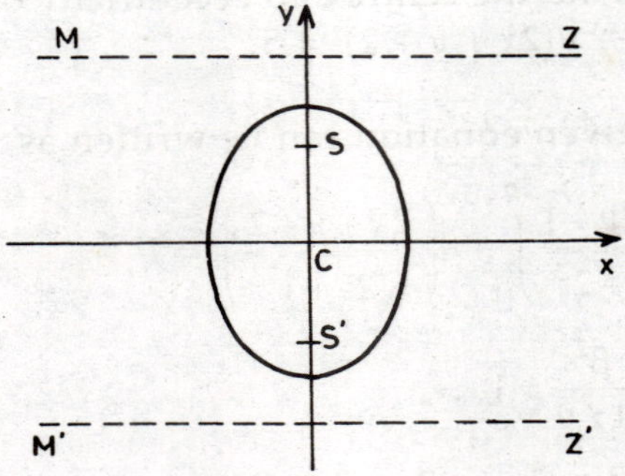


**Example 11.1.** Find the equation of the ellipse having its centre at the point (2, –3), one focus at (3, –3) and vertex at (4, –3).

*Solution*. Clearly the line joining the focus $S(3, -3)$ and the centre $C(2, -3)$ is parallel to the axis of $x$. Also one vertex, say $A$ is at (4, –3).

$$\therefore \quad a = \overline{CA} = 4 - 2 = 2$$

and $$ae = \overline{CS} = 3 - 2 = 1.$$

$$\Rightarrow \quad e = \frac{1}{2}$$

Also $$b^2 = a^2(1 - e^2) = 3.$$

As the centre of the ellipse is (2, –3), its equation is given by

$$\frac{(x-2)^2}{4} + \frac{(y+3)^2}{3} = 1$$

Its shape is shown as:

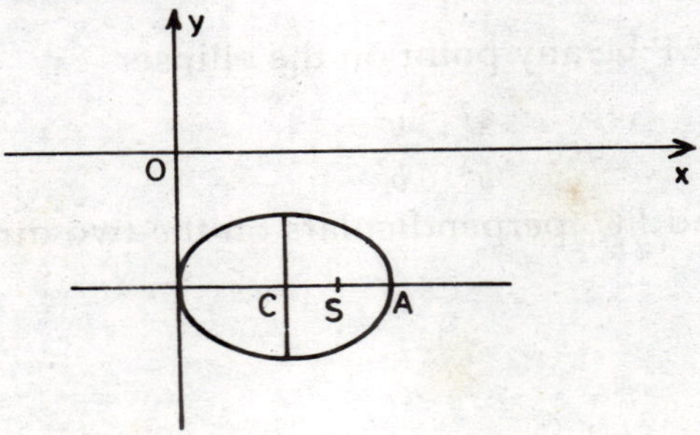

**Example 11.2** Find the centre and eccentricity of the ellipse, $4(x - 2xy + 1)^2 + 9(2x + y + 2)^2 = 5$.

*Solution.* The given equation can be written as

$$4\left(\frac{x-2y+1}{\sqrt{1^2+2^2}}\right)^2 + 9\left(\frac{2x+y+2}{\sqrt{2^2+1^2}}\right)^2 = 1$$

$$\Rightarrow \qquad \frac{\alpha^2}{1/4} + \frac{\beta^2}{1/9} = 1$$

where $\alpha = \dfrac{x-2y+1}{\sqrt{5}}$, $\beta = \dfrac{2x+2y+2}{\sqrt{5}}$

The semi-major axis is of length 1/2 and the semi-minor axis is of length 1/3. Their equations are $2x + y + 2 = 0$ and $x - 2y + 1 = 0$ respectively.

The co-ordinates of the centre are (–1, 0) and eccentricity is given by

$$b^2 = a^2 (1 - e)^2$$

$$\Rightarrow \qquad \frac{1}{9} = \frac{1}{4}(1-e^2)$$

$$\Rightarrow \qquad e^2 = 4\left(\frac{1}{4} - \frac{1}{9}\right) = \frac{5}{9}$$

$$\Rightarrow \qquad e = 5\sqrt{3}.$$

**Example 11.3.** Show that the sum of the focal distances of any point $P$ on the ellipse is constant and is equal to the major axis.

*Solution.* Let P be any point on the ellipse

$$\frac{x^2}{a^2} + \frac{y^2}{b^2} = 1$$

Draw $PQ$ and $PQ'$ perpendiculars on the two directrices.

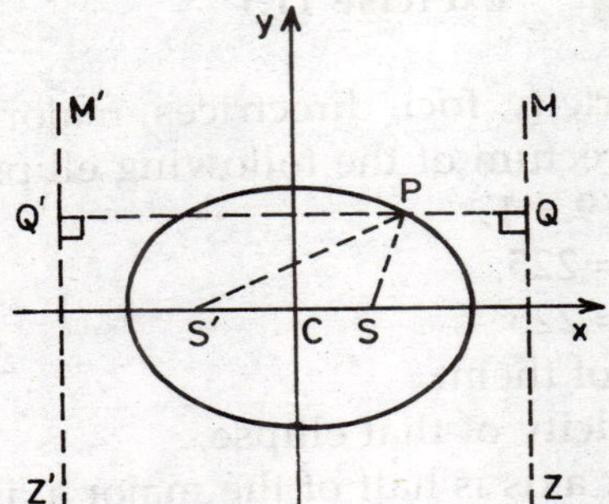


Then by definition

$$\overline{PS} = e.\overline{PQ}$$

and $$\overline{PS'} = e.\overline{Q'P}$$

$$\Rightarrow \quad \overline{PS} + \overline{PS'} = e.\left(\overline{Q'P} + \overline{PQ}\right) = e.\overline{Q'Q} = e.\frac{2a}{e} = 2a$$

Hence the result.

(Though we have taken $P$ in the first quadrant, yet it is easily seen that the result remains true for all positions of $P$ on the ellipse).

**Note**. From Example 2.4, we see that an ellipse can also be defined as the locus of a point, which moves such that the sum of its distances from two fixed points is always constant.

Also if we take the fixed points as $(ae, 0)$ and $(-ae, 0)$ and the sum of distances as $2a$, we arrive at the same equation,

$$x^2/a^2 + y^2/b^2 = 1 \text{ where } b^2 = a^2(1 - e^2),\ e < 1.$$

In fact this is how we can draw the diagram for an ellipse. Take a thread of fixed length and fasten its two ends to two nails attached at two fixed points $S$ and $S'$. Now roll the thread along a pencil and let the pencil move. The tip of the pencil will, then, describe an ellipse. This is known as the *gardener's method of tracing an ellipse.*

## Exercise 11.1

1. Find the eccentricity, foci, directrices, major axis, minor axis, and latus rectum of the following ellipses:
   (i) $x^2/16 + y^2/9 = 1,$
   (ii) $9x^2 + 25y^2 = 225,$
   (iii) $25x^2 + 9y^2 = 225.$
   Also trace each of them.
2. Find the eccentricity of that ellipse,
   (i) whose minor axis is half of the major axis;
   (ii) whose latus rectum is equal to half of its minor axis.
3. Find the equation of the ellipse referred to its axes as the axes of co-ordinates,
   (i) which passes through the points (2, 2) and (1, 4).
   (ii) with minor axis 4 and the distance between the foci 2.
   (iii) one vertex is (5, 0) and focus is (4, 0).
   (iv) one vertex is (8, 0) and the latus rectum is 9.
   (v) Minor axis = 12, eccentricity = 1/2
4. Find the equation of the ellipse (in standard form) whose latus rectum is 5 and whose eccentricity is 2/3.
5. Find the equation of the ellipse referred to the centre as the origin and its major and minor axes as the axes of $x$ and $y$ respectively and whose latus rectum is 3 and eccentricity is $1/\sqrt{2}$.
6. Find the equation of the ellipse referred to its centre as the origin and its major and minor axes as the axes of $x$ and $y$ respectively, such that its minor axis is equal to the distance between the foci and its latus rectum is 10.
7. Find the equation of the ellipse whose focus is (2, 2), the corresponding directrix is $x + y = 0$ and the eccentricity is 1/2. Also find its centre, the second focus and the equation of the second directrix .
   (*Hint*: See Chapter 13).
8. Find the equation of the ellipse, given that
   (i) foci (3, 1), (3, 7) and $e = 1/3$
   (ii) centre (0, –1), focus (0, 3), minor axis = 8.
9. If the angle between the lines joining the foci of an ellipse to an extremity of the minor axis is $\pi/2$; find the eccen-

tricity of the ellipse. Also find its equation if the major axis = $2\sqrt{2}$.

10. Find the centre, eccentricity, foci and lengths of the axes of the ellipse

$$\frac{(x-1)^2}{25}+\frac{(y+3)^2}{9}=1$$

(*Hint*. Shift the origin to (1, –3)).

11. A straight rod of given length slides between two fixed perpendicular bars. Show that the locus of a point on the rod which divides it in a given ratio is an ellipse. If this ratio is 1/2, find the eccentricity of the ellipse.

12. Find the equation of the ellipse whose major and minor axes are 10 and 8 and lie along the lines $3x + 4y - 7 = 0$ and $4x - 3y - 1 = 0$ respectively.

(*Hint*. If $P$ is any point on the ellipse and $PM$, $PN$ are perpendiculars on the major and minor axes respectively, the equation of the ellipse is

$$\frac{PN^2}{5^2}+\frac{PM^2}{4^2}=1)$$

13. Show that the triangle whose vertices are the points (3, –1), (1, 2) and (–2, –1) lies inside the ellipse, $\frac{x^2}{15}+\frac{y^2}{10}=1$.

## 11.2 SUMMARY OF THE RESULTS FROM CHAPTER 8.

For the benefit of the reader, we summarise here some of the results concerning the ellipse $x^2/a^2 + y^2/b^2 = 1$, already arrived at in Chapter 8.

1. The parametric equations of the ellipse or the co–ordinates of any point on the ellipse are

$$x = a\cos\theta,$$
$$y = b\sin\theta.$$

2. The equation of the tangent at the point $(x', y')$ to it is

$$\frac{xx'}{a^2}+\frac{yy'}{b^2}=1.$$

3. The equation of the tangent at $(a \cos \theta, b \sin \theta)$ to it is

$$\frac{x\cos\theta}{a}+\frac{y\sin\theta}{b}=1.$$

4. The equation of the normal at $(a \cos \theta, b \sin \theta)$ to it is

$$ax \sec \theta - by \operatorname{cosec} \theta = a^2 - b^2$$

or

$$\frac{ax}{\cos\theta}-\frac{by}{\sin\theta}=a^2-b^2$$

5. The condition that the line, $y = mx + c$ touches the ellipse is

$$c^2 = a^2m^2 + b^2.$$

Thus the equation of any tangent to it (not parallel to the axis of $y$) can be written as:

$$y = mx \pm \sqrt{a^2m^2+b^2}$$

6. The equation of the chord of contact of $(x', y')$ w.r.t. the ellipse is

$$\frac{xx'}{a^2}+\frac{yy'}{b^2}=1$$

7. The equation of the polar of $(x', y')$ w.r.t. the ellipse is

$$\frac{xx'}{a^2}+\frac{yy'}{b^2}=1$$

8. The equation of the chord of the ellipse whose middle point is $(x', y')$, is $T = S'$,

where
$$S=\frac{x^2}{a^2}+\frac{y^2}{b^2}-1$$

$$S'=\frac{x'^2}{a^2}+\frac{y'^2}{b^2}-1$$

and
$$T=\frac{xx'}{a^2}+\frac{yy'}{b^2}-1$$

9. The equation of the pair of tangents from a point $(x', y')$ to the ellipse is $T^2 = SS'$.

## 11.3 AUXILIARY CIRCLE

**Definition**. The circle described on the major axis of an ellipse is called the *auxiliary* circle of the ellipse. The equation of the auxiliary circle with centre (0, 0) and radius $a$, therefore, is

$$x^2 + y^2 = a^2.$$

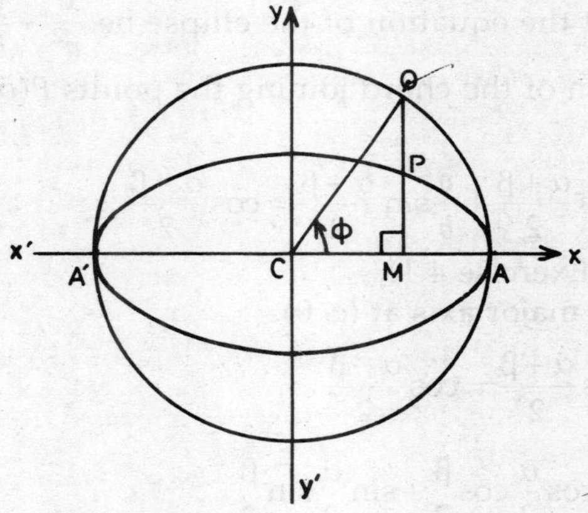


Let $P$ be any point on the ellipse. Let the ordinate $MP$ meet the auxiliary circle at $Q$.

Let $\angle xCQ = \phi$.

$\therefore$ $CM = a \cos \phi$, $MQ = a \sin \phi$

and $Q(a \cos \phi, a \sin \phi)$.

Now x-co-ordinate of $P$ is $a \cos \phi$. If its $y$-co-ordinate is $y'$, then $P(a \cos \phi, y')$ lies on

$$\frac{x^2}{a^2} + \frac{y^2}{b^2} = 1$$

whence $$\cos^2 \phi + \frac{y'^2}{b^2} = 1$$

$\Rightarrow$ $y' = b \sin \phi$.

Hence $P$ is $(a \cos \phi, b \sin \phi)$.

The parameter $\phi$ is called the *eccentric angle* of the point $P$. Note that $\phi$ is the angle which the line $CQ$ makes with the x-axis and not the angle which the line $CP$ makes with it.

**Example 11.4** If the chord joining two points whose eccentric angles are $\alpha$ and $\beta$, cut the major axis of an ellipse at *a* distance $c$ from its centre; show that

$\tan(\alpha/2) \tan(\beta/2) = (c - a)/(c + a)$,

where $a$ is the semi-major axis of the ellipse.

*Solution*. Let the equation of the ellipse be $\dfrac{x^2}{a^2} + \dfrac{y^2}{b^2} = 1$

The equation of the chord joining the points $P(\alpha)$ and $Q(\beta)$, is

$$\frac{x}{a}\cos\frac{\alpha+\beta}{2} + \frac{y}{b}\sin\frac{\alpha+\beta}{2} = \cos\frac{\alpha-\beta}{2}$$

(By $Q$ 2(ii), Exercise 4.1).

It meets the major axis at $(c, 0)$.

$$\therefore \quad \frac{c}{a}\cos\frac{\alpha+\beta}{2} = \cos\frac{\alpha-\beta}{2}$$

$$\Rightarrow \quad \frac{c}{a} = \frac{\cos\frac{\alpha}{2}\cos\frac{\beta}{2} + \sin\frac{\alpha}{2}\sin\frac{\beta}{2}}{\cos\frac{\alpha}{2}\cos\frac{\beta}{2} - \sin\frac{\alpha}{2}\sin\frac{\beta}{2}}$$

$$= \frac{1 + \tan\frac{\alpha}{2}\tan\frac{\beta}{2}}{1 - \tan\frac{\alpha}{2}\tan\frac{\beta}{2}}$$

$$\Rightarrow \quad \frac{c}{1 + \tan\frac{\alpha}{2}\tan\frac{\beta}{2}} = \frac{a}{1 - \tan\frac{\alpha}{2}\tan\frac{\beta}{2}} = k \text{ (say)}$$

$$\Rightarrow \quad c = k\left(1 + \tan\frac{\alpha}{2}\tan\frac{\beta}{2}\right)$$

and $$a = k\left(1 - \tan\frac{\alpha}{2}\tan\frac{\beta}{2}\right)$$

then $$\frac{c-a}{c+a} = \frac{k\left[2\tan\frac{\alpha}{2}\tan\frac{\beta}{2}\right]}{2k} = \tan\frac{\alpha}{2}\tan\frac{\beta}{2}$$

or $\tan\frac{\alpha}{2}\tan\frac{\beta}{2} = \frac{c-a}{c+a}$

**Corollary**. If any two chords be drawn through two points on the major axis equidistant from the centre, show that $\tan\frac{\alpha}{2}\tan\frac{\beta}{2}\tan\frac{\gamma}{2}\tan\frac{\delta}{2} = 1$, where $\alpha$, $\beta$, $\gamma$, $\delta$ are the eccentric angles of the extremities of the chords.

*Solution*. Let the chords meet the major axis at $(c, 0)$ and $(-c, 0)$.

$\therefore$ By the above example,

$$\tan\frac{\alpha}{2}\tan\frac{\beta}{2} = \frac{c-a}{c+a}$$

and
$$\tan\frac{\gamma}{2}\tan\frac{\delta}{2} = \frac{-c-a}{-c+a}$$

Multiply and get the result.

**Example 11.5**. A variable point $P$ on an ellipse of eccentricity $e$, is joined to its foci $S$, $S'$. Prove that the locus of the incentre of the triangle $PSS'$ is an ellipse whose eccentricity is $\sqrt{2e/(1+e)}$.

*Solution*. Let the given ellipse be

$$\frac{x^2}{a^2} + \frac{y^2}{b^2} = 1$$

By hypothesis, $e$ is given by

$$b^2 = a^2(1 - e^2) \tag{1}$$

Also $S(ae, 0)$; $S'(-ae, 0)$.

Let the co-ordinates of $P$ be

$(a \cos \phi, b \sin \phi)$.

$\therefore$ $\overline{SP}$ = Focal distance of the point $P$

$= a(1 - e \cos \phi)$

and $\overline{S'P} = a(1 + e \cos \phi)$

Also $$\overline{S'S} = 2ae$$

If$(x, y)$ be the incentre of the triangle $PSS'$, then by Example 1.4

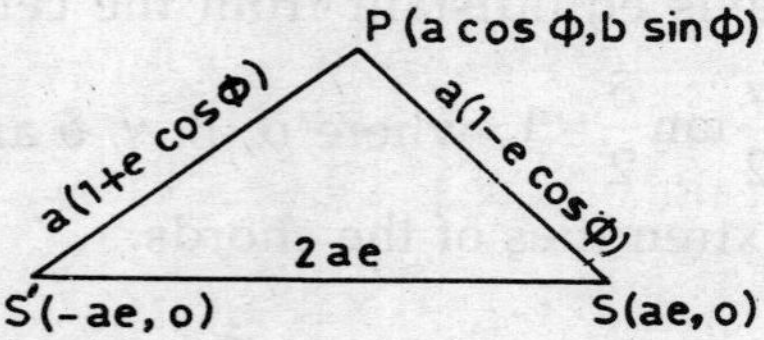


$$x = \frac{2ae.a\cos\phi + a(1-e\cos\phi)(-ae) + a(1+e\cos\phi).ae}{2ae + a\ (1-e\cos\ \phi) + a\ (1+e\cos\ \phi)}$$

$$= \frac{2a^2e.a\cos\phi + 2a^2e^2\cos\phi}{2ae+2a} = ae\cos\phi \tag{2}$$

and

$$y = \frac{2ae.b\sin\phi + a(1-e\cos\ \phi).0 + a(1+e\cos\phi).0}{2ae + a\ (1-e\cos\phi) + a\ (1+e\cos\phi)}$$

$$= eb\ \sin\ \phi/(e+1). \tag{3}$$

Eliminating $\phi$ from (2) and (3), we get

$$\left(\frac{x}{ae}\right)^2 + \left[\frac{y(e+1)}{be}\right]^2 = 1$$

$$\Rightarrow \qquad \frac{x^2}{ae^2} + \frac{y^2}{[be/(e+1)]^2} = 1$$

which represents an ellipse. Let $e_1$ be its eccentricity

$$\therefore \qquad \left(\frac{be}{e+1}\right)^2 = a^2e^2(1-e_1^2)$$

$$\Rightarrow \qquad e_1^2 = 1 - \frac{b^2}{a^2(e+1)^2}$$

$$= 1 - \frac{1-e^2}{(e+1)^2} \qquad \text{(by (1))}$$

$= 2e/(1 + e)$

$\Rightarrow e_1 = \sqrt{2e/(1+e)}$

## Exercise 11.2.

1. Show that the equation of the line joining the points $P(\alpha)$ and $Q(\beta)$ on the ellipse, $\frac{x^2}{a^2}+\frac{y^2}{b^2}=1$, is

$$\frac{x}{a}\cos\left(\frac{\alpha+\beta}{2}\right)+\frac{y}{b}\sin\left(\frac{\alpha+\beta}{2}\right)=\cos\left(\frac{\alpha-\beta}{2}\right)$$

Deduce the equation of tangent at the point $P(\alpha)$.

2. Let $\alpha$ and $\beta$ be the eccentric angles of two variable points $P$ and $Q$ on the ellipse, $\frac{x^2}{a^2}+\frac{y^2}{b^2}=1$

If $\alpha + \beta = 2\gamma$ (constant), prove that $PQ$ is always parallel to the tangent at "$\gamma$" and conversely.

3. If the chord joining the points "$\alpha$" and "$\beta$" on the ellipse $\frac{x^2}{a^2}+\frac{y^2}{b^2}=1$ is a focal chord, show that

(i) $\cos\left(\frac{\alpha-\beta}{2}\right)=\pm e\cos\left(\frac{\alpha+\beta}{2}\right)$

(ii) $\tan\frac{\alpha}{2}\tan\frac{\beta}{2}+\frac{1-ke}{1+ke}=0$, where $k = \pm 1$.

4. Prove that the equation of the chord joining two points $(x_1, y_1)$ and $(x_2, y_2)$ on the ellipse, $\frac{x^2}{a^2}+\frac{y^2}{b^2}=1$, is

$$\frac{x(x_1+x_2)}{a^2}+\frac{y(y_1+y_2)}{b^2}=\frac{x_1x_2}{a^2}+\frac{y_1y_2}{b^2}+1$$

Deduce the equation of the tangent at $(x_1, y_1)$.

5. Show that the chord joining the points $P(\alpha)$ and $Q(\beta)$ on the ellipse, $\frac{x^2}{a^2}+\frac{y^2}{b^2}=1$ subtends a right angle

(a) at a vertex if $\tan\frac{\alpha}{2}\tan\frac{\beta}{2}=-\frac{a^2}{b^2}$

(b) at the centre if $\tan\alpha\tan\beta = -a^2/b^2$.

6. Find the equation of the chord of the ellipse, $\frac{x^2}{a^2}+\frac{y^2}{b^2}=1$ joining two points whose eccentric angles are $\alpha+\beta$ and $\alpha-\beta$.
7. Prove that the line joining two points on an ellipse, the difference of whose eccentric angles is constant, touches another ellipse.
8. If three of the sides of a quadrilateral inscribed in an ellipse are parallel respectively to three given lines, show that the fourth side will also be parallel to a fixed line.
9. Show that the line, $lx + my + n = 0$ will cut the ellipse, $\frac{x^2}{a^2}+\frac{y^2}{b^2}=1$ in points whose eccentric angles differ *by* $\pi/2$ if $a^2l^2 + b^2m^2 = 2n^2$.
10. An ellipse slides between two perpendicular lines. Show that the locus of its centre is a circle. Find also the locus of its focus.

Now, we prove a result concerning the number of normals, which, in general, can be drawn from a point to an ellipse.

**Theorem 11.1.** In general, four normals can be drawn to an ellipse from any point and if $\alpha$, $\beta$, $\gamma$, $\delta$ be the eccentric angles of these four co-normal points; then

$\alpha+\beta+\gamma+\delta$ is an odd multiple of $\pi$.

*Proof.* Let $(h, k)$ be any given point and let $P(a\cos\phi, b\sin\phi)$ be any point on the ellipse,

$$\frac{x^2}{a^2}+\frac{y^2}{b^2}=1$$

Suppose further that the normal at $P$ passes through $(h, k)$.

Now the equation of the normal at $P(a \cos \phi, b \sin \phi)$ to the given ellipse is

$$ax \sec \phi - by \operatorname{cosec} \phi = a^2 - b^2.$$

It passes through (h, k),

$$\therefore \quad ah \sec \phi - bk \operatorname{cosec} \phi = a^2 - b^2 \tag{1}$$

Let $\tan \dfrac{\phi}{2} = t$; then

$$\cos \phi = \frac{1 - \tan^2 \phi/2}{1 + \tan^2 \phi/2} = \frac{1 - t^2}{1 + t^2}$$

and
$$\sin \phi = \frac{2 \tan \phi/2}{1 + \tan^2 \phi/2} = \frac{2t}{1 + t^2}$$

$\therefore$ (1) reduces to

$$ah.\frac{1+t^2}{1-t^2} - bk.\frac{1+t^2}{2t} = a^2 - b^2$$

$$bkt^4 + 2[ah + (a^2 - b^2)]t^3 + 2[ah - (a^2 - b^2)]t - bk = 0 \tag{2}$$

which is a fourth degree equation in $t$, and hence gives four values of $t$, say $t_i$ $(i = 1, 2, 3, 4)$.

Consequently, it has four values of $\phi$, say $\alpha, \beta, \gamma, \delta$ (since $t = \tan \phi/2$) the normals at which to the ellipse, pass through $(h, k)$.

$$\text{Now} \left(\frac{\alpha}{2} + \frac{\beta}{2} + \frac{\gamma}{2} + \frac{\delta}{2}\right) = \frac{s_1 - s_3}{1 - s_2 - s_4}$$

(from Trigonometry)

$$\cot\left(\frac{\alpha + \beta + \gamma + \delta}{2}\right) = \frac{1 - s_2 + s_4}{s_1 - s_3}$$

$$= \frac{1 - 0 - 1}{-\dfrac{2(ah - a^2 - b^2)}{bk} + \dfrac{2(ah - a^2 + b^2)}{bk}}$$

$$= \frac{0}{a^2 - b^2} = 0$$

using equation (2). (Note $a \neq b$ for an ellipse).

$\Rightarrow \dfrac{\alpha+\beta+\gamma+\delta}{2}$ = an odd multiple of $\pi/2$.

$\Rightarrow \alpha + \beta + \gamma + \delta$ = an odd multiple of $\pi$.

**Aliter**. Proceed as above, till the equation (1).

Set $\quad z = e^{i\phi} = \cos\phi + i\sin\phi$

$\Rightarrow \quad 1/z = e^{-i\phi} = \cos\phi - i\sin\phi$

$$\Rightarrow \qquad \cos\phi = \frac{1}{2}\left(z+\frac{1}{z}\right) = \frac{z^2+1}{2z}$$

and $$\sin\phi = \frac{1}{2i}\left(z-\frac{1}{z}\right) = \frac{z^2-1}{2iz}$$

Equation (1) now reduces to

$$ah.\frac{2z}{z+1} - bk.\frac{2iz}{z^2-1} = a^2-b^2$$

$$(a^2 - b^2)z^4 - 2(ah - ibk)z^3 + 2(ah + ibk)z - (a^2 - b^2) = 0 \qquad (2)$$

which yields four values of $z$, say $z_1, z_2, z_3, z_4$.

Consequently, $z = e^{i\phi}$, gives four values of $\phi$, say, $\alpha$, $\beta$, $\gamma$ and $\delta$, the normals at which pass through the point $(h, k)$.

Using this equation, we obtain,

$$z_1.z_2.z_3.z_4 = -1$$

$\Rightarrow e^{i\alpha}, e^{i\beta}, e^{i\gamma}, e^{i\delta} = -1$

$\Rightarrow e^{i(\alpha+\beta+\gamma+\delta)} = -1$

$\Rightarrow \cos(\alpha+\beta+\gamma+\delta) + i\sin(\alpha+\beta+\gamma+\delta) = -1$.

$\Rightarrow \cos(\alpha+\beta+\gamma+\delta) = -1$ and $\sin(\alpha+\beta+\gamma+\delta) = 0$.

$\Rightarrow \alpha+\beta+\gamma+\delta$ = an odd multiple of $\pi$.

**Example 11.6** Find the locus of the intersection of the normals at the extremities of a chord of the ellipse, $\dfrac{x^2}{a^2}+\dfrac{y^2}{b^2}=1$ draw parallel to a given line.

*Solution*. Let "$\alpha$" and "$\beta$" be the extremities of the chord. The equations of the normals at these points are

$$ax\sec\alpha - by\operatorname{cosec}\alpha = a^2-b^2 \qquad (1)$$

$$ax\sec\beta - by\operatorname{cosec}\beta = a^2-b^2 \qquad (2)$$

Let $(h, k)$ be the point of intersection of (1) and (2). Solving these equations, we get

$$h = \frac{a^2 - b^2}{a} \cos\alpha \cos\beta \frac{\cos\frac{\alpha+\beta}{2}}{\cos\frac{\alpha-\beta}{2}} \quad (3)$$

$$k = -\frac{a^2 - b^2}{a} \sin\alpha \sin\beta \frac{\sin\frac{\alpha+\beta}{2}}{\cos\frac{\alpha-\beta}{2}} \quad (4)$$

Now the slope of this chord

$$= \frac{b\sin\alpha - b\sin\beta}{a\cos\alpha - a\cos\beta} = -\frac{b}{a}\cot\frac{\alpha+\beta}{2}$$

By hypothesis, the chord is parallel to a given line.

$\therefore$ We have, $\alpha+\beta$ is a constant.

Let $\alpha + \beta = 2\gamma$ where $\gamma$ is a constant.

From (3) and (4)

$$\frac{ah}{\cos\gamma} + \frac{bk}{\sin\gamma} = \left(a^2 - b^2\right)\frac{\cos(\alpha+\beta)}{\cos\frac{\alpha-\beta}{2}}$$

$$= \left(a^2 - b^2\right)\frac{\cos 2\gamma}{\cos\frac{\alpha-\beta}{2}} \quad (5)$$

and $$\frac{ah}{\cos\gamma} - \frac{bk}{\sin\gamma} = \left(a^2 - b^2\right)\frac{\cos(\alpha-\beta)}{\cos\frac{\alpha-\beta}{2}}$$

$$= \left(a^2 - b^2\right)\left(2\cos\frac{\alpha-\beta}{2} - \sec\frac{\alpha-\beta}{2}\right) \quad (6)$$

Eliminating $\alpha-\beta$ from (5) and (6) we get

$a^2h^2 + 2abhk \text{ cosec } 2\gamma + b^2k^2 = (a^2 - b^2)^2 \cos^2 2\gamma$

$\therefore$ The locus of $(h, k)$ is

$a^2x^2 + 2abxy \text{ cosec } 2\gamma + b^2k^2 = (a^2 - b^2) \cos^2\gamma.$

**Example 11.7** Show that the eccentricity of the ellipse, in which the normal at one end of a latus rectum passes through an end of its minor axis, is given by $e^4 + e^2 - 1 = 0$.

*Solution.* Let $L$ be one end of the latus rectum of the ellipse

$$\frac{x^2}{a^2}+\frac{y^2}{b^2}=1;$$

then $L$ is $\left(ae, \dfrac{b^2}{a}\right)$

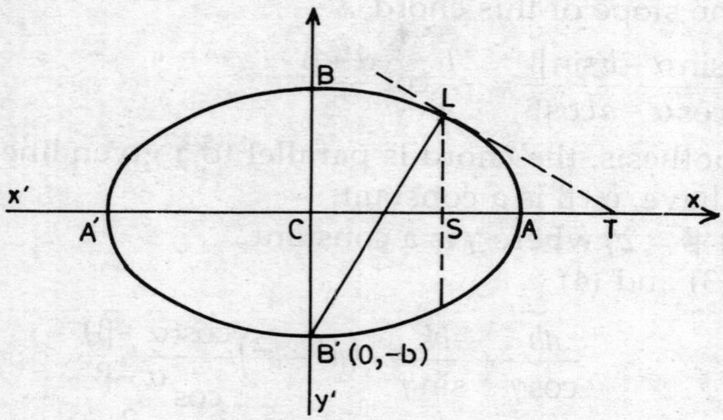


The equation of the tangent at $L$ to the ellipse is

$$\frac{xe}{a}+\frac{y}{a}=1,$$

whose slope is $-e$.

$\therefore$ the equation of the normal at $L$ is

$$y-\frac{b^2}{a}=\frac{1}{e}(x-ae)$$

By hypothesis, it passes through the end $B'(0, -b)$ of the minor axis (see the figure)

$$\therefore \quad -b-\frac{b^2}{a}=\frac{1}{e}(0-ae)=-a$$

$$\Rightarrow -b=\frac{b^2}{a}-a=\frac{b^2-a^2}{a}=-ae^2, \text{ as } b^2=a^2(1-e^2)$$

$$\Rightarrow b^2 = a^2e^4$$

$\Rightarrow a^2(1 - e^2) = a^2e^4$

$\Rightarrow e^4 + e^2 - 1 = 0.$

**Example 11.8** The tangent at a point $P(a \cos \theta, b \sin \theta)$ of an ellipse, $\frac{x^2}{a^2} + \frac{y^2}{b^2} = 1$, meets its auxiliary circle in two points, the chord joining which subtends a right angle at the centre. Show that the eccentricity of the ellipse is $(1 + \sin^2 \theta)^{-1/2}$.

*Solution.* The given ellipse is

$$\frac{x^2}{a^2} + \frac{y^2}{b^2} = 1 \tag{1}$$

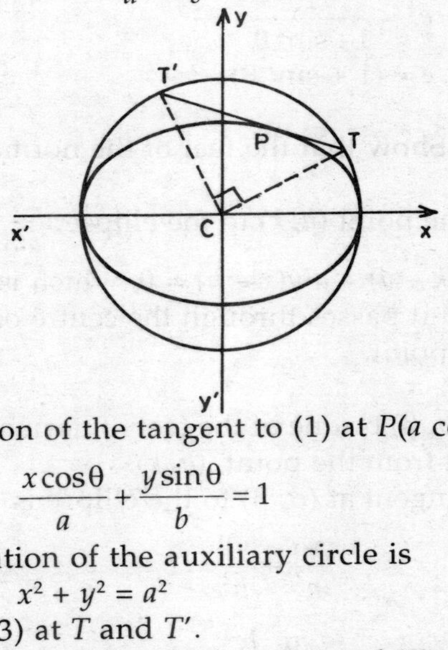


The equation of the tangent to (1) at $P(a \cos \theta, b \sin \theta)$ is

$$\frac{x\cos\theta}{a} + \frac{y\sin\theta}{b} = 1 \tag{2}$$

Also the equation of the auxiliary circle is

$$x^2 + y^2 = a^2 \tag{3}$$

Let (2) meet (3) at $T$ and $T'$.

We homogenise (3), with the help of (2) to get the joint equation of the line-pair and $CT$ and $CT'$,

$$x^2 + y^2 - a^2\left(\frac{x\cos\theta}{a} + \frac{y\sin\theta}{b}\right)^2 = 0 \tag{4}$$

Now by hypothesis

$$\angle TCT' = 90^0.$$

$\Rightarrow$ coefficient of $x^2$ + coefficient of $y^2$ in (4) is zero

$$\Rightarrow \quad (1-\cos^2\theta)+\left(1-\frac{a^2}{b^2}\sin^2\theta\right)=0$$

$$\Rightarrow \quad \sin^2\theta+1=\frac{a^2}{b^2}\sin^2\theta$$

$$=\frac{1}{1-e^2}.\sin^2\theta \text{ as } b^2=a^2(1-e^2).$$

$$\Rightarrow \quad (1-e^2)=\frac{\sin^2\theta}{1+\sin^2\theta}$$

$$\Rightarrow \quad e^2=\frac{1}{1+\sin^2\theta}$$

$$\Rightarrow \quad e=(1+\sin^2\theta)^{-1/2}$$

**Example 11.9** Show that the feet of the normals that can be drawn from the point $(h, k)$ to the ellipse, $\frac{x^2}{a^2}+\frac{y^2}{b^2}=1$, lie on the conic. $b^2x(k-y)+a^2y(x-h)=0$, which is a rectangular hyperbola and it passes through the centre of the ellipse as well as given point.

*Solution.* Let $(\alpha, \beta)$ be one of the feet of normals to the given ellipse, drawn from the point, $(h, k)$.

Now the tangent at $(\alpha, \beta)$ to the ellipse is

$$\frac{x\alpha}{a^2}+\frac{y\beta}{b^2}=1,$$

whose slope is $-\frac{\alpha}{a^2}\cdot\frac{b^2}{\beta}$

$\therefore$ the equation of the normal at $(\alpha, \beta)$ to it is

$$y-\beta=\frac{a^2\beta}{b^2\alpha}(x-\alpha)$$

$$\Rightarrow \quad b^2\alpha(y-\beta)-a^2\beta(x-\alpha)=0.$$

It passes through $(h, k)$, whence

$$b^2\alpha(k - \beta) - a^2\beta(h - \alpha) = 0$$

$\therefore$ the locus of $(\alpha, \beta)$, is

$$b^2x(k - y) + a^2y(x - h) = 0$$

$$\Rightarrow (a^2 - b^2)xy + b^2kx - a^2hy = 0$$

which represents a rectangular hyperbola passing through $(0, 0)$ and $(h, k)$.

**Example 11.10** If the normals at the four points $(x_1, y_1)$, $i = 1$, 2, 3, 4 on the ellipse, $\frac{x^2}{a^2} + \frac{y^2}{b^2} = 1$, are concurrent, show that

(i) $(x_1 + x_2 + x_3 + x_4)\left(\frac{1}{x_1} + \frac{1}{x_2} + \frac{1}{x_3} + \frac{1}{x_4}\right) = 4.$

(ii) $(y_1 + y_2 + y_3 + y_4)\left(\frac{1}{y_1} + \frac{1}{y_2} + \frac{1}{y_3} + \frac{1}{y_4}\right) = 4.$

*Solution*. Proceeding as in Example 11.9 we have

$$b^2\alpha(k - \beta) - a^2\beta(h - \alpha) = 0 \tag{1}$$

Now $(\alpha, \beta)$ lies on the ellipse,

$$\frac{x^2}{a^2} + \frac{y^2}{b^2} = 1,$$

$$\therefore \qquad \frac{\alpha^2}{a^2} + \frac{\beta^2}{b^2} = 1 \tag{2}$$

From (1), we have

$$\beta = \frac{b^2\alpha k}{a^2h - A\alpha}, \text{ where } A = a^2 - b^2 \tag{3}$$

Substituting for $\beta$ in equation (2), we obtain

$$\frac{\alpha^2}{a^2} + \frac{b^2\alpha^2k^2}{\left(a^2h - A\alpha\right)^2} = 1$$

$$\Rightarrow A^2\alpha^4 - 2Aa^2h\alpha^3 + (a^4h^2 + a^2b^2k^2 - a^2A^2)a^2 + 2a^4h\alpha - a^6h^2 = 0.$$

This is a fourth degree equation in $\alpha$, whose roots are $x_1$, $x_2$, $x_3$ and $x_4$.

$$\therefore \qquad \sum x_1 = \frac{2a^2h}{A}, \tag{4}$$

$$\sum x_1.x_2.x_3 = \frac{-2a^4h}{A} \tag{5}$$

and
$$x_1.x_2.x_3.x_4 = \frac{-a^6h^2}{A^2} \tag{6}$$

From (5) and (6) $\sum \frac{1}{x_1} = \frac{\sum x_1x_2x_3}{x_1x_2x_3x_4} = \frac{2A}{a^2h}$ (7)

But then (4) and (7) $\Rightarrow \sum x_1.\sum \frac{1}{x_1} = 4$

which proves (i)

Proof of (ii) is similar.

**Corollary.** If the normals at the four points($\phi_i$), $i = 1, 2, 3, 4$ on the ellipse, $\frac{x^2}{a^2} + \frac{y^2}{b^2} = 1$ are concurrent, prove that $\Sigma \cos \phi_i \, \Sigma \sec \phi_i = 4$.

Show that the mean position of these four points is $\left(\frac{1}{2}.\frac{a^2h}{a^2-b^2}, \frac{1}{2}.\frac{b^2k}{b^2-a^2}\right)$, where $(h, k)$ is the point of concurrency.

*Solution.* Here $x_i = a \cos \phi_i$ and $y_i = b \sin \phi_i$ $(i = 1, 2, 3, 4)$. Substitute these in

$\sum x_1.\sum \frac{1}{x_1} = 4$ to get the result.

Also $\quad \bar{x} = \frac{x_1 + x_2 + x_3 + x_4}{4} = \frac{1}{2}.\frac{a^2h}{a^2 - b^2}$

Similarly, $\bar{y} = \frac{y_1 + y_2 + y_3 + y_4}{4} = \frac{1}{2}.\frac{b^2k}{b^2 - a^2}$

**Example 11.11**.If the normals to $\frac{x^2}{a^2}+\frac{y^2}{b^2}=1$

at the ends of the chords

$$lx + my = 1 \text{ and } l'x + m'y = 1$$

be concurrent; show that

$$a^2ll' = b^2mm' = -1.$$

*Solution*. Let these normals meet in $(h, k)$.
Then their feet lie on the curve,

$$b^2x(k - y) + a^2y(x - h) = 0 \qquad \text{(Example 11.9)}$$

$$\Rightarrow (a^2 - b^2)\, xy + b^2kx - a^2hy = 0$$

$\therefore$ the locus passes through the points where the lines

$$lx + my - 1 = 0 \text{ and } l'x + m'y - 1 = 0$$

meet the given ellipse,

whence $$\left(\frac{x^2}{a^2}+\frac{y^2}{b^2}-1\right) - \lambda\,(lx + my - 1)(l'x + m'y - 1)$$

$$\equiv (a^2 - b^2)\, xy + b^2kx - a^2hy.$$

Comparing their respective coefficients, we get

$$\frac{1}{a^2}-\lambda ll'=0,\ \frac{1}{b^2}-\lambda mm'=0,\ -1-\lambda=0$$

$$\therefore \qquad a^2ll' = b^2mm' = -1.$$

**Exercise 11.3.**

1. If any tangent to the ellipse, $\frac{x^2}{a^2}+\frac{y^2}{b^2}=1$, intercepts lengths $\alpha$ and $\beta$ on the co-ordinate axes, show that

$$\frac{a^2}{\alpha^2}+\frac{b^2}{\beta^2}=1.$$

2. Given $\theta_1$ and $\theta_2$ are the inclinations of the tangents from a point $P$ to the ellipse, $\frac{x^2}{a^2}+\frac{y^2}{b^2}=1$. Find the locus of P if

(i) $\tan\theta_1 + \tan\theta_2 = c$ (ii) $\tan\theta_1 \tan\theta_2 = c$

(iii) $\tan^2\theta_1 + \tan^2\theta_2 = c$

3. If $p$ is the length of the perpendicular from the centre C

on the tangent at a point $P$ on the ellipse and $CP = r$; show that $a^2b^2/p^2 = a^2 + b^2 - r^2$.

4. (i) Find the common tangents of the ellipses, $\frac{x^2}{a^2}+\frac{y^2}{b^2}=1$ and $\frac{x^2}{b^2}+\frac{y^2}{a^2}=1$.

(ii) A circle of radius $r$ is concentric with the ellipse. Show that the common tangent is inclined to the major axis at an angle $\tan^{-1}[(r^2 - b^2)/(a^2 - r^2)]^{1/2}$ and find its length.

5. Find the point of intersection of the tangents at the points "$\alpha$" and "$\beta$" on the ellipse $\frac{x^2}{a^2}+\frac{y^2}{b^2}=1$.

6. P, Q, R are three points $\alpha, \beta, \gamma$ on the ellipse

$$\frac{x^2}{a^2}+\frac{y^2}{b^2}=1.$$

(i) Show that

$$\text{area } \Delta CPQ = \frac{1}{2}ab\sin(\alpha-\beta)$$

(ii) Show that

$$\text{area } \Delta PQR = 2ab\sin\left(\frac{\alpha-\beta}{2}\right).\sin\left(\frac{\beta-\gamma}{2}\right).\sin\left(\frac{\gamma-\alpha}{2}\right)$$

(iii) Show that the area of the traiangle formed by the tangents at P, Q, R is

$$ab\tan\left(\frac{\alpha-\beta}{2}\right)\tan\left(\frac{\beta-\gamma}{2}\right)\tan\left(\frac{\gamma-\alpha}{2}\right)$$

7. Prove that the tangents at the ends of a focal chord of an ellipse intersect on the directrix.

8. Find the equations to the tangents at the ends of the latera recta of the ellipse,

$$\frac{x^2}{a^2}+\frac{y^2}{b^2}=1$$

and show that they pass through the intersection of the major axis and the directrices.

9. Find the equations of the tangents and normals at the ends of latera recta of the ellipse,
$$9x^2 + 16y^2 = 144.$$

10. The tangent and normal to an ellipse at a point $T$ cut the minor axis at points $P$ and $Q$ respectively. Show that the circle described on $PQ$ as diameter passes through $T$ as well as the foci of the ellipse.

11. Any tangent to an ellipse is cut by the tangent at the extremities of the major axis in $T$ and $T'$. Prove that the circle drawn on $TT'$ as diameter passes through the foci.

12. Prove that the tangent and the normal at any point of an ellipse bisect the angles between the focal *radii* to that point.

13. If $\alpha$, $\beta$, $\gamma$ and $\delta$ be the eccentric angles of the four co–normal points on an ellipse, prove that
(i) $\Sigma \cos(\alpha + \beta) = 0$,
(ii) $\Sigma \sin(\alpha + \beta) = 0$.
(*Hint*: Use equation (2) of Theorem 11.1, (second proof)).

14. (i) The tangent at the point $(4 \cos \theta, (16/\sqrt{11}) \sin \theta)$ on the ellipse, $16x^2 + 11y^2 = 256$ is also the tangent to the circle, $x^2 + y^2 + 2x - 15 = 0$; find the value of $\theta$.
(ii) If $lx + my + n = 0$ is the tangent to the ellipse, $\frac{x^2}{a^2} + \frac{y^2}{b^2} = 1$, at the point whose normal passes through the point $(x_1, y_1)$; prove that
$$(a^2 - b^2)lm + mnx_1 - nly_1 = 0.$$

15. (i) The normal at any point $P$ of an ellipse cuts the major axis in $G$. Show that as $P$ varies, the mid-point of $PG$ traces another ellipse.
(ii) If the normals at the points $\alpha$, $\beta$, $\gamma$ on the ellipse, $\frac{x^2}{a^2} + \frac{y^2}{b^2} = 1$ are concurrent, prove that
$$\begin{vmatrix} \sec \alpha & \operatorname{cosec} \alpha & 1 \\ \sec \beta & \operatorname{cosec} \beta & 1 \\ \sec \gamma & \operatorname{cosec} \gamma & 1 \end{vmatrix} = 0$$
(*Hint*. See Q. 5 of Exercise 8.3).

16. (i) The length of the major axis intercepted between the tangent and normal at "θ" on the ellipse is equal to the semi–major axis. Prove that its eccentricity is $\sqrt{\sec\theta(\sec\theta - 1)}$

(ii) Prove that the eccentricity of the ellipse, $\frac{x^2}{a^2}+\frac{y^2}{b^2}=1$

is given by $e^2 \sin 2\theta = 2\sqrt{1-e^2}\cot w$.
where $w$ is one of the angles between the normals at the points whose eccentric angles are θ and $\frac{\pi}{2}+\theta$.

17. Show that the locus of the point of intersection of tangents at two points on the ellipse, $\frac{x^2}{a^2}+\frac{y^2}{b^2}=1$, is $\frac{x^2}{a^2}+\frac{y^2}{b^2}=\sec^2\alpha$ where $2\alpha$ is the difference of the eccentric angles of the two points.

18. If the tangent and the normal to the ellipse, $\frac{x^2}{a^2}+\frac{y^2}{b^2}=1$ at a point $(a\cos\theta, b\sin\theta)$ meet the major axis in $T_1, T_2$ and the minor axis in $G_1, G_2$ respectively, prove that
(i) $e^2\cos^2\theta = \cos\theta - 1 = 0$, where $T_1T_2 = a$.
(ii) $G_1G_2$ subtends a right angle at each foci of the ellipse.

19. If $P, Q, R$ be the points on the ellipse, $\frac{x^2}{a^2}+\frac{y^2}{b^2}=1$ and $p$, $q$, $r$ be the corresponding points on the auxiliary circle, prove that
(i) $\Delta CPQ / \Delta cpq = b/a$
(ii) $\Delta PQR / \Delta pqr = b/a$.

20. Prove that the normals at the points on the ellipse, whose eccentric angles are α, β, γ; are concurrent if and only if $\sin(\alpha+\beta) + \sin(\beta+\gamma) + \sin(\gamma+\alpha) = 0$.

21. If $p$ be the length of the perpendicular from a focus on

the tangent at any point $P$ of the ellipse and $r$ be the distance of $P$ from the focus, prove that $\frac{b^2}{p^2} = \frac{2a}{r} - 1$.

22. If $SY$ and $S'Y'$ are perpendiculars from the foci upon a tangent to the ellipse, $\frac{x^2}{a^2} + \frac{y^2}{b^2} = 1$ at $P$, show that $CY$ and $CY'$ are respectively parallel to $S'P$ and $SP$.

23. If the tangent any point $P$ on the ellipse, $\frac{x^2}{a^2} + \frac{y^2}{b^2} = 1$, whose centre is $C$ meets the major and minor axes in $T_1$ and $T_2$, prove that

(i) $\frac{a^2}{CT_1^2} + \frac{b^2}{CT_2^2} = 1$

(ii) $\overline{CT_1}.\overline{CN} = a^2$, where $PN$ is the perpendicular on the major axis.

(iii) $\overline{CT_2}.\overline{CM} = b^2$, where $PM$ is the perpendicular on the minor axis.

24. If the normal at any point $P$ on the ellipse, $\frac{x^2}{a^2} + \frac{y^2}{b^2} = 1$, meets the major and minor axes in $G_1$ and $G_2$ respectively and if $CD$ be the perpendicular from $C$ on this normal, then show that

(i) $\overline{PD}.\overline{PG_1} = b^2$; $\overline{PD}.\overline{PG_2} = a^2$

(ii) $\overline{PG_1}.\overline{PG_2} = \overline{SP}.\overline{S'P}$

(iii) $\overline{CG_1}.\overline{CT} = CS^2$, where $T$ is the point where tangent at $P$ meets the major axis.

(iv) $CG_1 = e^2\, CN$, where $PN$ is the perpendicular on the major axis.

(v) $SG_1/S'G_1 = SP/S'P$

where $S$ and $S'$ are the foci of the ellipse.

**Example 11.12**. If the chord of contact of the tangents from $P$ to the ellipse, $\frac{x^2}{a^2}+\frac{y^2}{b^2}=1$ always touches the circle on the join of foci as diameter, show that $P$ moves on the ellipse, $\frac{x^2}{a^4}+\frac{y^2}{b^4}=\frac{1}{a^2-b^2}$

*Solution*. Let the co-ordinates of $P$ be $(x_1, y_1)$. Its chord of contact w.r.t. the given ellipse is

$$\frac{xx_1}{a^2}+\frac{yy_1}{b^2}=1 \qquad (1)$$

Equation of the circle on the foci $(\pm ae, 0)$ as diameter is

$$(x-ae)(x+ae)+(y-0)(y-0)=0$$

$$\Rightarrow \quad x^2+y^2=a^2e^2 \qquad (2)$$

By hypothesis, line (1) touches the circle (2).

$$\therefore \quad \frac{-1}{\sqrt{\frac{x_1^2}{a^4}+\frac{y_1^2}{b^4}}}=\pm\, ae$$

$$\Rightarrow \quad \frac{x_1^2}{a^4}+\frac{y_1^2}{b^4}=\frac{1}{a^2e^2}=\frac{1}{a^2-b^2} \quad (\because\ b^2=a^2(1-e^2))$$

$\therefore$ The locus of $(x_1, y_1)$ is

$$\frac{x^2}{a^4}+\frac{y^2}{b^4}=\frac{1}{a^2-b^2} \quad \text{(See also Example 12.8)}$$

**Example 11.13** Prove that the locus of mid-points of chords of the ellipse, $\frac{x^2}{a^2}+\frac{y^2}{b^2}=1$, which touch the minor auxiliary circle, is $\left(\frac{x^2}{a^2}+\frac{y^2}{b^2}\right)^2=b^2\left(\frac{x^2}{a^4}+\frac{y^2}{b^4}\right)$

*Solution*. The equation of the minor auxiliary circle is $x^2+y^2=b^2$.

Any tangent to it is given by

$$y = mx + b\sqrt{1+m^2} \tag{1}$$

Now (1) is the chord of the ellipse,

$$\frac{x^2}{a^2} + \frac{y^2}{b^2} = 1 \tag{2}$$

Let $(x_1, y_1)$ be its mid-point.

$\therefore$ The equation of the chord of (2) with mid–point $(x_1, y_1)$ is

$$T = S_1 \Rightarrow \frac{xx_1}{a^2} + \frac{yy_1}{b^2} = \frac{x_1^2}{a^2} + \frac{y_1^2}{b^2} \tag{3}$$

Now (1) and (3) represent the same line, comparing their coefficients, we get

$$\frac{-m}{x_1 / a^2} = \frac{1}{y_1 / b^2} = \frac{b\sqrt{1+m^2}}{\frac{x_1^2}{a^2} + \frac{y_1^2}{b^2}}$$

$$\Rightarrow \qquad -m = \frac{x_1 / a^2}{y_1 / b^2}$$

$$\text{and} \qquad b\sqrt{1+m^2} = \left(\frac{x_1^2}{a^2} + \frac{y_1^2}{b^2}\right) \Big/ \frac{y_1}{b^2}$$

$$\Rightarrow \qquad b^2\left[1 + \left(\frac{x_1 / a^2}{y_1 / b^2}\right)^2\right] \equiv \left[\left(\frac{x_1^2}{a^2} + \frac{y_1^2}{b^2}\right) \Big/ \frac{y_1}{b^2}\right]^2$$

$$\Rightarrow \qquad \left(\frac{x_1^2}{a^2} + \frac{y_1^2}{b^2}\right)^2 = b^2\left(\frac{x_1^2}{a^4} + \frac{y_1^2}{b^4}\right)$$

Now, we take the locus of $(x_1, y_1)$ and get the desired result.

**Example 11.14.** Show that the rectangle under the perpendiculars drawn to the normal at a point $P$ of an ellipse from the centre and from the pole of the normal is equal to the rectangle under the focal distances of $P$.

*Solution*. Let the given ellipse be

$$\frac{x^2}{a^2}+\frac{y^2}{b^2}=1 \tag{1}$$

Let the co-ordinates of $P$ on (1) be $(a\cos\theta, b\sin\theta)$

The equation of the normal to (1) at $P$ is

$$ax\sec\theta - by\operatorname{cosec}\theta = a^2-b^2 \tag{2}$$

Let its pole w.r.t. (1) be $(x_1, y_1)$

$\therefore$ The polar of $(x_1, y_1)$ w.r.t. (1) is

$$\frac{xx_1}{a^2}+\frac{yy_1}{b^2}=1 \tag{3}$$

As (2) and (3) represent the same line, comparing their co-efficients, we get

$$\frac{x_1}{a^3\sec\theta}=\frac{y_1}{-b^3\operatorname{cosec}\theta}=\frac{1}{a^2-b^2}$$

$$\Rightarrow \qquad x_1=\frac{a^3\sec\theta}{a^2-b^2};\ y_1=\frac{-b^3\operatorname{cosec}\theta}{a^2-b^2} \tag{4}$$

Now $p_1 = \perp$ from C(0, 0) on (2)

$$=\frac{a^2-b^2}{\sqrt{a^2\sec^2\theta+b^2\operatorname{cosec}^2\theta}}$$

and $p_2 = \perp$ from $(x_1, y_1)$ on (2)

$$=\frac{ax_1\sec\theta-by_1\operatorname{cosec}\theta-(a^2-b^2)}{\sqrt{a^2\sec^2\theta+b^2\operatorname{cosec}^2\theta}}$$

$$=\frac{a^4\sec^2\theta-b^4\operatorname{cosec}^2\theta-(a^2-b^2)^2}{(a^2-b^2)\sqrt{a^2\sec^2\theta+b^2\operatorname{cosec}^2\theta}}$$

$$\therefore\ p_1p_2=\frac{a^4\sec^2\theta+b^4\operatorname{cosec}^2\theta-(a^2-b^2)^2}{a^2\sec^2\theta+b^2\operatorname{cosec}^2\theta}$$

$$=\frac{a^4\sin^2\theta+b^4\cos^2\theta-(a^2-b^2)^2\sin^2\theta\cos^2\theta}{a^2\sin^2\theta+b^2\cos^2\theta}$$

$$=\frac{a^4\sin^2\theta(1-\cos^2\theta)+b^4\cos^2\theta(1-\sin^2\theta)+2a^2b^2\sin^2\theta\cos^2\theta}{a^2\sin^2\theta+b^2\cos^2\theta}$$

$$=a^2\sin^2\theta+b^2\cos^2\theta$$

$$=a^2\sin^2\theta+a^2(1-e^2)\cos^2\theta$$

$= a^2(1 - e^2 \cos^2 \theta)$
$= a(1 - e \cos \theta).\, a(1 + e \cos \theta)$
$= \overline{SP}.\overline{S'P}$ (See Example 11.5).

$\therefore$ The result follows.

**Example 11.15.** *PQ, PR* is a pair of perpendicular tangents to the ellipse, $\frac{x^2}{a^2}+\frac{y^2}{b^2}=1$. Prove that *QR* always touches the ellipse $\frac{x^2}{a^4}+\frac{y^2}{b^4}=\frac{1}{a^2+b^2}$

*Solution*. Since $PQ \perp PR$ and *PQ, PR* are tangents to the ellipse,

$$\frac{x^2}{a^2}+\frac{y^2}{b^2}=1 \qquad (1)$$

$\therefore$ P lies on the director circle of (1), whose equation is

$$x^2 + y^2 = a^2 + b^2 \qquad (2)$$

(See Q. 5(ii), Exercise 8.7)

Thus we can take the co-ordinates of *P* as $(r \cos \theta, r \sin \theta)$ where $r=\sqrt{a^2+b^2}$.

Now *QR*, being the chord of contact of *P* w.r.t. (1), has the equation

$$\frac{x.r\cos\theta}{a^2}+\frac{y.r\sin\theta}{b^2}=1$$

$$\Rightarrow \quad \frac{x\cos\theta}{a^2/r}+\frac{y\sin\theta}{b^2/r}=1$$

which touches the ellipse, whose equation is

$$\frac{x^2}{\left(a^2/r\right)^2}+\frac{y^2}{\left(b^2/r\right)^2}=1$$

(cf. $\frac{x\cos\theta}{a}+\frac{y\sin\theta}{b}=1$ touches the ellipse $\frac{x^2}{a^2}+\frac{y^2}{b^2}=1$).

$$\Rightarrow \frac{x^2}{a^4}+\frac{y^2}{b^4}=\frac{1}{r^2}=\frac{1}{a^2+b^2}.$$

## Exercise 11.4

1. If $lx + my = 1$ meets the ellipse $\frac{x^2}{a^2}+\frac{y^2}{b^2}=1$ in $P$ and $Q$, prove that the mid-point of $PQ$ is

$$\left(\frac{a^2l}{a^2l^2+b^2m^2}, \frac{b^2m}{a^2l^2+b^2m^2}\right).$$

2. A chord of $x^2/a^2 + y^2/b^2 = 1$ has $(\alpha, \beta)$ as its mid-point. Prove that the co-ordinates of its pole are

$$\left(\frac{\alpha}{\alpha a^{-2}+\beta b^{-2}}, \frac{\beta}{\alpha a^{-2}+\beta b^{-2}}\right).$$

3. Show that the points $(x_i, y_i)$ $(i = 1, 2, 3)$ are collinear if their polars w.r.t. the ellipse, $\frac{x^2}{a^2}+\frac{y^2}{b^2}=1$ are concurrent.
4. Tangents are drawn from any point on $\frac{x^2}{a^2}+\frac{y^2}{b^2}=1$, to the circle, $x^2 + y^2 = ab$, prove that chords of contact always touch the ellipse, $\frac{x^2}{a^2}+\frac{y^2}{b^2}=1$.
5. From a point $P$ of an ellipse, two tangents are drawn to the circle on the minor axis as diameter. Prove that these tangents will meet the diameter at right angles to $CP$ in points lying on two fixed lines parallel to the major axis.
6. Prove that the locus of the mid-points of the chords of the ellipse, $\frac{x^2}{a^2}+\frac{y^2}{b^2}=1$, tangent at the ends of which intersect on the circle, $x^2 + y^2 = a^2$ is

$$\left(\frac{x^2}{a^2}+\frac{y^2}{b^2}\right)^2=\frac{x^2+y^2}{a^2}.$$

7. If $l_1$ and $l_2$ be the lengths of two tangents to the ellipse, $\frac{x^2}{a^2}+\frac{y^2}{b^2}=1$, at right angles to one another, prove that

$$\frac{4\left(a^2+b^2\right)^3}{l_1^2+l_1^2}=[a^2+b^2+a^2b^2\,(l_1^{-2}+l_2^{-2})]^2.$$

8. A variable chord subtends a right angle at the centre of the ellipse. Find the locus of the point of intersection of tangents at the ends.
9. Triangles are formed by pair of tangents drawn from any point on the ellipse, $a^2x^2+b^2y^2=(a^2+b^2)^2$ to the ellipse, $\frac{x^2}{a^2}+\frac{y^2}{b^2}=1$, and their chord of contact. Prove that the orthocentre of each such triangle lies on the second ellipse.
10. $RP$, $RQ$ are the tangents drawn from an external point $R(x_1, y_1)$ to the ellipse, $\frac{x^2}{a^2}+\frac{y^2}{b^2}=1$; show that

    (i) if $S$ be a focus, $\frac{RS^2}{SP.SQ}=\frac{x_1^2}{a^2}+\frac{y_1^2}{b^2}$

    (ii) if $C$ be the centre of the ellipse, area of $\Delta CPQ$ is

    $$\frac{a^2b^2\left(b^2x_1^2+a^2y_1^2-a^2b^2\right)^{1/2}}{\left(b^2x_1^2+a^2y_1^2\right)}$$

    and area of $\Delta PQR$ is $\frac{\left(b^2x_1^2+a^2y_1^2-a^2b^2\right)^{1/2}}{\left(b^2x_1^2+a^2y_1^2\right)}$.
11. Find the locus of the point of intersection of a pair of tangents to an ellipse, the bisectors of the angles between which are equally inclined to the axes of the ellipse.

12. (i) Prove that the perpendicular from the focus of an ellipse whose centre is $C$, on the polar of any point $P$ will meet $CP$ on the directrix.

(ii) If the product of perpendiculars from the foci of the ellipse, $\frac{x^2}{a^2}+\frac{y^2}{b^2}=1$ upon the polar of $P$ be always $c^2$; prove that the locus of $P$ is the ellipse,

$$b^4(c^2 + a^2 - b^2)x^2 + c^2a^4y^2 = a^4b^4.$$

13. (i) Show that the locus of middle points of focal chords of the ellipse,

$$\frac{x^2}{a^2}+\frac{y^2}{b^2}=1 \text{ is } \frac{x^2}{a^2}+\frac{y^2}{b^2}\pm\frac{ex}{a}$$

(ii) Prove that the locus of the middle points of the portion of a tangent to the ellipse,

$$\frac{x^2}{a^2}+\frac{y^2}{b^2}=1$$

included between the axes is $a^2x^{-2} + b^2y^{-2} = 4$.

## 11.4 TO FIND THE LOCUS OF THE MIDDLE POINTS OF A SYSTEM OF PARALLEL CHORDS OF AN ELLIPSE.

Let $M(\alpha, \beta)$ be the middle point of one of these parallel chords of the ellipse,

$$\frac{x^2}{a^2}+\frac{y^2}{b^2}=1$$

and let their common slope be m.

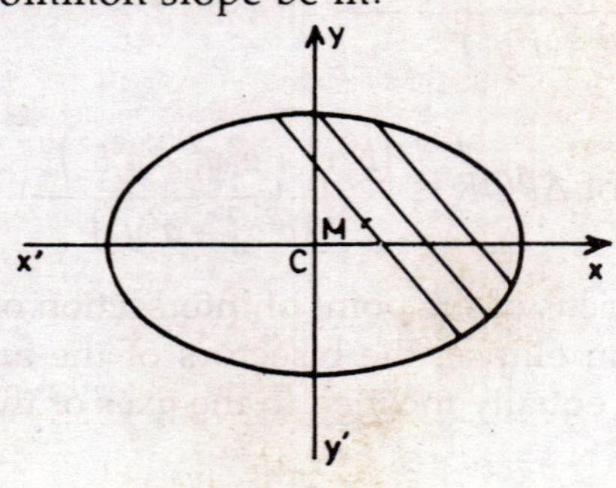

The equation of this chord of the ellipse

is $$T = S' \Rightarrow \frac{x\alpha}{a^2} + \frac{y\beta}{b^2} = \frac{\alpha^2}{a^2} + \frac{\beta^2}{b^2}$$

$$\text{Its slope} = -\frac{\alpha / a^2}{\beta / b^2} = m$$

$$\Rightarrow \qquad \beta = -\frac{b^2}{a^2 m}\alpha$$

$\therefore$ the locus of $(\alpha, \beta)$ is

$$y = -\frac{b^2}{a^2 m}x.$$

Clearly, it passes through the centre of the ellipse.

**Definition**. The locus of the middle points of a system of parallel chords of an ellipse (or any conic) is called a diameter of the ellipse (the conic).

From above, we see that $y = -\dfrac{b^2}{a^2 m}x$ is a diameter of the ellipse. If $m'$ is its slope then, clearly, $mm' = -\dfrac{b^2}{a^2}$.

The symmetry of this result in $m$ and $m'$ shows that the diameter, $y = mx$, will bisect all parallel chords of the ellipse having the common slope, $m'$ and vice–versa.

### 11.4.1. Conjugate diameters

**Definition**. Two diameters of an ellipse are called *conjugate* if each bisects all the chords of the ellipse parallel to the other. Thus for two diameters, $y = mx$ and $y = m'x$ of the ellipse to be conjugate, $mm' = -b^2/a^2$.

### 11.4.2. To find the co–ordinates of the extremities of a pair of conjugate diameters of an ellipse.

Let $P'CP$ and $D'CD$ be a pair of conjugate diameters.

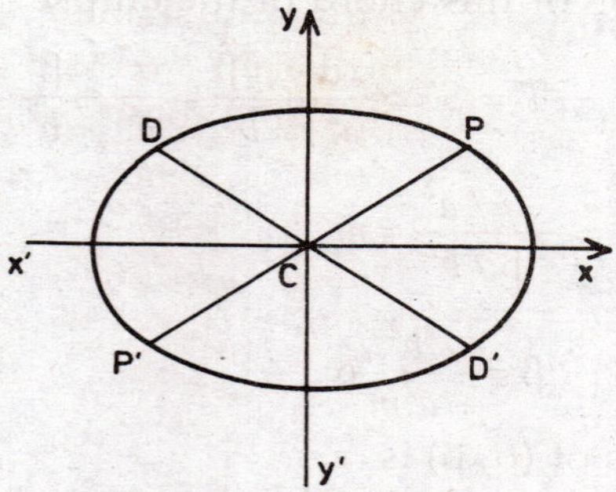


Let $P$ be $(a \cos \varphi, b \sin \varphi)$
and $D$ be $(a \cos \theta, b \sin \theta)$.

$\Rightarrow$ the slope of $CP = m = \dfrac{b \sin\varphi}{a \cos\varphi}$

and the slope of $CD = m' = \dfrac{b \sin\theta}{a \cos\theta}$

Now $CP$ and $CD$ are conjugate diameters,

$$\therefore \quad mm' = \frac{-b^2}{a^2}$$

$$\Rightarrow \quad \frac{b^2 \sin\phi \sin\theta}{a^2 \cos\phi \cos\theta} = \frac{-b^2}{a^2}$$

$$\Rightarrow \quad \sin \varphi \sin \theta = -\cos \varphi \cos \theta$$

$$\Rightarrow \quad \cos (\theta - \varphi) = 0.$$

$$\therefore \quad \theta - \varphi = \pi/2 \Rightarrow \theta = \pi/2 + \varphi$$

Thus $D$ is $[a \cos (\pi/2 + \phi), b \sin (\pi/2 + \varphi)]$
$= (-a \sin \varphi, b \cos \varphi)$.

Thus, if $CP$ and $CD$ are a pair of conjugate diameters and if the eccentric angle of the point $P$ is $\varphi$, then the eccentric angle of the point $D$ is $\pi/2 + \varphi$. Further the eccentric angles of $P'$ and $D'$ are respectively $\pi + \phi$ and $3\pi/2 + \phi$ and their co-ordinates are

$$(-a \cos \varphi, -b \sin \varphi)$$

and

$$(a \sin \varphi, -b \cos \varphi)$$

respectively.

### 11.4.3. Properties of Conjugate Diameters

We now prove certain results about conjugate diameters.

*Proposition 1.* If $CP$ and $CD$ be a pair of conjugate diameters of an ellipse, prove that $CP^2 + CD^2$ is constant and is equal to sum of squares of its semi-major and semi-minor axes.

*Proof.* We have

$$CP^2 = a^2 \cos^2 \varphi + b^2 \sin^2 \varphi$$

and

$$CD^2 = a^2 \sin^2 \varphi + b^2 \cos^2 \varphi.$$

$$\therefore \quad CP^2 + CD^2 = a^2 + b^2$$

Hence the result.

*Proposition 2.* The tangents at the extremities of a pair of conjugate diameters of an ellipse enclose a parallelogram whose area is constant and is equal to the area of the rectangle contained under the major and the minor axes of the ellipse.

*Proof.* Let $P'CP$ and $D'CD$ be a pair of conjugate diameters and let the tangents at $P$, $D$, $P'$, $D'$ meet in $Q$, $R$, $Q'$ and $R'$.

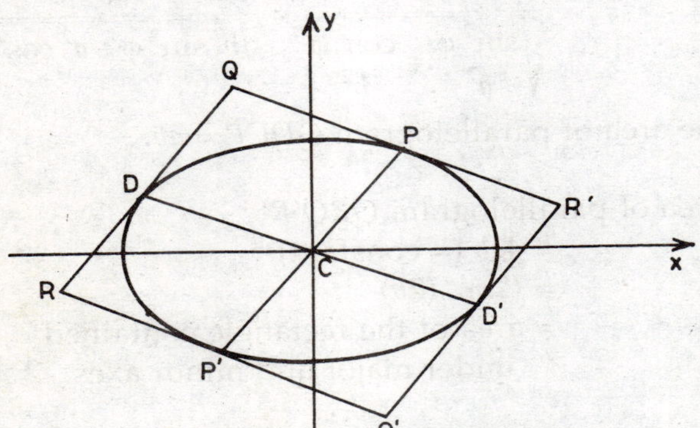


The equations of the tangents at P and P' are

$$\frac{x \cos \phi}{a} + \frac{y \sin \phi}{b} = 1$$

and $$\frac{-x\cos\phi}{a}-\frac{y\sin\phi}{b}=1$$

which are both parallel to the line $DCD'$ whose equation is

$$y=-\frac{b\cos\phi}{a\sin\phi}.x$$

Similarly the tangents at $D$ and $D'$ are both parallel to the line $P'CP$.

$QRQ'R'$ is a parallelogram.

By symmetry, the area of parallalogram $QRQ'R'$

$= 4.$(the area of parallelogram $QDCP$)

$= 4.|QD|.[|\perp$ from $C$ on $QD|$

$= 4.|CP|.[|\perp$ from $C$ on $QD|$.

Now $$|CP|=\sqrt{a^2\cos^2\varphi+b^2\sin^2\varphi}$$

Also the equation of $QD$, the tangent to the ellipse at the point $D$ $(-a\sin\phi, b\cos\phi)$ is

$$-\frac{x\sin\phi}{a}+\frac{y\cos\phi}{b}=1$$

$\therefore$ The perpendicular distance from $C(0,0)$ on this tangent, is

$$\frac{1}{\sqrt{\frac{\sin^2\varphi}{a^2}+\frac{\cos^2\varphi}{b^2}}}=\frac{ab}{\sqrt{b^2\sin^2\varphi+a^2\cos^2\varphi}}$$

$\therefore$ The area of parallelogram $QDCP = ab$.

Thus

the area of parallelogram $QRQ'R'$

$= 4ab$ (= constant)

$= (2a) . (2b)$

= area of the rectangle contained under major and minor axes.

*Remark.* We know that polar of a point and the chord of contact of tangents from it to the ellipse are same. So the pole of the diameter $P'CP$ will be the point of intersection of the tangents at $P$ and $P'$, which are parallel (See Proposition 2, above). Thus the pole lies at infinity on the conjugate diameter $D'CD$. In the same way, the pole of the diameter $D'CD$ lies on $P'CP$. Consequently $P'CP$ and $D'CP$ are conjugate lines i.e. the conjugate diameters are a special case of conjugate lines.

### 11.4.4. Equi-conjugate diameters.

**Definition.** Two conjugate diameters of an ellipse are said to be *equi-conjugate* if their lengths are equal.

**Example 11.16.** Prove that equi-conjugate diameters of an ellipse lie along the diagonals of the rectangle formed by the tangents at the ends of its axes.

*Solution.* Let $P'CP$ and $D'CD$ be two conjugate diameters of the ellipse, $\frac{x^2}{a^2}+\frac{y^2}{b^2}=1$.

Let the co-ordinates of $P$ be $(a\cos\theta, b\sin\theta)$ then the co-ordinates of $D$ are

$(-a\sin\theta, b\cos\theta)$. Also $C$ is $(0, 0)$

$\therefore \quad CP^2 = a^2\cos^2\theta + b^2\sin^2\theta$

and $\quad CD^2 = a^2\sin^2\theta + b^2\cos^2\theta$.

For equi-conjugate diameters,

$$\overline{CP} = \overline{CD} \Rightarrow CP^2 = CD^2$$

$\Rightarrow a^2\cos^2\theta + b^2\sin^2\theta = a^2\sin^2\theta + b^2\cos^2\theta$

$\Rightarrow \tan^2\theta = 1 \Rightarrow \tan\theta = \pm 1$

$\Rightarrow \theta = \pi/4$ or $3\pi/4$.

When $\theta = \pi/4$, the equation of the diameter $CP$ is

$$y-0=\frac{b\sin\theta}{a\cos\theta}(x-0)\Rightarrow y=\frac{b}{a}x$$

and that of CD is

$$y-0=\frac{b\cos\theta}{-a\sin\theta}(x-0)\Rightarrow y=-\frac{b}{a}x$$

When $\theta = 3\pi/4$, we again get these two equations.

$\therefore$ The equations of the equi–conjugate diameters are $y=\pm\frac{b}{a}x$

which are the equations of the diagonals of the rectangle formed by the tangents at the four vertices of the ellipse.

**Example 11.17.** Prove that the acute angle between two conjugate diameters is least when these are of equal lengths i.e. when the diameters are equi-conjugate.

*Solution.* Let $CP$ and $CD$ be the conjugate semi-diameters, then

$$\left(\overline{CP}-\overline{CD}\right)^2 = CP^2 + CD^2 - 2\overline{CP}.\overline{CD}$$

$$\Rightarrow \quad 2\overline{CP}.\overline{CD} = a^2 + b^2 - \left(\overline{CP}-\overline{CD}\right)^2$$

Now area of parallelogram $CPQD$

$$= ab = \overline{CP}.\overline{CD}\sin(\angle PCD)$$

$$\Rightarrow \quad 2ab = [a^2 + b^2 - (\overline{CP}-\overline{CD})^2]\sin(\angle PCD)$$

$$\Rightarrow \quad \sin(\angle PCD) = 2ab/[a^2 + b^2 - (\overline{CP}-\overline{CD})^2]$$

For $\angle PCD$ to be least, $[a^2 + b^2 - (\overline{CP}-\overline{CD})^2]$ must be greatest, which is possible only when $\overline{CP} = \overline{CD}$
i.e. when the diameters are equi–conjugate.

### 11.4.5. Supplemental Chords

**Definition**. The two chords drawn from any point on an ellipse to the extremities of any diameter are called *supplemental chords*.

If $P'CP$ be any diameter of an ellipse and $Q$ is any point of it, then $QP'$ and $QP$ are supplemental chords.

**Example 11.18.** Prove that any two supplemental chords of the ellipse are parallel to a pair of conjugate diameters.

*Solution.* Let the given ellipse be

$$\frac{x^2}{a^2} + \frac{y^2}{b^2} = 1 \qquad (1)$$

Let $P'CP$ be any diameter of (1). Suppose the eccentric angle of $P$ is $\theta$, then the eccentric angle of $P'$ is $\pi + \theta$ i.e. their co-ordinates respectively are $(a \cos \theta, b \sin \theta)$ and $(-a \cos \theta, -b \sin \theta)$.

Let any point $Q$ on (1) have co–ordinates $(a \cos \phi, b \sin \phi)$. Now the slope of $QP$

$$= \frac{b \sin\phi - b \sin\theta}{a\cos\phi - a\cos\theta}$$

and the slope of Q'P

$$= \frac{b \sin\phi + b \sin\theta}{a\cos\phi + a\cos\theta}$$

$\therefore$ The product of their slopes

$$= \frac{b(\sin\phi - \sin\theta)}{a(\cos\phi - \cos\theta)} \times \frac{b(\sin\phi + \sin\theta)}{a(\cos\phi + \cos\theta)}$$

$$= \frac{b^2}{a^2} \cdot \frac{\sin^2\phi - \sin^2\theta}{\cos^2\phi - \cos^2\theta} = -\frac{b^2}{a^2}$$

$\Rightarrow QP$ and $QP'$ (the supplemental chords) are parallel to a pair of conjugate diameters.

**Example 11.19**. Find the locus of the point of intersection of the tangents drawn at the extremities of a pair of conjugate diameters, of an ellipse.

*Solution*. Let $(\alpha, \beta)$ be one such point of intersection. The tangents at $P(a \cos \phi, b \sin \phi)$ and $D(-a \sin \phi, b \cos \phi)$to the ellipse,

$$\frac{x^2}{a^2} + \frac{y^2}{b^2} = 1$$

are
$$\frac{x\cos\phi}{a} + \frac{y\sin\phi}{b} = 1$$

and
$$-\frac{x\sin\phi}{a} + \frac{y\cos\phi}{b} = 1$$

Both of them pass through $(\alpha, \beta)$,

$$\therefore \qquad \frac{\alpha\cos\phi}{a}+\frac{\beta\sin\phi}{b}=1 \tag{1}$$

and
$$-\frac{\alpha\sin\phi}{a}+\frac{\beta\cos\phi}{b}=1 \tag{2}$$

Now, to get the locus of (α, β), we have to eliminate φ from (1) and (2). To do this, we square and add these equations and obtain

$$\frac{\alpha^2}{a^2}+\frac{\beta^2}{b^2}=2$$

∴ The locus of (α, β) is

$$\frac{x^2}{a^2}+\frac{y^2}{b^2}=2$$

**Example 11.20**. If θ and θ′ be the angles subtended by the major axis of an ellipse at the extremities of a pair of conjugate diameters, show that $\cot^2\theta + \cot^2\theta'$ is constant.

*Solution*. Let

$$P(a\cos\phi,\ b\sin\phi)$$

and
$$D\left(a\cos\left(\frac{\pi}{2}+\phi\right),\ b\sin\left(\frac{\pi}{2}+\phi\right)\right)$$

be the extremities of a pair of conjugate diameters of the ellipse,

$$\frac{x^2}{a^2}+\frac{y^2}{b^2}=1$$

with $A'(-a, 0)$ and $A(a, 0)$ as the end points of its major axis.

Then
$$\angle A'PA = \theta$$
and
$$\angle A'DA = \theta'$$

Now $\quad m_1 = \text{slope of } AP = \dfrac{b\sin\varphi}{a(\cos\varphi-1)}$

and $\quad m_2 = \text{slope of } A'P = \dfrac{b\sin\varphi}{a(\cos\varphi+1)}$

$$\therefore\ \tan^2\theta = \left(\frac{m_1 - m_2}{1 + m_1 m_2}\right)^2 = \frac{\left[\frac{b}{a}\frac{\sin\phi}{(\cos\phi - 1)} - \frac{b}{a}\frac{\sin\phi}{(\cos\phi + 1)}\right]^2}{\left[1 + \frac{b^2}{a^2}\frac{\sin^2\phi}{(-\sin^2\phi)}\right]^2}$$

$$= \frac{4a^2b^2}{(a^2 - b^2)^2 \sin^2\phi}$$

$$\Rightarrow \qquad = \cot^2\theta = \frac{\left(a^2 - b^2\right)^2}{4a^2b^2}\sin^2\phi .$$

Replacing $\phi$ by $\frac{\pi}{2} + \phi$ in it, for $D$, to get

$$\cot^2\theta' = \frac{\left(a^2 - b^2\right)^2}{4a^2b^2}\cos^2\varphi$$

Thus $\quad \cot^2\theta + \cot^2\theta' = \dfrac{\left(a^2 - b^2\right)^2}{4a^2b^2}$

which is a constant.

**Example 11.21** (a) Find the condition that the lines, $Ax^2 + 2Hxy + By^2 = 0$ may be a pair of conjugate diameters of the ellipse, $\dfrac{x^2}{a^2} + \dfrac{y^2}{b^2} = 1$.

(b) If the points of intersection of the ellipses $\dfrac{x^2}{a^2} + \dfrac{y^2}{b^2} = 1$ and $\dfrac{x^2}{A^2} + \dfrac{y^2}{B^2} = 1$ be the points of the conjugate diameters of the former; prove that $\dfrac{a^2}{A^2} + \dfrac{b^2}{B^2} = 2$.

*Solution.* (a) We know that

$$Ax^2 + 2Hxy + By^2 = 0 \qquad (1)$$

representes two lines through the origin. Let their equations be

$$y = m_1 x \text{ and } y = m_2 x \quad (2)$$

$$\therefore \quad m_1 + m_2 = -2H/B \text{ and } m_1 m_2 = A/B \quad (3)$$

For the lines (2) to be a pair of conjugate diameters,

$$m_1 . m_2 = -\, b^2/a^2 \quad (4)$$

From (3) and (4), we get

$$A/B = -b^2/a^2 \Rightarrow a^2A + b^2B = 0$$

which is the required condition.

(b) The equation of a conic, passing through the points of intersection of the ellipses

$$\frac{x^2}{a^2} + \frac{y^2}{b^2} = 1 \quad (1)$$

and

$$\frac{x^2}{A^2} + \frac{y^2}{B^2} = 1 \quad (2)$$

is

$$\left(\frac{x^2}{a^2} + \frac{y^2}{b^2} - 1\right) + \lambda\left(\frac{x^2}{A^2} + \frac{y^2}{B^2} - 1\right) = 0$$

Taking $\lambda = -1$, in it, the equation reduces to

$$\left(\frac{1}{a^2} - \frac{1}{A^2}\right)x^2 + \left(\frac{1}{b^2} - \frac{1}{B^2}\right)y^2 = 0 \quad (3)$$

which represents two lines through the points of intersection of (1) and (2) and also through the origin (centre).

i.e. (3) represents a pair of diameters of the ellipse (1) (by hypothesis).

Applying part (a), we get

$$a^2A + b^2B = 0$$

$$\Rightarrow \quad a^2\left(\frac{1}{a^2} - \frac{1}{A^2}\right) + b^2\left(\frac{1}{b^2} - \frac{1}{B^2}\right) = 0$$

$$\Rightarrow \quad \frac{a^2}{A^2} + \frac{b^2}{B^2} = 2$$

## Exercise 11.5.

1. $CP$ and $CD$ are two conjugate diameters of the ellipse
$$\frac{x^2}{a^2}+\frac{y^2}{b^2}=1;$$
show that the equation of $PD$ is
$$\frac{x}{a}\cos\left(\phi+\frac{\pi}{4}\right)+\frac{y}{b}\sin\left(\phi+\frac{\pi}{4}\right)=\frac{1}{\sqrt{2}}$$
where $\phi$ is the eccentric angle of the point $P$.

2. Prove that
$$\overline{SP}.\overline{S'P}=CD^2,$$
where $CP$ and $CD$ are conjugate diameters of the ellipse and $S$, $S'$ are its foci.

3. (i) Prove that the polar of any point on a diameter of an ellipse w.r.t. it is parallel to the conjugate diameter.
(ii) Deduce that the tangent lines drawn at the ends of a diameter of an ellipse are parallel to the conjugate diameter.

4. Let $CP$ and $CD$ be a pair of conjugate semi-diameters of an ellipse and let $p$ and $d$ be the points on the auxiliary circle corresponding to $P$ and $D$ respectively. Prove that $\angle pCd$ is a right angle.

5. Prove that the tangents to the ellipse
$$\frac{x^2}{a^2}+\frac{y^2}{b^2}=1$$
at the points whose eccentric angles differ by $\pi/2$, intersect on the ellipse
$$\frac{x^2}{a^2}+\frac{y^2}{b^2}=2$$
whose eccentricity is the same as that of the given ellipse. (*Hint*: See Example 11.19).

6. Let $CP$ and $CQ$ be a pair of conjugate diameters of an ellipse and let the tangents at $P$ and $Q$ meet at $R$. Show that $CR$ and $PQ$ bisect each other.

7. Find the condition for the diameters of

$$\frac{x^2}{a^2}+\frac{y^2}{b^2}=1.$$

through its points of intersection with the line

$$lx + my + n = 0$$

to be conjugate.

(*Hint*: $P$ and $D$ lie on this line and eliminate $\phi$).

8. (i) Prove that

$$b^2x^2 + 2hxy - a^2y^2 = 0$$

represents conjugate diameters of the ellipse

$$\frac{x^2}{a^2}+\frac{y^2}{b^2}=1$$

for all values of $h$.

(ii) Prove that

$$ax^2 + 2hxy - by^2 = 0$$

represents conjugate diameters of the ellipse

$$ax^2 + by^2 = 1$$

for all values of $h$.

(iii) Show that the lines joining the centre to the points of intersection of the line, $y = mx + \sqrt{(a^2m^2+b^2)/2}$ with the ellipse, $\frac{x^2}{a^2}+\frac{y^2}{b^2}=1$ are a pair of conjugate diameters.

9. Find the condition for the lines

$$l_1x + m_1y = 0$$

and

$$l_2x + m_2y = 0$$

to be conjugate diameters of

$$\frac{x^2}{a^2}+\frac{y^2}{b^2}=1$$

10. If $PP'$ is a diameter and $Q$ is any point on the ellipse, prove that $QP$, $QP'$ are parallel to a pair of conjugate diameters of the ellipse.

11. Find the locus of the foot of the perpendicular from the centre of an ellipse on the chord joining the ends of a variable pair of conjugate diameters.

12. $CP$ and $CQ$ are conjugate semi-diameters of the ellipse

$$\frac{x^2}{a^2}+\frac{y^2}{b^2}=1.$$

A tangent parallel to $PQ$ meets $CP$ and $CQ$ in $R$ and $S$ respectively. Show that $R$ and $S$ lie on the ellipse,

$$\frac{x^2}{a^2}+\frac{y^2}{b^2}=2$$

13. If $CP$ and $CD$ be any two conjugate semi-diameters of the ellipse,

$$\frac{x^2}{a^2}+\frac{y^2}{b^2}=1$$

and circles with $CP$ and $CD$ as diameters intersect in $R$, show that $R$ lies on the curve

$$2(x^2+y^2)^2=a^2x^2+b^2y^2.$$

14. Show that the locus of the middle point of the line joining the extremities of a pair of conjugate diameters of the ellipse

$$\frac{x^2}{a^2}+\frac{y^2}{b^2}=1$$

is

$$\frac{x^2}{a^2}+\frac{y^2}{b^2}=\frac{1}{2}$$

15. $P$ and $D$ are two extremities of a pair of conjugate diameters of the ellipse, $\frac{x^2}{a^2}+\frac{y^2}{b^2}=1$. Show that $PD$ touches the ellipse, $\frac{x^2}{a^2}+\frac{y^2}{b^2}=\frac{1}{2}$.

16. If $\alpha+\beta=\gamma$ (a constant), prove that tangents at '$\alpha$' and '$\beta$' on the ellipse, $\frac{x^2}{a^2}+\frac{y^2}{b^2}=1$, intersect on the diameter through '$\gamma$'.

17. (i) Show that the line joining the extremities of any two diameters of an ellipse, which are at right angles to one another will always touch a fixed circle.

(ii) Show that the sum of the reciprocals of the squares of any two diameters of an ellipse, which are at right angles to one another is constant.

18. If $P'CP$ and $D'CD$ are two conjugate diameters of the ellipse, $\frac{x^2}{a^2}+\frac{y^2}{b^2}=1$ and $Q$ is a variable point on the circle, $x^2 + y^2 = c^2$; prove that $PQ^2 + DQ^2 + P'Q^2 + D'Q^2 = 2(a^2 + b^2 + 2c^2)$.

19. If $\lambda$, $\lambda'$ be the angles which any two conjugate diameters subtend at any fixed point on an ellipse, prove that $\cot^2 \lambda + \cot^2 \lambda'$ is constant.

20. Through a fixed point P, a pair of lines is drawn parallel to a variable pair of conjugate diameters of a given ellipse. The lines meet the principal axes in $Q$ and $R$ respectively. Show that the mid-point of $QR$ lies on a fixed line.

21. Lines are drawn through the foci of an ellipse perpendicular respectively to a pair of conjugate diameters and intersect in $R$. Show that the locus of $R$ is a concentric ellipse.

22. In an ellipse, a pair of conjugate diameters is produced to meet a directrix. Show that the orthocentre of the triangle so formed is a focus.

23. Show that the locus of the point of intersection of normals at the extremities of the conjugate diameters of the ellipse, $\frac{x^2}{a^2}+\frac{y^2}{b^2}=1$ is the curve

$$2(a^2x^2 + b^2y^2)^3 = (a^2 - b^2)^2(a^2x^2 - b^2y^2)^2.$$

24. If $CP$, $CD$ be a pair of conjugate semi-diameters of an ellipse whose semi-axes are $a$ and $b$, find $CD$ given that $CP = a \sin \phi + b \cos \phi$.

25. If there is a system of parallel chords of the ellipse, $\frac{x^2}{a^2}+\frac{y^2}{b^2}=1$, prove that the sum of the eccentric angles of the ends of each chord is same. If the constant value of the sum of the eccentric angles is $\lambda$, find the equa-

tion of the diameter of the ellipse, which bisects the system of parallel chords.

26. If two conjugate diameters $CP$ and $CQ$ of an ellipse cut the director circle in $L$ and $M$, prove that $LM$ touches the ellipse.

27. Perpendiculars $PM$ and $PN$ are drawn from any point $P$ (of an ellipse) on the equi-conjugate diameters of the ellipse. Prove that the perpendicular from $P$ on its polar line bisects $MN$.

28. Two conjugate diameters of the ellipse, $\frac{x^2}{a^2}+\frac{y^2}{b^2}=1$ cut the circle $x^2 + y^2 = r^2$ at $P$ and $Q$. Show that the locus of the mid-point of $PQ$ is
$$a^2[(x^2 + y^2)^2 - r^2x^2] + b^2[(x^2 + y^2)^2 - r^2y^2] = 0.$$

**Example 11.22.** Show that a circle and an ellipse intersect in four points and the sum of eccentric angles of these four points is equal to an even multiple of π.

*Solution.* Let the given circle be
$$x^2 + y^2 + 2gx + 2fy + c = 0 \tag{1}$$
and the given ellipse be
$$\frac{x^2}{a^2}+\frac{y^2}{b^2}=1 \tag{2}$$
Let $(a \cos \theta, b \sin \theta)$ be a point of intersection of (1) and (2). As it lies on the circle (1).

$\therefore$ $a^2 \cos^2 \theta + b^2 \sin^2 \theta + 2g\, a \cos \theta + 2fb \sin \theta + c = 0$ (3)

Put: $\tan (\theta/2) = t$
$$\Rightarrow \quad \cos\theta = \frac{1-\tan^2(\theta/2)}{1+\tan^2(\theta/2)} = \frac{1-t^2}{1+t^2}$$
and
$$\sin\theta = \frac{2\tan(\theta/2)}{1+\tan^2(\theta/2)} = \frac{2t}{1+t^2}$$

Substituting these values of cos θ and sin θ in (3), we get

$$a^2\left(\frac{1-t^2}{1+t^2}\right)^2+b^2\left(\frac{2t}{1+t^2}\right)^2+2ga\left(\frac{1-t^2}{1+t^2}\right)+2fb\left(\frac{2t}{1+t^2}\right)+c=0$$

$$\Rightarrow \quad (a^2-2ga+c)t^4+4bft^3+(4b^2-2a^2+2c)t^2$$
$$+4bft+(a^2+2ga+c)=0 \quad (4)$$

which being a fourth degree equation, has four values of $t$ ($t_i$, $i = 1, 2, 3, 4$)

i.e. it has four values of tan (θ/2)

i.e. it has four values of θ, the eccentric angle.

Thus these are, in general, four points of intersection of (1) and (2).

Let these four points have eccentric angles $\theta_i$, $i = 1, 2, 3, 4$).

Now $\tan\frac{1}{2}(\theta_1+\theta_2+\theta_3+\theta_4)$

$$=\frac{\Sigma\tan\frac{\theta_1}{2}-\Sigma\tan\frac{\theta_1}{2}\tan\frac{\theta_2}{2}\tan\frac{\theta_3}{2}}{1-\Sigma\tan\frac{\theta_1}{2}\tan\frac{\theta_2}{2}+\tan\frac{\theta_1}{2}\tan\frac{\theta_2}{2}\tan\frac{\theta_3}{2}\tan\frac{\theta_4}{2}}$$

$$=\frac{\Sigma t_1-\Sigma t_1t_2t_3}{1-\Sigma t_1t_2+t_1t_2t_3t_4}=0$$

$$\Rightarrow \quad \frac{1}{2}(\theta_1+\theta_2+\theta_3+\theta_4)=n\pi \quad (5)$$

$$\Rightarrow \quad \theta_1+\theta_2+\theta_3+\theta_4=2n\pi;\ \text{where } n \text{ is any integer.}$$

**Corollary 1.** Prove that the common chords of a circle and an ellipse are equally inclined to the axes of the ellipse.

*Solution.* If the points of intersection of (1) and (2) be $P_i$ having eccentric angle $\theta_i$ ($i = 1, 2, 3, 4$); then the equation of the chord $P_1P_2$ is

$$\frac{x}{a}\cos\frac{\theta_1+\theta_2}{2}+\frac{y}{b}\sin\frac{\theta_1+\theta_2}{2}=\cos\frac{\theta_1-\theta_2}{2}$$

$\therefore$ Slope of $P_1P_2 = -\frac{b}{a}\cot\frac{\theta_1+\theta_2}{2}$

$= -\frac{b}{a}\cot\left(n\pi - \frac{\theta_3+\theta_4}{2}\right)$ by (5)

$= \frac{b}{a}\cot\frac{\theta_3+\theta_4}{2}$

$= -$ (Slope of $P_3P_4$).

$\therefore$ $P_1P_2$ and $P_3P_4$ are equally inclined to the axis of $x$, which is an axis of the ellipse. One can similarly prove this fact about the other two pairs of chords.

**Corollary 2**. If the normals to an ellipse at $P_i$ ($i = 1, 2, 3, 4$) be concurrent, then the circle through $P_1$, $P_2$, $P_3$ meets the ellipse again in a point $P_4'$ with is the other end of the diameter through $P_4$.

*Solution*. Let $\theta_i$ be the eccentric angle of $P_i$ ($i = 1, 2, 3, 4$) and $\theta'_4$ be that of $P'_4$. Then

$\theta_1 + \theta_2 + \theta_3 + \theta_4 = (2n + 1)\pi$, by Theorem 11.1.

Again $P_1$, $P_2$, $P_3$, $P'_4$ are concyclic,

$\theta_1 + \theta_2 + \theta_3 + \theta'_4 = 2m\pi$

(where $n$ and $m$ are integers)

$\therefore$ $\theta_4 - \theta'_4 = [2(n - m)+1]\pi$

$= (2k + 1)\pi$, $k = n - m$

= an odd multiple of $\pi$

$\Rightarrow$ $P_4$ and $p'_4$ are the opposite ends of the diameter of the ellipse.

**Corollary 3**. Find centre of the circle passing through the three points, on an ellipse, whose eccentric angles are $\theta_1$, $\theta_2$, $\theta_3$.

*Solution*. Let the fourth point of intersection of the ellipse and the circle through $P_1(\theta_1)$, $P_2(\theta_2)$, $P_3(\theta_3)$ be $P_4(\theta_4)$.

Let the equation of this circle be

$$x^2 + y^2 + 2gx + 2fy + c = 0. \quad (1)$$

Then as in Example 11.22,

$$\theta_1 + \theta_2 + \theta_3 + \theta_4 = 2n\pi \qquad (2)$$

(where $n$ is an integer).

If '$\theta$' be any of these points, then

$x = a\cos\theta$, $y = b\sin\theta$ lies on (1).

$$\therefore\ a^2\cos^2\theta + b^2\sin^2\theta + 2ga\cos\theta + 2fb\sin\theta + c = 0 \qquad (3)$$

$$\Rightarrow [(a^2-b^2)\cos^2\theta + 2ga\cos\theta + (b^2+c)]^2 = 4f^2b^2(1-\cos^2\theta)$$

$$\Rightarrow (a^2-b^2)^2\cos^4\theta + 4ga(a^2-b^2)\cos^3\theta$$

$$+ [2(a^2-b^2)(b^2+c) + 4g^2a^2 + 4f^2b^2]\cos^2\theta$$

$$+ 4ga(b^2+c)\cos\theta + [(b^2+c^2) - 4f^2b^2] = 0$$

This is a fourth degree equation in $\cos\theta$; it has four roots, $\cos\theta_i$ $(i = 1, 2, 3, 4)$

$$\therefore\ \cos\theta_1 + \cos\theta_2 + \cos\theta_3 + \cos\theta_4 = -4ga/(a^2-b^2) \qquad (4)$$

Similarly changing (3) into an equation in $\sin\theta$, we get

$$\sin\theta_1 + \sin\theta_2 + \sin\theta_3 + \sin\theta_4 = -4fb/(b^2-a^2) \qquad (5)$$

From (2), $\theta_4 = 2n\pi - (\theta_1 + \theta_2 + \theta_3)$

$\Rightarrow \cos\theta_4 = \cos(\theta_1 + \theta_2 + \theta_3)$

and $\sin\theta_4 = -\sin(\theta_1 + \theta_2 + \theta_3)$.

$\therefore$ (4) and (5), along with these, imply

$$-g = \frac{a^2-b^2}{4a}[\cos\theta_1 + \cos\theta_2 + \cos\theta_3 + \cos(\theta_1 + \theta_2 + \theta_3)]$$

and $$-f = \frac{b^2-a^2}{4b}[\sin\theta_1 + \sin\theta_2 + \sin\theta_3 - \sin(\theta_1 + \theta_2 + \theta_3)]$$

which give the co-ordinates of the centre of the circle through $P_1$, $P_2$, $P_3$.

**Corollary 4**. $P'CP$ and $D'CD$ are conjugate diameters of an ellipse and $\theta$ is the eccentric angle of $P$. Prove that the eccentric angle of the point where the circle through $P$, $P'$, $D$ again cuts the ellipse is $\frac{\pi}{2} - 3\theta$.

*Solution*. The eccentric angles of $P$, $P'$ and $D$ respectively are $\theta$, $\pi + \theta$ and $\frac{\pi}{2} + \theta$. Let $\delta$ be the eccentric angle of the fourth point.

As above

$$\theta + (\pi + \theta) + \left(\frac{\pi}{2} + \theta\right) + \delta = 2n\pi$$

$$\Rightarrow \delta = 2n\pi - \left(\frac{3\pi}{2} + 3\theta\right)$$

$$= \frac{\pi}{2} - 3\theta \qquad \text{when } n = 1$$

(Note that any other value of $n$ gives the same point on the ellipse).

*Remark.* We show below that both circle and parabola are limiting cases of an ellipse.

The equation of an ellipse is $\frac{x^2}{a^2} + \frac{y^2}{b^2} = 1$ where $b^2 = a^2(1 - e^2)$.

(i) Let $e \to 0$, $a$ remaining finite, then $b \to a$ and the foci $(\pm ae, 0)$ approach the centre and the ellipse becomes the circle, $x^2 + y^2 = a^2$.

(ii) Let us shift the origin to $(-a, 0)$; the equation of the ellipse then becomes $\frac{(x-a)^2}{a^2} + \frac{y^2}{b^2} = 1$, referred to the new axes.

$$\Rightarrow \qquad \frac{x^2}{a^2} + \frac{y^2}{b^2} = \frac{2x}{a}$$

$$\Rightarrow \qquad y^2 + \frac{b^2}{a^2}x^2 = \frac{2b^2}{a}x$$

Let now $a \to \infty$, $b \to \infty$ such that

$$\text{Lim } \frac{b^2}{a} = k \text{ (a finite number)}.$$

$$\therefore \quad \lim \frac{b^2}{a^2} = \lim \frac{k}{a} = 0$$

Now the above equation reduces to

$$y^2 = 2kx,$$

which is the equation of a parabola.

Thus a parabola is the limiting case of an ellipse of which the axes are of infinite length but the latus rectum is finite.

It may be noted that it is a routine problem in calculus to find numbers $a$ and $b$ both tending to infiinity but Lim $\frac{b^2}{a}$ is finite e.g. take $b = \sqrt{n}$, $a = n$.

**Example 11.23.** $T$ is the foot of the perpendicular from the centre $C$ of an ellipse, $\frac{x^2}{a^2} + \frac{y^2}{b^2} = 1$, on the tangent at a point P on it. Show that the locus of $T$ is $(x^2 + y^2)^2 = a^2x^2 + b^2y^2$.

*Solution.* Let the co-ordinates of $P$ be $(a \cos\theta, b \sin\theta)$. Now the equation of the tangent at $P$ to the ellipse

$$\frac{x^2}{a^2} + \frac{y^2}{b^2} = 1$$

is
$$\frac{x\cos\theta}{a} + \frac{y\sin\theta}{b} = 1$$

Let the co–ordinates of $T$ be $(\alpha, \beta)$.

By hypothesis

$T$ lies on the tangent at $P$ and $CT \perp PT$.

$$\therefore \qquad \frac{\alpha\cos\theta}{a} + \frac{\beta\sin\theta}{b} = 1 \tag{1}$$

and
$$\frac{\beta - 0}{\alpha - 0} \cdot \left(\frac{-\cos\theta / a}{\sin\theta / b}\right) = -1$$

$$\Rightarrow \qquad \cos\theta = \frac{a\alpha\sin\theta}{b\beta} \tag{2}$$

Substituting for cos θ in (1), we have

$$\sin\theta = \frac{b\beta}{\alpha^2 + \beta^2}$$

and then (2) ⇒

$$\cos\theta = \frac{a\alpha}{\alpha^2 + \beta^2}$$

Eliminating θ, we obtain,

$$a^2 \alpha^2 + b^2 \beta^2 = (\alpha^2 + \beta^2)^2.$$

Thus the locus of $T$ is $(x^2 + y^2)^2 = a^2x^2 + b^2y^2$.

**Example 11.24.** If $SY$ and $S'Y'$ be the perpendiculars from the foci $S$ and $S'$ upon any tangent to the ellipse,

$$\frac{x^2}{a^2} + \frac{y^2}{b^2} = 1,$$

then prove that

(i) $\overline{SY}.\overline{S'Y'} = b^2$,

(ii) $Y$ and $Y'$ lie on the auxiliary circle.

*Solution*. We know that any tangent to the given ellipse is

$$y = mx + \sqrt{a^2m^2 + b^2} \tag{1}$$

(i) The points $S$ and $S'$ are $(\pm\, ae, 0)$.

$$\therefore \quad \overline{SY} = \left| \frac{aem + \sqrt{a^2m^2 + b^2}}{\sqrt{1 + m^2}} \right|$$

and

$$\overline{S'Y'} = \left| \frac{-aem + \sqrt{a^2m^2 + b^2}}{\sqrt{1 + m^2}} \right|$$

$$\therefore \quad \overline{SY}.\overline{S'Y'} = \left| \frac{a^2m^2 + b^2 - a^2e^2m^2}{1 + m^2} \right| = \frac{b^2(1 + m^2)}{(1 + m^2)} = b^2$$

(ii) Equation of the line $SY$ is $y = \dfrac{-1}{m}(x - ae)$

$$\Rightarrow \quad my + x + ae \tag{2}$$

To find the locus of $Y$, we eliminate $m$ from (1) and (2). We rewrite (1) as

$$y - mx = \sqrt{a^2m^2 + b^2} \tag{3}$$

Squaring (2) and (3), and then adding them, we obtain

$$(x^2 + y^2)(1 + m^2) = a^2m^2 + b^2 + a^2e^2 = a^2\,(1 + m^2)$$

$$\Rightarrow \quad x^2 + y^2 = a^2.$$

Similarly $Y'$ also lies on $x^2 + y^2 = a^2$.
Thus $Y$ and $Y'$ lie on the auxiliary circle.

**Example 11.25.** If $ABC$ be a maximum triangle inscribed in an ellipse, show that the eccentric angles of the vertices differ by $2\pi/3$ and the normals at $A, B, C$ are concurrent.

*Solution.* Let $\alpha, \beta, \gamma$ be the eccentric angles of the vertices $A$, $B$, $C$ lying on the ellipse, $\frac{x^2}{a^2} + \frac{y^2}{b^2} = 1$.
If $A', B', C'$ be the corresponding points on the auxiliary circle, the points have co-ordinates

$A(a \cos \alpha, b \sin \alpha)$; $A'(a \cos \alpha, a \sin \alpha)$ etc.

Let $\Delta(\Delta')$ denote the area of the triangle $ABC(A'B'C')$.

$$\therefore \quad \Delta = \frac{1}{2}\begin{vmatrix} a\cos\alpha & b\sin\alpha & 1 \\ a\cos\beta & b\sin\beta & 1 \\ a\cos\gamma & b\sin\gamma & 1 \end{vmatrix},$$

$$\Delta' = \frac{1}{2}\begin{vmatrix} a\cos\alpha & a\sin\alpha & 1 \\ a\cos\beta & a\sin\beta & 1 \\ a\cos\gamma & a\sin\gamma & 1 \end{vmatrix}$$

$$\Rightarrow \qquad \frac{\Delta}{\Delta'} = \frac{b}{a}$$

Thus $\Delta$ will be maximum when $\Delta'$ is so (and conversely).

Also $\Delta'$(the area of $\Delta\, A'B'C'$, inscribed in a circle) is maximum when $A'B'C'$ is an equilateral triangle.

$\therefore$ The consecutive eccentric angles differ by $2\pi/3$; so that we can take these as $\alpha$, $\alpha + \frac{2}{3}\pi$, $\alpha + \frac{4}{3}\pi$

Now $\sin(\alpha + \beta) + \sin(\beta + \gamma) + \sin(\gamma + \alpha)$

$$= \sin\left(2\alpha + \frac{2}{3}\pi\right) + \sin(2\alpha + 2\pi) + \sin\left(2\alpha + \frac{4}{3}\pi\right)$$

$= \sin 2\alpha + 2\sin(2\alpha + \pi)\cos\dfrac{\pi}{3}$

$= \sin 2\alpha - \sin 2\alpha = 0.$

$\therefore$ The normals at $A$, $B$, $C$ are concurrent (by Q. 20 of Exercise 11.3).

**Example 11.26**. If $w$ is the difference of the eccentric angles of two points on an ellipse, the tangents at which are at right angles, prove that $ab \sin w = \lambda\mu$, where $\lambda$, $\mu$ are the semi–diameters parallel to the tangents at the points and $a$, $b$ are the semi–axes of the ellipse.

*Solution*. Let the given ellipse be

$$\frac{x^2}{a^2} + \frac{y^2}{b^2} = 1 \qquad (1)$$

Let $P(\alpha)$and $Q(\beta)$ be two points on (1) such that

$$w = \alpha - \beta \qquad (2)$$

By hypothesis, the tangents at $P$ and $Q$ are at right angles.

$$\therefore \left(-\frac{b}{a}\cot\alpha\right)\left(-\frac{b}{a}\cot\beta\right) = -1$$

$$\Rightarrow a^2 \sin\alpha \sin\beta + b^2 \cos\alpha \cos\beta = 0 \qquad (3)$$

Now the diameter parallel to the tangent at $P(\alpha)$ will be conjugate to the diameter $CP$ and therefore one of its extremities will be

$$(-a \sin\alpha, b\cos\alpha).$$

$\therefore \lambda^2 = a^2 \sin^2\alpha + b^2 \cos^2\alpha$

Similarly, $\mu^2 = a^2 \sin^2\beta + b^2\cos^2\beta$

$\Rightarrow \lambda^2\mu^2 = (a^2 \sin^2\alpha + b^2\cos^2\alpha)(a^2\sin^2\beta + b^2\cos^2\beta)$

$= (a^2 \sin\alpha \sin\beta + b^2 \cos\alpha\cos\beta)^2$

$\qquad + a^2b^2(\sin\alpha\cos\beta - \cos\alpha\sin\beta)^2$

$= a^2b^2 \sin^2 w$; using (3) and (2)

$\Rightarrow \lambda\mu = ab \sin w.$

**Miscellaneous Excercise on Chapter 11.**

1. $P(a\cos\theta, b\sin\theta)$ is any point on the ellipse $\dfrac{x^2}{a^2} + \dfrac{y^2}{b^2} = 1$ and $S$ and $S'$ are the foci. Show that

(i) $\overline{SP} = a(1 - e\cos\theta)$,
(ii) $\overline{S'P} = a(1 + e\cos\theta)$,
(iii) $\overline{SP} + \overline{S'P} = 2a$.

2. Let P be any point on the ellipse, $\frac{x^2}{a^2} + \frac{y^2}{b^2} = 1$ and let $Q$ be the corresponding point on the auxiliary circle. Show that the perpendicular distance of the focus $S$ on the tangent at $Q$ to the auxiliary circle is equal to $\overline{SP}$.
3. Show that the focus of an ellipse is the pole of the corresponding directrix.
4. Show that the system of ellipses, $\frac{x^2}{a^2+p} - \frac{b^2y^2}{b^2+p} = 1$, for different values of $p$, have the same foci.
5. Let $P(\theta)$ be any point on the ellipse, $\frac{x^2}{a^2} + \frac{y^2}{b^2} = 1$, with centre $C$ and $Q$ be the corresponding point on the auxiliary circle. Let the normal at $P$ to the ellipse and the line $CQ$ meet at $R$. Show that
(i) $\overline{CR} = (a + b)$,
(ii) the locus of $R$ is, $x^2 + y^2 = (a + b)^2$.
6. Let the tangent at a point $P$ of an ellipse meet the $x$-axis at $T$. Let $Q$ be the corresponding point on the auxiliary circle. Show that $QT$ is the tangent at $Q$, to the auxiliary circle.
7. (i) Show that the locus of the middle points of the chords of the ellipse, $\frac{x^2}{a^2} + \frac{y^2}{b^2} = 1$, which pass through the positive end of the major axis is $\frac{x^2}{a^2} + \frac{y^2}{b^2} = \frac{x}{a}$.
(ii) Find the locus of the middle points of chords of an ellipse which are drawn through the positive end of the minor axis.

8. Let $PCP'$ be a diameter of the ellipse, $\frac{x^2}{a^2}+\frac{y^2}{b^2}=1$. Let the normal at $P$ meet the ordinate at $P'$ at $T$. Show that the locus of $T$ is $\frac{x^2}{a^2}+\frac{b^2y^2}{(2a^2-b^2)^2}=1$..

9. The tangent and the normal to the ellipse, $x^2 + 4y^2 = 2$, at the point $P$ meet the major axis in $Q$ and $R$ respectively. If $\overline{QR} = \sqrt{2}$, show that the eccentric angle of $P$ is $\cos^{-1}(2/3)$

10. $CP$ and $CD$ are semi-conjugate diameters of the ellipse,
$$\frac{x^2}{a^2}+\frac{y^2}{b^2}=1$$
Prove that the locus of the orthocentre of the triangle $CPD$ is
$$2(a^2x^2 + b^2y^2)^3 = (a^2 - b^2)^2(a^2x^2 - b^2y^2)^2.$$

11. Lines through the origin are drawn, perpendicular to the tangents from a point $P$ to the ellipse, $\frac{x^2}{a^2}+\frac{y^2}{b^2}=1$. Find the locus of $P$ if the lines are its conjugate diameters. (*Hint*: If $P$ is $(\alpha, \beta)$, equation of pair of tangents is $SS' = T^2$. Now use Theorem 5.4 and Question 2, Exercise 5.1).

12. Prove that the minimum angle between a pair of conjugate diameters of the ellipse, $\frac{x^2}{a^2}+\frac{y^2}{b^2}=1$, is $\tan^{-1}\frac{2ab}{a^2-b^2}$.

13. Prove that the minimum intercept made by the axes on the tangents to the ellipse, $\frac{x^2}{a^2}+\frac{y^2}{b^2}=1$ is $(a + b)$.

(*Hint*: Intercept $= \sqrt{a^2\sec^2\theta + b^2\operatorname{cosec}^2\theta}$
$= \sqrt{(a+b)^2 + (a\tan\theta - b\cot\theta)^2}$

14. Tangents are drawn to the ellipse, $\frac{x^2}{a^2}+\frac{y^2}{b^2}=1$ and the circle, $x^2 + y^2 = a^2$, at the points where a common ordinate meets them. Show that if $\theta$ be the greatest inclina-

tion of these tangents, then

$$\tan\theta = \frac{a-b}{2\sqrt{ab}}$$

15. Prove that the square of the perpendicular from any point on the ellipse, on its major (minor) axis varies as the rectangle of the segments of the major (minor) axis made by the perpendicular.

16. (i) Show that the equation of the circle on the chord intercepted by $\frac{x^2}{a^2}+\frac{y^2}{b^2}=1$, on $y = mx + c$ as a diameter is $(a^2m^2 + b^2)(x^2 + y^2) + 2\,a^2mcx - 2b^2cy + c^2(a^2 + b^2) - a^2b^2\,(1 + m^2) = 0.$

(ii) Prove that the circle whose diameter is the chord, $\frac{lx}{a}+\frac{my}{b}=1$ cuts the ellipse $\frac{x^2}{a^2}+\frac{y^2}{b^2}=1$ in two other points whose join is the line, $\frac{lx}{a}-\frac{my}{b}=\frac{a^2+b^2}{a^2-b^2}$

(iii) Prove that the circle on the join of $P(\theta_1)$ and $Q(\theta_2)$ as diameter cuts the ellipse in two other points whose join is the line.

$$\frac{x}{a}\cos\frac{\theta_1+\theta_2}{2}-\frac{y}{b}\sin\frac{\theta_1+\theta_2}{2}=\frac{a^2-b^2}{a^2+b^2}\cos\frac{\theta_1+\theta_2}{2}$$

17. (i) $P$ is a point on a chord $RS$ of the ellipse, $\frac{x^2}{a^2}+\frac{y^2}{b^2}=1$, such that $r_1 r_2 = 1$ where $r_1, r_2$ are the algebraic distances of R, S from P. Show that if the chord moves parallel to itself, the locus of $P$ is an ellipse.

(ii) Any two perpendicular lines are drawn through a fixed point $O$, meeting on ellipse on $P_1$, $Q_1$ and $P_2$, $Q_2$.

Prove that $\frac{1}{\overline{OP_1}.\overline{OQ_1}}+\frac{1}{\overline{OP_2}.\overline{OQ_2}}$ is constant.

(iii) An ellipse is cut by a variable secant through a fixed point $O$ in $P$ and $Q$. A point $R$ is taken on $PQ$ such that

(a) $\frac{1}{\overline{OP}}+\frac{1}{\overline{OQ}}=\frac{1}{\overline{OR}}$

(b) $\overline{OP}.\overline{OQ}=OR^2$

Find the locus of $R$ in each case.

18. Prove that the subtangent and subnormal of a point $(x_1,y_1)$ on the ellipse, $\frac{x^2}{a^2}+\frac{y^2}{b^2}=1$ are respectively $\left(\frac{a^2}{x_1}-x_1\right)$ and $\left(\frac{b^2x_1}{a}\right)$

19. A straight line moves in such a manner that the product of the perpendiculars from two fixed points on it is a positive constant. Show that it always touches an ellipse with fixed points as foci.

20. If the normals at four points on the ellipse, $\frac{x^2}{a^2}+\frac{y^2}{b^2}=1$ are concurrent, and if two of the points lie on the line, $lx + my = 1$, show that the other two points lie on the line, $\frac{x}{a^2l}+\frac{y}{b^2m}+1=0$. Hence show that if the feet of the two of the normals from a point P to this ellipse are coincident, the locus of the mid-point of the chords joining the feet of the other normals is $\left(\frac{xy}{ab}\right)^2=\left(\frac{x^2}{a^2}+\frac{y^2}{b^2}\right)^3$

21. A point $P$ on the ellipse, $\frac{x^2}{a^2}+\frac{y^2}{b^2}=1$ whose eccentric angle is $\theta$, is joined to the foci $S$ and $S'$ and $PS$, $PS'$ meet the curve again in $Q$ and $Q'$. Show that the equation of $QQ'$ is $\frac{x}{a}\cos\theta\left(1-e^2\right)+\frac{y}{b}\sin\theta\left(1+e^2\right)=e^2-1$

22. A parallelogram circumscribes the ellipse, $\frac{x^2}{a^2}+\frac{y^2}{b^2}=1$

and two of its angular points are on the lines, $x^2 - c^2 = 0$; prove that the other two angular points lie on the conic,

$$\frac{x^2}{a^2} + \frac{y^2}{b^2}\left(1 - \frac{a^2}{c^2}\right) = 1$$

23. Two conjugate diameters of the ellipse, $\frac{x^2}{a^2} + \frac{y^2}{b^2} = 1$ meet a fixed line, $lx + my = 1$ in $P$ and $Q$ and the lines through $P$ and $Q$ respectively perpendicular to these diameters meet in $R$. Find the locus of $R$.
24. If two concentric ellipses be such that the foci of one lie on the other, prove that the angle between their axes is $\left(\sqrt{e_1^2 + e_2^2 - 1} / e_1 e_2\right)$ where $e_1$, $e_2$ are their eccentricities.

25. The eccentric angles of two points $P$ and $Q$ on the ellipse, $\frac{x^2}{a^2} + \frac{y^2}{b^2} = 1$ are $\alpha$ and $\beta$. Prove that the area of the parallelogram formed by the tangents at the ends of the diameters through $P$ and $Q$ is $4ab/\sin(\alpha \sim \beta)$ and hence that it is least when $P$ and $Q$ are the extremities of a pair of conjugate diameters.

# 12. HYPERBOLA

## INTRODUCTION

We have defined in Chapter 9, that a hyperbola is the locus of a point wich moves such that its distance from a fixed point, called *focus*, is always $e$ times $(e > 1)$ its distance from a fixed line, called *directrix*. Here, we find its equation in the *Standard Form* and study some of its properties.

## 12.1 TO FIND THE EQUATION OF A HYPERBOLA IN THE STANDARD FORM.

Let $S$ be a focus and $ZM$ be a directrix of the given hyperbola. Draw $ST \perp ZM$. Divide $ST$ internally and externally in the ratio $e : 1$, at the points, $A$ and $A'$ respectively.

$$\therefore \qquad \overline{AS} = e.\overline{TA} \text{ and } \overline{A'S} = e.\overline{A'T} \tag{1}$$

whence the points, $A$ and $A'$ lie on the hyperbola.

Set $\qquad \overline{A'A} = 2a, \ (a > 0).$

Let $C$ be the middle point of $A'A$. Take C to be the origin, the line $CS$ as the axis of $x$ and a line through $C$, perpendicular to it as the axis of $y$.

Now the equations (1) imply

$$\overline{A'S} + \overline{AS} = e.\left(\overline{A'T} + \overline{TA}\right) \tag{2}$$

and $$\overline{A'S} - \overline{AS} = e.\left(\overline{A'T} - \overline{TA}\right) \tag{3}$$

We express each side of these equations in terms of the origin, $C$.

Now (2) $\Rightarrow \overline{A'C} + \overline{CS} + \overline{CS} - \overline{CA}$

$$= e.\left(\overline{A'C} + \overline{CT} + \overline{CA} - \overline{CT}\right)$$

$\Rightarrow CS = ae.$ $\quad (\because\ \overline{CA} = \overline{A'C} = a)$

Thus, the co-ordinates of $S$ are $(ae, 0)$.

Again (3) $\Rightarrow \left(\overline{A'C} + \overline{CS}\right) - \left(\overline{CS} - \overline{CA}\right)$

$$= e\left[\left(\overline{A'C} + \overline{CT}\right) - \left(\overline{CA} - \overline{CT}\right)\right]$$

$$\Rightarrow CT = a/e.$$

Hence the equation of the directrix is,

$$x = a/e.$$

Let, now, $P(x, y)$ be any point on the hyperbola.

Then $\quad \overline{PS} = e.\overline{PQ}$

where $PQ \perp ZM$, the directrix, whence

$$\sqrt{(x - ae)^2 + y^2} = e.|x - a/e| = |ex - a|$$

$$\Rightarrow \quad (x - ae)^2 + y^2 = (ex - a)^2$$

$$\Rightarrow \quad (e^2 - 1)x^2 - y^2 = a^2(e^2 - 1)$$

$$\Rightarrow \quad \frac{x^2}{a^2} - \frac{y^2}{a^2(e^2 - 1)} = 1$$

Set $\quad a^2(e^2 - 1) = b^2 \quad$ (note: $e > 1$).

Hence, the required equation of the hyperbola, in the *standard form* is $\dfrac{x^2}{a^2} - \dfrac{y^2}{b^2} = 1.$

By the method of tracing a curve (see Chapter 7) the shape of the above hyperbola is:

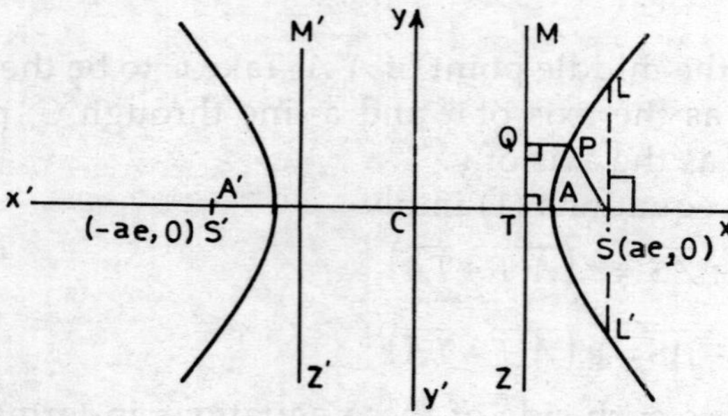


$x = -\dfrac{a}{e}$ $\qquad\qquad$ $x = \dfrac{a}{e}$

**Definitions**

1. *Centre.* The point C is called the *centre* of the hyperbola.
2. *Axes.* The hyperbola, given by

$$\frac{x^2}{a^2} - \frac{y^2}{b^2} = 1$$

meets the $x$-axis at $A(a, 0)$ and $A'(-a, 0)$. These two points are called its *vertices*. Further $A'A$ is caled the *transverse axis* of the hyperbola and its length is $2a$.

The above hyperbola does not meet the $y$-axis in real points. Take the points $B'(0, -b)$ and $B(0, b)$ on the $y$-axis; then $B'B$ is called the *conjugate axis* and its length is $2b$.

These two axes together are called the *pincipal axes* of the hyperbola.
3. *Latus Rectum.* A line through the focus, $S$, of the hyperbola, perpendicular to its transverse axis is called its *latus rectum*.

If we denote it by $L'L$ (see the above figure), then its equation is

$$x = ae.$$

This meets the hyperbola,

$$\frac{x^2}{a^2} - \frac{y^2}{b^2} = 1$$

where $$e^2 - \frac{y^2}{b^2} = 1$$

$$\Rightarrow \quad y^2 = b^2(e^2 - 1) = b^2 \cdot \frac{b^2}{a^2} \quad (\because \; b^2 = a^2(e^2 - 1))$$

$$\Rightarrow \quad y = \pm \frac{b^2}{a}$$

$\therefore$ the co–ordinates of $L$ and $L'$ are $\left(ae, \frac{b^2}{a}\right)$ and $\left(ae, -\frac{b^2}{a}\right)$ respectively.

Also, Latus Rectum $= L'L = \dfrac{2b^2}{a}$.

4. *Eccentricity.* Since , $b^2 = a^2(e^2 - 1)$, on solving this, we get

$$e = \sqrt{a^2 + b^2} / a$$

It may be noted $e <, = \text{ or } > \sqrt{2}$ according as $a >, = \text{ or } < b$. The case, $e = \sqrt{2}$ i.e. when $a = b$ is of special interest to us. In this case, equation of the hyperbola is $x^2 - y^2 = a^2$ and it is known as *rectangular* (or *equilateral*) *hyperbola*.

**Note 1.** By the symmetry of the curve or one can see directly that if the point $S'(-ae, 0)$ is taken as the focus and the line, $x = -a/e$ as the directrix, we get the same hyperbola. Hence, we see that a hyperbola has two foci $(\pm ae, 0)$ and also two directrices: $x = \pm a/e$. Consequently, it has two latera recta, $x = \pm ae$.

**Note 2.** Clearly, if we consider the equation,

$$\frac{y^2}{b^2} - \frac{x^2}{a^2} = 1,$$

this hyperbola lies along the $y$-axis and is as shown below:

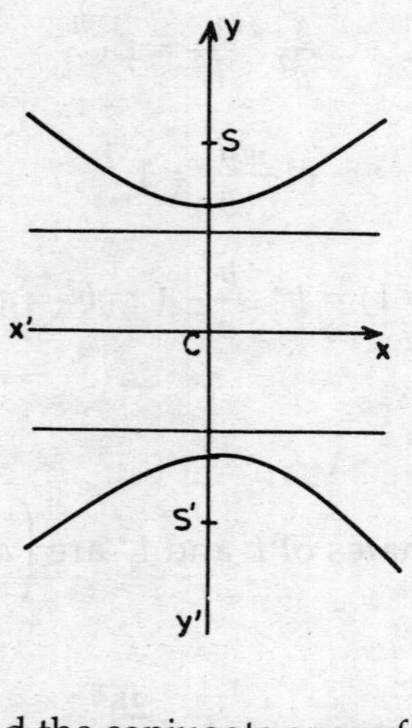


The transverse and the conjugate axes of this hyperbola are, then, the conjugate and transverse axes of the hyperbola,

$$\frac{x^2}{a^2} - \frac{y^2}{b^2} = 1 \tag{1}$$

The hyperbola,

$$\frac{y^2}{b^2} - \frac{x^2}{a^2} = 1 \Leftrightarrow \frac{x^2}{a^2} - \frac{y^2}{b^2} = -1 \tag{2}$$

is called the *conjugate hyperbola* of the hyperbola given by equation (1).

It may be noted further that the foci of (2) are, $(0, \pm be)$, the directrices are, $y = \pm b/e$ and its eccentricity is given by

$$a^2 = b^2(e^2 - 1).$$

Also the transverse axis = $2b$,
the conjugate axis = $2a$,

and the latus rectum = $\dfrac{2a^2}{b}$.

**Note 3.** The equation,

$$\frac{\left[\dfrac{lx + my + n}{\sqrt{l^2 + m^2}}\right]^2}{a^2} - \frac{\left[\dfrac{mx - ly + n'}{\sqrt{l^2 + m^2}}\right]^2}{b^2} = 1$$

represents a hyperbola whose transverse axis is of length $2a$ and it lies along the line $mx - ly + n' = 0$ and whose conjugate axis is of length $2b$ and it lies along the line $lx + my + n = 0$. The point of intersection of these lines is the centre of the hyperbola.

**Example 12.1.** Find the equation of the hyperbola (in the standard form) whose eccentricity is $\sqrt{2}$ and the distance between the foci is 16.

*Solution.* Let the equation of the hyperbola be

$$\frac{x^2}{a^2} - \frac{y^2}{b^2} = 1$$

Here $e = \sqrt{2}$.

Also distance between the foci = $2ae$ = 16.

$$\Rightarrow 2a\sqrt{2} = 16 \Rightarrow a = 4\sqrt{2}$$

$$\therefore \quad b^2 = a^2(e^2 - 1) = 32(2 - 1) = 32.$$

Hence, the equation of the hyperbola is

$$\frac{x^2}{32} - \frac{y^2}{32} = 1, \text{ i.e. } x^2 - y^2 = 32.$$

**Example 12.2**. Show that the difference of focal distances of any point on the hyperbola is constant and is equal to its transverse axis.

*Solution*. (See the figure below).

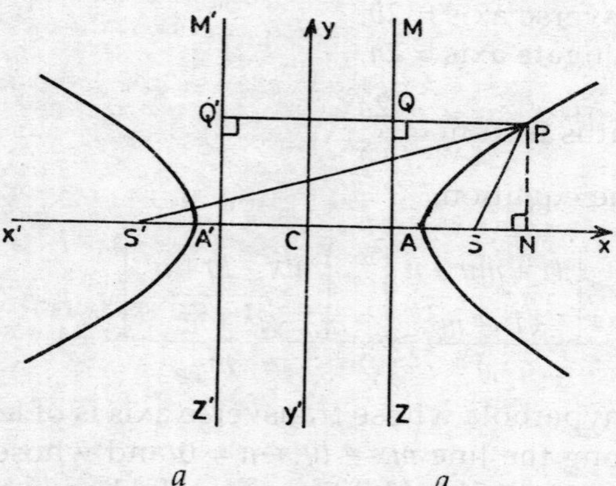


$$x = -\frac{a}{e} \qquad\qquad x = \frac{a}{e}$$

Let $P(\alpha, \beta)$ by any point on the hyperbola.

Let $PQ$ and $PQ'$ be the perpendiculars on the directrices $ZM$ and $Z'M'$ of the hyperbola.

By definition of the hyperbola,

$$\overline{PS} = e.\overline{PQ}$$

and

$$\overline{PS'} = e.\overline{PQ'}$$

$$\Rightarrow \quad \left[\overline{PS'} - \overline{PS}\right] = e.\left[\overline{PQ'} - \overline{PQ}\right]$$

$$= e.|QQ'| = e.\left|\frac{2a}{e}\right| = 2a$$

The result is not affected if the point $P$ is taken on the other branch of the hyperbola.

**Note**. From *Question 9, Exercise 2.1* we see that a hyperbola can also be defined as the locus of a point, which moves such that the difference of its distances from two fixed points is always constant. In fact, if we take the fixed points as $(ae, 0)$ and $(-ae, 0)$ and the constant difference of the distances of the moving point as $2a$, then the equation of its locus is again $\frac{x^2}{a^2} - \frac{y^2}{b^2} = 1$, where $b^2 = a^2(e^2 - 1)$, $e > 1$.

**Example 12.3**. Given the base of a triangle and the ratio of the tangents of half the base angles, show that the vertex moves on a hyperbola whose foci are the extremities of the base.

*Solution*. Let $ABC$ be the given triangle with base $= BC$ and $A$ its vertex. Take $x$-axis along $BC$. The axis of $y$ can be chosen arbitrarily. Using elementary Trigonometric formulae (with the usual notation), $\tan\frac{B}{2} = \frac{r}{s-b}$; $\tan\frac{C}{2} = \frac{r}{s-c}$

$$\Rightarrow \quad \frac{\tan(B/2)}{\tan(C/2)} = \frac{s-c}{s-b} = \frac{a+b-c}{c+a-b} = k$$

($a$ constant, by hypothesis)

$$\Rightarrow \quad \frac{k-1}{k+1} = \frac{b-c}{a}$$

$$\Rightarrow \quad b - c = a\frac{k-1}{k+1} = \lambda \text{ (say)}$$

($\lambda$ is a constant, as both $a$ and $k$ are constants)

$$\Rightarrow \quad \overline{CA} - \overline{BA} = \lambda$$

$\Rightarrow$ by the focal property, mentioned in Example 12.2, the locus of $A$ is a hyperbola, whose foci are $B$ and $C$.

**Example 12.4**. Prove that the square on the perpendicular from any point of the hyperbola, $\frac{x^2}{a^2} - \frac{y^2}{b^2} = 1$, on its transverse axis, varies as the rectangle of the segments of the transverse axis made by this perpendicular.

*Solution.* See the figure of Example 12.2. Let $P(x, y)$ be any point on

$$\frac{x^2}{a^2} - \frac{y^2}{b^2} = 1 \tag{1}$$

Let N be the foot of the perpendicular from $P$, on the transverse axis $A'CA$ of (1).

Now (1) yields

$$\frac{y^2}{b^2} = \frac{x^2}{a^2} - 1$$

$$\Rightarrow \quad \frac{NP^2}{b^2} = \frac{CN^2}{a^2} - 1$$

$$= \frac{CN^2 - a^2}{a^2}$$

$$= \frac{(\overline{CN} + a)(\overline{CN} - a)}{a^2}$$

$$= \frac{\overline{A'N}.\overline{AN}}{a^2} \quad (\because \ \overline{A'C} = \overline{CA} = a)$$

$$\Rightarrow \quad = \frac{NP^2}{\overline{AN}.\overline{A'N}} = \frac{b^2}{a^2} \tag{2}$$

$\Rightarrow$ $NP^2$ varies as $\overline{AN}.\overline{A'N}$.

**Note.** Equation (2) in this example is known as the *Geometrical Interpretation* of the hyperbola.

**Example 12.5**. Show that the minimum length of the radius vector from the centre of the hyperbola, $\frac{x^2}{a^2} - \frac{y^2}{b^2} = 1$ is $a$, the semi-tranverse axis.

*Solution.* Let $P$ be any point on the hyperbola, distant $r$ from the centre $C$. Suppose $\angle XCP = \theta$.

$\therefore$ The polar coordinates of $P$ are $(r, \theta)$ and we have

$$x = r\cos\theta,\ y = r\sin\theta.$$

Substituting these values of $x$ and $y$ in $\frac{x^2}{a^2} - \frac{y^2}{b^2} = 1$, we get

$$r^2\left[\frac{\cos^2\theta}{a^2} - \frac{\sin^2\theta}{b^2}\right] = 1$$

$$\Rightarrow \qquad r^2 = \frac{a^2b^2}{b^2\cos^2\theta - a^2\sin^2\theta}$$

$$= \frac{2a^2b^2}{\left[b^2(1+\cos 2\theta) - a^2(1-\cos 2\theta)\right]}$$

$$= \frac{2a^2b^2}{[(b^2 - a^2) + (b^2 + a^2)\cos 2\theta]}$$

For the minimum value of $r$, $\cos 2\theta$ should be maximum. As the maximum value of $\cos 2\theta$ is 1.

$\Rightarrow$ The minimum value of $r^2$

$$= \frac{2a^2b^2}{(b^2 - a^2) + (b^2 + a^2).1} = a^2$$

$\Rightarrow$ the minimum value of $r$ is $a$.

**Example 12.6**. If $e$ and $e'$ be the eccentricities of a hyperbola and its conjugate, prove that $\frac{1}{e^2} + \frac{1}{e'^2} = 1$.

*Solution*. Let the equation of the hyperbola be

$$\frac{x^2}{a^2} - \frac{y^2}{b^2} = 1,$$

$\therefore$ the equation of the conjugate hyperbola is

$$\frac{y^2}{b^2} - \frac{x^2}{a^2} = 1,$$

Now $e$ and $e'$ are given by

$$b^2 = a^2(e^2 - 1) \text{ and } a^2 = b^2(e'^2 - 1).$$

Multiplying these equations, we get

$$1 = (e^2 - 1)(e'^2 - 1)$$

$$\Rightarrow \quad e^2 + e'^2 = e^2 e'^2$$

$$\Rightarrow \quad \frac{1}{e'^2} + \frac{1}{e^2} = 1.$$

## Exercise 12.1

1. Find the equation of the hyperbola with foci $(\pm 2, 0)$ and eccentricity, $\frac{3}{2}$.
2. Find the equations of the hyperbolas having distance between directrices, $\frac{4}{3}\sqrt{3}$ and passing through the point (2, 1).
3. Find the equation of the hyperbola referred to its axes as the axes of co-ordinates where distance between the foci is 1 and eccentricity $\sqrt{2}$.
4. Find the equation of the hyperbola in the standard form, given that (i) vertex is (4, 0); focus is (6, 0).
(ii) latus rectum = 4; eccentricity = 3.
5. Find the equation of the hyperbola, the equations of whose transverse and conjugate axes are along the lines, $x + 2y - 3 = 0$ and $2x - y + 4 = 0$ with their respective lengths $\sqrt{2}$ and $2/\sqrt{3}$.
6. Find the equation of the hyperbola, given that:
   (i) Centre is (0, 3); focus is (0, 5); semi-conjugate axis = 1.
   (ii) Conjugate axis = 4 and the distance between the foci = 12.
7. In the hyperbola, $9x^2 - 16y^2 = 144$, find the eccentricity, the co-ordinates of foci and length of the latus rectum.
8. Find the eccentricity, foci, equation of the directrix, equations and lengths of the latera recta of the hyperbola,
   (i) $y^2 - 3x^2 = 4$.
   (ii) $x^2 - y^2 = 4$.
9. Find the eccentricity of the hyperbola conjugate to the

hyperbola, $x^2 - 3y^2 = 3$.

10. Find the centre, lengths of the axes, eccentricity and foci of the hyperbola:

$$\frac{(x-1)^2}{9} - \frac{(y-2)^2}{16} = 1$$

## 12.2. SUMMARY OF THE RESULTS FROM CHAPTER 8

For the benefit of the reader we summarise here some of the results concerning the hyperbola, already derived in Chapter 8.

1. The parametric equations of the hyperbola, or the co-ordinates of any point on the hyperbola, $\frac{x^2}{a^2} - \frac{y^2}{b^2} = 1$

   are $x = a \sec \theta$, $y = b \tan \theta$,
   or $x = a \cosh \theta$, $y = b \sinh \theta$.

   **Note**. It is because of this reason that cosh θ, sinh θ etc. are called the hyperbolic functions.

2. The equation of the tangent to the hyperbola, $\frac{x^2}{a^2} - \frac{y^2}{b^2} = 1$

   at the point $(x', y')$ is
$$\frac{xx'}{a^2} - \frac{yy'}{b^2} = 1$$

3. The equation of the tangent to it at $(a \sec \theta, b \tan \theta)$ is
$$\frac{x}{a}\sec\theta - \frac{y}{b}\tan\theta = 1.$$

4. The equation of the normal to it at $(a \sec \theta, b \tan \theta)$ is
$$ax \cos \theta + by \cot \theta = a^2 + b^2$$

   or
$$\frac{ax}{\sec\theta} + \frac{by}{\tan\theta} = a^2 + b^2$$

5. The condition that the line $y = mx + c$ may touch the hyperbola, $x^2/a^2 - y^2/b^2 = 1$, is $c^2 = a^2m^2 - b^2$ and hence the equation of any tangent to it (not parallel to the $y$-axis) is

$$y = mx \pm \sqrt{a^2m^2 - b^2}$$

6. The equation of the chord of contact of $(x', y')$ w.r.t. the hyperbola is

$$\frac{xx'}{a^2} - \frac{yy'}{b^2} = 1$$

7. The equation of the polar of the point $(x', y')$ w.r.t. the hyperbola is

$$\frac{xx'}{a^2} - \frac{yy'}{b^2} = 1$$

8. The exchange of the chord of the hyperbola, whose middle point is $(x', y')$, is $T = S'$,

where $$S = \frac{x^2}{a^2} - \frac{y^2}{b^2} - 1; \quad S' = \frac{x'^2}{a^2} - \frac{y'^2}{b^2} - 1$$

and $$T = \frac{xx'}{a^2} - \frac{yy'}{b^2} - 1$$

9. The equation of the pair of tangents from the point $(x', y')$ to the hyperbola is $SS' = T^2$.

**Remark**. Most of the formulae for the hyperbola $\frac{x^2}{a^2} - \frac{y^2}{b^2} = 1$ can be derived from the corresponding results for the ellipse, $\frac{x^2}{a^2} + \frac{y^2}{b^2} = 1$, by changing $b^2$ to $-b^2$. In the above summary of the Results, Results 2 and 5 to 9 can be derived in this way. The reader will find many more results in the sequel, which can be proved for the hyperbola, by simply imitating the proofs of the corresponding results for the ellipse, replacing $b^2$ by $-b^2$.

## 12.3 GEOMETRICAL INTERPRETATION OF THE PARAMETER θ, WHEN THE EQUATIONS OF THE HYPERBOLA ARE TAKEN AS $x = a \sec \theta, y = b \tan \theta$.

Let the equation of the hyperbola be

$$\frac{x^2}{a^2} - \frac{y^2}{b^2} = 1 \tag{1}$$

It meets the tranverse axis at $A$ and $A'$.

Draw the auxiliary circle, $x^2 + y^2 = a^2$, with $A'A$ as its diameter.

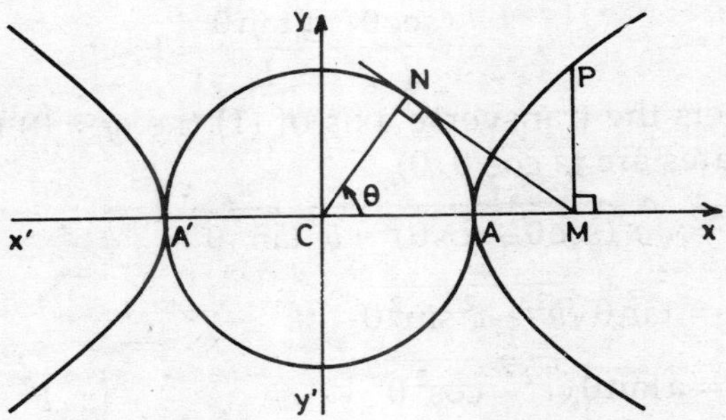


Let $P(x, y)$ be any point on the hyperbola. Draw $PM \perp Cx$. From $M$, draw a tangent, $MN$, to touch the auxiliary ecircle at the point $N$ and let $\angle MCN = \theta$.

Then $x = CM = CN \sec \theta = a \sec \theta$.

Substituting this value of $x$ in (1), we get

$$= \frac{a^2 \sec^2 \theta}{a^2} - \frac{y^2}{b^2} = 1 \Rightarrow \sec^2 \theta - 1 = \frac{y^2}{b^2}$$

$$\Rightarrow y = b \tan \theta.$$

Thus, we have explained the meaning of the parameter "θ"

**Note**. Throughout this chapter, a point "θ" on the hyperbola means the point ($a \sec \theta$, $b \tan \theta$).

**Example 12.7**. Prove that the part of the tangent at any point of a hyperbola intercepted between the point of contact and the transverse axis is a harmonic mean between the lengths of the perpendiculars drawn from the foci on the normal at the same point.

*Solution*. Let the coordinates of any point $P$ on the hyperbola

$$\frac{x^2}{a^2} - \frac{y^2}{b^2} = 1 \qquad (1)$$

be $P(a \sec \theta, b \tan \theta)$.

Now the equation of the tangent to (1) at $P$ is

$$\frac{x \sec \theta}{a} - \frac{y \tan \theta}{b} = 1.$$

This meets the transverse axis of (1) (i.e. $y = 0$) at $T$ whose coordinates are $(a \cos \theta, 0)$.

$$\text{Now } PT = \sqrt{a^2 (\sec \theta - \cos \theta)^2 + b^2 \tan^2 \theta}$$

$$= \tan \theta \sqrt{b^2 + a^2 \sin^2 \theta}$$

$$= \text{a} \tan \theta \sqrt{e^2 - \cos^2 \theta} \qquad \left(\because\ b^2 = a^2 (e^2 - 1)\right)$$

Again the equation of the normal to (1) at $P$ is

$$ax \cos \theta + by \cot \theta = a^2 + b^2 \qquad (2)$$

Let $p_1$ and $p_2$ be the perpendicular distances from $S(ae, 0)$ and $S'(-ae, 0)$ to (2). Then

$$p_1 = \frac{a^2 + b^2 - a^2 e \cos \theta}{\sqrt{a^2 \cos^2 \theta + b^2 \cot^2 \theta}}$$

$$= \frac{a^2 + a^2(-1 + e^2) - a^2 e \cos \theta}{\cos \sqrt{a^2 + a^2 (-1 + e^2) \operatorname{cosec}^2 \theta}}$$

$$= \frac{ae \,.\, (e - \cos \theta) \tan \theta}{\sqrt{e^2 - \cos^2 \theta}}$$

In the same way,

$$p_2 = \frac{ae(e+\cos\theta)\tan\theta}{\sqrt{e^2-\cos^2\theta}}$$

$$\frac{1}{p_1}+\frac{1}{p_2} = \frac{\sqrt{e^2-\cos^2\theta}}{ae\tan\theta}\left(\frac{1}{e-\cos\theta}+\frac{1}{e+\cos\theta}\right)$$

$$= \frac{2}{a\tan\theta\sqrt{e^2-\cos^2\theta}} = \frac{2}{PT}$$

$\Rightarrow$ $PT$ is H.M. between $p_1$ and $p_2$.

**Example 12.8**. A series of chords of the hyperbola, $\frac{x^2}{a^2}-\frac{y^2}{b^2}=1$ are tangents to the circle described on the line joining the foci of the hyperbola as diameter. Show that the locus of their poles with respect to the hyperbola is

$$\frac{x^2}{a^4}+\frac{y^2}{b^4}=\frac{1}{a^2+b^2}$$

*Solution*. The hyperbola, $\frac{x^2}{a^2}-\frac{y^2}{b^2}=1$ has the foci, $S(ae, 0)$ and $S'(-ae, 0)$. The equation of the circle on $SS'$ as diameter is

$$(x-ae)(x+ae)+(y-0)(y-0)=0$$

$$\Rightarrow \quad x^2+y^2=a^2e^2 \tag{1}$$

If $(h, k)$ be the pole of a chord of the above hyperbola, then the equation of its polar is

$$\frac{xh}{a^2}-\frac{yk}{b^2}=1 \tag{2}$$

Now the line (2) touches (1).

$\therefore$ Perpendicular from (0, 0) on (2)

= Radius of (1)

$$\Rightarrow \quad \frac{-1}{\sqrt{\frac{h^2}{a^4}+\frac{k^2}{b^4}}} = \pm\, ae$$

$$\Rightarrow \quad \frac{h^2}{a^4}+\frac{k^2}{b^4}=\frac{1}{a^2e^2}=\frac{1}{a^2+b^2}$$

$\therefore$ The locus of $(h, k)$ is

$$\frac{x^2}{a^4}+\frac{y^2}{b^4}=\frac{1}{a^2+b^2}$$

(See also Example 11.12).

**Exercise 12.2.**

1. (i) Show that the point

$$P(t)=\left(\frac{a(1+t^2)}{1-t^2},\frac{2bt}{1-t^2}\right)$$

lies on the hyperbola,

$$\frac{x^2}{a^2}-\frac{y^2}{b^2}=1$$

for all values of $t$.

(ii) Show that the point

$$P(t)=\left(\frac{a}{2}\left(t+\frac{1}{t}\right),\frac{b}{2}\left(t-\frac{1}{t}\right)\right)$$

lies on the hyperbola,

$$\frac{x^2}{a^2}-\frac{y^2}{b^2}=1,$$

for all values of $t$.

2. For the hyperbola, $x^2-y^2 = a^2$, determine the
   (i) eccentricity,
   (ii) polar of the point $P(\alpha, \beta)$,
   (iii) pole of the line, $Ax + By + C = 0$.
3. For the hyperbola, $x^2 - 2y^2 = 1$,
   (i) eccentricity,
   (ii) find $\overline{SA}.\overline{S'A}$, where $A$ is a vertex.
4. Prove that the equation of the chord joining two points

$(x_1, y_1)$ and $(x_2, y_2)$ on the hyperbola, $\frac{x^2}{a^2}-\frac{y^2}{b^2}=1$ is

$$\frac{x(x_1+x_2)}{a^2}-\frac{y(y_1+y_2)}{b^2}=\frac{x_1x_2}{a^2}-\frac{y_1y_2}{b^2}+1$$

5. Show that the equation of the chord joining the points 'α', 'β' on the hyperbola $x^2/a^2 - y^2/b^2 = 1$ is

$$\frac{x}{a}\cos\frac{\alpha-\beta}{2}-\frac{y}{b}\sin\frac{\alpha+\beta}{2}=\cos\frac{\alpha+\beta}{2}$$

Deduce the equation of the tangent at 'α'.

6. If the chord joining the points 'α' and 'β' on the hyperbola $x^2/a^2 - y^2/b^2 = 1$ is a focal chord, prove that

(i) $\pm e\cos\frac{\alpha-\beta}{2}=\cos\frac{\alpha+\beta}{2}$

(ii) $\tan\frac{\alpha}{2}\tan\frac{\beta}{2}+\frac{ke-1}{ke+1}=0$ where $k = \pm 1$.

7. A normal to the hyperbola,

$$\frac{x^2}{a^2}-\frac{y^2}{b^2}=1,$$

meets the axes in $M$ and $N$ and the lines $MP$ and $NP$ are drawn at right angles to the axes. Show that the locus of $P$ is the hyperbola

$$a^2x^2 - b^2y^2 = (a^2 + b^2)^2.$$

8. Find the point of intersection of the tangents at the points "α" and "β" on the hyperbola, $\frac{x^2}{a^2}-\frac{y^2}{b^2}=1$.

9. Let the tangent and the normal at a point $P$, on the hyperbola meet the transverse axis at $T$ and $G$ respectively. Prove that

$$\overline{CG}.\overline{CT} = a^2 + b^2.$$

10. Prove that the chords of a hyperbola which touch the conjugate hyperbola are bisected at their point of contact.

11. Show that the tangent and the normal at any point of a hyperbola bisect the focal distances of the point.

12. If the tangent at the point $(h, k)$ to the hyperbola cuts the auxiliary circle in points whose ordinates are $y_1$ and $y_2$, show that

$$\frac{1}{y_1}+\frac{1}{y_2}=\frac{2}{k}$$

13. Find the locus of the foot of the perpendicular drawn from the centre of the hyperbola, $\frac{x^2}{a^2}-\frac{y^2}{b^2}=1$ upon (i) any tangent (ii) any normal to it.

14. If $SY$ and $S'Y'$ be perpendiculars from the foci, $S$ and $S'$, on the tangent to a hyperbola at any point $P$, prove that $Y$ and $Y'$ will be on a circle with $AA'$ as its diameter, where $A$ and $A'$ are the vertices of the hyperbola. Show also that

$$\overline{SY}.\overline{S'Y'}$$

is constant.
(*Hint*. See Example 11.24).

15. The tangent at any point $P$ on the hyperbola, $\frac{x^2}{a^2}-\frac{y^2}{b^2}=1$ whose centre is $C$, meets the transverse and conjugate axes in $T_1$ and $T_2$; prove that
(i) $\overline{CN}.\overline{CT_1}=a^2$, where $PN$ is the perpendicular on the transverse axis.
(ii) $\overline{CM}.CT_2=-b^2$ where $PM$ is the perpendicular on the conjugate axis.

16. The normal at any point $P$ on the hyperbola, $\frac{x^2}{a^2}-\frac{y^2}{b^2}=1$, with centre $C$, meets the transverse and conjugate axes in $G_1$ and $G_2$; and if $CF$ be the perpendicular upon the normal, prove that
(i) $CG_1 = e^2\,CN$, where $PN$ is the perpendicular on the transverse axis.
(ii) $SG/S'G_1 = SP/S'P$.
(iii) $\overline{PF}.\overline{PG_1}=b^2$, $\overline{PF}.\overline{PG_2}=a^2$

17. The chord of the hyperbola, $\frac{x^2}{a^2} - \frac{y^2}{b^2} = 1$ whose equation is $x \cos \alpha + y \sin \alpha = p$ subtends a right angle at its centre. Prove that it always touches a circle of radius = $ab / \sqrt{b^2 - a^2}$

18. Prove that the sum of the squares of the reciprocals of two perpendicular radii vectors of the hyperbola is constant.

19. If $p$ is the length of the perpendicular from $C$, the centre of the hyperbola, $\frac{x^2}{a^2} - \frac{y^2}{b^2} = 1$ on the tangent at a point $P$ on it and $CP = r$; prove that $(a^2b^2/p^2) = r^2 + b^2 - a^2$.

20. An ellipse and a hyperbola have the same principal axes. Show that the polar of any point on either curve w.r.t. the other, touches the first curve.

21. Show that the area of the triangle formed by the lines $x - y = 0$, $x + y = 0$ and any tangent to the hyperbola, $x^2 - y^2 = a^2$ is constant.

22. A line is drawn parallel to the conjugate axis of a hyperbola to meet it and the conjugate hyperbola in the points $P$ and $Q$. Show that the tangent at $P$ and $Q$ meet on the curve,

$$\left(\frac{y^2}{b^2} - \frac{x^2}{a^2}\right)\frac{y^4}{b^4} = \frac{4x^2}{a^2}$$

Show also that the normals at $P$ and $Q$ meet on the transverse axis of the hyperbola.

## 12.4 FOUR NORMALS FROM A POINT TO A HYPERBOLA

**Definition**. *Conormal Points.* The points on the hyperbola, the normals at which pass through a common point are called conormal points of the hyperbola.

We now prove a theorem, concerning the number of normals which can be drawn from a point to a hyperbola.

**Theorem 12.1**. In general, four normals can be drawn to a hyperbola from any point in its plane, and if "$\alpha$", "$\beta$", "$\gamma$" and "$\delta$" be the feet of these four conormal points,

then $\alpha + \beta + \gamma + \delta$ = an odd multiple of $\pi$.

*Proof*. Let the hyperbola be $x^2/a^2 - y^2/b^2 = 1$ and let $(h, k)$ be any point in its plane. Let $P(a \sec\theta, b\tan\theta)$ be any point on the hyperbola, the normal at which passes through $(h, k)$.

The equation of the normal at $(a \sec\theta, b \tan\theta)$ to the hyperbola is $ax \cos\theta + by \cot\theta = a^2 + b^2$.

It passes through $(h, k)$,

$$\therefore\ ah \cos\theta + bk \cot\theta = a^2 + b^2 \tag{1}$$

Put $\tan\theta/2 = t$.

Now $\cos\theta = \dfrac{1-\tan^2\theta/2}{1+\tan^2\theta/2} = \dfrac{1-t^2}{1+t^2}$

and $\tan\theta = \dfrac{2\tan\theta/2}{1-\tan^2\theta/2} = \dfrac{2t}{1-t^2}$

Equation (1), then, becomes

$$\frac{ah(1-t^2)}{1+t^2} + \frac{bk(1-t^2)}{2t} = a^2 + b^2$$

$$\Rightarrow\ ah.2t(1-t^2) + bk(1-t^4) = (a^2+b^2).2t(1+t^2)$$

$$\Rightarrow\ bkt^4 + 2(a^2+b^2+ah)\,t^3 + 2(a^2+b^2-ah)\,t - bk = 0$$

which is a fourth degree equation in $t$, and hence there are four values of $t$, and then $\tan\theta/2 = t$, gives four values of $\theta$, say $\alpha$, $\beta$, $\gamma$ and $\delta$, the normals at which to the given hyperbola pass through the point $(h, k)$. Thus, in general, four normals can be drawn to a hyperbola from any point in its plane.

Also, $\tan\left(\dfrac{\alpha+\beta+\gamma+\delta}{2}\right) = \tan\left(\dfrac{\alpha}{2}+\dfrac{\beta}{2}+\dfrac{\gamma}{2}+\dfrac{\delta}{2}\right)$

$= \dfrac{s_1 - s_3}{1 - s_2 - s_4}$ (from Trigonometry)

$$\text{Now } s_1 - s_2 = \frac{-2(a^2 + b^2 + ah)}{bk} + \frac{2(a^2 + b^2 - ah)}{bk}$$

$$= \frac{-4ah}{bk} \neq 0$$

and $1 - s_2 + s_4 = 1 - 0 + (-1) = 0$.

$$\therefore \quad \frac{\alpha + \beta + \gamma + \delta}{2} = \text{an odd multiple of } \pi/2.$$

$\Rightarrow \alpha + \beta + \gamma + \delta =$ an odd multiple of $\pi$

**Example 12.9** The normal at $P$ to a hyperbola of eccentricity $e$, intersects its transverse and conjugate axes at $L$ and $M$ respectively. Show that the locus of the middle point of $LM$ is a hyperbola of eccentricity $\dfrac{e}{\sqrt{e^2 - 1}}$.

*Solution.* The equation of the normal at $P(a \sec \theta, b \tan \theta)$ to the hyperbola,

$$\frac{x^2}{a^2} - \frac{y^2}{b^2} = 1$$

is $\qquad ax \cos \theta + by \cot \theta = a^2 + b^2 = k$ (say).

It meets the transverse axis at $L$, where $y = 0$.

$$\therefore L \text{ is } \left(\frac{k \sec \theta}{a}, 0\right)$$

Similarly $M$ is $\left(0, \dfrac{k \tan \theta}{b}\right)$.

If $(\alpha, \beta)$ be the middle point of $LM$, then

$$\alpha = \frac{k \sec \theta}{2a},$$

$$\beta = \frac{k \tan \theta}{2b}$$

$$\Rightarrow 1 = \sec^2 \theta - \tan^2 \theta = \left(\frac{2a\alpha}{k}\right)^2 - \left(\frac{2b\beta}{k}\right)^2$$

$\therefore$ the locus of $(\alpha, \beta)$ is

$$\frac{x^2}{k^2/4a^2} - \frac{y^2}{k^2/4b^2} = 1.$$

Writing this as

$$\frac{x^2}{A^2} - \frac{y^2}{B^2} = 1$$

and using $B^2 = A^2(e'^2 - 1)$,
its eccentricity,

$$e' = \sqrt{\frac{A^2 + B^2}{A^2}} = \sqrt{\frac{k^2/4a^2 + k^2/4b^2}{k^2/4a^2}} = \sqrt{\frac{a^2 + b^2}{b^2}}$$

$$= \sqrt{\frac{a^2 + a^2(e^2 - 1)}{a^2(e^2 - 1)}} = \frac{e}{\sqrt{e^2 - 1}}$$

**Example 12.10**. A chord of the hyperbola $\frac{x^2}{a^2} - \frac{y^2}{b^2} = 1$ touches the circle, $x^2 + y^2 = c^2$. Show that the mid-point of the chord lies on the curve

$$\left(\frac{x^2}{a^2} - \frac{y^2}{b^2}\right)^2 = c^2\left(\frac{x^2}{a^4} + \frac{y^2}{b^4}\right).$$

*Solution*. Let $x \cos \theta + y \sin \theta = C$ (1)
be a tangent to

$$x^2 + y^2 = c^2 \tag{2}$$

Now (1) is a chord of the hyperbola

$$\frac{x^2}{a^2} - \frac{y^2}{b^2} = 1 \tag{3}$$

Let the mid-point of (1) be $(x_1, y_1)$.
Then the equation of chord of (3) with mid-point $(x_1, y_1)$, is $T = S_1$

$$\Rightarrow \qquad \frac{xx_1}{a^2} - \frac{yy_1}{b^2} = \frac{x_1^2}{a^2} - \frac{y_1^2}{b^2} \qquad (4)$$

As (1) and (4) represent the same chord; on comparing their coefficients, we get

$$\frac{\cos\theta}{x_1 / a^2} = \frac{\sin\theta}{-y_1 / b^2} = \frac{c}{\dfrac{x_1^2}{a^2} - \dfrac{y_1^2}{b^2}}$$

Eliminating θ, we get

$$c^2\left(\frac{x_1^2}{a^4} + \frac{y_1^2}{b^4}\right) = \left(\frac{x_1^2}{a^2} - \frac{y_1^2}{b^2}\right)^2$$

Now we take the locus of $(x_1, y_1)$ and get the required result.

**Example 12.11**. Find the locus of the point of intersection of the tangents to the hyperbola, $\frac{x^2}{a^2} - \frac{y^2}{b^2} = 1$ which include an angle whose tangent is $b/a$.

*Solution.* Let $(x_1, y_1)$ be the point from where tangents are drawn to the hyperbola,

$$\frac{x^2}{a^2} - \frac{y^2}{b^2} = 1 \qquad (1)$$

Suppose these tangents intersect at an angle θ.
By hypothesis, $\tan\theta = b/a$
Now the joint equation of these tangents through $(x_1, y_1)$ to (1) is

$$T^2 = SS_1$$

$$\Rightarrow \qquad \left(\frac{xx_1}{a^2} - \frac{yy_1}{b^2} - 1\right)^2 = \left(\frac{x^2}{a^2} - \frac{y^2}{b^2} - 1\right)\left(\frac{x_1^2}{a^2} - \frac{y_1^2}{b^2} - 1\right) \qquad (2)$$

Here coefficient of $x^2$

$$= \frac{x_1^2}{a^4} - \frac{1}{a^2}\left(\frac{x_1^2}{a^2} - \frac{y_1^2}{b^2} - 1\right)$$

$$= \frac{y_1^2 + b^2}{a^2b^2}$$

Coefficient of $y^2 = \frac{y_1^2}{b^4} + \frac{1}{b^2}\left(\frac{x_1^2}{a^2} - \frac{y_1^2}{b^2} - 1\right) = \frac{x_1^2 - a^2}{a^2b^2}$

and $\frac{1}{2}$. coefficient of $xy$

$$= -\frac{x_1y_1}{a^2b^2}$$

Now $\qquad \frac{b}{a} = \tan\theta = \frac{2\sqrt{h^2 - ab}}{a+b}$

$$= \frac{2\sqrt{\frac{x_1^2y_1^2}{a^4b^4} - \left(\frac{y_1^2 + b^2}{a^2b^2}\right)\left(\frac{x_1^2 - a^2}{a^2b^2}\right)}}{\frac{y_1^2 + b^2}{a^2b^2} + \frac{x_1^2 - a^2}{a^2b^2}}$$

$$= \frac{2\sqrt{a^2y_1^2 - b^2x_1^2 + a^2b^2}}{x_1^2 + y_1^2 + b^2 - a^2}$$

$\Rightarrow = b^2(x^2 + y^2 + b^2 - a^2)^2 = 4a^2(a^2y^2 - b^2x^2 + a^2b^2)$,
which is the required locus.

### Exercise 12.3.

1. Find the equations of the common tangents to the hyperbolas, $\frac{x^2}{a^2} - \frac{y^2}{b^2} = 1$ and $\frac{y^2}{a^2} - \frac{x^2}{b^2} = 1$. Also find the length of this common tangent.
2. Show that the common tangents of the conics

   $$\frac{x^2}{a^2} + \frac{y^2}{b^2} = 1 \text{ and } \frac{x^2}{a^2} - \frac{y^2}{b^2} + \frac{a^2 + b^2}{a^2 - b^2} = 0$$

   intercepted between the points of contact subtends a right angle at their centre.

3. Find the condition of concurrence of the normals to the hyperbola, $\frac{x^2}{a^2} - \frac{y^2}{b^2} = 1$ at the points whose parameters are $\theta_1, \theta_2, \theta_3, \theta_4$.
4. Prove that the feet of the normals to,

$$\frac{x^2}{a^2} - \frac{y^2}{b^2} = 1,$$

from $(h, k)$ lie on,
$a^2y(x - h) + b^2x(y - k) = 0$
(*Hint*: See Example 11.9).
5. If the normals at four points
$(x_i, y_i), i = 1, 2, 3, 4$
on the hyperbola

$$\frac{x^2}{a^2} - \frac{y^2}{b^2} = 1$$

are concurrent, show that

$$(\Sigma x_i)\left(\Sigma \frac{1}{x_i}\right) = (\Sigma y_i)\left(\Sigma \frac{1}{y_i}\right) = 4$$

6. Find the condition that the normals to the hyperbola,

$$\frac{x^2}{a^2} - \frac{y^2}{b^2} = 1$$

at the ends of the chords given by

$$lx + my = 1, \qquad l'x + m'y = 1$$

be concurrent.
(Hint: See Example 11.11).
7. Show that the locus of the mid-points of normal chords of the rectangular hyperbola, $x^2 - y^2 = a^2$ is $(y^2 - x^2)^3 = 4a^2x^2y^2$.
8. Show that the locus of the middle points of the portions of tangents to the hyperbola,

$$\frac{x^2}{a^2} - \frac{y^2}{b^2} = 1$$

intercepted between the axes, is

$$4x^2y^2 = a^2y^2 - b^2x^2.$$

9. Prove that the locus of the mid-points of the chords of the hyperbola, $\frac{x^2}{a^2} - \frac{y^2}{b^2} = 1$, passing through the fixed point $(\alpha, \beta)$ is a hyperbola whose centre is $(\alpha/2, \beta/2)$.
10. Prove that the two tangents drawn from any point on the hyperbola, $x^2 - y^2 = a^2 - b^2$ to the ellipse, $\frac{x^2}{a^2} + \frac{y^2}{b^2} = 1$ make complementary angles with the axes.
11. Tangents are drawn from any point on the hyperbola, $x^2 - y^2 = a^2 + b^2$ to the hyperbola, $\frac{x^2}{a^2} - \frac{y^2}{b^2} = 1$. Prove that they meet the axes in concyclic points.

## 12.5 ASYMPTOTES

Let $f(x, y) = 0$ be a curve having a branch at infinity and $l = 0$ be a given line. Let $P(x, y)$ be a point on this curve and suppose its distance from the line, $l = 0$, be $p$. If $p \to 0$, as the point $P$ recedes to infinity along the curve, then we say that the line, $l = 0$, is an asymptote to the given curve.

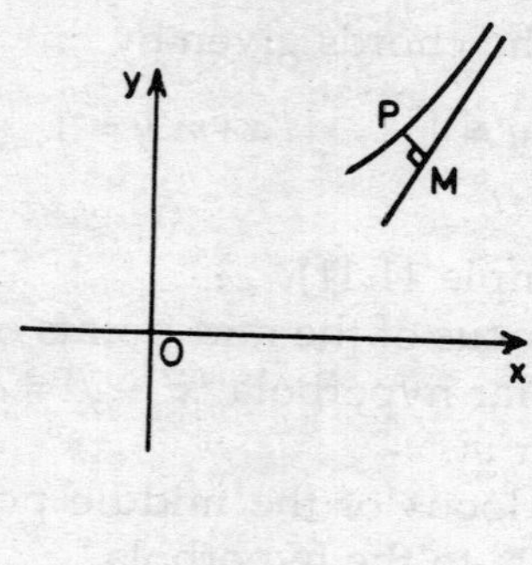

**12.5.1. To find the asymptotes of the hyperbola,**

$$\frac{x^2}{a^2} - \frac{y^2}{b^2} = 1.$$

First note that as $y \to \infty$, so does $x$ and vice versa. This means that the hyperbola has only oblique asymptotes.

Let $y = mx + c$ be any asymptote to it, and let $p$ denote the perpendicular distance from a point $P(x, y)$ on the hyperbola to this line; then

$$p = \left|\frac{y - mx - c}{\sqrt{1+m^2}}\right|.$$

Let now $P$ recede to infinity; then $x \to \infty$ and $p \to 0$.

$$0 = \lim_{x\to\infty} p = \lim_{x\to\infty} \left|\frac{y - mx - c}{\sqrt{1+m^2}}\right|$$

$$\Rightarrow \quad \lim_{x\to\infty} (y - mx - c) = 0$$

$$\Rightarrow \quad c = \lim_{x\to\infty} (y - mx) \tag{1}$$

Further

$$\lim_{x\to\infty} \left(\frac{y}{x} - m\right) = \lim_{x\to\infty} \left[(y - mx).\frac{1}{x}\right] = c.0 = 0$$

$$\Rightarrow \quad m = \lim_{x\to\infty} \left(\frac{y}{x}\right) \tag{2}$$

We are given the equation of the hyperbola as

$$\frac{x^2}{a^2} - \frac{y^2}{b^2} = 1. \tag{3}$$

$$\Rightarrow \quad \frac{1}{a^2} - \frac{1}{b^2}\left(\frac{y}{x}\right)^2 = \frac{1}{x^2}.$$

Let $x \to \infty$. By (2), $\frac{y}{x} \to m$

$$\therefore \quad \frac{1}{a^2} - \frac{1}{b^2} m^2 = 0 \Rightarrow m = \pm\frac{b}{a}$$

Let, first, $m = \dfrac{b}{a}$

Then $y - mx = y - \dfrac{b}{a}x = d$ (say).

Again when $x \to \infty$, $d \to c$ (by (1))

Substituting for $y$ in (3), we get

$$\frac{x^2}{a^2} - \frac{1}{b^2}\left(\frac{bx}{a} + d\right)^2 = 1$$

$$\Rightarrow \quad \frac{2bd}{a}x + d^2 + b^2 = 0$$

$$\Rightarrow \quad \frac{2bd}{a} + \frac{d^2 + b^2}{x} = 0$$

Proceeding to limits as $x \to \infty$,

$$\frac{2bc}{a} + 0 = 0 \Rightarrow c = 0$$

Hence $\quad y = mx + c \Rightarrow y = \dfrac{b}{a}x$

This gives one asymptote of the hyperbola.
Similarly taking

$$m = -\frac{b}{a},$$

we get the other asymptotes to it as

$$y = -\frac{b}{a}x.$$

Thus the two asymptotes are given by

$$\frac{x}{a} - \frac{y}{b} = 0$$

and

$$\frac{x}{a} + \frac{y}{b} = 0$$

Their joint equation is

$$\frac{x^2}{a^2} - \frac{y^2}{b^2} = 0.$$

**Note 1.** The reader familiar with calculus can proceed as follows:

Write the equation of the hyperbol as

$$\phi(x, y) = b^2x^2 - a^2y^2 - a^2b^2 = 0$$

Here $\phi_2(m) = b^2 - a^2m^2$

and $\phi_2(m) = 0 \Rightarrow m = \pm \dfrac{b}{a}$

Also $c = -\dfrac{\phi_1(m)}{\phi_2{}'(m)} = -\dfrac{0}{-2a^2m} = 0$

for $m = \pm \dfrac{b}{a}$.

Hence the two asymptotes to the hyperbola are

$$y = \pm \frac{b}{a}x$$

(These are the only two asymptotes, as a second degree equation can have at most two asymptotes).

**Note 2**. Let $S \equiv \dfrac{x^2}{a^2} - \dfrac{y^2}{b^2} - 1 = 0$

$$A \equiv \frac{x^2}{a^2} - \frac{y^2}{b^2} = 0$$

and $C \equiv \dfrac{x^2}{a^2} - \dfrac{y^2}{b^2} + 1 = 0$

be the equations of the hyperbola, asymptotes and the conjugate hyperbola respectively; then clearly

$$C + S = 2A.$$

**Note 3.** The asymptotes of the conjugate hyperbola are again

$$\frac{x^2}{a^2} - \frac{y^2}{b^2} = 0.$$

Its proof is left to the reader.

**Note 4**. The equation of the asymptotes differs from the equation of the hyperbola by a constant and the equation of the conjugate hyperbola differs from the equation of the asymptotes by the same constant. (*For proof of this result*, in a general case, see: *Theorem 13.3*). Thus if $S = 0$ is the equation of a hyperbola, the equation of the asymptotes is, $S + k = 0$ provided it represents a pair of lines and the equation of the conjugate hyperbola is $S + 2k = 0$.

Also if $S = 0$ represents a pair of lines, the general equations of a hyperbola and its conjugate are of the form $S + k = 0$ and $S - k = 0$, respectively.

**Note 5.** As both circle and ellipse are bounded curves, (i.e. they have no branches at infinity), they cannot have asymptotes. A parallel, although it has a branch at infinity, yet it has no asymptote. The reader can easily verify this.

**Note 6**. The asymptotes of the hyperbola, $\frac{x^2}{a^2} - \frac{y^2}{b^2} = 1$ are tangents from the centre (0, 0) of the hyperbola. This can be very easily seen by using the formula, $T^2 = SS_1$.

**Example 12.12** (a) Prove that any line parallel to either of the asymptotes of a hyperbola shall meet it in one point at infinity.

(b) Prove that the asymptotes of a hyperbola are the diagnols of the rectangle formed by the lines drawn parallel to the axes at the vertices of the hyperbola (i.e. at $(\pm a, 0)$ and $(0, \pm b)$).

*Solution*. Let the given hyperbola be

$$\frac{x^2}{a^2} - \frac{y^2}{b^2} = 1 \tag{1}$$

∴ Its asymptotes are

$$\frac{x}{a} + \frac{y}{b} = 0 \text{ and } \frac{x}{a} - \frac{y}{b} = 0$$

(a) Let us take the asymptote

$$\frac{x}{a}-\frac{y}{b}=0 \text{ i.e. } y=\frac{b}{a}x$$

Any line parallel to it is

$$y=\frac{b}{a}x+c \qquad (2)$$

Now (1) and (2) intersect, where

$$\frac{x^2}{a^2}-\left(\frac{b}{a}x+c\right)^2\cdot\frac{1}{b^2}=1$$

This equation has one of its roots to be infinite as coefficient of $x^2$ in it is zero.

Also from (2), $y\to\infty$ as $x\to\infty$

$\therefore$ (1) and (2) meet in one point at infinity.

(b) Lines through $(a, 0)$, $(-a, 0)$ and $(0, b)$, $(0, -b)$ parallel to the principal axes enclose a rectangle whose vertices are $O(a, b)$, $P(-a, b)$, $Q(-a, -b)$ and $R(a, -b)$.

Now the equation of $OQ$ is

$$y-b=\frac{b-(-b)}{a-(-a)}(x-a)$$

$$\Rightarrow \quad y=\frac{b}{a}x \Rightarrow \frac{x}{a}-\frac{y}{b}=0$$

ie. the diagonal $OQ$ is one of the asymptotes of the hyperbola. In the same way, diagonal $PR$ is the other asymptote.

**Example 12.13**. The tangent at any point $P$ on the hyperbola meets the asymptotes at $Q$ and $R$. Prove that $\overline{CQ}.\overline{CR}$ = constant.

*Solution*. Let the hyperbola be

$$\frac{x^2}{a^2}-\frac{y^2}{b^2}=1$$

and any point $P$ on it be $(a \sec\theta, b\tan\theta)$

The tangent to the hyperbola at $P$ is given by

$$\frac{x\sec\theta}{a}-\frac{y\tan\theta}{b}=1 \tag{1}$$

Also the asymptotes to it are given by

$$y=\pm\frac{b}{a}x \tag{2}$$

Now (1) and asymptotes (2) meet at $Q$ and $R$, whose coordinates, respectively are

$$\left(\frac{a}{\sec\theta-\tan\theta},\frac{b}{\sec\theta-\tan\theta}\right)\text{and}\left(\frac{a}{\sec\theta+\tan\theta},\frac{b}{\sec\theta+\tan\theta}\right)$$

Now $CQ^2.CR^2=\dfrac{a^2+b^2}{(\sec\theta-\tan\theta)^2}.\dfrac{a^2+b^2}{(\sec\theta+\tan\theta)^2}=(a^2+b^2)^2$

$\Rightarrow \overline{CQ}.\overline{CR}$ = constant.

**Example 12.14**. Prove that the locus of the centre of the circle circumscribing the triangle formed by the asymptotes of the hyperbola, $\dfrac{x^2}{a^2}-\dfrac{y^2}{b^2}=1$ and a variable tangent is

$$4(a^2x^2-b^2y^2)=a^2+b^2.$$

*Solution*. The asymptotes of the given hyperbola are given by

$$\frac{x}{a}+\frac{y}{b}=0 \tag{1}$$

and

$$\frac{x}{a}-\frac{y}{b}=0 \tag{2}$$

Let any tangent to the hyperbola be

$$\frac{x\sec\theta}{a}-\frac{y\tan\theta}{b}=1 \tag{3}$$

Now (1) and (3) meet at

$P(a(\sec\theta-\tan\theta), -b(\sec\theta-\tan\theta))$

and (2) and (3) meet at

$Q(a(\sec\theta+\tan\theta), -b(\sec\theta+\tan\theta))$

We want to find the equation of the circle through $C$, $P$ and $Q$. Now any circle passing through $C(0,0)$ is given by the equation,

$$x^2 + y^2 + 2gx + 2fy = 0 \tag{4}$$

As $P$ and $Q$, lie on (4),

$$\therefore \quad (a^2+b^2)(\sec\theta - \tan\theta) + 2(ag - bf) = 0 \tag{5}$$

and

$$(a^2 + b^2)(\sec\theta + \tan\theta) + 2(ag + bf) = 0 \tag{6}$$

Now (5) + (6) and (6) – (5) imply

$$\sec\theta = \frac{-2ag}{a^2+b^2} \text{ and } \tan\theta = \frac{-2bf}{a^2+b^2}$$

$$\Rightarrow \quad 1 = \sec^2\theta - \tan^2\theta$$

$$= \frac{4a^2g^2}{(a^2+b^2)^2} - \frac{4b^2f^2}{(a^2+b^2)^2}$$

$$\Rightarrow \quad 4(a^2g^2 - b^2f^2) = (a^2 + b^2)^2$$

$\therefore$ Locus of $(-g, -f)$, the centre of the circumscribing circle, is $4(a^2x^2 - b^2y^2) = (a^2 + b^2)^2$.

**Example 12.15**. A parallelogram is constructed with its sides parallel to the asymptotes of a hyperbola and one of its diagonals is a chord of the hyperbola, show that the other diagonal passes through the centre.

*Solution*. Let the given hyperbola be

$$\frac{x^2}{a^2} - \frac{y^2}{b^2} = 1 \tag{1}$$

As the parallelogram to be constructed has one diagonal a chord of (1), the extremities $(x_1, y_1)$ and $(x_2, y_2)$ of it lie on (1).

$$\therefore \quad \frac{x_1^2}{a^2} - \frac{y_1^2}{b^2} = 1 \text{ and } \frac{x_2^2}{a^2} - \frac{y_2^2}{b^2} = 1 \tag{2}$$

Also asymptotes of (1) are given by

$$\frac{x}{a} - \frac{y}{b} = 0 \text{ and } \frac{x}{a} + \frac{y}{b} = 0 \tag{3}$$

Now the equations of two adjacent sides of the parallelogram passing through $(x_1, y_1)$ and $(x_2, y_2)$ and parallel to (3) are

$$\frac{x}{a} - \frac{y}{b} = \frac{x_1}{a} - \frac{y_1}{b} \tag{4}$$

and

$$\frac{x}{a} + \frac{y}{b} = \frac{x_2}{a} + \frac{y_2}{b} \tag{5}$$

Now (4) and (5) intersect in the point

$$P\left(\frac{x_1 + x_2}{2} - \frac{a}{2b}(y_1 - y_2), \frac{y_1 + y_2}{2} - \frac{b}{2a}(x_1 - x_2)\right)$$

The other diagonal passes through $P$ and the mid-point $\left(\frac{x_1 + x_2}{2}, \frac{y_1 + y_2}{2}\right)$ of the first diagonal.

$\therefore$ Its equation is

$$y - \frac{y_1 + y_2}{2} = \frac{\frac{b}{2a}(x_1 - x_2)}{\frac{a}{2b}(y_1 - y_2)}\left(x - \frac{x_1 + x_2}{2}\right)$$

$$= \frac{b^2(x_1 - x_2)}{a^2(y_1 - y_2)}\left(x - \frac{x_1 + x_2}{2}\right)$$

$$\Rightarrow \quad a^2 y(y_1 - y_2) - \frac{a^2}{2}(y_1^2 - y_2^2) = b^2 x(x_1 - x_2) - \frac{b^2}{2}(x_1^2 - x_2^2)$$

$$\Rightarrow \quad a^2 y(y_1 - y_2) = b^2 x(x_1 - x_2); \qquad \text{(using(2))}$$

This clearly passes through the centre (0, 0) of the hyperbola.

**Example 12.16**. Prove that the locus of the poles of a given line w.r.t. the circles of a coaxal system is, in general, a hyperbola having one asymptote parallel to their radical axis and the other asymptote perpendicular to the given line.

*Solution*. Let the coaxal system of circles be given by

$$x^2 + y^2 + 2gx + C = 0 \tag{1}$$

where $g$ is a parameter and its radical axis is $x = 0$.

Let the given line be $lx + my + 1 = 0$ (2)

If $(x_1, y_1)$ be the pole of (2) w.r.t. (1), then its polar is

$$xx_1 + yy_1 + g(x + x_1) + C = 0$$

$$\Rightarrow \quad (x_1 + g)x + yy_1 + (gx_1 + C) = 0 \tag{3}$$

Now (2) and (3) represent the same line; comparing their coefficients, we get

$$\frac{x_1 + g}{l} = \frac{y_1}{m} = \frac{gx_1 + C}{1}$$

$$\Rightarrow \quad g = \frac{ly_1}{m} - x_1 = \frac{1}{x_1}\left(\frac{y_1}{m} - C\right)$$

$$\Rightarrow \quad lx_1y_1 - x_1^2 m = y_1 - mc$$

$\therefore$ The locus of $(x_1, y_1)$ is

$$mx^2 - lxy + y - mc = 0 \tag{4}$$

which represents a hyperbola

(You can easily verify: $\Delta \neq 0$ and $ab - h^2 < 0$; see Chapter 13) and its asymptotes are parallel to the lines given by $mx^2 - lxy = 0$ (*see Cor. 1 of Theorem 13.3*)

$$\Rightarrow \quad x(mx - ly) = 0$$

$$\Rightarrow \quad x = 0;\ mx - ly = 0.$$

Thus one asymptote of the hyperbola, (4) is parallel to radical axis ($x = 0$) of (1) and the other is perpendicular to the line (2).

**Exercise 12.4.**

1. (i) Find the general equation of the hyperbolas having the same asymptotes as the hyperbola,

$$\frac{x^2}{a^2} - \frac{y^2}{b^2} = 1.$$

(ii) Find the asymptotes and the conjugate hyperbola of the hyperbola,

$$(px + qy + r)(p'x + q'y + r') + A = 0.$$

2. Show that the angle between the asymptotes of the hyperbola

$$\frac{x^2}{a^2} - \frac{y^2}{b^2} = 1 \text{ is } 2\tan^{-1}\left(\frac{b}{a}\right).$$

3. Find the equation of the asymptotes of the hyperbola $9y^2 - 4x^2 = 36$ and obtain the product of the perpendicular distance of any point on the hyperbola from the asymptotes.
4. Show that the eccentricity of a hyperbola in which a normal at one end of its latus rectum is parallel to one of its asymptotes, is given by

$$e^2 = \frac{\sqrt{5}+1}{2}.$$

5. Show that the points of intersection of the asymptotes with the directrices of the hyperbola,

$$\frac{x^2}{a^2} - \frac{y^2}{b^2} = 1,$$

lie on the circle, $x^2 + y^2 = a^2$.
6. Show that the portion of a tangent to a hyperbola intercepted between its asymptotes is bisected at the point of contact.
7. (i) Prove that the area of a triangle formed by the asymptotes and any tangent to a hyperbola is constant and is equal to $ab$.
   (ii) Find the eccentricity of the hyperbola, such that the area of a triangle formed by the asymptotes and any tangent to the hyperbola,

$$\frac{x^2}{a^2} - \frac{y^2}{b^2} = 1, \text{ is } a^2 \tan \lambda.$$

8. A line through the focus $S$ drawn parallel to the asymptote $y = (b/a)x$ of the hyperbola

$$\frac{x^2}{a^2} - \frac{y^2}{b^2} = 1$$

meets the hyperbola and the corresponding directrix at $P$ and $Q$. Show that $\overline{SQ} = 2\overline{SP}$
9. A line drawn, through $P$ (any point on the hyperbola),

with slope $m$, meets its asymptotes at $Q$ and $R$. Show that $\overline{PQ} = \overline{PR}$ = constant.

10. Prove that the polarofany point on an asymptote of a hyperbola w.r.t. it is parallel to that asymptote.
11. Prove that the tangent to the hyperbola, $x^2 - 3y^2 = 3$ at the point $(\sqrt{3}, 0)$ when associated with the two asymptotes forms an equilateral triangle. Find also the area of this triangle.
12. A line has its extremities on two fixed lines and cuts off from them a triangle of constant area. Prove that the locus of the mid–point of the line is a hyperbola having the two lines as asymptotes.
13. From any point on one hyperbola, tangents are drawn to another which has the same asymptotes. Show that the chord of contact cuts off constant area from the asymptote.
14. Prove that the centroid of the triangle formed by a tangent to a hyperbola with its asymptotes, lies on another hyperbola.
15. A series of hyperbolas is drawn having a common transverse axis of length $2a$. Prove that the locus of a point $P$ on each hyperbola, such that its distance from the tranverse axis is equal to its distance from an asymptote, is the curve,
$$(x^2 - y^2)^2 = 4x^2(x^2 - a^2).$$
16. Find the locus of a point such that the angle between the tangents from it to a hyperbola is equal to the angle between the asymptotes of the hyperbola.
17. $PN$ is the ordinate of a point $P$ on a hyperbola, the normal at which meets the axis in $G$. If $NP$ be produced to meet the asymptote in $Q$, prove that $QG$ is at right angles to it.
18. Let $C$ be the centre of a hyperbola. The tangent at $P$ meets the axes in $T_1$ and $T_2$ and the asymptotes in $L_1$, $L_2$; the normal at P meets the axes in $G_1$ and $G_2$. Prove that $L_1$, $L_2$ lie on the circle $OG_1G_2$ and $T_1$,$T_2$ are conjugate w.r.t. the circle.

## 12.6 TO FIND THE LOCUS OF THE MIDDLE POINTS OF A SYSTEM OF PARALLEL CHORDS, HAVING SLOPE $m$, OF THE HYPERBOLA.

Let the hyperbola be

$$\frac{x^2}{a^2} - \frac{y^2}{b^2} = 1$$

Let $(\alpha, \beta)$ be the middle point of its any chord, then the equation of this chord is

$$T = S' \text{ i.e.} \frac{x\alpha}{a^2} - \frac{y\beta}{b^2} = \frac{\alpha^2}{a^2} - \frac{\beta^2}{b^2}$$

By hypothesis, its slope is $m$.

$$\therefore \quad m = \frac{\alpha}{a^2} \cdot \frac{b^2}{\beta} \quad \Rightarrow \quad \beta = \frac{b^2}{a^2 m}\alpha$$

Thus the locus of $(\alpha, \beta)$ is

$$y = \frac{b^2}{a^2 m} x$$

which is a line passing through the centre (0, 0) of the hyperbola. It is called a diameter of the hyperbola.

**Remark**. Let its slope be $m'$.

$$\therefore \quad m' = \frac{b^2}{a^2 m} \Rightarrow mm' = \frac{b^2}{a^2}$$

The symmetry of the result in $m$ and $m'$ shows that the locus of the middle points of parallel chords of the hyperbola having slope $m'$, is the diameter, $y = mx$; i.e. each diameter bisects all the chords of the hyperbola, parallel to the other diameter. Two diameters of the hyperbola having this property are called *conjugate diameters*.

**Note**. If $y = mx$ and $y = m'x$ be a pair of conjugate diameters w.r.t. a hyperbola, $\frac{x^2}{a^2} - \frac{y^2}{b^2} = 1$; one can easily see that they are also conjugate w.r.t. the conjugate hyperbola.

$$\frac{x^2}{a^2}-\frac{y^2}{b^2}=-1 \text{ and } mm'=\frac{b^2}{a^2}.$$

We now prove a result concerning a pair of conjugate diameters of a hyperbola and its conjugate hyperbola.

**Theorem 12.2**. If a diameter meets a hyperbola in real points, it will meet the conjugate hyperbola in imaginary points and its conjugate diameter will meet the hyperbola in imaginary points and the conjugate hyperbola in real points and vice versa.

*Proof*. The equations of the hyperbola and the conjugate hyperbola are

$$\frac{x^2}{a^2}-\frac{y^2}{b^2}=1 \qquad (1)$$

and $$\frac{x^2}{a^2}-\frac{y^2}{b^2}=-1, \text{ respectively.} \qquad (2)$$

Let $y = mx$ and $y = m'x$ be a pair of conjugate diameters. Then

$$mm'=\frac{b^2}{a^2}$$

The points of intersection of $y = mx$ and (1) are given by

$$\frac{x^2}{a^2}-\frac{x^2}{b^2}m^2=1$$

$$\Rightarrow \quad x^2=\frac{1}{1/a^2-m^2/b^2}=\frac{a^2b^2}{b^2-a^2m^2}$$

$$\Rightarrow \quad x=\pm\frac{ab}{\sqrt{b^2-a^2m^2}}$$

By hypothesis, $y = mx$ meets (1) in real points. This requires $b^2 - a^2m^2 > 0$.

Again $y = mx$ meets (2), where $x^2 = -\dfrac{a^2b^2}{b^2-a^2m^2}$ and hence, these points of intersection are imaginary.

Now, the conjugate diameter, $y = m'x$, meets (1) in points given by

$$x^2 = \frac{a^2b^2}{b^2 - a^2m'^2} = \frac{a^2b^2}{b^2 - a^2\dfrac{b^4}{a^4m^2}} = -\frac{a^4m^2}{b^2 - a^2m^2}$$

$$\left(\because mm' = \frac{b^2}{a^2}\right)$$

which are imaginary.

Also, the conjugate diameter $y = m'x$ will meet the conjugate hyperbola (2), in points given by

$$x^2 = -\frac{a^4m^2}{b^2 - a^2m^2} \text{ which are real}$$

Converse follows on the same lines.

**Note 1**. If $CD$ is the conjugate diameter of a diameter $CP$, of th

e hyperbola, $\frac{x^2}{a^2} - \frac{y^2}{b^2} = 1$; where $P$ is ($a$ sec θ, $b$ tan θ), the

the equation of $CP$ is

$$y = mx = \frac{b\tan\theta}{a\sec\theta}x = \frac{b}{a}\sin\theta.x$$

As $mm' = b^2/a^2$, we see that the equation of the conjugate diameter is $y = m'x = (b/a)$ cosec $\theta.x$. On solving $y = (b/a)$ cosec $\theta.x$ and

$$\frac{x^2}{a^2} - \frac{y^2}{b^2} = -1,$$

we see that the coordinates of $D$, where the conjugate diameter meets the conjugate hyperbola, are ($a$ tan θ, $b$ sec θ).

**Note 2.** It follows immediately from Theorem 12.2 that only one of a pair of conjugate diameters of a hyperbola meets the hyperbola in real points.

**Theorem 12.3.** If a pair of conjugate diameters meet the hyperbola and its conjugate in $P$ and $D$ respectively, then

$$CP^2 - CD^2 = a^2 - b^2.$$

*Proof.* Let $P$ and $D$ be $(a \sec \theta, b \tan \theta)$ and $(a \tan \theta, b \sec \theta)$ respectively.
Then

$$CP^2 - CD^2 = (a^2 \sec^2 \theta + b^2 \tan^2 \theta) - (a^2 \tan^2 \theta + b^2 \sec^2 \theta)$$
$$= a^2 - b^2.$$

**Theorem 12.4.** The parallelogram formed by the tangents at the extremities of conjugate diameters of a hyperbola has its vertices lying on the asymptotes and is of constant area.
*Proof.*

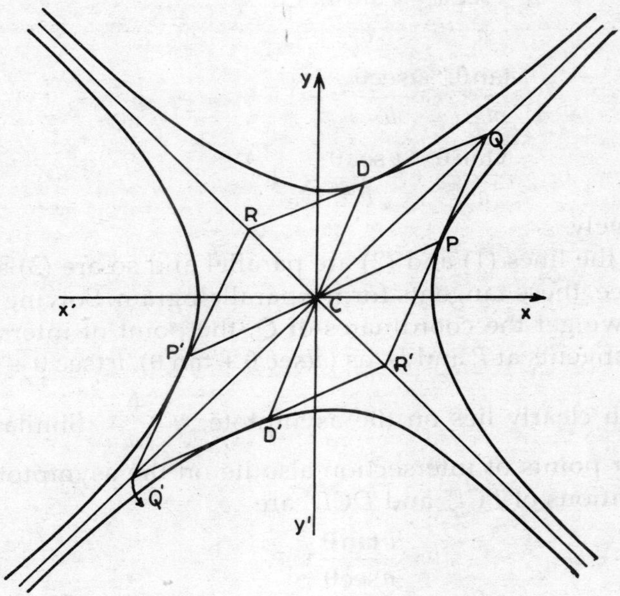


As in Theorem 12.3, let $P$ and $D$ be the points $(a \sec \theta, b \tan \theta)$ and $(a \tan \theta, b \sec \theta)$ on the hyperbola $\frac{x^2}{a^2} - \frac{y^2}{b^2} = 1$ and its conjugate, $\frac{x^2}{a^2} - \frac{y^2}{b^2} = -1$, respectively.

Then $D'$ is $(-a \tan \theta, -b \sec \theta)$ and $P'$ is $(-a \sec \theta, -b \tan \theta)$.

The equations of the asymptotes $CQ$ and $CR$ are $y = \frac{b}{a}x$ and $y = -\frac{b}{a}x$, respectively.

Further the equations of the tangents at $P$, $P'$, $D$ and $D'$ are

$$\frac{x\sec\theta}{a} - \frac{y\tan\theta}{b} = 1 \quad (1)$$

$$\frac{-x\sec\theta}{a} + \frac{y\tan\theta}{b} = 1 \quad (2)$$

$$\frac{x\tan\theta}{a} - \frac{y\sec\theta}{b} = -1 \quad (3)$$

and

$$\frac{-x\tan\theta}{a} + \frac{y\sec\theta}{b} = -1 \quad (4)$$

respectively.

Now the lines (1) and (2) are parallel and so are (3) and (4). Hence, these tangents form a parallelogram. Solving (1) and (3) we get the coordinates of $Q$, the point of intersection of tangents at $P$ and $D$, as $(a(\sec\theta + \tan\theta), b(\sec\theta + \tan\theta))$ which clearly lies on the asymptote, $y = \frac{b}{a}x$. Similarly, the other points of intersection also lie on the asymptotes. The equations of $PCP'$ and $DCD'$ are

$$y = \frac{b\tan\theta}{a\sec\theta}x \quad (5)$$

and

$$y = \frac{b\sec\theta}{a\tan\theta}x \quad (6)$$

respectively.

Clearly the lines (3), (4) and (5) are parallel and so are the lines (1), (2) and (6).

Hence by symmetry, area parallellogram $QRQ'R'$

$= 4.$ area parallelogram $QDCP$

$= 4. CP. (\perp$ from C on $QD)$

$$= 4.\sqrt{a^2 \sec^2\theta + b^2 \tan^2\theta}.\frac{1}{\sqrt{\frac{\tan^2\theta}{a^2} + \frac{\sec^2\theta}{b^2}}}$$

$= 4ab$. Hence the result.

**Example 12.17** Prove that the chord which joins the points in which a pair of conjugate diameters meets the hyperbola and its conjugate, is parallel to one asymptote and is bisected by the other.

*Solution*. Let the hyperbola be

$$\frac{x^2}{a^2} - \frac{y^2}{b^2} = 1 \qquad (1)$$

The asymptotes of (1) are

$$\frac{x}{a} - \frac{y}{b} = 0 \qquad (2)$$

and

$$\frac{x}{a} + \frac{y}{b} = 0 \qquad (3)$$

Now the conjugate hyperbola is given by

$$\frac{x^2}{a^2} - \frac{y^2}{b^2} = -1 \qquad (4)$$

Let $CP$ and $CD$ be a pair of conjugate semi-diameters meeting (1) at $P$ ($a \sec\theta$, $b \tan\theta$) and (4) at $D$ ($a \tan\theta$, $b \sec\theta$). Now the slope of the chord $PD$

$$= \frac{b\tan\theta - b\sec\theta}{a\sec\theta - a\tan\theta} = -\frac{b}{a}$$

$\Rightarrow$ $PD$ is parallel to the asymptote (3).

Also mid-point of $PD$ is

$$\left(\frac{a}{2}(\sec\theta + \tan\theta), \frac{b}{2}(\tan\theta + \sec\theta)\right)$$

which clearly lies on the asymptote (2).

Hence the result follows.

**Example 12.18**. If any tangent to the hyperbola, $\frac{x^2}{a^2} - \frac{y^2}{b^2} = 1$, with centre $C$, meets its director circle in $P$ and $Q$; show that $CP$ and $CQ$ are conjugate semi–diameters of the hyperbola.

*Solution.* We have the equation of the hyperbola as

$$\frac{x^2}{a^2}-\frac{y^2}{b^2}=1 \qquad (1).$$

The equation of the tangent to (1) at $(a \sec\theta, b\tan\theta)$ is

$$\frac{x\sec\theta}{a}-\frac{y\tan\theta}{b}=1 \qquad (2)$$

Also the equation of the director circle of (1) is

$$x^2+y^2=a^2-b^2 \qquad (3)$$

(See Q. 5(iii) of Exercise 8.7).

Now (2) meets (3) at $P$ and $Q$.

$\therefore$ The joint equation of the lines $CP$ and $CQ$ is

$$x^2+y^2-(a^2-b^2)\left(\frac{x\sec\theta}{a}-\frac{y\tan\theta}{b}\right)^2=0$$

(on homogenising (3) with the help of (2)).

$$\Rightarrow \left(1-\frac{a^2-b^2}{a^2}\sec^2\theta\right)x^2+\frac{2(a^2-b^2)}{ab}\sec\theta\tan\theta.xy$$

$$+\left(1-\frac{a^2-b^2}{a^2}\tan^2\theta\right)y^2=0 \qquad (4)$$

Let $m$, $m'$ represent the slopes of the lines given by (4);

$$\text{then } mm' = \frac{'A'}{B}=\frac{1-\dfrac{a^2-b^2}{a^2}\sec^2\theta}{1-\dfrac{a^2-b^2}{b^2}\tan^2\theta}$$

$$=\frac{b^2}{a^2}.\frac{a^2-(a^2-b^2)\sec^2\theta}{b^2-(a^2-b^2)\tan^2\theta}$$

$$=\frac{b^2}{a^2}.\frac{b^2\sec^2\theta-a^2\tan^2\theta}{b^2\sec^2\theta-a^2\tan^2\theta}=\frac{b^2}{a^2}$$

$\therefore$ The lines $CP$ and $CQ$ are conjugate semi-diameters of the hyperbola.

## Exercise 12.5.

1. Find the conjugate diameter of the diameter, $y = tx$, of the hyperbola, $x^2/a^2 - y^2/b^2 = 1$.
2. In the hyperbola, $16x^2 - 9y^2 = 144$, find the equation of the diameter, conjugate to the diameter, $x = 2y$.
3. Find the condition for $l_1x + m_1y = 0$ and $l_2x + m_2y = 0$ to be a pair of conjugate diameters of the hyperbola,
$$x^2/a^2 - y^2/b^2 = 1.$$
4. For the hyperbola, $x^2 - y^2 = a^2$, prove that the triangle $CPD$ is isosceles and has constant area, where $CP$ and $CD$ are a pair of its conjugate diameters.
5. Find the condition for the lines, $Ax^2 + 2Hxy + By^2 = 0$, to be conjugate diameters of $x^2/a^2 - y^2/b^2 = 1$.
6. Show that an asymptote of a hyperbola is its own conjugate diameter.
7. Show that the equation of the conjugate diameter of the diameter $y = mx$ of the curve $xy = c^2$ is $y = -mx$. (or alternatively, show that $y = mx$ and $y = -mx$ are conjugate diameters of the hyperbola, $xy = c^2$, whatever $m$ may be).
(*Hint*: Proceed as in § 12.6).
8. If the line $lx + my + n = 0$, passes through the extremities of a pair of conjugate diameters of the hyperbola, $x^2/a^2 - y^2/b^2 = 1$; show that $a^2l^2 - b^2m^2 = 0$.
9. Prove that the tangents at the ends of a system of parallel chords of a hyperbola intersect on the diameter which bisects the chords.
10. If $P'CP$ and $D'CD$ be a pair of conjugate diameters of a hyperbola; show that
    (i) $\overline{SP}.\overline{S'P} = CD^2$, where $S$ and $S'$ are foci of the hyperbola.
    (ii) Asymptotes of the hyperbola, bisect the chords, $PD$, $PD'$, $P'D$ and $P'D'$.
11. (i) Prove that the polar of any point on a hyperbola w.r.t. its conjugate hyperbola touches the original hyperbola at the other end of the diameter through

the given point.

(ii) If from any point $P$ on a hyperbola, tangents $PQ$, $PR$ be drawn to the conjugate hyperbola, show that $QR$ will touch the original hyperbola at the other end of the diameter through $P$.

12. If the normal at a point $P$ on a rectangular hyperbola meets the curve again in $Q$, show that the circle on $PQ$ as diameter cuts the curve again at the other end of the diameter through $P$.

## 12.7 RECTANGULAR HYPERBOLA

A hyperbola is called *rectangular* if its asymptotes are at right angles. The asymptotes of the hyperbola $\frac{x^2}{a^2} - \frac{y^2}{b^2} = 1$ are $y = \pm \frac{b}{a} x$. These are at right angles if $\frac{b}{a}\left(\frac{-b}{a}\right) = -1 \Rightarrow b^2 = a^2 \Rightarrow$ $a = b$. (Here positive sign is taken as $a$ and $b$ are semi–axes). Thus, the equation of the hyperbola reduces to $x^2 - y^2 = a^2$ and the asymptotes are $y = \pm x$. Its eccentricity is given by,

$$a^2 = a^2(e^2 - 1) \Rightarrow e = \sqrt{2}.$$

Now we find its equation in the simplest form.

### 12.7.1 To find the equation of a rectangular hyperbola referred to asymptotes as the coordinate axes.

The equation of the rectangular hyperbola is $x^2 - y^2 = a^2$ and its asymptotes are: $y = \pm x$.

Call them by $CX$ and $CY$. These are to be taken as the coordinate axes. Clearly $CX(y = -x)$ makes an angle $\theta$ with the $x$-axis, where $\tan\theta = -1$, i.e. $\theta = -45^0$.

Rotate the axes through this angle $\theta$, origin remaining the same; then

$$x = X\cos\theta - Y\sin\theta = \frac{X+Y}{\sqrt{2}}$$

and $\quad y = X \sin \theta + Y \cos \theta = \dfrac{-X+Y}{\sqrt{2}}.$

Substituting for $x, y$ in $x^2 - y^2 = a^2$ , we get

$$\left(\frac{X+Y}{\sqrt{2}}\right)^2 - \left(\frac{-X+Y}{\sqrt{2}}\right)^2 = a^2$$

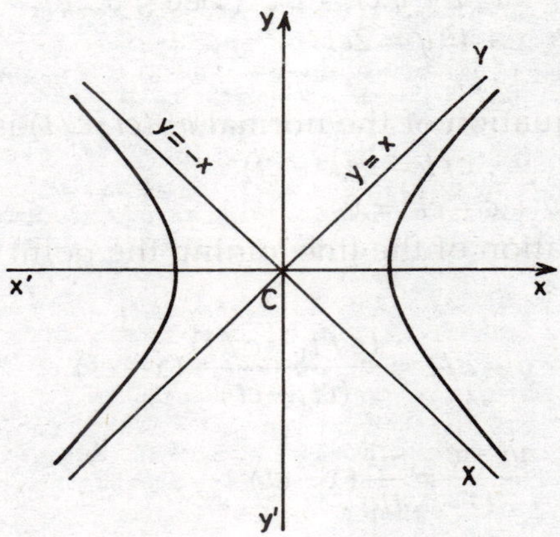


$$\Rightarrow \qquad XY = \frac{a^2}{2} = c^2. \text{ (say)}$$

Thus the equation of rectangular (also called sometimes, *equilateral*) *hyperbola*, referred to its asymptotes as the axes is $XY = c^2$.

**Note 1**. The equations of the asymptotes and the conjugate hyperbola of the rectangular hyperbola $xy = c^2$, where the axes are the asymptotes, are, $xy = 0$ and $xy = -c^2$, respectively.

**Note 2**. The parametric equations of the rectangular hyperbola (in the case, $xy = c^2$) are $x = ct$ and $y = c/t$ $(t \neq 0)$ and by the point "$t$" on $xy = c^2$, we shall mean the point $(ct, c/t)$. See Chapter 7 for the diagram of the rectangular hyperbola, $xy = c^2$.

**Some solved examples**

**Example 12.19.** If the normal to the rectangular hyperbola $xy = c^2$ at the point "$t$" on it intersects the hyperbola at "$t_1$", prove that $t^3 t_1 = -1$.

*Solution*. The equation of the rectangular hyperbola is $xy = c^2$.

$\therefore$ the equation of the tangent at the point "$t$" = $(ct, c/t)$ is

$$x.c/t + y.ct = 2c^2 \text{ ( See § 8.2)}$$

$$\Rightarrow \quad x + t^2 y = 2ct.$$

Its slope = $-1/t^2$.

$\therefore$ the equation of the normal *at* $(ct, c/t)$ is

$$y - c/t = t^2(x - ct)$$

$$\Rightarrow \quad t^3x - ty + c - ct^4 = 0 \qquad (1)$$

Now the equation of the line joining the point "$t$" and "$t_1$", is

$$y - c/t = \frac{(c/t_1 - c/t)}{(ct_1 - ct)}(x - ct)$$

$$\Rightarrow \quad \frac{yt - c}{t} = \frac{-1}{tt_1}(x - ct)$$

$$\Rightarrow \quad x + tt_1\, y - c(t + t_1) = 0. \qquad (2)$$

But (1) and (2) are same; so, we compare the coefficients of $x$ and $y$. This gives

$$\frac{1}{t^3} = \frac{tt_i}{-t}$$

$$\Rightarrow \quad t^3 t_1 = -1.$$

Hence the result.

**Example 12.20**. If the normals at $(x_i, y_i)$, $i = 1, 2, 3, 4$ on the rectangular hyperbola, $xy = c^2$, meet at the point $(h, k)$; show that

(i) $\Sigma x_i = h$,
(ii) $\Sigma y_i = k$,
(iii) $x_1 x_2 x_3 x_4 = y_1 y_2 y_3 y_4 = -c^4$,
(iv) $\Sigma x^2_i = h^2$,
(v) $\Sigma y^2_i = k^2$.

*Solution*. Let $(x_i, y_i) = \left(ct_i, \dfrac{c}{t_i}\right)$, $i = 1, 2, 3, 4$.

Now the equation of the normal to the hyperbola, $xy = c^2$, at $\left(ct, \dfrac{c}{t}\right)$ is

$$ty = t^3x + c - ct^4 \qquad \text{(See: Example 12.19)}$$

It passes through $(h,k)$ whence

$$ct^4 - ht^3 + kt - c = 0$$

The roots of this equation are $t_1, t_2, t_3, t_4$.

$$\Rightarrow \quad \Sigma t_i = h/c,\ \Sigma t_1t_2 = 0,\ \Sigma t_1t_2t_3 = -k/c,\ t_1t_2t_3t_4 = -1$$

Now (i) $\quad \Sigma x_i = c(\Sigma t_1) = h.$

(ii) $\quad \Sigma y_i = c(\Sigma \dfrac{1}{t_i}) = c(\dfrac{\sum t_1t_2t_3}{t_1t_2t_3t_4}) = k.$

(iii) $\quad x_1x_2x_3x_4 = c^4t_1t_2t_3t_4 = -c^4$

$$y_1y_2y_3y_4 = c^4 \frac{1}{t_1t_2t_3t_4} = -c^4.$$

(iv) $\quad \Sigma x_1^2 = c^2 (\sum t_i^2) = c^2[(\sum t_i)^2 - 2\sum t_1t_2] = h^2$

(v) $\quad \Sigma y_i^2 = (\Sigma y_i)^2 - 2\Sigma y_1y_2 = k^2 - 2c^2 \Sigma (1/t_1t_2)$

$$= k^2 - 2c^2 \frac{\sum t_1t_2}{t_1t_2t_3t_4} = k^2.$$

**Example 12.21**. A triangle has its vertices on a rectangular hyperbola. Prove that the orthocentre of the triangle also lies on the same hyperbola.

*Solution*. Let "$t_1$", "$t_2$", "$t_3$" be the vertices of the triangle $ABC$, described on the rectangular hyperbola, $xy = c^2$, where the point "$t$" is $(ct, c/t)$.

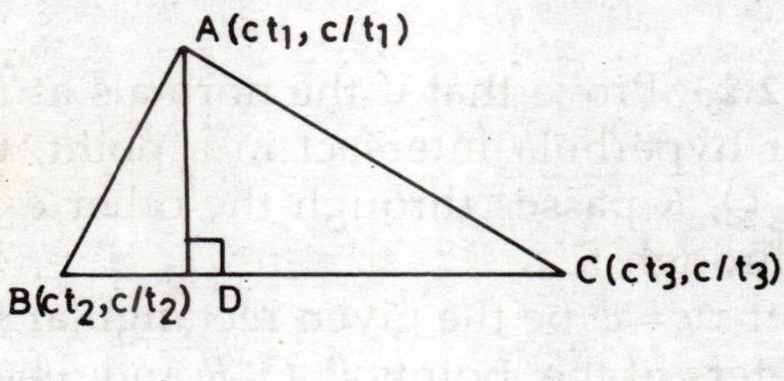

Now the slope of $BC$ is $\dfrac{(c/t_2 - c/t_3)}{(ct_2 - ct_3)} = \dfrac{-1}{t_2 t_3}$

$\therefore$ the equation of the altitude, $AD$, is

$$y - c/t_1 = t_2 t_3(x - ct_1)$$

$$\Rightarrow \quad y = t_2 t_3 x + c/t_1 - ct_1 t_2 t_3.$$

Similarly, the equation of the altitude through $B$, is

$$y = t_3 t_1 x + c/t_2 - ct_1 t_2 t_3.$$

Solving these, the orthocentre of the $\Delta\, ABC$, is

$$\left(\frac{-c}{t_1 t_2 t_3}, -ct_1 t_2 t_3\right)$$

which clearly lies on, $xy = c^2$.

**Example 12.22**. A circle cuts a rectangular hyperbola, $xy = c^2$, in $A, B, C, D$ and the parameters of these four points are $t_1, t_2, t_3, t_4$ respectively. Prove that $t_1 t_2 t_3 t_4 = 1$. Also find the value of $\Sigma(1/t_1)$.

*Solution*. Let the equation of the circle be,

$$x^2 + y^2 + 2gx + 2fy + d = 0.$$

Any point on the hyperbola is $(ct, c/t)$. If it is also on the circle, then

$$c^2t^2 + \frac{c^2}{t^2} + 2gct + 2f\frac{c}{t} + d = 0$$

$$\Rightarrow \quad c^2t^4 + 2gct^3 + dt^2 + 2fct + c^2 = 0. \qquad (1)$$

The roots of this equation are $t_1, t_2, t_3, t_4$ and $t_1 t_2 t_3 t_4 = c^2/c^2 = 1$.

Also from equation (1), we get

$$\Sigma\, t_1 t_2 t_3 = -2f/c$$

$$\therefore \quad \Sigma\frac{1}{t_1} = \frac{\Sigma t_1 t_2 t_3}{t_1 t_2 t_3 t_4} = -\frac{2f}{c}$$

**Example 12.23**. Prove that if the normals at $P, Q, R, S$ on a rectangular hyperbola intersect in a point, then the circle through $P, Q, R$ passes through the other extremity of the diameter through $S$.

*Solution*. Let $xy = c^2$ be the given rectangular hyperbola. Let the parameters of the points $P, Q, R$ and $S$ be $t_1, t_2, t_3$ and $t_4$

respectively and suppose the normals at these points meet at $(h, k)$. Then, by Example 12.20 $t_1, t_2, t_3, t_4$ are the roots of

$$ct^4 - ht^3 + kt - c = 0 \tag{1}$$

Also, the parameters of the points of intersection of the circle through $P, Q, R$ and the rectangular hyperbola are, by Example 12.22, given by

$$c^2t^4 + 2get^3 + dt^2 + 2fct + c^2 = 0.$$

Equation (2) gives four values of $t$, of which $t_1, t_2, t_3$ correspond to the points $P, Q, R$ and let $t_0$, the fourth root, correspond to the point $T$.

Now, by (1), $\quad t_1t_2t_3t_4 = -1$

and *by* (2), $\quad t_1t_2t_3t_0 = 1.$

$\Rightarrow \quad t_0 = -t_4.$

$\therefore$ the point $T = (ct_0, c/t_0) = (-ct_4 - c/t_4)$ which is the other extremity of the diameter of the rectangular hyperbola through the point $S(ct_4, c/t_4)$. Hence the result.

**Example 12.24**. From the points $(x_1, y_1)$ and $(x_2, y_2)$, tangents are drawn to the rectangular hyperbola, $xy = c^2$. If the conic passing through the two given points and the four points of contact is a circle, then show that $x_1x_2 = y_1y_2$ and $x_1y_2+x_2y_1 = 4\,c^2$.

*Solution*. The given rectangular hyperbola is

$$xy = c^2 \tag{1}$$

Now the equations of the chords of contact of$(x_1, y_1)$ and $(x_2, y_2)$ w.r.t. (1) are

$$xy_1 + yx_1 = 2\,c^2 \tag{2}$$

$$xy_2 + yx_2 = 2\,c^2 \tag{3}$$

The equation of a conic passing through the intersection of (1) with (2) and (3) is

$$(xy_1 + yx_1 - 2c^2)(xy_2 + yx_2 - 2c^2) + \lambda(xy - c^2) = 0 \tag{4}$$

Now (4) represents a circle.

$\therefore$ Coefficient of $xy$ in (4) is zero

$$\Rightarrow \quad x_1y_2 + x_2y_1 + \lambda = 0 \tag{5}$$

and coefficient of $x^2$ = coefficient of $y^2$

$$\Rightarrow \quad y_1y_2 = x_1x_2$$

This is one of the conditions to be found.
Again the conic (4) is to pass through $(x_1, y_1)$ and $(x_2, y_2)$.
$\therefore\ 2(x_1y_1 - c^2)(x_1y_2 + x_2y_1 - 2c^2) + \lambda(x_1y_1 - c^2) = 0$
and $(x_2y_1 + x_1y_2 - 2c^2).\ 2(x_2y_2 - c^2) + \lambda(x_2y_2 - c^2) = 0$

$$\Rightarrow \quad 2(x_1y_2 + x_2y_1 - 2c^2) + \lambda = 0 \qquad (6)$$

(as $x_1y_1 \neq c^2$ ; $x_2y_2 \neq c^2$ )
Now (6) – (5) implies

$$x_1y_2 + x_2y_1 = 4c^2$$

which is the other desired condition.

**Example 12.25**. Show that an infinite number of triangles can be inscribed in the rectangular hyperbola, $xy = c^2$, whose sides touch the parabola, $y^2 = 4ax$.

*Solution*. Consider three points $A(t_1)$, $B(t_2)$ and $C(t_3)$ on the rectangular hyperbola,

$$xy = c^2 \qquad (1)$$

Suppose $AB$ and $AC$ touch the parallel, $y^2 = 4ax$ (2)
We shall show: $BC$ also touches (2).
Now the equation of $AB$ is

$$y - \frac{c}{t_1} = \frac{\frac{c}{t_1} - \frac{c}{t_2}}{ct_1 - ct_2}(x - ct_1)$$

$$\Rightarrow \quad y = -\frac{x}{t_1t_2} + \frac{c(t_1 + t_2)}{t_1t_2}$$

This is to be tangent to (2)

$$\therefore \quad c\left(\frac{t_1 + t_2}{t_1t_2}\right) = a \Big/ \left(\frac{-1}{t_1t_2}\right) \qquad (\because\ 'c = a/m')$$

$$\Rightarrow \quad at_1^2t_2^2 + c(t_1 + t_2) = 0 \qquad (3)$$

Similarly, the chord $AC$ touches (2) implies

$$at_1^2t_3^2 + c(t_1 + t_3) = 0 \qquad (4)$$

Now (3) and (4) imply that $t_2$, $t_3$ are the roots of the equation

$$at_1^2t^2 + ct + ct_1 = 0.$$

$$\Rightarrow \quad t_2 + t_3 = -c/at_1^2;\ t_2t_3 = c/at_1 \qquad (5)$$

Again the equation of the chord $BC$ is $x + t_2t_3y = c(t_2 + t_3)$

$$\Rightarrow \qquad x + \frac{c}{at_1} y = -\frac{c^2}{at_1^2} \qquad \text{(using (5))}$$

$$\Rightarrow \qquad y = \left(-\frac{at_1}{c}\right)x - \frac{c}{t_1}$$

which touches (2)

(as $m = -\dfrac{at_1}{c}$ and $a/m = -c/t_1$).

$\Rightarrow$ $BC$ also touches (2).

We have got one triangle $ABC$ whose sides touch the parabola, (2). Since $A$ is arbitrary point on (1), there are infinitely many such triangles.

### Exercise 12.6.

1. (i) Find the coordinate of the foci and the equations of the directrices of the rectangular hyperbola, $xy = c^2$.
   (ii) Prove that the director circle of a rectangular hyperbola is a point circle.
2. Show that the length of the rectangular hyperbola, $xy = c^2$ which is bisected at $(p, q)$ is
   $$2\sqrt{(p^2+q^2)(pq-c^2)}\,/\,pq\cdot$$
3. If the normal and the tangent to a rectangular hyperbola cut off intercepts $a_1, a_2$ on one axis and $b_1, b_2$ on the other; show that $a_1a_2 + b_1b_2 = 0$.
4. Prove that the distance of any point from the centre of a rectangular hyperbola varies inversely as the perpendicular distance of its polar from the centre.
5. Prove that the locus of the poles of normal chords of the rectangular hyperbola, $xy = c^2$ is the curve $(x^2 - y^2)^2 + 4c^2xy = 0)$
6. Prove that the feet of normals to the rectangular hyperbola $xy = c^2$, which concur at $(\alpha, \beta)$ lie on the rectangular hyperbola, $x^2 - y^2 - \alpha x + \beta y = 0$.
7. Show that the coordinates of the point of intersection of two tangents to a rectangular hyperbola, are harmonic

means between the coordinates of the point of contact.

8. $PQ$ and $PR$ are two perpendicular chords of a rectangular hyperbola; show that $QR$ is parallel to the normal at $P$.
9. Prove that a circle and a hyperbola meet in four points. If $P_i(\theta_i)$ ($i = 1,2,3,4$) be these four points, common to them, show that
   (i) $\theta_1 + \theta_2 + \theta_3 + \theta_4$ = an even multiple of $\pi$.
   (ii) The common chords are equally inclined to the axes of the hyperbola.
10. Through a point $(h, k)$, a line is drawn perpendicular to any diameter of an ellipse, $\frac{x^2}{a^2} + \frac{y^2}{b^2} = 1$ to meet the conjugate diameter in $Q$. Prove that the locus $Q$ is a rectangular hyperbola passing through the given point as well as the centre of the ellipse.
11. Show that the conjugate diameters of a rectangular hyperbola are equally inclined to the asymptotes.
12. Given the base of a triangle in position and magnitude, and the difference of the base angles to be a constant; prove that the locus of the vertex is a rectangular hyperbola having the given base as a diameter.
13. If a line cuts a hyperbola at $P$, $P'$ and its asymptotes at $Q$, $Q'$; show that $PQ = P'Q'$. Deduce that the portion of any tangent to a hyperbola included between its asymptotes is bisected at the point of contact.
14. A triangle is inscribed in the hyperbola, $xy = c^2$ and two of its sides are parallel to $y = m_1x$ and $y = m_2x$. Prove that the third side touches the hyperbola.
    $4m_1m_2\, xy = c^2(m_1 + m_2)^2$.
15. (i) Prove that there are four normals from a point to a rectangular hyperbola, $xy = c^2$.
    (ii) If the normals at the points "$t_i$" ($i$= 1, 2, 3, 4) on the rectangular hyperbola, $xy = c^2$ be concurrent, prove that $\Sigma\, t_1t_2 = 0$ and $t_1t_2t_3t_4 = -1$.
    (iii) Find the coordinates of the point of concurrence of the four normals in (i) in terms of $t_i$, the parameters of their feet.

(iv) Prove that if the normals at the points "$t_i$" ($i = 1, 2, 3$) on the rectangular hyperbola, $xy = c^2$ are concurrent, then

$$t_1t_2 + t_2t_3 + t_3t_1 = \frac{1}{t_1t_2} + \frac{1}{t_2t_3} + \frac{1}{t_3t_1}$$

16. The normals at $A, B, C, D$ to the rectangular hyperbola, $xy = a^2$ meet in $P(h, k)$. Prove that $PA^2 + PB^2 + PC^2 + PD^2 = 3(h^2 + k^2)$.
17. Prove that in any rectangular hyperbola, the rectangle under the distances of any point on it from two fixed tangents is to the square on the distance from their chord of contact as $\cos\theta : 1$, where $\theta$ is the angle between the tangents.

## Miscellaneous Exercise on Chapter 12

1. If $P(a \sec\theta, b \tan\theta)$ be any point on the hyperbola $x^2/a^2 - y^2/b^2 = 1$, show that
   (i) $\overline{SP} = a(e \sec\theta - 1)$,
   (ii) $\overline{S'P} = a(e \sec\theta + 1)$,
   (iii) $\overline{S'P} - \overline{SP} = 2a$.
   Prove further that $\overline{SP}.\overline{S'P} = CP^2$, in case of a rectangular hyperbola.
2. Obtain the equation of the hyperbola in the form $x^2/a^2 - y^2/b^2 = 1$ and prove that the lines, $x/a - y/b = m$ and $x/a + y/b = 1/m$ always meet on the hyperbola.
3. $P$ is a variable point on the hyperbola, $x^2/a^2 - y^2/b^2 = 1$, whose vertex $A$ is $(a, 0)$. Show that the locus of the mid-point of $AP$ is $\dfrac{(2x-a)^2}{a^2} - \dfrac{4y^2}{b^2} = 1$.
4. $P$ and $Q$ are two points on the hyperbola, $x^2/a^2 - y^2/b^2 = 1$, such that the product of the slopes of the tangents at $P$ and $Q$ is $k^2$. Show that the locus of the point of intersection of the tangents at $P$ and $Q$ is $y^2 + b^2 = k^2(x^2 - a^2)$.
5. The normals at three points $P$, $Q$, and $R$ on a rectangular hyperbola intersect at a point $S$ on the curve. Prove that

the centre of the hyperbola is the centroid of triangle $PQR$.

6. $P, Q, R, S$ are four points on a rectangular hyperbola such that the chord $PQ$ is perpendicular to $RS$.
   (i) Prove that each of the points is the orthocentre of the triangle formed by the other three.
   (ii) Further if $C$ be the centre of the above hyperbola and $CP, CQ, CR, CS$ make angles $\alpha, \beta, \gamma, \delta$ with an asymptote, show that
   $$\tan \alpha \tan \beta \tan \gamma \tan \delta = 1.$$
7. If the normal at $P$ to the rectangular hyperbola, $xy = c^2$, meets the curve again in $Q$, show that the circle on $PQ$ as diameter cuts the curve at the other end of the diameter of the rectangular hyperbola through $P$.
   (*Hint*. Use Example 12.19).
8. A rectangular hyperbola $xy = c^2$ whose centre is $C$, is cut by a circle of radius $r$ in four points $P, Q, R$ and $S$. Prove that
   $$CP^2 + CQ^2 + CR^2 + CS^2 = 4r^2.$$
9. If a circle cuts a rectangular hyperbola, $xy = 1$, in four points $(x_i, y_i)$, $i = 1, 2, 3, 4$, prove that
   $$x_1x_2x_3x_4 = y_1y_2y_3y_4 = 1.$$
   (*Hint*: See Example 12.22).
10. If a circle cuts a rectangular hyperbola, $xy = c^2$, in $P, Q, R, S$ and the parameters of these four points be $t_1, t_2, t_3, t_4$ respectively, prove that

    (i) the centre of the mean position of these four points, i.e.
    $$\left(\frac{x_1 + x_2 + x_3 + x_4}{4}, \frac{y_1 + y_2 + y_3 + y_4}{4}\right)$$
    bisects the distance between the centres of the two curves.
    (ii) the centre of the circle through the points $t_1, t_2, t_3$ is
    $$\left[\frac{1}{2}c\left(t_1 + t_2 + t_3 + \frac{1}{t_1t_2t_3}\right), \frac{1}{2}c\left(\frac{1}{t_1} + \frac{1}{t_2} + \frac{1}{t_3} + t_1t_2t_3\right)\right]$$
    (iii) if $H$ be the orthocentre of the triangle $PQR$, then $H$

and $S$ are the extremities of a diameter of the rectangular hyperbola.

11. Three tangents are drawn to the rectangular hyperbola, $xy = a^2$ at the points $(x_i, y_i)$; $i = 1, 2, 3$ and form a triangle whose circumcircle passes through the centre of the hyperbola. Show that

$$\frac{\Sigma x_1}{x_1x_2x_3}+\frac{\Sigma y_1}{y_1y_2y_3}=0,$$

and that the centre of the circle lies on the hyperbola.

12. A circle with fixed centre $(3p, 3q)$ and of variable radius cuts the rectangular hyperbola, $x^2-y^2 = 9a^2$ at the points $P, Q, R$ and $S$. Prove that the locus of the centroid of the triangle $PQR$ is given by

$$(x - 2p)^2 - (y - 2q)^2 = a^2 .$$

13. Show that the area of the triangle formed by the asymptotes of the rectangular hyperbola $xy = c^2$ and $e$ the normal at $(x', y')$ is

$$\frac{1}{2}\left(\frac{x'^2 - y'^2}{c}\right)^2$$

14. From any point $P$ on the hyperbola

$$\frac{x^2}{a^2}-\frac{y^2}{b^2}=1,$$

three normals other than that at $P$ are drawn. Show that the centroid of the triangle tormed by their feet lies on the hyperbola,

$$\frac{x^2}{a^2}-\frac{y^2}{b^2}=\frac{(a^2-b^2)^2}{9(a^2+b^2)^2}$$

[*Hint*: Take any point as $\{a/2(t+1/t), b/2(t-1/t)\}$].

15. The tangent to the hyperbola, $xy = c^2$ , at the point $(ct, c/t)$ intersects the hyperbola, $xy = - c^2$, at the points $M$ and $N$. Show that the tangents at $M$ and $N$ to the hyperbola $xy = - c^2$ intersect on the hyperbola $xy = c^2$.

16. The line, $ax + by = 1$ meets the hyperbola, $xy = c^2$ in $P$ and $Q$. If $C$ is the centre of this, show that $CP$ and $CQ$

are perpendicular if $c^2e^2(a^2 + b^2) - (2 - e^2)(2abc^2 - 1) = 4$ where $e$ is the eccentricity of the hyperbola.

17. Prove that the sum of the eccentric angles of the four points of intersection of an ellipse and a rectangular hyperbola whose asymptotes are parallel to the axes of the ellipse is an odd multiple of $\pi$.

18. $A$, $B$, $C$ are three points on the rectangular hyperbola, $xy = d^2$ whose abscissae are $a$, $b$, $c$ respectively. Prove that the area of $\Delta ABC$ is

$$\frac{1}{2}d^2 . \frac{(a-b)(b-c)(c-a)}{abc}$$

and the area of the triangle enclosed by tangents at these points is

$$2d^2 . \frac{(a-b)(b-c)(c-a)}{(a+b)(b+c)(c+a)}$$

19. $P$ and $Q$ are two variable points on the rectangular hyperbola, $xy = c^2$ such that the tangent at $Q$ passes through the foot of the ordinate of $P$. Show that the locus of the point of intersection of tangents at $P$ and $Q$ is a hyperbola with the same asymptotes as the given hyperbola.

20. From any point $P$ on the line, $lx + my + 1 = 0$, tangents $PQ$, $PR$ are drawn to the rectangular hyperbola, $2xy = c^2$ and the circle $PQR$ cuts the hyperbola again in $T$, $T'$. Prove that $TT'$ touches the parabola, $(l^2 + m^2)(x^2 + y^2) = (lx + my + 1)^2$.

21. A circle cuts two fixed perpendicular lines so that each intercept is of given length. Prove that the locus of the centre of the circle is a rectangular hyperbola.

22. A circle and a rectangular hyperbola intersect in the points $P$, $Q$, $R$ and $S$. If $PQ$ is a diameter of the circle, then show that $RS$ is a diameter of the rectangular hyperbola.

# 13. TRACING OF THE GENERAL CONIC

## INTRODUCTION

In the previous chapters, we always considered the equation of a conic in a standard form (see Chapter 8). In this chapter, we take up the study of curves given in a general form. Let $S = 0$ stand for the general equation of second degree where

$$S = ax^2 + 2hxy + by^2 + 2gx + 2fy + c$$

and we shall discuss how to transform, $S = 0$, to one of the standard forms and thus be able to trace it.

## 13.1 THE NATURE OF CONIC REPRESENTED BY $S = 0$, where $S = ax^2 + 2hxy + by^2 + 2gx + 2fy + c$.

Set $\Delta = abc + 2fgh - af^2 - bg^2 - ch^2$.

**Case 1**. $\Delta = 0$ then by § 5.4, $S = 0$ represents a pair of lines.

**Case 2**. $\Delta \neq 0$, then a rotation of the axes by an angle $\theta$, given by $\tan 2\theta = \dfrac{2h}{a-b}$ (§ 3.5), without changing the origin, renders $S = 0$, free of $xy$ term and the transformed equation is

$$AX^2 + BY^2 + 2GX + 2FY + C = 0. \qquad (1)$$

Now $ab - h^2$ is an invariant of $S$, by § 3.6; whence, $AB = ab - h^2$.

**Subcase 2.1**. Let $ab - h^2 = 0 \Rightarrow AB = 0 \Rightarrow A = 0$ or $B = 0$. Also both $A$ and $B$ cannot be zero, because then the transformed equation reduces to first degree.

We assume $A = 0$, $B \neq 0$. Now, equation (1) becomes,

$$BY^2 + 2GX + 2FY + C = 0 \qquad (2)$$

Also $G \neq 0$, otherwise (2) will represent two lines.

$\therefore$ Equation (2) can be rewritten as

$$B\left(Y + \frac{F}{B}\right)^2 = -2G\left(X + \frac{C}{2G} - \frac{F^2}{2BG}\right) \qquad (3)$$

Shifting the origin to

$$\left(\frac{F^2}{2BG}-\frac{C}{2G},\frac{-F}{B}\right).$$

The axis remaining parallel to the original axes, equation (3) takes the form,

$$Y'^2=\frac{-2G}{B}X'$$

which now represents a parabola, whose

(i) axis is given by $Y' = 0$,

(ii) the tangent at the vertex is given by $X' = 0$,

(iii) the vertex is given by $X' = 0$, $Y' = 0$,

(iv) the latus rectum $= 4a = \left|\frac{-2G}{B}\right|$

(v) (a) if $\frac{G}{B}<0$, its equation is $Y'^2 = 4\,aX'$, then its focus is given by $X' = a$, $Y' = 0$, and the directrix is given by $X' = -a$.

(b) if $\frac{G}{B}>0$, its equation is $Y'^2 = -4\,aX'$, then its focus is given by $X' = -a$, $Y' = 0$, and the directrix is given by $X' = a$.

Also in the case, when $B = 0$, $A \neq 0$; we similarly get a parabola.

**Subcase 2.2.** $ab - h^2 \neq 0 \Rightarrow AB \neq 0 \Rightarrow A \neq 0, B \neq 0$.

Equation (1) can now be rewritten as

$$A\left(X+\frac{G}{A}\right)^2+B\left(Y+\frac{F}{B}\right)^2=\frac{G^2}{A}+\frac{F^2}{B}-C=D\text{ (say)}.$$

Shifting the origin to $(-G/A, -F/B)$, the axes remaining parallel to the original axes, the above equation transforms to

$$AX'^2 + BY'^2 = D. \tag{4}$$

Clearly $D \neq 0$, otherwise this will again represent a pair of lines.

Now (4) $\Rightarrow$
$$\frac{X'^2}{D/A}+\frac{Y'^2}{D/B}=1. \quad (5)$$

then (5) represents an ellipse or a hyperbola according as $A$ and $B$ are of the same or of opposite signs, i.e. according as

$$ab - h^2 = AB > \text{ or } < 0.$$

The co-ordinates of the centre, are given by $X' = 0$, $Y' = 0$.

**Note**. A circle is a special case of the ellipse,

$$\frac{x^2}{a^2}+\frac{y^2}{b^2}=1$$

when $a = b$.

We have thus proved the following theorem:

**Theorem 13.1**. The general equation of second degree, $S = 0$, either represents a pair of lines, or a circle or a conic.

*Summary*. (a) If $\Delta = 0$, $S = 0$ represents a pair of lines.

(b) $\Delta \neq 0$.

(i) $ab - h^2 = 0$, i.e., second degree terms in $S = 0$, viz., $ax^2 + 2hxy + by^2$ form a perfect square, then $S = 0$ represents a parabola.

(ii) $ab - h^2 > 0$, $S = 0$, is an ellipse

(iii) $ab - h^2 < 0$, $S = 0$, is a hyperbola.

(iv) $a = b \neq 0$, $h = 0$; $S = 0$ is a circle.

(v) $a + b = 0$, obviously then, $ab - h^2 < 0$, $S = 0$ is a rectangular hyperbola.

**Definition**. $\Delta$ is known as the *discriminant* of $S = 0$.

**Example 13.1**. What curves are represented by the equation, $x^2 + 2\,xy + (1 + \lambda)y^2 + 2\lambda y - 1 = 0$ for different values of $\lambda$.

*Solution*. The given equation is

$$x^2 + 2xy + (1 + \lambda)\, y^2 + 2\lambda y - 1 = 0 \quad (1)$$

Here $a = 1$, $h = 1$, $b = 1+\lambda$, $g = 0$, $f = \lambda$, $c = -1$

$$\therefore\ \Delta = abc + 2fgh - af^2 - bg^2 - ch^2$$
$$= -(1 + \lambda) - \lambda^2 + 1 = -\lambda(\lambda + 1).$$

Thus for $\lambda = 0$ or $-1$, $\Delta = 0$ and (1) represents a pair of lines. For $\lambda = 0$, (1) $\Rightarrow (x + y)^2 = 1$

$\Rightarrow$ the lines are $x + y = \pm 1$.

For $\lambda = -1$, (1) $\Rightarrow x^2 + 2xy - 2y - 1 = 0$

$\Rightarrow (x - 1)(x + 2y + 1) = 0$

$\Rightarrow$ the lines are $x = 1$ and $x + 2y + 1 = 0$.

Suppose next, $\lambda \neq 0, -1$ (i.e. $\Delta \neq 0$).

Now $ab - h^2 = (1 + \lambda) - 1 = \lambda$.

(i) If $ab - h^2 > 0$ i.e. if $\lambda > 0$; then (1) represents an ellipse.

(ii) If $ab - b^2 < 0$ i.e. if $\lambda < 0$, then (1) represents a hyperbola.

(iii) For no value of $\lambda$, (1) can represent a parabola.

**Theorem 13.2**. If $S = 0$ represents a central conic with centre at (0, 0), then $g = 0 = f$. i.e. the equation of a conic referred to its centre is free from the first degree terms in $x$ and $y$.

*Proof*. Let $(x_1, y_1)$ by any point on the conic $S = 0$, whose centre is at (0, 0), where

$$S = ax^2 + 2hxy + by^2 + 2gx + 2fy + c,$$

then $(-x_1, -y_1)$ also lies on the conic

$\therefore \quad ax_1^2 + 2hx_1y_1 + by_1^2 + 2gx_1 + 2fy_1 + c = 0$

and $\quad ax_1^2 + 2hx_1y_1 + by_1^2 - 2gx_1 - 2fy_1 + c = 0$

Substracting, we get

$$gx_1 + fy_1 = 0$$

for all points $(x_1, y_1)$ which lie on the conic, $S = 0$. If any of $g$ or $f$ is non-zero, then all the points on $S = 0$, also lie on

$$gx + fy = 0,$$

which represents a straight line. This is absurd

$$\therefore f = 0 = g.$$

## 13.2 TO FIND THE CO-ORDINATES OF THE CENTRE OF THE CONIC, S = 0.

Suppose $(x_1, y_1)$ is the centre of the given central conic, $S = 0$, where $\quad S = ax^2 + 2hxy + by^2 + 2gx + 2fy + c$.

Shifting the origin to $(x_1, y_1)$ axes remaining parallel to the original axes, $S = 0$ transforms to

$$aX^2 + 2hXY + bY^2 + 2(ax_1 + hy_1 + g)X$$
$$+ 2(hx_1 + by_1 + f)Y + S_1 = 0,$$

where $S_1 = ax_1^2 + 2hx_1y_1 + by_1^2 + 2gx_1 + 2fy_1 + c.$

By Theorem 13.2,

$$ax_1 + hy_1 + g = 0, hx_1 + by_1 + f = 0 \quad (1)$$

and hence, $S_1 = (ax_1 + hy_1 + g)x_1 + (hx_1 + by_1 + f)y_1 + (gx_1 + fy_1 + c)$
$= gx_1 + fy_1 + c$

$$\therefore \text{ centre } (x_1, y_1) = \left(\frac{hf - bg}{ab - h^2}, \frac{gh - af}{ab - h^2}\right) \quad \text{(Solving (1))}.$$

(It may be noted that for a central conic, i.e. when the conic is an ellipse or a hyperbola, $ab - h^2 \neq 0$)

**Corollary**. Now

$$gx_1 + fy_1 + c = \frac{g(hf - bg) + f(gh - af) + c(ab - h^2)}{ab - h^2}$$

$$= \frac{abc + 2fgh - af^2 - bg^2 - ch^2}{ab - h^2} = \frac{\Delta}{ab - h^2}$$

Thus the equation of the conic referred to its centre as the origin is

$$aX^2 + 2hXY + bY^2 + d = 0, \quad (2)$$

where $\qquad d = S_1 = gx_1 + fy_1 + c = \dfrac{\Delta}{ab - h^2}$

Clearly for a central conic, both $\Delta$ and $ab - h^2$ are non-zero.

**Remark 1**. The centre giving equations (1) of the conic, $S = 0$ can also be obtained by differentiating $S = 0$, partially w.r.t. $x$ and $y$.

**Remark 2.** If $\Delta = 0$, the equation (2) becomes, $aX^2 + 2hXY + bY^2 = 0$ and the given equation $S = 0$ represens a pair of lines.

**Remark 3.** If $\Delta \neq 0$, the equation (2) reduces to the form, $a'X^2 + 2h'XY + b'Y^2 = 1$ ($a' = -a(ab - h^2)/\Delta$ etc); which is known as the *Standard Form* of the equation of a given conic.

**Example 13.2.** Find the centre and hence reduce it to the standard form, the conic represented by

$$x^2 - 3xy + y^2 + 10x - 10y + 50 = 0 \qquad (1)$$

**Solution.** Let

$$S \equiv x^2 - 3xy + y^2 + 1 - x - 10y + 50 = 0 \qquad (1)$$

Its centre is given by

$$\frac{\partial S}{\partial x} = 0 \text{ and } \frac{\partial S}{\partial y} = 0 \text{ (See Remark 1).}$$

$\Rightarrow \qquad 2x - 3y + 10 = 0$

and $\qquad -3x + 2y - 10 = 0$

$\Rightarrow \qquad x = -2;\ y = 2$

i.e. Centre of (1) has coordinates $(-2, 2)$

Now $\quad d = \text{'}gx_1 + fy_1 + c\text{'}$

$\quad = 5 \times (-2) - 5 \times 2 + 50 = 30$

$\therefore$ (1) reduces to

$$x^2 - 3xy + y^2 + 30 = 0$$

$$\Rightarrow \qquad \left(-\frac{1}{30}\right)x^2 + \left(\frac{1}{10}\right)xy + \left(\frac{-1}{30}\right)y^2 = 1.$$

which is the required standard form.

## 13.3 TO FIND THE DIRECTION AND LENGTHS OF THE AXES OF A CENTRAL CONIC, $S = 0$.

By the preceding corollary, the equation of the conic, $S = 0$, with its centre as the origin and the axes remaining parallel to the original axes, transforms to

$$aX^2 + 2hXY + bY^2 + d = 0 \qquad (1)$$

where $$d = \frac{\Delta}{ab - h^2}$$

We remove from this equation $XY$-term by rotating the axes (keeping the origin fixed) through an angle $\theta$, given by,

$$\tan 2\theta = \frac{2h}{a - b} \qquad \text{(see § 3.5)}$$

$$\Rightarrow \qquad \frac{2\tan\theta}{1-\tan^2\theta} = \frac{2h}{a-b}$$

$$\Rightarrow \qquad h\tan^2\theta + (a-b)\tan\theta - h = 0,$$

which yields two values of $\tan\theta$, say, $\tan\theta_1$ and $\tan\theta_2$. These two give the slopes of the two axes of the given conic.

Clearly $\tan\theta_1 . \tan\theta_2 = -1$.

$\therefore$ If $m = \tan\theta_1$ then $\tan\theta_2 = \dfrac{-1}{m}$ and the *equations of the axes of the conic* are given by $y - y_1 = m(x - x_1)$ and $y - y_1 = (-1/m)(x - x_1)$ where $(x_1, y_1)$ are the coordinates of its centre.

Now changing to polar coordinates (by $X = r\cos\theta$, $Y = r\sin\theta$), equation (1) yields

$$r^2 = \frac{-d}{a\cos^2\theta + 2h\sin\theta\cos\theta + b\sin^2\theta}$$

$$= \frac{-d(1+\tan^2\theta)}{a + 2h\tan\theta + b\tan^2\theta}$$

Substituting for $\tan\theta = \tan\theta_1$ and $\tan\theta_2$, in above, we get two values of $r^2$, say $r_1^2$ and $r_2^2$.

In case both $r_1^2$ and $r_2^2$ are positive, the conic is an ellipse (in fact, if both $r_1^2$ and $r_2^2$ and negative, even then, the conic is an ellipse, but it is an imaginary one); the lager and the smaller values of $r^2$, giving the squares of the semi-major and the semi-minor axes respectively.

In case one of $r_1^2$ and $r_2^2$ is positive and the other negative, then the conic is a hyperbola; the positive and negative values of $r^2$, respectively giving the squares of the semi-transverse and semi-conjugate axes.

**13.3.1. To find the lengths and positions of the axes of the conic** $\quad a'x^2 + 2h/xy + b'y^2 = 1 \qquad (1)$

Consider the circle:

$$x^2 + y^2 = r^2 \Leftrightarrow \frac{x^2+y^2}{r^2} = 1 \tag{2}$$

The equation of the lines joining the origin to the points of intersection of (1) and (2) is obtained by homogenising (1) with the help of (2). This is

$$a'x^2 + 2h'xy + b'y^2 - \left(\frac{x^2+y^2}{r^2}\right) = 0$$

$$\Rightarrow \quad (a'-1/r^2)\, x^2 + 2h'xy + (b'-1/r^2)y^2 = 0 \tag{3}$$

The lines given by (3) will coincide if it is any of the semi-axes, because in that case the circle and the central conic touch each other; this requires that the discriminant of (3) must be zero;

$$\Rightarrow \quad (a'-1/r^2)(b'-1/r^2) = h'^2 \tag{4}$$

$$\Rightarrow \quad 1/r^4 - (a'+b')1/r^2 + (a'b' - h'^2) = 0 \tag{5}$$

The conic is an ellipse or a hyperbola according as the two values of $r^2$, obtained from (5) are both positive or are of opposite signs (the conic is an imaginary ellipse if both the values of $r^2$ are negative).

Using (4), the equation (3) can be written as

$$[(a' - 1/r^2)x + h'y]^2 = 0.$$

Thus, the equation of either axis is

$$(a'-1/r^2)x + h'y = 0. \tag{6}$$

If $r_1^2$, $r_2^2$ be the roots of the equation (5), the equations of the semi-axes of the conic are

$$\left(a - \frac{1}{r_1^2}\right)x + h'y = 0 \tag{7}$$

and

$$\left(a - \frac{1}{r_2^2}\right)x + h'y = 0 \tag{8}$$

If $r_1^2 > r_2^2 > 0$, the conic is an ellipse and, (7) and (8) represent its major and minor axes with lengths $2r_1$ and $2r_2$ respectively. These very equations respectively represent transverse and conjugate axes of lengths $2r_1$ and $2\sqrt{-r_2^2}$ if

$r_1^2 > 0$ and $r_2^2 < 0$ (i.e. when the conic is a hyperbola).

**Note**. The equations (5) and (6) should be remembered; they yield the lengths and the positions, respectively of the axes of the central conic given by (1).

## 13.4. TO FIND THE ECCENTRICITY OF A CENTRAL CONIC

We can find the principal axes of the given central conic as in § 13.3.1. Two cases arise:

**Case I**. *The central conic is an ellipse.*

Let $r_1^2$, $r_2^2$ denote the squares of its semi-major and semi-minor axes; its eccentricity, $e$, is given by

$$r_2^2 = r_1^2 (1 - e^2)$$

$$\Rightarrow e = \sqrt{\frac{r_1^2 - r_2^2}{r_1^2}}$$

**Case II**. *The central conic is a hyperbola.*

Let $r_1^2$, $r_2^2$ denote the squares of its semi-transverse and semi-conjugates axes; its eccentricity, $e$, is given by

$$r_2^2 = r_1^2 (e^2 - 1)$$

$$\Rightarrow e = \sqrt{\frac{r_1^2 + r_2^2}{r_1^2}}$$

**Example 13.3**. Find the centre of the conic

$$5x^2 - 2xy + 5y^2 + 2x - 10y - 7 = 0$$

and show that it represents an ellipse. Also find its eccentricity, the lengths and the equations of its principal axes.

*Solution.*

**Method I**. Here

$$\Delta = abc + 2fgh - af^2 - bg^2 - ch^2$$
$$= -175 + 10 - 125 - 5 + 7 = -288 \neq 0,$$

and $\quad ab - h^2 = 24 > 0.$

Thus the given equation represents an ellipse.

Its centre $(x_1, y_1)$ is given by the equations:

$$ax_1 + hy_1 + g = 0$$

and $$hx_1 + by_1 + f = 0 \quad \text{(see § 13.2)}$$

$$5x_1 - y_1 + 1 = 0$$

and

$$-x_1 + 5y_1 - 5 = 0$$

$\Rightarrow$ the centre $(x_1, y_1) = (0, 1)$.

Shifting the origin to (0, 1), the axes remaining parallel to the original axes, the equation

$$5x^2 - 2xy + 5y^2 + 2x - 10y - 7 = 0,$$

transforms to

$$5X^2 - 2XY + 5Y^2 + \frac{\Delta}{ab - h^2} = 0 \text{(by Cor. to § 13.2)}$$

$$\Rightarrow \quad 5X^2 - 2XY + 5Y^2 = 12 \tag{1}$$

$$\Rightarrow \quad \frac{5}{12}X^2 - 2.\frac{1}{12}XY + \frac{5}{12}Y^2 = 1$$

Let $r$ be any of the semi-axis of this conic, then it is given by

$$\frac{1}{r^4} - (a' + b')\frac{1}{r^2} + (a'b' - h'^2) = 0$$

(by (5) of § 13.3.1)

$$\Rightarrow \quad \frac{1}{r^4} - \left(\frac{5}{12} + \frac{5}{12}\right)\frac{1}{r^2} + \left(\frac{5}{12}.\frac{5}{12} - \left(\frac{1}{12}\right)^2\right) = 0$$

$$\Rightarrow \quad r^4 - 5r^2 + 6 = 0$$

$$\Rightarrow \quad r^2 = 3, 2.$$

$\therefore$ the semi–axes of the ellipse are $\sqrt{3}$ and $\sqrt{2}$. Clearly, its eccentricity is given by

$$e = \sqrt{\frac{r_1^2 - r_2^2}{r_1^2}} = \sqrt{\frac{3-2}{3}} = \frac{1}{\sqrt{3}}$$

Again using (6) of § 13.3.1, the equation of the semi-axis of the ellipse is

$$\left(a' - \frac{1}{r^2}\right)X + h'Y = 0$$

$\Rightarrow$ the equations of the major and the minor axes of the ellipse are

$$\left(\frac{5}{12}-\frac{1}{3}\right)X-\frac{1}{12}Y=0$$

and $$\left(\frac{5}{12}-\frac{1}{2}\right)X-\frac{1}{12}Y=0$$

$\Rightarrow$ $X - Y = 0$

and $X + Y = 0$

But $X = x$

and $Y = y - 1.$

$\therefore$ these axes are given by

$$x - y + 1 = 0$$

and $$x + y - 1 = 0.$$

**Method II**. Proceed as in Method I and obtain equation (1).

Rotating the axes (keeping the origin fixed) through an angle $\theta$ (to get rid of $XY$–term) given by

$$\cot 2\theta = \frac{a-b}{2h} \qquad (\S\ 3.5)$$

$$= 0 \qquad \text{(Here } a = b = 5 \text{ and } h = -1)$$

$$\Rightarrow \qquad \theta = \frac{\pi}{4}, \frac{3\pi}{4}$$

$$\Rightarrow \qquad \tan\theta = \pm 1.$$

Changing (1) to polar coordinates, we get

$$r^2 = \frac{12}{(5\cos^2\theta - 2\cos\theta\sin\theta + 5\sin^2\theta)}$$

$$= \frac{12(1+\tan^2\theta)}{5-2\tan\theta+5\tan^2\theta}$$

$$\Rightarrow \qquad r^2 = 3\ ,2$$

for $\tan\theta = 1, -1$ respectively.

$\Rightarrow$ the major and the minor axes lie along the lines which make angles $\frac{\pi}{4}$ and $\frac{3\pi}{4}$ respectively with the $x$-axis and they pass through the centre (0, 1).

$\therefore$ their equations are given by

$$y - 1 = 1(x - 0) \qquad \left(m = \tan\frac{\pi}{4} = 1\right)$$

and $$y - 1 = -1(x - 0) \qquad \left(m' = \tan\frac{3\pi}{4} = -1\right)$$

$\Rightarrow$ $$x - y + 1 = 0$$

and $$x + y - 1 = 0$$

Eccentricity of the ellipse can be determined, as in Method I.

**Example 13.4**. Show that the joint equation of the axes of the central conic

$$ax^2 + 2hxy + by^2 = 1$$

is $$h(x^2 - y^2) = (a - b)xy.$$

*Solution*. Let $r_1, r_2$ denote the semi-axes of the given conic

$$ax^2 + 2hxy + by^2 = 1$$

The equations of the semi-axes are

$$\left(a - \frac{1}{r_1^2}\right)x + hy = 0$$

and $$\left(a - \frac{1}{r_2^2}\right)x + hy = 0 \qquad \text{(by § 13.3.1)}$$

where $\frac{1}{r_1^2}, \frac{1}{r_2^2}$ are the roots of

$$\frac{1}{r^4} - (a + b)\frac{1}{r^2} + (ab - h^2) = 0$$

$$\left.\begin{aligned} \Rightarrow \quad & \frac{1}{r_1^2} + \frac{1}{r_2^2} = a + b \\ \text{and} \quad & \frac{1}{r_1^2} \cdot \frac{1}{r_2^2} = ab - h^2 \end{aligned}\right\} \qquad (1)$$

Now the joint equation of the axes is

$$\left[\left(a-\frac{1}{r_1^2}\right)x+hy\right].\left[\left(a-\frac{1}{r_2^2}\right)x+hy\right]=0$$

$$\Rightarrow \quad \left[a^2-\left(\frac{1}{r_1^2}+\frac{1}{r_2^2}\right)a+\frac{1}{r_1^2}.\frac{1}{r_2^2}\right]x^2$$

$$+h\left[2a-\left(\frac{1}{r_1^2}+\frac{1}{r_2^2}\right)\right]xy+h^2y^2=0$$

$\Rightarrow \quad -h^2x^2 + h(a-b)xy + h^2y^2 = 0,$ using (1).

$\Rightarrow \quad h(x^2 - y^2) = (a - b)xy,$

(when $h \neq 0$. However, if $h = 0$, the joint equation of the axes of the conic, $ax^2 + by^2 = 1$ is $xy = 0$).

**Example 13.5** Show that the product of the semi-axes of the conic whose equation is, $x^2 - xy + 2y^2 - 2x - 6y + 7 = 0$ is $2/\sqrt{7}$ and that the equation of its axes is $x^2 - y^2 - 2xy + 8y - 8 = 0$.

*Solution.* The equation of the given conic is

$$S = x^2 - xy + 2y^2 - 2x - 6y + 7 = 0 \qquad (1)$$

The centre giving equations are

$$\frac{\partial S}{\partial x}=0 \text{ and } \frac{\partial S}{\partial y}=0$$

$\Rightarrow 2x - y - 2 = 0$ and $-x + 4y - 6 = 0$

$\Rightarrow$ Centre of (1) has coordinates (2, 2).

Also $d = gx_1 + fy_1 + c$

$= (-1) \times 2 + (-3) \times 2 + 7 = -1.$

$\therefore$ (1) transforms to

$$X^2 - XY + 2Y^2 = 1 \qquad (2)$$

Now the lengths of the semi-axes are given by

$$\frac{1}{r^4}-(a'+b').\frac{1}{r^2}+\left(a'b'-h'^2\right)=0 \qquad (\S\ 13.3.1).$$

$$\Rightarrow \quad \frac{1}{r^4}-\frac{3}{r^2}+\frac{7}{4}=0 \qquad (\because a' = 1, b' = 2 \text{ and } h' = -1/2)$$

$\Rightarrow \quad 7r^4 - 12r^2 + 4 = 0$

$\Rightarrow \quad r_1^2.r_2^2 = 4/7 \Rightarrow r_1r_2 = 2/\sqrt{7}$

Note that the above equation yields both $r_1^2$ and $r_2^2$ to be positive and consequently the given conic is an ellipse.

Now the joint equation of the axes, by Example 13.4, is given by $h'(X^2 - Y^2) = (a' - b')XY$

$$\Rightarrow \quad -\frac{1}{2}(X^2 - Y^2) = (1 - 2)XY$$

$$\Rightarrow \quad X^2 - 2XY - Y^2 = 0$$

$$\Rightarrow \quad (x-2)^2 - 2(x-2)(y-2) - (y-2)^2 = 0$$

$$(\because\ x = X + 2 \text{ and } y = Y + 2)$$

$$\Rightarrow \quad x^2 - 2xy - y^2 + 8y - 8 = 0.$$

## Exercise 13.1

1. Show that the following equation represents an ellipse. Find its centre and the axes of the ellipse,
$$11x^2 - 4xy + 14y^2 - 58x - 44y + 71 = 0.$$
2. Show that the equation
$$22x^2 - 12xy + 17y^2 - 112x + 92y + 78 = 0$$
represents an ellipse. Find the centre of this ellipse and the lengths of its axes.
3. What conic does the equation
$$36x^2 + 24xy + 29y^2 - 72x + 126y + 81 = 0$$
represent? Find the coordinates of its centre, lengths of its axes and the equations of its axes.
4. Find the position of the centre and the lengths of the axes of the conic
$$97x^2 - 60xy + 72y^2 - 314x + 348y + 37 = 0.$$
5. Find the centre of the conic section
$$13x^2 - 18xy + 37y^2 + 2x + 14y - 2 = 0.$$
Also, find the transformed equation of the conic section with its centre as origin.
6. Find the lengths and equations of the axes of the conic
$$8x^2 + 4xy + 5y^2 = 24(x + y).$$

7. Show that the lengths of the semi-axes of the conic

$$ax^2 + 2hxy + ay^2 = k$$

are $\left(\dfrac{k}{a \pm h}\right)^{1/2}$ and their joint equation is $x^2 - y^2 = 0$.

8. Prove that the principal axes of the conic

$$ax^2 + 2hxy + by^2 + 2gx + 2fy + c = 0$$

are parallel to the lines

$$h(x^2 - y^2) = (a - b)xy.$$

Hence find the principal axes of the conic, $2x^2 + 12xy - 7y^2 - 16x + 2y - 3 = 0$.

9. Show that the eccentricity e of the conic

$$ax^2 + 2hxy + by^2 = 1$$

is given by

$$\frac{e^2}{2 - e^2} = \pm \frac{\left[(a-b)^2 + 4h^2\right]^{1/2}}{a + b}$$

Also show that

$$e^4 + \frac{(a-b)^2 + 4h^2}{ab - h^2} e^2 - \frac{(a-b)^2 + 4h^2}{ab - h^2} = 0$$

10. If $r$ be the length of any of the semi-axis of the conic

$$ax^2 + 2hxy + by^2 + 2gx + 2fy + c = 0,$$

show that

$$(ab - h^2)^3 r^4 + \Delta(a + b)(ab - h^2)r^2 + \Delta^2 = 0,$$

where $D = abc + 2fgh - af^2 - bg^2 - ch^2$.

(*Hint*. Shift the origin to the centre of the conic).

11. (i) Show that the squares of the lengths of the semi-axes of the conic, $ax^2 + 2hxy + by^2 = 1$ are the reciprocals of the roots of

$$x^2 - (a + b)x + (ab - h^2) = 0.$$

(ii) Prove that the squares of the semi-axes of the conic

$$ax^2 + 2hxy + by^2 + 2gx + 2fy + c = 0$$

are $\dfrac{-\Delta\left[(a+b) \pm \sqrt{(a-b)^2 + 4h^2}\right]}{2(ab - h^2)^2}$

12. If $a$ and $b$ are both of the same sign, prove that the equa-

tion, $ax^2 + by^2 + 2gx + 2fy + c = 0$ represents an ellipse.

## 13.5. TO FIND THE COORDINATES OF THE FOCUS (OR FOCI) OF A CONIC, $S = 0$.

We find the coordinates of the focus (or foci) when the given conic is a parabola, an ellipse or a hyperbola.

**Case I**. Let $S = 0$ be a *parabola*.
Let $(x', y')$ be its vertex, $m = \tan\theta$ be the slope of its axis and $4a$ be the length of its latus rectum. Then the equation of the axis of the parabola, is

$$\frac{x - x'}{\cos\theta} = \frac{y - y'}{\sin\theta} = r \tag{1}$$

We know that the focus of the parabola lies on its axis (i.e. on (1) and is at a distance $a$ from its vertex.)

Thus the coordinates of the focus of the parabola are

$$(x' + a\cos\theta, y' + a\sin\theta).$$

**Case II**. Let $S = 0$, be an *ellipse*.
Let its semi-major axis be of length $a$ and its slope be, $m = \tan\theta$.
Let $(x', y')$ denote its centre. Proceeding as in *Case I*, the coordinates of the foci of the ellipse are

$$(x' \pm ae\cos\theta, y' \pm ae\sin\theta);$$

as its foci are at a distance, $\pm ae$, from its centre.
(It may be *noted* that we have to take either both upper signs or both lower signs).

**Case III** Let $S = 0$ be a *hyperbola*.
This case is similar to Case II; the only difference being that here $a$ stands for semi-traverse axis.

**Corollary**. If the equation of the conic is $ax^2 + 2hxy + by^2 = 1$, then the foci are given by
(i) $(\pm ae\cos\theta, \pm ae\sin\theta)$, where $a$ is semi-major axis, $e$ is the eccentricity and (0, 0) is the centre of the ellipse.
(ii) $(\pm ae\cos\theta, \pm ae\sin\theta)$, where $a$ is semi–tranverse axis, $e$ is

the eccentricity and (0, 0) is the centre of the hyperbola.

## 13.6 TO FIND THE EQUATION OF THE DIRECTRIX (OR DIRECTRICES) OF A CONIC, $S = 0$.

**Case I. Let** $S = 0$ *be a parabola.*
Let the equation of the axis of the parabola be

$$\frac{x-x'}{\cos\theta} = \frac{y-y'}{\sin\theta} = r \qquad \text{(see § 13.5.)}$$

$$\Rightarrow x \sin \theta - y \cos \theta - x' \sin \theta + y' \cos \theta = 0 \qquad (1)$$

We know that the point $T$, where the directrix of the parabola meets its axis is at a distance - $a$ from its *vertex*. Thus the coordinates of $T$ are,

$$(x' - a \cos \theta, y' - a \sin \theta).$$

Now the directrix being perpendicular to the axis, given by (1), of the parabola, we have its equation as

$$x \cos \theta + y \sin \theta + k = 0 \qquad (2)$$

It passes through $T(x' - a \cos \theta, y' - a \sin \theta)$.

$$\therefore \quad (x' - a \cos \theta) \cos \theta + (y' - a \sin \theta)\sin \theta + k = 0 \qquad (3)$$

Now $\quad (2) - (3) \Rightarrow$

$$[x - (x' - a \cos \theta)]\cos \theta + [y - (y' - a \sin \theta)]\sin \theta = 0,$$

$$\Rightarrow \quad (x - x') \cos \theta + (y - y') \sin \theta + a = 0,$$

which is the required equation of the directrix of the parabola.

**Case II**. *Let S be an ellipse or a hyperbola.*
Proceeding exactly as in Case I, above; the equations of their directrices are

$$(x - x') \cos \theta + (y - y') \sin \theta \pm a/e = 0.$$

Note that the distance between the centre and the point where directrix meets the major (transverse) axis is $\pm (a/e)$.

**Corollary**. If the equation of the conic is $ax^2 + 2hxy + by^2 = 1$, then the directrices are given by

$$x \cos \theta + y \sin \theta \pm a/e = 0,$$

where $a$ is the semi-major (semi-transverse) axis, $e$ is the eccentricity and the centre is (0, 0) of the ellipse (hyperbola).

## 13.7 TRACING OF AN ELLIPSE.

We explain this by the following example.

**Example 13.6.** Find the centre, the lengths and the equations of the principal axes, the eccentricity, the latus rectum, the foci and the directrices of the conic

$$5x^2 - 6xy + 5y^2 + 22x - 26y + 29 = 0.$$

Also trace it.

*Solution.* Here

$$\Delta = abc + 2fgh - af^2 - bg^2 - ch^2 = -128 \neq 0$$

and $$ab - h^2 = 16 > 0.$$

$\Rightarrow$ The given curve is an ellipse.

Its centre $(x_1, y_1)$ is given by the equations

$$ax_1 + hy_1 + g = 0 \text{ and } hx_1 + by_1 + f = 0$$

i.e. by $5x_1 - 3y_1 + 11 = 0$ and $-3x_1 + 5y_1 - 13 = 0$

whence $x_1 = -1,\ y_1 = 2$

i.e. centre of the ellipse is (–1, 2).

Now shifting the origin to (–1, 2), the axes remaining parallel to the original axes, (i.e. $x = X - 1$ and $y = Y + 2$) the given equation transforms to

$$5X^2 - 6XY + 5Y^2 + \frac{\Delta}{ab - h^2} = 0 \qquad \text{(by Cor. § 13.2)}$$

$$\Rightarrow \quad 5X^2 - 6XY + 5Y^2 - 8 = 0$$

$$\Rightarrow \quad \frac{5}{8}X^2 - \frac{3}{4}XY + \frac{5}{8}Y^2 = 1.$$

Now if $r$ represents any of its semi-axis, then by equation (5) of § 13.3.1, we have

$$1/r^4 - (a' + b')1/r^2 + (a'b' - h'^2) = 0$$

where $a' = b' = 5/8$ and $h' = -3/8$

$\Rightarrow \quad r^4 - 5r^2 + 4 = 0 \quad \Rightarrow \quad r^2 = 4, 1.$

$\therefore$ the semi–axes of the ellipse are 2 and 1.

$\therefore$ the lengths of its axes are 4 and 2.

Again by equation (6) of § 13.3.1, the equation of the axis is given by

$$(a' - 1/r^2)X + h'Y = 0$$

$\Rightarrow$ $X - Y = 0$ when $r^2 = 4$

and $X + Y = 0$ when $r^2 = 1$

$\Rightarrow$ the equations of major and minor axes respectively are

$$x - y + 3 = 0$$

and $x + y - 1 = 0$ (as $X = x + 1$, $Y = y - 2$)

Further $b^2 = a^2(1 - e^2) \Rightarrow 1 = 4(1 - e^2) \Rightarrow e = \sqrt{3}/2$.

Latus Rectum $= \dfrac{2b^2}{a} = 1.$

Now the slope of the major axis, $\tan\theta = 1$, whence $\theta = \pi/4$

$\therefore$ The foci of the ellipse are

$(x_1 \pm ae\cos\theta, y_1 \pm ae\sin\theta)$ (See § 13.5)

$$= \left(-1 \pm 2.\frac{\sqrt{3}}{2}.\frac{1}{\sqrt{2}},\ 2 \pm 2.\frac{\sqrt{3}}{2}.\frac{1}{\sqrt{2}}\right)$$

$$= \left(-1 \pm \sqrt{\frac{3}{2}}, 2 \pm \sqrt{\frac{3}{2}}\right).$$

Also the equations of its directrices are given by

$(x - x_1)\cos\theta + (y - y_1)\sin\theta \pm a/e = 0$ (See § 13.6)

$\Rightarrow (x + 1) + (y - 2) \pm 4\sqrt{2/3} = 0.$

We now come to tracing of the ellipse;

It will be enough to find the intersection of the ellipse

$$5x^2 - 6xy + 5y^2 + 22x - 26y + 29 = 0$$

with the coordinate axes.

Putting $y = 0$, in this equation, we have

$$5x^2 + 22x + 29 = 0$$

$\Rightarrow$ $x$ is imaginary

$\Rightarrow$ the ellipse does not intersect the $x$-axis.

Next putting, $x = 0$, in the same equation, we have

$$5y^2 - 26y + 29 = 0$$

$$\Rightarrow \quad y = \frac{26 \pm 4\sqrt{6}}{10} = 3.58 \text{ and } 1.62 \text{ (approximately)}$$

$\Rightarrow$ the ellipse meets the $y$-axis at $P(3.58, 0)$ and $Q(1.62, 0)$.
The shape of the ellipse is:

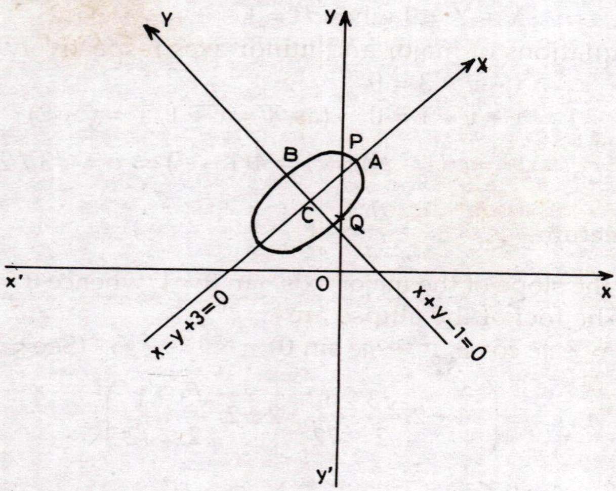


(Note: $CA = 2$, $CB = 1$).

**Example 13.7**. Trace the conic $(x - y + 1)^2 + 4(x + y + 3)^2 = 18$.

*Solution*. The lines $x - y + 1 = 0$ and $x + y + 3 = 0$ are perpendicular to each other. We take these lines as the new axes of coordinates.

Let $X = \dfrac{x - y + 1}{\sqrt{2}}$ and $Y = \dfrac{x + y + 3}{\sqrt{2}}$ (1)

so that $x - y + 1 = 0$ is the new $y$-axis,
and $x + y + 3 = 0$ is the new $x$-axis.

Note that the origin lies on the positive side of both these lines. Hence with respect to the new axes, the origin lies in the positive quadrant.
Now the given equation is $(x - y + 1)^2 + 4(x + y + 3)^2 = 18$

$$\Rightarrow \qquad \left(\frac{x-y+1}{\sqrt{2}}\right)^2 + 4\left(\frac{x+y+3}{\sqrt{2}}\right)^2 = 9$$

$$\Rightarrow \qquad \frac{X^2}{9} + \frac{Y^2}{9/4} = 1 \qquad \text{(using (1))}$$

which is the transformed equation.

This is clearly an ellipse whose

(i) major axis is of length = 6 and it lies along the new $x$-axis i.e. along the line, $x + y + 3 = 0$)

(ii) minor axis is of length = 3 and it lies along the new $y$-axis i.e. along the line, $x - y + 1 = 0$.

The diagram of this is shown below:

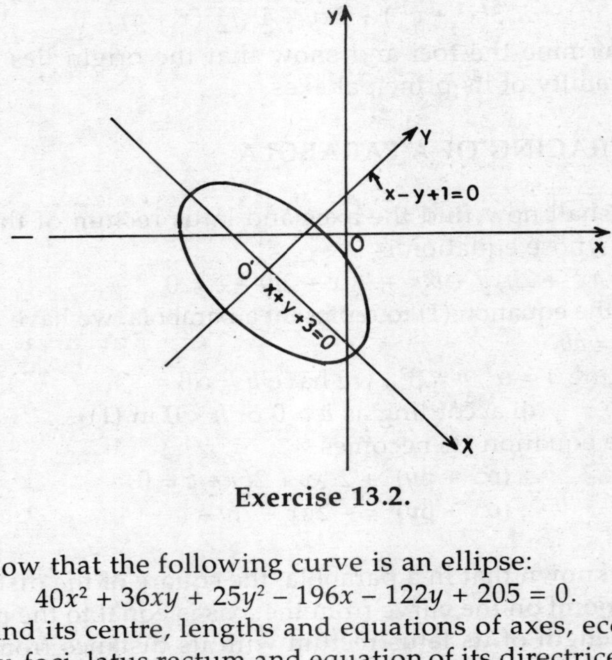


**Exercise 13.2.**

1. Show that the following curve is an ellipse:

$$40x^2 + 36xy + 25y^2 - 196x - 122y + 205 = 0.$$

Find its centre, lengths and equations of axes, eccentricity, foci, latus rectum and equation of its directrices. Also trace the conic.

2. Trace the conic:

$$5x^2 - 2xy + 5y^2 + 2x - 10y - 7 = 0.$$

3. Trace the conic given by the equation
$$22x^2 - 12xy + 17y^2 - 112x + 92y + 178 = 0.$$
4. Find the nature of the conic represented by the equation
$$14x^2 - 4xy + 11y^2 - 44x - 58y + 71 = 0$$ and trace it.
5. Trace the conic
$$17x^2 - 12xy + 8y^2 + 46x - 28y + 17 = 0.$$
Also find the equations of its directrices.
6. Find the eccentricity, foci and the directrices of the ellipse,
$$3x^2 + 4y^2 - 12x - 8y + 4 = 0.$$
7. Show that the curve
$$x^2 + xy + y^2 - x + 4y + 3 = 0.$$
is an ellipse. Find its centre, foci and the equations of its principal axes. Also trace it.
8. Draw a rough sketch of the conic
$$3(x^2 + y^2) + 2xy - 4\sqrt{2}(x + y) = 0.$$
Determine the foci and show that the origin lies at an extremity of its principal axes.

## 13.8. TRACING OF A PARABOLA

We shall now find the axes and latus rectum of the parabola whose equation is
$$ax^2 + 2hxy + by^2 + 2gx + 2fy + c = 0 \tag{1}$$
For the equation (1) to represent a parabola, we have $\Delta \neq 0$ and $h^2 = ab$.

Putting $a = \alpha^2$, $b = \beta^2$, we have $h = \alpha\beta$
(or $h = -\alpha\beta$ according as $h > 0$ or $h < 0$ in (1)).
and the equation (1) becomes
$$(\alpha x + \beta y)^2 + 2gx + 2fy + c = 0.$$
$$\Rightarrow \quad (\alpha x + \beta y)^2 = -2gx - 2fy - c \tag{2}$$

It is known that in a parabola, the square of the distance of any point on the curve from the axis is equal to the product of length of its latus rectum with its distance from the tangent at its vertex. Also the axis of the parabola and the tangent at its vertex are perpendicular, therefore, we write equation (2) in a form which satisfies the aforesaid properties as

$$(\alpha x + \beta y + \lambda)^2 - (2\alpha\lambda x + 2\beta\lambda y + \lambda^2) = -2gx - 2fy - c$$

$$\Rightarrow = (\alpha x + \beta y + \lambda)^2 = 2(\alpha\lambda - g)x + 2(\beta\lambda - f)y + (\lambda^2 - c) \quad (3)$$

Now choose $\lambda$ such that the lines given by

$$\alpha x + \beta y + \lambda = 0 \quad (4)$$

and $$2(\alpha\lambda - g)x + 2(\beta\lambda - f)y + (\lambda^2 - c) = 0 \quad (5)$$

are perpendicular

This requires

$$\left(-\frac{\alpha}{\beta}\right)\cdot\left[-\frac{2(\alpha\lambda - g)}{2(\beta\lambda - f)}\right] = -1$$

$$\Rightarrow \alpha(\alpha\lambda - g) = -\beta(\beta\lambda - f)$$

$$\lambda = (\alpha g + \beta f)/(\alpha^2 + \beta^2) \quad (6)$$

With this value of $\lambda$, equation (4) represents the axis and equation (5) represents the tangent at the vertex to the parabola.

For finding the length of the latus rectum, we write equation (3) as

$$\left[\frac{\alpha x + \beta y + \lambda}{\sqrt{\alpha^2 + \beta^2}}\right]^2 .(\alpha^2 + \beta^2)$$

$$= 2\sqrt{(\alpha\lambda - g)^2 + (\beta\lambda - f)^2}\cdot\left[\frac{2(\alpha\lambda - g)x + 2(\beta\lambda - f)y + (\lambda^2 - c)}{2\sqrt{(\alpha\lambda - g)^2 + (\beta\lambda - f)^2}}\right]$$

Compare it with '$Y^2 = 4p\,X$', the length of the latus rectum

$$= 4p = \frac{2\sqrt{(\alpha\lambda - g)^2 + (\beta\lambda - f)^2}}{\alpha^2 + \beta^2}$$

$$= \frac{2}{\alpha^2 + \beta^2}\cdot\sqrt{\left[\alpha\left(\frac{\alpha g + \beta f}{\alpha^2 + \beta^2}\right) - g\right]^2 + \left[\beta\left(\frac{\alpha g + \beta f}{\alpha^2 + \beta^2}\right) - f\right]^2}$$

$$= \frac{2}{\alpha^2 + \beta^2}\cdot\sqrt{\beta^2(\alpha f - \beta g)^2 + \alpha^2(\beta g - \alpha f)^2}$$

$$= 2\,|\alpha f - \beta g|\,/\,(\alpha^2 + \beta^2)^{3/2}$$

***Note***. Equations (4) and (5) above represent the axis and the tangent at the vertex of the parabola, where $\lambda$ is given by (6). Their point cf intersection is the vertex of the parabola.

We now illustrate the tracing of a parabola with the help of the following examples.

**Example 13.8**. Show that $x^2 + 2xy + y^2 - 2x - 1 = 0$ represents a parabola. Find the equations of its axis and the tangent at the vertex. Also find the length of its latus rectum.

*Solution*. **Method I**.

Here
$$\Delta = abc + 2fgh - af^2 - bg^2 - ch^2$$
$$= -1 + 0 - 0 - 1 + 1 = -1 \neq 0$$
and $ab - h^2 = 1 - 1 = 0$, so the given conic is a parabola. Rotate the axes by an angle $\theta$ (origin remaining the same), given by

$$\cot 2\theta = \frac{a-b}{2h} = 0 \Rightarrow \theta = \frac{\pi}{4}$$

so as to remove the $xy$-term.

The transforming equations are given by

$$x = X\cos\theta - Y\sin\theta = \frac{X-Y}{\sqrt{2}}$$

and
$$y = X\sin\theta + Y\cos\theta = \frac{X+Y}{\sqrt{2}}$$

$\therefore$ the equation of the conic transforms to

$$\left(\frac{X-Y}{\sqrt{2}}\right)^2 + 2\left(\frac{X-Y}{\sqrt{2}}\right)\left(\frac{X+Y}{\sqrt{2}}\right) + \left(\frac{X+Y}{\sqrt{2}}\right)^2 - 2\left(\frac{X-Y}{\sqrt{2}}\right) - 1 = 0$$

$\Rightarrow 2X^2 - \sqrt{2}\,X = 1 - \sqrt{2}\,Y$

After completing the square on L.H.S., this can be rewritten as

$$\left(X - \frac{1}{\sqrt{8}}\right)^2 = -\frac{1}{\sqrt{2}}\left(Y - \frac{5}{8}\sqrt{2}\right)$$

Shifting the origin to $\left(\frac{1}{\sqrt{8}}, \frac{5}{8}\sqrt{2}\right)$, this equation becomes

$$X'^2 = -\frac{1}{\sqrt{2}}Y'$$

which is the equation of the parabola in the standard form.

(i) Its latus rectum $= \frac{1}{\sqrt{2}}$

(ii) Its axis is:

$$0 = X' = X - \frac{1}{\sqrt{8}} = x\cos\theta + y\sin\theta - \frac{1}{\sqrt{8}}$$

$$= \frac{x+y}{\sqrt{2}} - \frac{1}{\sqrt{8}}$$

$\Rightarrow \quad 2x + 2y - 1 = 0.$

(iii) The equation of the tangent at the vertex is

$$0 = Y' = Y - \frac{5\sqrt{2}}{8} = -x\sin\theta + y\cos\theta - \frac{5\sqrt{2}}{8}$$

$$= \frac{-x+y}{\sqrt{2}} - \frac{5\sqrt{2}}{8}$$

$\Rightarrow \quad 4x - 4y + 5 = 0.$

**Method II**. We first make an *observation*.
Consider the equation: $y^2 = 4x$, of the parabola. This implies (see the figure)

$(MP)^2 =$ (Latus Rectum), $NP$

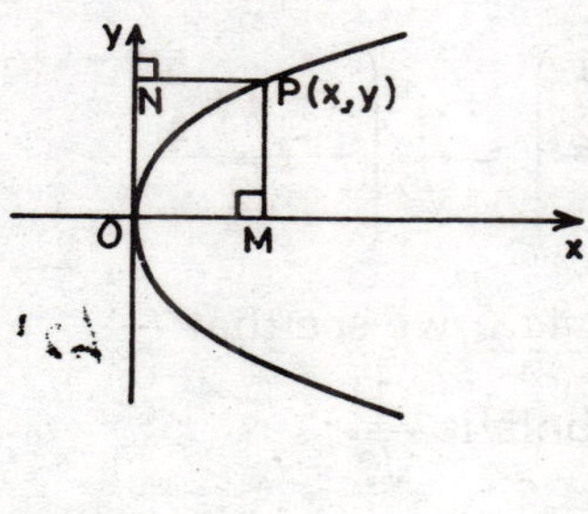

i.e., if $P$ is any point on the given parabola, then

(the distance of $P$ from its axis)$^2$

= (Latus Rectum) . (the distance of $P$ from the tangent at its vertex).

Also, the axis of the parabola and the tangent at its vertex are at right angles.

We now come to Method II.

Verify as in Method I, that the conic represented by the given equation

$$x^2 + 2xy + y^2 - 2x - 1 = 0 \qquad (1)$$

is a parabola.

Now (1) can be written as

$$(x + y)^2 = 2x + 1.$$

Rewrite this equation as

$$(x + y + \lambda)^2 = 2(\lambda + 1)x + 2\lambda y + (\lambda^2 + 1), \qquad (2)$$

where $\lambda$ is a constant to be so chosen that the lines

$$x + y + \lambda = 0 \text{ and } 2(\lambda + 1)x + 2\lambda\, y + (\lambda^2 + 1) = 0$$

are perpendicular.

This requires:

$$(-1).\left[\frac{-2(\lambda+1)}{2\lambda}\right] = -1 \Leftrightarrow \lambda = -\frac{1}{2}$$

$\therefore$ the equation (2) becomes,

$$(x + y - \frac{1}{2})^2 = (x - y + \frac{5}{4}) \qquad (3)$$

Arrange this equation, so that the expression inside the bracket on each side, appears as a perpendicular distance from $(x, y)$, as:

$$\left(\frac{x+y-\frac{1}{2}}{\sqrt{2}}\right)^2 = \frac{1}{\sqrt{2}}\left(\frac{x-y+\frac{5}{4}}{\sqrt{2}}\right) \qquad (4)$$

Comparing it with $Y^2 = 4aX$, we see that

(i) latus rectum of parabola is $\frac{1}{\sqrt{2}}$,

(ii) the axis of the parabola is

$$x + y - \frac{1}{2} = 0, \text{ i.e. } 2x + 2y - 1 = 0,$$

(iii) the equation of the tangent to it at the vertex is

$$x - y + \frac{3}{4} = 0, \text{ i.e. } 4x - 4y + 5 = 0.$$

**Example 13.9.** Trace the conic

$$x^2 + 2xy + y^2 - 2x - 1 = 0.$$

Also find the coordinates of the vertex, the focus and equation of directrix.

*Solution.* **Method I**. Proceeding as in Method I, in Example 13.8, the transformed equation is

$$X'^2 = -\frac{1}{\sqrt{2}}Y'$$

The points of intersection of the given parabola with the axes are given by $x^2 - 2x - 1 = 0$ and $y^2 - 1 = 0$, respectively.

$\therefore$ the parabola meets the axes at

$$A(1 + \sqrt{2}, 0) = (2, 4, 0), B(1 - \sqrt{2}, 0) = (-0.4, 0)$$

and $C(0, 1), D(0, -1)$.

Thus the shape of the parabola is as shown below.

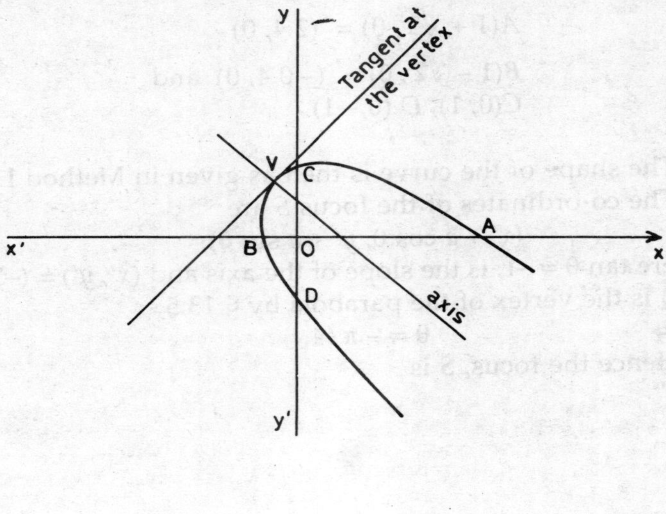

(i) Vertex is the intersection of the lines $X' = 0$, $Y' = 0$

i.e. of $2x + 2y - 1 = 0$ and $4x - 4y + 5 = 0$ whence the vertex is $(-3/8, 7/8)$.

(ii) Coordinates of the focus are given by

$$Y' = -a,\ X' = 0 \text{ where } 4a = \frac{1}{\sqrt{2}},$$

is the latus rectum. Transfering back, the coordinates of the focus are given by

$$\frac{-x+y}{\sqrt{2}} - \frac{5\sqrt{2}}{8} = \frac{-1}{4\sqrt{2}} \text{ and } 2x + 2y - 1 = 0$$

$$\Rightarrow \quad x - y + 1 = 0 \text{ and } 2x + 2y - 1 = 0,$$

whence the focus $S$ is $(-1/4, 3/4)$.

(iii) The equation of the directrix is

$$Y' = a$$

$$\Rightarrow \frac{-x+y}{\sqrt{2}} - \frac{5\sqrt{2}}{8} = \frac{1}{4\sqrt{2}} \Rightarrow 2x - 2y + 3 = 0.$$

**Method II**. Proceeding as in Method II, in the previous example, we get the equation of the axis of the parabola as

$$2x + 2y - 1 = 0,$$

and that of the tangent at its vertex as $4x - 4y + 5 = 0$, and solving these, the vertex is $V(-3/8, 7/8)$.

The given parabola meets the axes at

$$A(1 + \sqrt{2}, 0) = (2.4, 0)$$

$$B(1 - \sqrt{2}, 0) = (-0.4, 0), \text{ and}$$

$$C(0, 1),\ D(0, -1).$$

and

The shape of the curve is then as given in Method I.

The co-ordinates of the focus $S$ are

$$(x' + a\cos\theta,\ y' + a\sin\theta),$$

where $\tan\theta = -1$, is the slope of the axis and $(x', y') = (-3/8, 7/8)$ is the vertex of the parabola by § 13.5.

$$\Rightarrow \qquad \theta = -\pi/4.$$

Hence the focus, $S$ is

$$\left(-\frac{3}{8}+\frac{1}{4\sqrt{2}}\cdot\frac{1}{\sqrt{2}},\frac{7}{8}+\frac{1}{4\sqrt{2}}\cdot\frac{-1}{\sqrt{2}}\right)=\left(-\frac{1}{4},\frac{3}{4}\right)$$

Replacing $a$ by $-a$ in the coordinates of $S$, we find the coordinates of $T$, the point of intersection of the directrix with the axis of parabola as $(-\frac{1}{2}, 1)$.

$\therefore$ the equation of the directrix passing through $(-\frac{1}{2}, 1)$ and perpendicular to the axis:

$$2x + 2y - 1 = 0$$

is $\qquad y - 1 = 1.(x + \frac{1}{2}) \Rightarrow 2x - 2y + 3 = 0$

**Example 13.10.** Show that the curve

$$(a^2 + b^2)(x^2 + y^2) = (bx + ay - ab)^2$$

is a parabola and find its latus rectum.

*Solution.* The given equation, after simplification, is

$$a^2x^2 + b^2y^2 - 2abxy = -2ab^2x - 2a^2by + a^2b^2,$$

which can be further written as

$$(ax - by)^2 = -2ab(bx + ay - \frac{1}{2}ab). \qquad (1)$$

As the lines

$$ax - by = 0$$

and $\qquad bx + ay - \frac{1}{2}ab = 0$

are perpendicular, there is no need to introduce, $\lambda$, (in fact, if we do so, it is a routine matter to verify that $\lambda = 0$),

Now (1) $\Rightarrow$

$$\left(\frac{ax-by}{\sqrt{a^2+b^2}}\right)^2 = -\frac{2ab}{\sqrt{a^2+b^2}}\cdot\left(\frac{bx+ay-\frac{1}{2}ab}{\sqrt{a^2+b^2}}\right)$$

which is of the form,

$$Y^2 = -4AX$$

and thus represents a parabola whose

$$\text{latus rectum} = \frac{2ab}{\sqrt{a^2+b^2}}.$$

**Exercise 13.3**

1. Find the vertex, directrix and the focus of the parabola
$$16x^2 - 24xy + 9y^2 - 104x - 172y + 44 = 0$$
and sketch the curve.
2. Sketch the curve
$$9x^2 + 24xy + 16y^2 - 2x + 14y + 1 = 0$$
and find its vertex, directrix and focus.
3. Trace the parabola
$$9x^2 + 24xy + 16y^2 - 44x + 8y + 92 = 0$$
and find its vertex, length of the latus rectum and equations of the axis and the tangent at the vertex.
4. Trace the conic represented by the equation
$$4x^2 - 4xy + y^2 - 8x - 6y + 5 = 0$$
5. Trace th conic
$$9x^2 - 24xy + 16y^2 - 18x - 101y + 19 = 0.$$
6. Find the axis, tangent at the vertex, length of the latus rectum and the coordinates of the vertex and focus of the parabola
$$16x^2 - 24xy + 9y^2 + 77x - 64y + 95 = 0$$
and trace the curve.
7. Sketch the curve
$$4x^2 + 4xy + y^2 + 3x + 4y + 1 = 0.$$
8. If
$$S = ax^2 + 2hxy + by^2 + 2gx + 2fy + c = 0$$
represents a parabola, find the equation of its axis and prove that its latus rectum is
$$\frac{2\left|\sqrt{a}f - \sqrt{b}g\right|}{(a+b)^{3/2}}$$
9. The Cartesian coordinates $(x, y)$ of a point on a curve are given by

$$x = at^2 + bt + c,$$
$$y = \alpha t^2 + \beta t + \gamma$$

where $t$ is a parameter. Show that the curve is a parabola of latus rectum

$$\frac{(a\beta - b\alpha)^2}{\left(a^2 + \alpha^2\right)^{3/2}}$$

10. Prove that the equations

$$y^2 + 2Ax + 2By + C = 0$$

and $$x^2 + 2Ax + 2By + C = 0,$$

represent parabolas, whose axes are parallel to $x$ and $y$ axes respectively. Also find the latus rectum in each case.

11. Trace the curve given by:

$$(x + y)^2 = 4(x - y - 1).$$

12. Find the focus, directrix and axis of the parabola,

$$\sqrt{ax} + \sqrt{by} = 1.$$

Prove further that if its axis passes through a fixed point, the locus of its focus is a rectangular hyperbola.

13. Show that the curve given by

$$x = \frac{at^2 + bt + c}{\alpha t^2 + \beta t + \gamma};\ y = \frac{a't^2 + b't + c'}{\alpha' t^2 + \beta' t + \gamma'}$$

is a conic section. What is the condition that it may be a parabola?

(*Hint*: Eliminate $t$ and get a second degree equation in $x$ and $y$).

## 13.9 TRACING OF A HYPERBOLA

First we prove a theorem concerning the asymptotes of a hyperbola.

**Theorem 13.3**. The equations of the asymptotes and its conjugate hyperbola differ from the equation of a given hyperbola only by a constant. Further if

$$S = ax^2 + 2hxy + by^2 + 2gx + 2fy + c$$

is the equation of a hyperbola, the equation of its asymptotes and the conjugate hyperbola are respectively

$$S = S_1$$

and

$$S = 2S_1$$

where

$$S_1 = ax_1^2 + 2hx_1y_1 + by_1^2 + 2gx_1 + 2fy_1 + c = 0,$$

and $(x_1, y_1)$ is the centre of the hyperbola.

*Proof.* Let

$$S = ax^2 + 2hxy + by^2 + 2gx + 2fy + c = 0$$

be the given hyperbola with centre

$$(x_1, y_1) = \left(\frac{hf - bg}{ab - h^2}, \frac{gh - af}{ab - h^2}\right) \qquad \text{(see § 13.2)}$$

Shifting the origin to $(x_1, y_1)$ the equation of a hyperbola becomes

$$aX^2 + 2hXY + bY^2 + S_1 = 0 \qquad (1)$$

where

$$S_1 = ax_1^2 + 2hx_1y_1 + by_1^2 + 2gx_1 + 2fy_1 + c = 0.$$

Rotating the axes by an angle θ, given by

$$\tan 2\theta = \frac{2h}{a - b}$$

the equation (1) becomes

$$a(X' \cos\theta - Y' \sin\theta)^2 + 2h(X' \cos\theta - Y' \sin\theta)(X' \sin\theta + Y' \cos\theta)$$
$$+ b(X' \sin\theta + Y' \cos\theta)^2 + S_1 = 0.$$

As in § 3.5, we get

$$a'X'^2 + b'Y'^2 + S_1 = 0,$$

which can be written as

$$\frac{X'^2}{\left(-\frac{S_1}{a'}\right)} - \frac{Y'^2}{\left(\frac{S_1}{b'}\right)} = 1$$

Thus, the equation of the asymptotes and the conjugate hyperbola, then are

$$\frac{X'^2}{\left(-\frac{S_1}{a'}\right)} - \frac{Y'^2}{\left(\frac{S_1}{b'}\right)} = 0.$$

and $$\frac{X'^2}{\left(-\frac{S_1}{a'}\right)} - \frac{Y'^2}{\left(\frac{S_1}{b'}\right)} = -1$$

respectively which can be written as

$$a'X'^2 + b'Y'^2 = k$$

where $k = 0$ or $S_1$.

Rotating the axes back, this equation becomes

$$aX^2 + 2hXY + bY^2 - k = 0$$

Substituting,

$$X = x - x_1, Y = y - y_1,$$

we get

$$ax^2 + 2hxy + by^2 + 2gx + 2fy + c - S_1 - k = 0.$$

Substituting,

$$k = 0 \text{ and } S_1,$$

the equations of the asymptotes and conjugate hyperbola of

$$S = 0$$

are $$S = S_1$$

and $$S = 2S_1$$

respectively which also proves the first part of the theorem.

**Corollary 1.** $S_1 = ax_1^2 + 2hx_1y_1 + by_1^2 + 2gx_1 + 2fy_1 + c$
$= gx_1 + fy_1 + c$
$= \dfrac{\Delta}{ab - h^2}$ (by § 13.2)

Thus the equations of the asymptotes and the conjugate hyperbola are

$$S = \frac{\Delta}{ab - h^2}$$

and $$S = \frac{2\Delta}{ab - h^2}$$

respectively, where

$$\Delta = abc + 2fgh - af^2 - bg^2 - ch^2.$$

**Corollary 2.** If

$$S = 0$$

represents a pair of lines, the general equations of a hyperbola and its conjugate hyperbola, whose asymptotes are these lines, are of the forms

$$S + k = 0$$

and $$S - k = 0$$

respectively.

We can find the asymptotes of the conic

$$ax^2 + 2hxy + by^2 + 2gx + 2fy + c = 0. \quad (1)$$

in the following way also.

Since the equation of the asymptotes differ from that of the curve by a constant only, the equation of the asymptotes will be

$$ax^2 + 2hxy + by^2 + 2gx + 2fy + c + \lambda = 0 \quad (2)$$

Choose $\lambda$ so that equation (2) represents a pair of lines which requires

$$\Rightarrow \quad \begin{vmatrix} a & h & g \\ h & b & f \\ g & f & c+\lambda \end{vmatrix} = 0$$

$$\Rightarrow \quad \begin{vmatrix} a & h & g \\ h & b & f \\ g & f & c \end{vmatrix} + \begin{vmatrix} a & h & 0 \\ h & b & 0 \\ g & f & \lambda \end{vmatrix} = 0$$

$\Rightarrow \quad \Delta + \lambda(ab - h^2) = 0.$

$\Rightarrow \quad \lambda = -\Delta/(ab - h^2).$

Substituting for $\lambda$ in (2), the required equation of the asymptotes is

$$ax^2 + 2hxy + by^2 + 2gx + 2fy + c - \Delta/(ab - h^2) = 0 \quad (3)$$

**Corollary 1**. The asymptotes of the conic (1) are parallel to the lines, $ax^2 + 2hxy + by^2 = 0$.

**Corollary 2**. The asymptotes of the conic $ax^2 + 2hxy + by^2 = 1$ are given by the equation, $ax^2 + 2hxy + by^2 = 0$.

One can easily verify that $\lambda = 1$.

**Corollary 3**. The conic represented by $ax^2 + 2hxy + by^2 = 1$ is an ellipse or a hyperbola according as $ab - h^2 > 0$ or $< 0$.

By Corollary 2, the asymptotes are given by $ax^2 + 2hxy + by^2 = 0$, which are imaginary (i.e. when $ab - h^2 > 0$) in case of ellipse and real (i.e. when $ab - h^2 < 0$) in case of hyperbola.

**Corollary 4**. The equation of the asymptotes of the conic (1) $(ab - h^2 \neq 0)$ is $bX^2 - 2hXY + aY^2 = 0$, where

$$X = ax + hy + g \text{ and } Y = hx + by + f.$$

The asymptotes are given by equation (3) which implies

$$(ab - h^2)(ax^2 + 2hxy + by^2 + 2gx + 2fy + c) - \Delta = 0$$

$$\Rightarrow (ab - h^2)(ax^2 + 2hxy + by^2 + 2gx + 2fy) + (ab - h^2)c$$
$$- (abc + 2fgh - af^2 - bg^2 - ch^2) = 0$$

$$\Rightarrow (ab - h^2)(ax^2 + 2hxy + by^2 + 2gx + 2fy) + af^2 + bg^2 - 2fgh = 0$$

which can be simplified to

$$bX^2 - 2hXY + aY^2 = 0, \text{ where}$$
$$X = ax + hy + g \text{ and } Y = hx + by + f.$$

**Example 13.11**. Find the asymptotes of the curve, $2x^2 + 5xy + 2y^2 + 4x + 5y = 0$, and find the general equation of all hyperbolas having the same asymptotes.

*Solution*. Since the equations of the curve and the asymptotes differ only by a constant, the equation of the asymptotes of the given curve is of the form

$$2x^2 + 5xy + 2y^2 + 4x + 5y + \lambda = 0$$

This must represent a pair of lines, i.e.

$$\Delta = 0$$

$$\Rightarrow \quad 4\lambda + 25 - \frac{25}{2} - 8 - \frac{25}{4}\lambda = 0 \Rightarrow \lambda = 2$$

$\therefore$ The required equation of the asymptotes is

$$2x^2 + 5xy + 2y^2 + 4x + 5y + 2 = 0$$

On factorising this, we get

$$(2x + y + 2)(x + 2y + 1)=0 \tag{1}$$

$\therefore$ The general equation of all hyperbolas having the asymptotes given by (1) is

$$2x^2 + 5xy + 2y^2 + 4x + 5y + k = 0 \qquad (k \neq 2).$$

**Example 13.12** (i) Find the co-ordinates of the centre, the eccentricity, the equations of the principal axes and the equations of the asymptotes of the hyperbola, represented by the equation

$$3x^2 + 10xy + 3y^2 - 26x - 22y + 43 = 0.$$

(ii) Also find the co-ordinates of the foci, equations of its directrices, conjugate hyperbola, latus rectum and trace the curve.

*Solution.* Here $\Delta = abc + 2fgh - af^2 - bg^2 - ch^2$

$$= 387 + 1430 - 363 - 507 - 1075$$

$$= -128 \neq 0$$

and $ab - h^2 = -16 < 0$.

$\therefore$ the given conic is a hyperbola.

Its centre $(x_1, y_1)$ is given by the equations

$$ax_1 + hy_1 + g = 0$$

and

$$hx_1 + by_1 + f = 0$$

i.e. by $3x_1 + 5y_1 - 13 = 0$ and $5x_1 + 3y_1 - 11 = 0$

$\Rightarrow x_1 = 1, y_1 = 2.$

i.e. the centre is (1, 2).

Shifting the origin to (1, 2), with the axes remaining parallel to the original axes, the equation of the hyperbola

$$3x^2 + 10xy + 3y^2 - 26x - 22y + 43 = 0 \qquad (1)$$

transforms to

$$3X^2 + 10XY + 3Y^2 + (gx_1 + fy_1 + c) = 0 \qquad \text{(by § 13.2)}$$

$$\Rightarrow \quad 3X^2 + 10XY + 3Y^2 + 8 = 0 \qquad (2)$$

Rotating the axes, (keeping the origin fixed), by an angle $\theta$ given by

$$\cot 2\theta = \frac{a-b}{2h} = 0 \Rightarrow \theta = \frac{\pi}{4}$$

$$\therefore \quad X = X' \cos\theta - Y' \sin\theta = \frac{X' - Y'}{\sqrt{2}}$$

and $\quad Y = X' \sin\theta + Y' \cos\theta = \dfrac{X' + Y'}{\sqrt{2}}$

The equation(2) now, transforms to

$$3\left(\frac{X'-Y'}{\sqrt{2}}\right)^2 + 10\left(\frac{X'-Y'}{\sqrt{2}}\right).\left(\frac{X'+Y'}{\sqrt{2}}\right) + 3\left(\frac{X'+Y'}{\sqrt{2}}\right)^2 + 8 = 0$$

$$\Rightarrow \quad \frac{Y'^2}{4} - \frac{X'^2}{1} = 1$$

$$\Rightarrow \quad \frac{X''^2}{4} - \frac{Y''^2}{1} = 1 \qquad (3)$$

where $X'' = Y'$ and $Y'' = X'$.

Compare (3) with the standard form,

$$\frac{x^2}{a^2} - \frac{y^2}{b^2} = 1,$$

of the hyperbola.

Here $a = 2$ and $b = 1$.

(i) Eccentricity $e$ is given by

$$b^2 = a^2(e^2 - 1) \Rightarrow 1 = 4(e^2 - 1) \Rightarrow e = \frac{\sqrt{5}}{2}.$$

(ii) Latus rectum $= \dfrac{2b^2}{a} = 1$.

(iii) The equation of the transverse axis is

$$0 = Y'' = X' = X\cos\theta + Y\sin\theta = \frac{(x-1)+(y-2)}{\sqrt{2}}$$

$$\Rightarrow x + y - 3 = 0.$$

(iv) The equation of the conjugate axis is

$$0 = X'' = Y' = -X\sin\theta + Y\cos\theta = \frac{-(x-1)+(y-2)}{\sqrt{2}}$$

$$\Rightarrow x - y + 1 = 0.$$

(v) The equation of the asymptotes is

$$\frac{X''^2}{4} - \frac{Y''^2}{1} = 0$$

$$\Rightarrow \qquad \frac{(-x+y-1)^2}{8} - \frac{(-x+y-3)^2}{2} = 0$$

which simplifies to

$$3x^2 + 10xy + 3y^2 - 26x - 22y + 35 = 0.$$

(vi) The equations of the asymptotes and the hyperbola are

$$3x^2 + 10xy + 3y^2 - 26x - 22y + 35 = 0.$$

and $\quad (3x^2 + 10xy + 3y^2 - 26x - 22y + 35) + 8 = 0.$

Thus, the equation of the conjugate hyperbola is

$$(3x^2 + 10xy + 3y^2 - 26x - 22y + 35) - 8 = 0.$$

$$\Rightarrow \quad 3x^2 + 10xy + 3y^2 - 26x - 22y + 27 = 0.$$

(vii) The slope of the transverse axis is

$$\tan \theta' = -1 \Rightarrow \theta' = \frac{3\pi}{4}$$

$\therefore$ the co-ordinates of its foci are

$$(x_1 \pm ae \cos \theta' \; y_1 \pm ae \sin \theta') \qquad \text{(see § 13.5)}$$

$$= \left(1 \mp \sqrt{\frac{5}{2}}, 2 \pm \sqrt{\frac{5}{2}}\right)$$

(viii) The equations of its directrices are

$$X'' = \pm \frac{a}{e}$$

$$\Rightarrow \qquad \frac{-x+y-1}{\sqrt{2}} = \pm \frac{4}{\sqrt{5}}$$

$$\Rightarrow \qquad \sqrt{5}\,(x - y + 1) \pm 4\sqrt{2} = 0$$

(ix) Now, we trace the hyperbola.

The points of intersection of the hyperbola (1) with the co-ordinate axes are given by

$$3x^2 - 26x + 43 = 0$$

and $\quad 3y^2 - 22y + 43 = 0.$

The roots of $3y^2 - 22y + 43 = 0$ are imaginary, so the hyperbola does not meet the $y$-axis.

Also, by solving

$$3x^2 - 26x + 43 = 0,$$

we see that the curve meets the $x$-axis at $\left(\dfrac{13 \pm 2\sqrt{10}}{3}\right)$

which are approximately (6, 4, 0) and (2, 3, 0). Thus the hyperbola is as shown here:

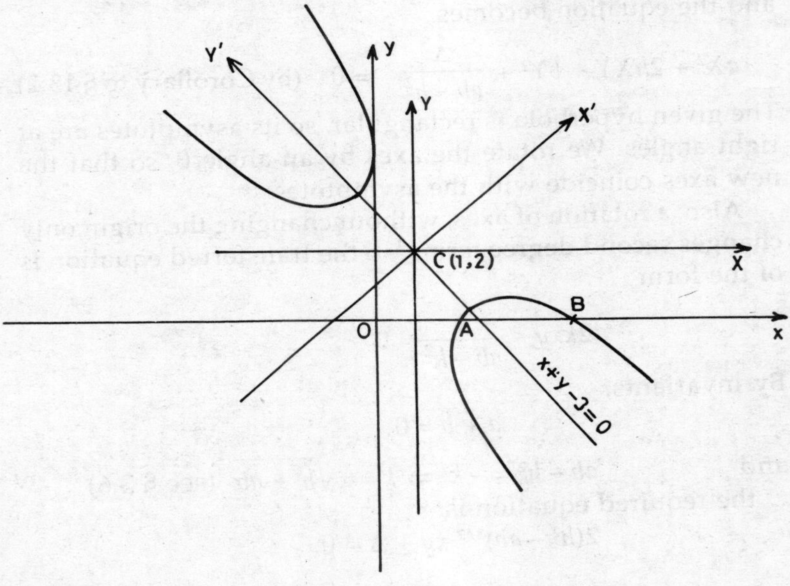


**Example 13.13**. The asymptotes of a hyperbola are

$$x + 2y + 3 = 0 \text{ and } 3x + 4y + 5 = 0$$

and it passes through the point (1, –1). Find its equation and the equation of the conjugate hyperbola.

*Solution*. Let the equation of the hyperbola be

$$(x + 2y + 3)(3x + 4y + 5) + k = 0.$$

It passes through (1, –1), so $k = -8$.

$\therefore$ the equation of the hyperbola is

$$(x + 2y + 3)(3x + 4y + 5) - 8 = 0$$

and that of the conjugate hyperbola is

$$(x + 2y + 3)(3x + 4y + 5) - k = 0.$$

$$\Rightarrow \quad (x + 2y + 3)(3x + 4y + 5) + 8 = 0$$

**Example 13.14**. If the conic

$$ax^2 + 2hxy + by^2 + 2gx + 2fy + c = 0$$

represents a rectangular hyperbola, show that its equation referred to its asymptotes as the co-ordinate axes is

$$2(h^2 - ab)^{3/2}\, xy \pm \Delta = 0$$

*Solution*. We shift the origin to the centre of the hyperbola and the equation becomes

$$aX^2 + 2hXY + bY^2 + \frac{\Delta}{ab - h^2} = 0. \quad \text{(by Corollary to § 13.2)}$$

The given hyperbola is rectangular, so its asymptotes are at right angles. We rotate the axes by an angle $\theta$, so that the new axes coincide with the asymptotes.

Also, a rotation of axes, without changing the origin only changes second degree terms, so the transferred equation is of the form

$$2kxy + \frac{\Delta}{ab - h^2} = 0$$

By invariants,

$$a + b = 0$$

and

$$ab - h^2 = -k^2 \Rightarrow k = \pm\sqrt{h^2 - ab} \quad \text{(see § 3.6)}$$

$\therefore$ the required equation is

$$2(h^2 - ab)^{3/2}\, xy \pm \Delta = 0.$$

## Exercise 13.4

1. Show that the equation
$$x^2 - 3xy + y^2 + 10x - 10y + 21 = 0$$
represents a hyperbola and trace the same.
2. Determine the type of the conic
$$y^2 + 2x + xy = 2x^2 + y.$$
Find its centre.
3. Find the foci and the eccentricity of the conic
$$x^2 + 4xy + y^2 - 2x + 2y - 6 = 0$$
4. Find the centre and the axes of the conic
$$x^2 - 4xy - 2y^2 + 10x + 4y = 0$$

and trace the same.

5. Find the asymptotes of

$$x^2 + xy - 2y^2 + 2x + y + 4 = 0$$

and the general equation of all hyperbolas having the same asymptotes.

6. Find the equation to the asymptotes of the hyperbola

$$3x^2 - 5xy - 2y^2 + 5x + 11y - 8 = 0$$

Hence obtain the equation of its conjugate hyperbola.

7. Find the asymptotes of

$$\text{(i) } xy - 3x - 2y = 0, \text{ (ii) } xy + 4x - 2y + 1 = 0$$

What is the equation of its conjugate hyperbola?

8. Show that the equation

$$6x^2 - 7xy - 3y^2 - 2x - 8y - 6 = 0$$

represents a hyperbola. Find the asymptotes of the hyperbola and also find the equation of the conjugate hyperbola.

9. Find the equation of the hyperbola which has,

$$3x - 4y + 7 = 0 \text{ and } 4x + 3y + 1 = 0$$

for asymptotes and passes through the origin.

10. The asymptotes of a hyperbola are parallel to

$$2x + 3y = 0 \text{ and } 3x + 2y = 0$$

Its centre is (1, 2) and it passes through (5, 3). Find its equation.

11. Prove that if the centre of the hyperbola whose equation is $S = 0$ be $(\alpha, \beta)$, then the equation of the asymptotes can be written as

$$ax^2 + 2hxy + by^2 + 2gx + 2fy - (g\alpha + f\beta) = 0$$

12. Find the equations of the asymptotes and the conjugate hyperbola to the hyperbola, $(ax + by + c)(a'x + b'y + c') = \lambda^2$.

13. If $(\alpha, \beta)$ be the centre of the hyperbola, whose equation is $f(x, y) = 0$, show that the equations of the asymptotes and the conjugate hyperbola are respectively given by $f(x, y) = f(\alpha, \beta)$ and $f(x, y) = 2f(\alpha, \beta)$.

14. Trace the conic

$$(3x - 4y + 1)(4x + 3y - 7) = 50.$$

15. Trace the conic given by

$$4x^2 - 9y^2 - 24x - 36y - 36 = 0,$$

giving complete details.

(*Hint*. The given equation can be written as

$$4(x-3)^2 - 9(y+2)^2 = 36$$

$\Rightarrow \dfrac{(x-3)^2}{9} - \dfrac{(y+2)^2}{4} = 1$ One can easily trace this hyperbola now).

16. Find the centre, lengths of the axes, eccentricity and foci of the hyperbola:

(i) $\dfrac{(3x+4y-7)^2}{64} - \dfrac{(4x-3y-1)^2}{36} = 25$

(ii) $x^2 - y^2 - 8x - 8y - 4 = 0$.

17. If $a$ and $b$ are of opposite signs, prove that the equation,

$$ax^2 + by^2 + 2gx + 2fy + c = 0$$

represents a hyperbola with its transverse axis parallel to one of the coordinate axes, provided

$$\frac{g^2}{a} + \frac{f^2}{b} - c \neq 0$$

## Miscellaneous Exercise on Chapter 13

1. Find the magnitude and the equations of the axes of the central conic

$$ax^2 + 2hxy + by^2 = 1.$$

2. Find the condition that

$$ax^2 + 2hxy + by^2 + 2gx + 2fy + c = 0$$

may represent

(i) an ellipse,

(ii) a hyperbola,

(iii) a parabola.

3. What do you understand by

(i) Centre?

(ii) Axis of a conic?

Does every conic have (i) a unique centre (ii) a unique axis?

4. Show that an equation of second degree in $x$ and $y$ is either a void locus or represents one of the following:

(i) a pair of straight lines,

(ii) a circle,
(iii) an ellipse,
(iv) a parabola,
(v) a hyperbola.

5. Obtain the expressions giving the position and the squares of the lengths of the axes of the central conic
$$ax^2 + 2hxy + by^2 = 1$$

6. Obtain the equation of a conic referred to the axes through the centre, parallel to the original axes.

7. Find the co–ordinates of the centre of the conic
$$ax^2 + 2hxy + by^2 + 2gx + 2fy + c = 0$$

8. Find the asymptotes of the conic
$$ax^2 + 2hxy + by^2 + 2gx + 2fy + c = 0$$

9. Show that
$$ax^2 + 2hxy + by^2 + 2gx + 2fy + c = 0$$
represents
(i) an ellipse,
(ii) a parabola, or
(iii) a hyperbola according as $ab - h^2$ is positive, zero or negative.

10. Prove that the locus of the vertices of the rectangular hyperbola, $x^2 - y^2 + 2\lambda xy = a^2$ for different vaues of $\lambda$, is the curve, $(x^2 + y^2)^2 = a^2(x^2 - y^2)$.

11. If the conics
$$S \equiv ax^2 + 2hxy + by^2 + 2gx + 2fy + c = 0$$
and $S' \equiv a'x^2 + 2h'xy + b'y^2 + 2g'x + 2f'y + c' = 0$
intersect in concyclic points, prove that $h'(a - b) = h(a' - b')$ and the coordinates of the centre of the circle are
$$\left(\frac{hg' - h'g}{ah' - a'h}, \frac{hf' - h'f}{ah' - a'h}\right)$$

12. Find the locus of the centres of the conics through the points of intersection of the conics, $S = 0$ and $S' = 0$ ($S$ and $S'$ as in Q. 11 above).

# 14. POLAR CO-ORDINATES

## INTRODUCTION

In § 1.5, we briefly introduced the polar co-ordinates of a point and found also the relations between its polar and the cartesian co-ordinates. In this chapter, we derive the various important formulae e.g. distance between two points, area of a triangle, equations of a line, of a circle and also of a conic when its focus is taken as pole.

## 14.1. TO FIND THE DISTANCE BETWEEN TWO POINTS $(r_1, \theta_1)$ and $(r_2, \theta_2)$

Let $P(r_1, \theta_1)$ and $Q(r_2, \theta_2)$ be the two given points. Applying cosine formula to the $\Delta OPQ$, we have

$$PQ^2 = OP^2 + OQ^2 - 2.\overline{OP}.\overline{OQ} \cos \angle POQ$$

$$\Rightarrow PQ^2 = r_1{}^2 + r_2{}^2 - 2r_1r_2 \cos(\theta_2 - \theta_1)$$

$$\Rightarrow |PQ| = \sqrt{r_1^2 + r_2^2 - 2r_1r_2 \cos(\theta_2 - \theta_1)}$$

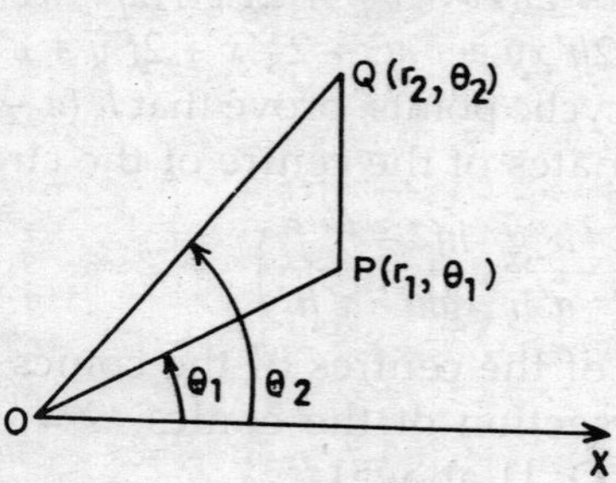


### 14.1.1. To find the area of a triangle whose vertices are $(r_i, \theta_i)$, $i = 1, 2, 3$.

Let $P(r_1, \theta_1)$, $Q(r_2, \theta_2)$ and $R(r_3, \theta_3)$ be the vertices of the given triangle.

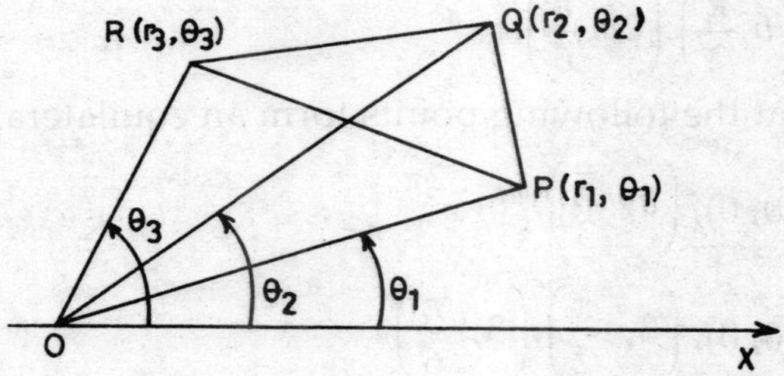


Now, (taking care of the directions for the area),

area $\Delta PQR$ = area $\Delta OPQ$ + area $\Delta OQR$ − area $\Delta OPR$

$$\Rightarrow \text{area } \Delta PQR = \frac{1}{2} r_1 r_2 \sin(\theta_2 - \theta_1) + \frac{1}{2} r_2 r_3 \sin(\theta_3 - \theta_2) - \frac{1}{2} r_1 r_3 \sin(\theta_3 - \theta_1)$$

$$= \frac{1}{2} [r_1 r_2 \sin(\theta_2 - \theta_1) + r_2 r_3 \sin(\theta_3 - \theta_2) + r_3 r_1 \sin(\theta_1 - \theta_3)]$$

which is the required formula for the area. We generally take the absolute value of the area.

**Corollary 1**. Area of the triangle $OPQ$ where $O$ is the pole is $\frac{1}{2} r_1 r_2 \sin(\theta_2 - \theta_1)$.

**Corollary 2.** Three points $P(r_1, \theta_1)$, $Q(r_2, \theta_2)$ and $R(r_3, \theta_3)$ are collinear iff,

$$r_1 r_2 \sin(\theta_2 - \theta_1) + r_2 r_3 \sin(\theta_3 - \theta_2) + r_3 r_1 \sin(\theta_1 - \theta_3) = 0$$

$$\Leftrightarrow \frac{1}{r_3} \sin(\theta_2 - \theta_1) + \frac{1}{r_1} \sin(\theta_3 - \theta_2) + \frac{1}{r_2} \sin(\theta_1 - \theta_3) = 0$$

## Exercise 14.1.

1. Find the distance between the points

(i) $(a, 115^\circ)$, $(3a, 25^\circ)$

(ii) $\left(b, \frac{\pi}{2}\right), \left(2b, \frac{\pi}{6}\right)$

2. Show that the following points form an equilateral triangle

(i) $(0, 0), \left(b, \frac{\pi}{3}\right), \left(b, \frac{2\pi}{3}\right).$

(ii) $(0, 0), \left(3, \frac{\pi}{2}\right), \left(3, \frac{\pi}{6}\right).$

3. Find the area of the triangle whose vertices are

(i) $(b, \alpha), \left(3b, \alpha + \frac{\pi}{3}\right), \left(4b, \alpha + \frac{2\pi}{3}\right)$

(ii) $\left(1, \frac{\pi}{6}\right), \left(-1, \frac{\pi}{2}\right), \left(1, \frac{4\pi}{3}\right)$

$\left[\text{Hint: } \left(-1, \frac{\pi}{2}\right) = \left(1, \frac{3\pi}{2}\right)\right]$

4. The vectorial angle of a point $P$ on the join of the points $P_1(r_1, \theta_1)$ and $P_2(r_2, \theta_2)$ is $\frac{\theta_1 + \theta_2}{2}$. Show that the length of the radius vector of $P$ is

$$\frac{2r_1r_2}{r_1 + r_2}\cos\left(\frac{\theta_1 - \theta_2}{2}\right).$$

## 14.2. TO FIND THE EQUATION OF A LINE IN POLAR CO-ORDINATES

**Case 1**. *Let the line pass through the pole.*

Let this line $AB$, through the pole, make an angle $\theta_1$ with the initial line. Then for every point $P$ $(r, \theta)$ on $AB$, we have, $\theta = \theta_1$ which is, therefore, the required equation of the line $AB$, passing through the pole, $O$.

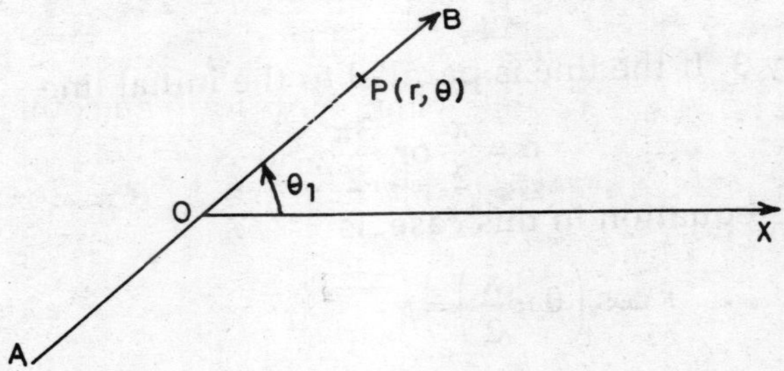


**Case 2**. *Let the line not pass through the pole.*

Suppose that such a line is at a distance $p$, from the pole. Also let the perpendicular, $OL$, from the pole on this line make an angle, $\alpha$ with the initial line. Then, if $P(r, \theta)$ is any point on this line, from $\Delta OLP$, we have

$$\cos(\theta - \alpha) = \frac{p}{r}$$

$\Rightarrow \qquad r\cos(\theta - \alpha) = p,$

which is the required equation of the line.

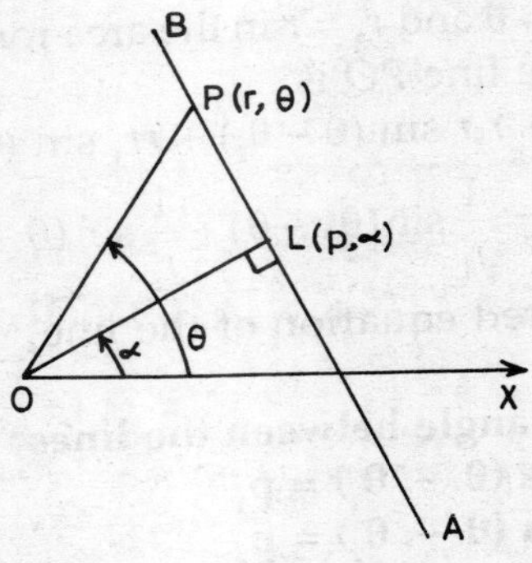


**Corollary 1**. The co-ordinates of the foot of the perpendicular, from the pole on the line

$$r\cos(\theta - \alpha) = p \text{ are } (p, \alpha).$$

**Corollary 2.** If the line is perpendicular to the initial line, then $\quad \alpha = 0$ or $\pi$, so that its equation is

$$r \cos \theta = p \text{ or } r \cos \theta = -p.$$

**Corollary 3**. If the line is parallel to the initial line,

$$\alpha = \frac{\pi}{2} \text{ or } \frac{3\pi}{2},$$

so that its equation in this case, is

$$r \cos\left(\theta - \frac{\pi}{2}\right) = p$$

or $$r \cos\left(\theta - \frac{3\pi}{2}\right) = p;$$

i.e $$r \sin \theta = p$$

or $$r \sin \theta = -p.$$

**14.2.1 To find the equation of the line joining two points P($r_1$, $\theta_1$) and Q($r_2$, $\theta_2$).**

Let $R(r, \theta)$ be any other point on the line $PQ$.
Then area $\Delta PQR = 0$.

Substituting $\theta_3 = \theta$ and $r_3 = r$ in the area formula in § 14.1.1, the equation of the line $PQ$ is

$$r_1 r_2 \sin(\theta_2 - \theta_1) + r_2 r \sin(\theta - \theta_2) + r r_1 \sin(\theta_1 - \theta) = 0$$

$$\Rightarrow \frac{1}{r}\sin(\theta_2 - \theta_1) = \frac{1}{r_1}\sin(\theta_2 - \theta) + \frac{1}{r_2}\sin(\theta - \theta_1),$$

which is the required equation of the line.

**14.2.2. To find the angle between the lines $r \cos(\theta - \theta_1) = p_1$ and $r \cos(\theta - \theta_2) = p_2$**

Let $AB$ and $CD$ be the given lines as shown in the figure.

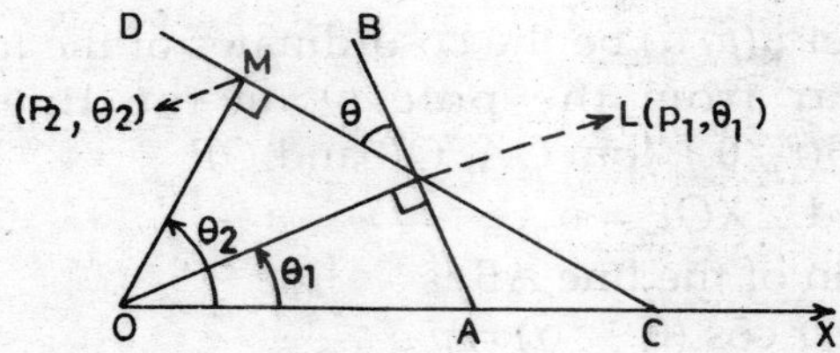


Now, if $\theta$ is the angle between the lines $AB$ and $CD$, then $\theta$ = angle between a pair of perpendicular lines to $AB$ and $CD$, through any point, say $O$.

$$= |\theta_2 - \theta_1|$$

**Corollary 1**. The lines are parallel iff $\theta_2 - \theta_1 = 0$ i.e. iff $\theta_1 = \theta_2$.

**Corollary 2**. The lines are perpendicular iff $\theta = \dfrac{\pi}{2}$

i.e. $\quad$ iff $\theta_2 - \theta_1 = \pm\dfrac{\pi}{2}$

i.e. $\quad$ iff $\theta_2 = \theta_1 \pm \dfrac{\pi}{2}$

**Corollary 3**. The equation of any line, parallel to

$r \cos(\theta - \theta_1) = p$, is, $r \cos(\theta - \theta_1) = p'$,

and the equation of any line, perpendicular to

$$r \cos(\theta - \theta_1) = p \text{ is,}$$

$$r \cos\left(\theta - \left(\theta_1 \pm \frac{\pi}{2}\right)\right) = p''$$

i.e. $\quad r \sin(\theta - \theta_1) = A.$

where $p', p''$ and $A$ are certain constants.

**Example 14.1** (i) Find the co-ordinates of the foot of the perpendicular from the pole on the line joining the points $(r_1, \theta_1)$ and $(r_2, \theta_2)$.

(ii) Find the length of the perpendicular drawn from the point $(r_1, \theta_1)$ on the line $r \cos(\theta - \alpha) = p$.

*Solution*. (i) Let $P(p, \alpha)$ be the co-ordinates of the foot of the perpendicular from the pole $O$ on the line joining $A(r_1, \theta_1)$ and $B(r_2, \theta_2)$. Join $OA$, $OB$ and $OP$.

$\therefore$ $\overline{OP} = p$, and $\angle XOP = \alpha$.

$\therefore$ the equation of the line $AB$ is

$$r \cos (\theta - \alpha) = p \qquad (1)$$

Now $A$ and $B$, both lie on it.

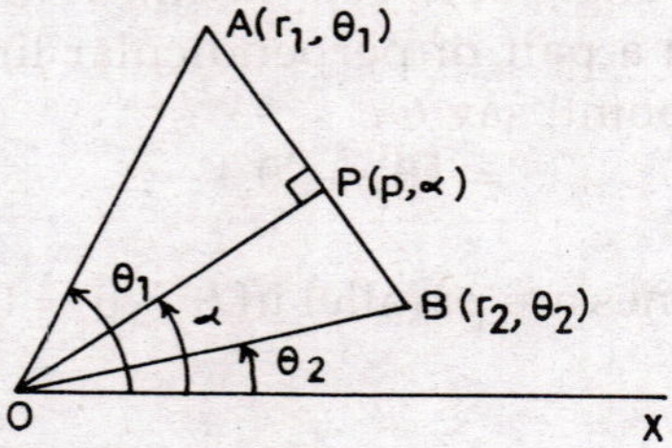


$\therefore$ $r_1 \cos (\theta_1 - \alpha) = p$

and $\quad r_2 \cos (\theta_2 - \alpha) = p$

$\Rightarrow r_1 \cos (\theta_1 - \alpha) = r_2 \cos (\theta_2 - \alpha)$

$\Rightarrow r_1 \cos \theta_1 \cos \alpha + r_1 \sin \theta_1, \sin \alpha = r_2 \cos \theta_2 \cos \alpha +$
$r_2 \sin \theta_2 \sin \alpha$

$\Rightarrow (r_1 \sin \theta_1 - r_2 \sin \theta_2) \sin \alpha = (r_2 \cos \theta_2 - r_1 \cos \theta_1) \cos \alpha$

$$\Rightarrow \qquad \tan\alpha = \frac{r_2 \cos \theta_2 - r_1 \cos \theta_1}{r_1 \sin \theta_1 - r_2 \sin \theta_2}$$

$$\Rightarrow \qquad \alpha = \tan^{-1}\left[\frac{r_2 \cos \theta_2 - r_1 \cos \theta_1}{r_1 \sin \theta_1 - r_2 \sin \theta_2}\right]$$

Also area $\Delta OAB = \frac{1}{2}\overline{AB}.p = \frac{1}{2}p.\sqrt{r_1^2 + r_2^2 - 2r_1r_2 \cos (\theta_1 - \theta_2)}$

and $\quad \Delta OAB = \frac{1}{2}\overline{OA}.\overline{OB}\sin(\angle AOB) = \frac{1}{2}r_1r_2 \sin(\theta_1 - \theta_2)$

$$\therefore \quad \frac{1}{2}p\sqrt{r_1^2 + r_2^2 - 2r_1r_2 \cos (\theta_1 - \theta_2)} = \frac{1}{2}r_1r_2 \sin (\theta_1 - \theta_2)$$

$$\Rightarrow p = \frac{r_1r_2 \sin (\theta_1 - \theta_2)}{\sqrt{r_1^2 + r_2^2 - 2r_1r_2 \cos (\theta_1 - \theta_2)}}$$

Hence the polar co-ordinates of $P$, the foot of perpendicular from $O$ on $AB$, are $(p, \alpha)$ i.e.

$$\left[\frac{r_1 r_2 \sin(\theta_1-\theta_2)}{\sqrt{r_1^2+r_2^2-2r_1r_2\cos(\theta_1-\theta_2)}}, \tan^{-1}\left(\frac{r_2\cos\theta_2-r_1\cos\theta_1}{r_1\sin\theta_1-r_2\sin\theta_2}\right)\right]$$

(ii) Let $AB$ be the line whose equation is

$$r\cos(\theta - a) = p \quad (1)$$

Draw $OL \perp$ on $AB$. Then $OL = p$ and $\angle XOL = \alpha$

Let $P$ $(r_1, \theta_1)$ be the given point.

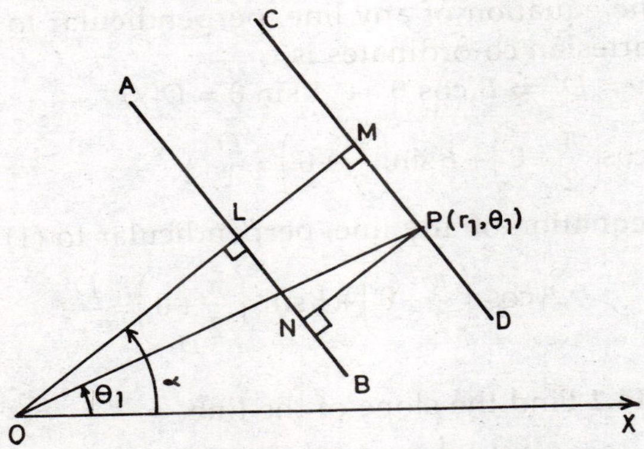


Let $PN = d$ be the length of the perpendicular from $P$ on the line $AB$. Through $P$, draw a line $CPD$ parallel to $AB$ and produce $OL$ to meet it in $M$. Clearly $\angle XOM = \alpha$. Also let $OM = p'$.

$\therefore$ the equation of $CD$ is

$$r\cos(\theta - \alpha) = p'.$$

As $CD$ passes through $P(r_1, \theta_1)$, we have

$$r_1\cos(\theta_1 - \alpha) = p' = p + d$$

$$\Rightarrow \quad d = r_1\cos(\theta_1 - \alpha) - p,$$

which is the required perpendicular distance from $(r_1, \theta_1)$ on the given line. If this distance is negative, we take its absolute value.

### 14.2.3 To find the general equation of a line.

Transforming

$$Ax + By = C$$

(the general equation of a line in Cartesian co-ordinates) to polar co-ordinates, we see that the general equation of any line, in polar co-ordinates, is

$$r(a \cos \theta + B \sin \theta) = C$$

i.e. $$A \cos \theta + B \sin \theta = C/r \quad (1)$$

Hence the equation of any line, parallel to (1) is

$$A \cos \theta + B \sin \theta = D/r.$$

Also the equation of any line, perpendicular to $Ax + By = C$, in Cartesian co-ordinates is,

$$Bx - Ay = D' \Rightarrow B \cos \theta - A \sin \theta = D'/r$$

$$\Rightarrow \quad A\cos\left(\frac{\pi}{2}+\theta\right) + B\sin\left(\frac{\pi}{2}+\theta\right) = \frac{D'}{r}$$

Thus the equation of any line, perpendicular to (1) is

$$A\cos\left(\frac{\pi}{2}+\theta\right) + B\sin\left(\frac{\pi}{2}+\theta\right) = \frac{D'}{r}$$

**Example 14.2** Find the slope of the line

$$\frac{1}{r} = \cos(\theta - \alpha) + e\cos\theta$$

*Solution.* The given equation, when transformed to Cartesian co-ordinates, is

$$l = x(e + \cos \alpha) + y \sin \alpha$$

Hence, the slope of this line is $-(e + \cos \alpha)/\sin \alpha$.

**Example 14.3.** Find the point of intersection of the two lines

$$\frac{l}{r} = \cos(\theta - \theta_1) + e\cos\theta$$

and $$\frac{l}{r} = \cos(\theta - \theta_2) + e\cos\theta$$

*Solution.* Solving these equations simultaneously, we get

$$\cos(\theta - \theta_1) = \cos(\theta - \theta_2).$$

$$\Rightarrow \qquad \theta - \theta_i = \pm (\theta - \theta_2).$$

$$\Rightarrow \qquad \theta = \frac{\theta_2 + \theta_1}{2} \qquad (\text{as } \theta_1 \neq \theta_2)$$

Also then,

$$\frac{l}{r} = \cos\left(\frac{\theta_2 - \theta_1}{2}\right) + e\cos\left(\frac{\theta_2 + \theta_1}{2}\right)$$

$$\Rightarrow r = \frac{l}{\cos\left(\frac{\theta_2 - \theta_1}{2}\right) + e\cos\left(\frac{\theta_2 + \theta_1}{2}\right)}$$

Thus, the point of intersection of the given lines, is

$$\left(\frac{l}{\cos\left(\frac{\theta_2 - \theta_1}{2}\right) + e\cos\left(\frac{\theta_2 + \theta_1}{2}\right)}, \frac{\theta_2 + \theta_1}{2}\right)$$

**Exercise 14.2.**

1. Find the angle between the following pairs of lines:

   (i) $r\cos\left(\theta - \frac{\pi}{2}\right) = p$, $r\cos\left(\theta - \frac{\pi}{3}\right) = p'$.

   (ii) $r\cos\theta = p$, $r\sin\theta = p'$.

   (iii) $r\cos\theta = 2$, $r\cos\left(\theta - \frac{\pi}{6}\right) = 3$.

2. Find the point of intersection of the lines:
   $r\cos(\theta - \theta_1) = p_1$; $r\cos(\theta - \theta_2) = p_2$.

3. Find the equation of the line joining the points $\left(2, \frac{\pi}{3}\right)$ and $\left(3, \frac{\pi}{6}\right)$. Deduce that the points $\left(2, \frac{\pi}{3}\right)$ $\left(3, \frac{\pi}{6}\right)$ and $\left(\frac{6}{3\sqrt{3} - 2}, \frac{\pi}{2}\right)$ are collinear.

4. Show that the equation of the line perpendicular to

$$\frac{l}{r} = \cos(\theta - \alpha) + e\cos\theta$$

is
$$\frac{l'}{r} = \sin(\theta - \alpha) + e\sin\theta$$

and hence find the equation of that perpendicular line which passes through $(r_1, \theta_1)$.

5. Show that the equation of any line, parallel to

$$\frac{l}{r} = \cos(\theta - \alpha) + e\cos\theta,$$

through the pole, is

$$\theta = \tan^{-1}\left(\frac{(\cos\alpha + e)}{\sin\alpha}\right)$$

6. Show that the four lines

$$r(\cos\theta + \sin\theta) = \pm 1$$

and
$$r(\cos\theta - \sin\theta) = \pm 1$$

enclose a square and calculate the length of the side of this square.

(*Hint*: Change to Cartesian co-ordinates).

7. Prove that the equation of any line passing through the intersection of the lines

$$\frac{l}{r} = A\cos\theta + B\sin\theta$$

and
$$\frac{l'}{r} = A'\cos\theta + B'\sin\theta$$

is
$$\frac{l + \lambda l'}{r} = (A + \lambda A')\cos\theta + (B + \lambda B')\sin\theta.$$

(*Hint*: Change to Cartesian co-ordinates).

## 14.3 TO FIND THE EQUATION OF A CIRCLE

**General case**. Let $C(r_1, \theta_1)$ be the centre of a given circle and let its radius be $a$. If now $P(r, \theta)$ is any point on it, applying cosine formula to $\Delta OCP$, we get

$$CP^2 = OC^2 + OP^2 - 2\,\overline{OC}.\overline{OP}\,\cos(\theta - \theta_1)$$

$\Rightarrow \quad a^2 = r_1^2 + r^2 - 2rr_1 \cos(\theta - \theta_1)$

$\Rightarrow \quad r^2 - 2rr_1 \cos(\theta - \theta_1) + (r_1^2 - a^2) = 0, \qquad (1)$

which is the required equation of the circle.

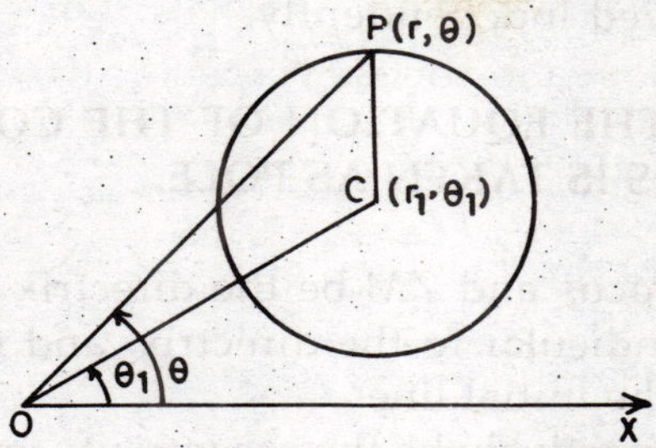


**Corollary 1**. When the pole lies on the circle. In this case, the radius,

$$a = r_1.$$

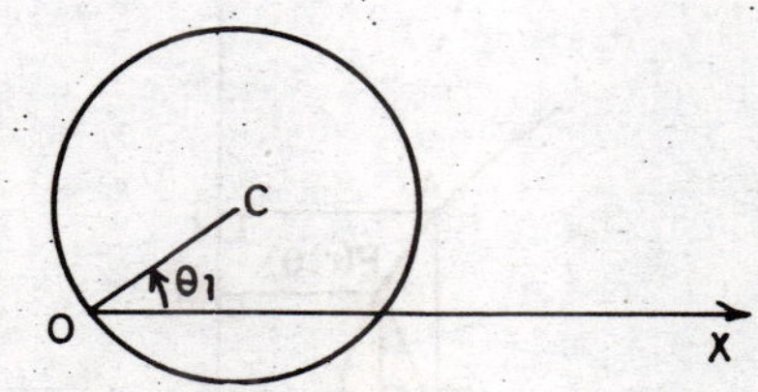


Substituting this in (1), the equation of the circle reduces to

$$r^2 = 2ra \cos(\theta - \theta_1)$$

$$\Rightarrow \quad r = 2a \cos(\theta - \theta_1)$$

**Corollary 2**. When the initial line is a diameter of the circle and the pole lies on the circle. In this case, the radius

$$a = r_1$$

and

$$\theta_1 = 0 \text{ or } \pi.$$

The equation (2) then reduces to

$$r = 2a \cos\theta$$

or

$$r = -2a \cos\theta.$$

**Corollary 3**. When the centre is at the pole. Here the centre

$$(r_1, \theta_1) = (0, \theta_1).$$

Substituting $r_1 = 0$, in (1) we get

$$r^2 = a^2 \Rightarrow r = a$$

as the required equation of the circle.

**Note**. The equations of the circles in the above corollaries can also be derived independently.

## 14.4 TO FIND THE EQUATION OF THE CONIC WHEN THE FOCUS IS TAKEN AS POLE.

Let $S$ be the focus and $ZM$ be the directrix of the conic. Draw $SX$, perpendicular to the directrix and take $S$ as the pole and $SX$ as the initial line.

Let $e$ and $l$ respectively be the eccentricity and the length of the semi – latus rectum of the conic.

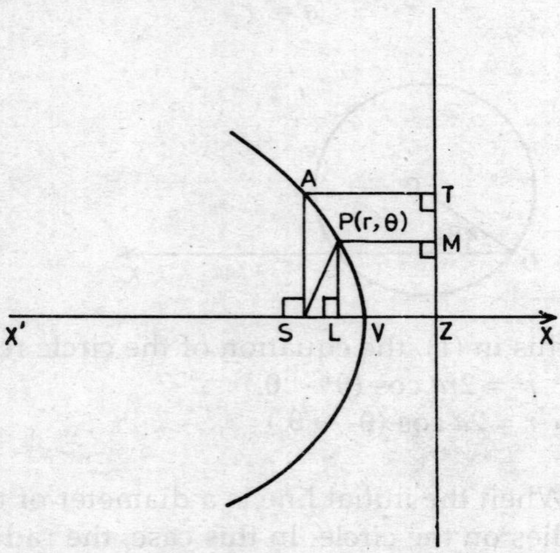


Let $P(r, \theta)$ be any point on the conic, then

$$\overline{SP} = e.\overline{PM},$$

where $PM$ is the perpendicular on the directrix .

$$\Rightarrow \quad r = e.\overline{LZ} = e.\left(\overline{SZ} - \overline{SL}\right)$$

$$= e.\left(\overline{AT} - \overline{SL}\right) = e\left(\frac{\overline{SA}}{e} - \overline{SL}\right)$$

$$= e.\left(\frac{l}{e} - r\cos\theta\right)$$

$$= l - r\,e\cos\theta,$$

$$\Rightarrow \quad r(1 + e\cos\theta) = l$$

$$\Rightarrow \quad \frac{l}{r} = 1 + e\cos\theta$$

which is the required equation of the conic, whose focus is at the pole.

**Note 1**. If instead of taking $SX$ as the initial line, we take $SX'$ as the initial line, $\theta$ is replaced by $(\pi + \theta)$ and the equation of the conic becomes

$$\frac{l}{r} = 1 - e\cos\theta$$

which is taken as the standard equation in some text books on Co-ordinate Geometry.

However, if the reader has to deal with the conic,

$$\frac{l}{r} = 1 - e\cos\theta,$$

we suggest to him to deal with,

$$\frac{l}{r} = 1 + e\cos\theta,$$

taking opposite direction of the initial line and finally replacing $\theta$, $\theta_1$ etc. by $(\theta + \pi)$, $(\theta_1 + \pi)$ respectively.

**Note 2**. In the sequel, we shall always denote the point $(r_1, \theta_1)$ on the conic as the point "$\theta_1$". Clearly its radius vector, $r_1$ is

$$\frac{l}{1 + e\cos\theta_1}$$

**Note 3**. If the axis of the conic is inclined at an angle $\alpha$ to the initial line, its equation can be easily seen to be

$$\frac{l}{r} = 1 + e\cos(\theta + \alpha)$$

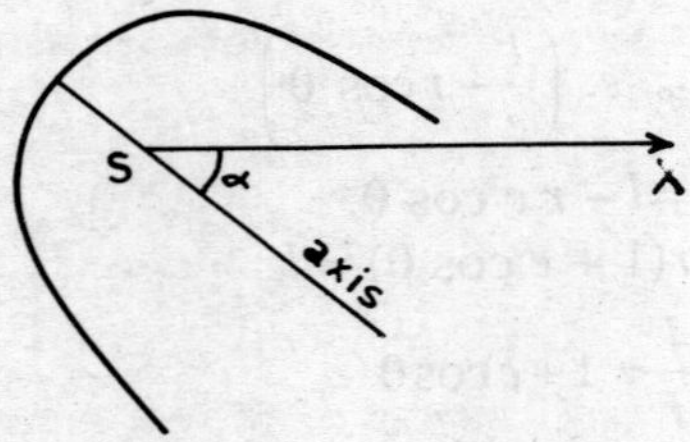


or
$$\frac{l}{r} = 1 + e\cos\ (\theta - \alpha)$$

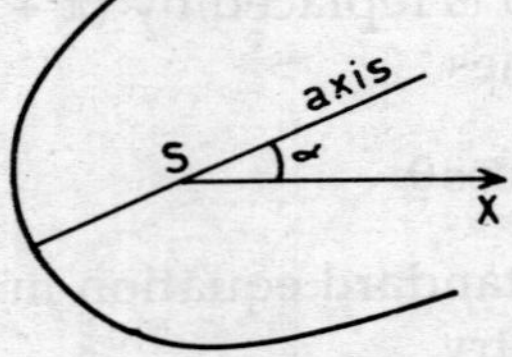


### 14.4.1. To trace the conic

$$\frac{l}{r} = 1 + e\cos\theta$$

We note that as cos θ is periodic with period 2π, it is enough to trace the conic as θ varies from − π to π . Further, as cos ( − θ) = cos θ, the conic is symmetrical w.r.t. the initial line. Hence it is enough to trace the conic as θ varies from 0 to π. We discuss the various cases for different values of the eccentricity, *e*.

**Case I**. $e = 0$. The equation of the conic becomes, $r = l$, which is a circle of radius, *l*, with its centre at the pole, *S*.

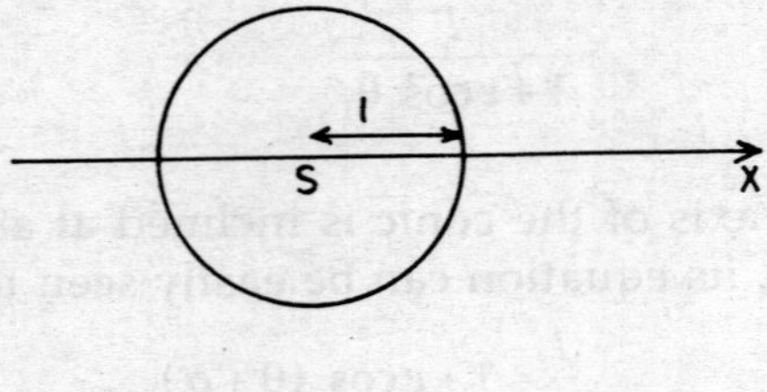

**Case II**. $e = 1$. The equation of the conic is

$$\frac{l}{r} = 1 + \cos\theta$$

$$\Rightarrow \qquad r = \frac{l}{1+\cos\theta}$$

As $\theta$ varies from 0 to $\pi$, $\cos\theta$ decreases from 1 to $-1$; $1 + \cos\theta$ decreases from 2 to 0 and then $r$ increases from $\frac{l}{2}$ to $\infty$.

The conic, which is a parabola, in this case is as shown here:

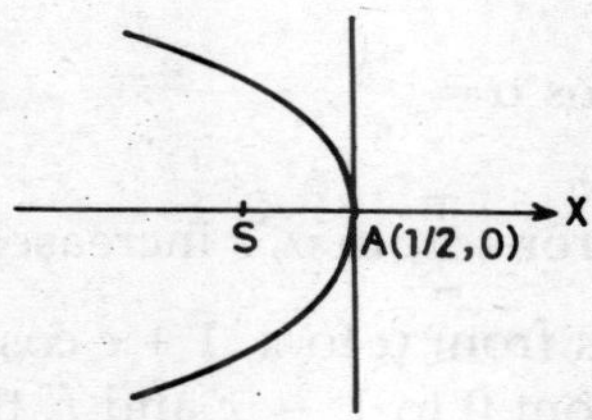


**Case III**. $e < 1$. Here

$$r = \frac{l}{1+e\cos\theta}$$

is the conic.

As $\theta$ varies from 0 to $\pi$, $1 + e\cos\theta$ decreases from $1 + e$ to $1 - e$, whence r increases from $\frac{l}{1+e}$ to $\frac{l}{1-e}$. The conic, an ellipse, in this case, is shown below:

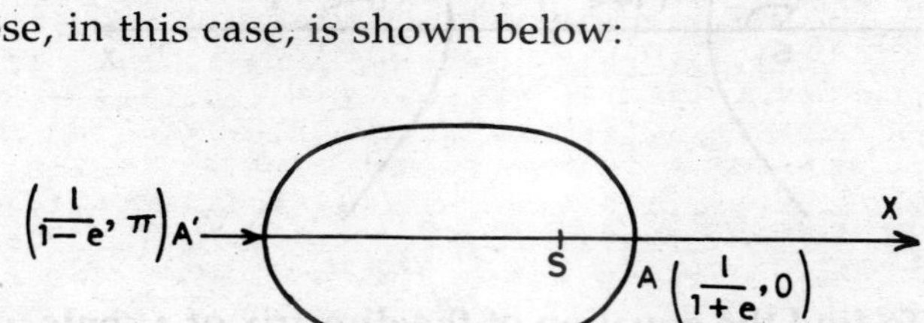


**Case IV.** $e > 1$. Here

$$r = \frac{l}{1 + e\cos\theta}$$

As $\theta$ varies from 0 to $\frac{\pi}{2}$, $1 + e\cos\theta$ decreases from $1 + e$ to 1 and so $r$ increases from $\frac{l}{1+e}$ to $l$. Further as $\theta$ varies from $\frac{\pi}{2}$ to $\pi$, $1 + e\cos\theta$ decreases from 1 to $1 - e(< 0)$. Hence, there exists an angle $\alpha$, $\frac{\pi}{2} < \alpha < \pi$, such that

$$1 + e\cos\alpha = 0$$

$$\text{i.e. } \cos\alpha = \frac{-1}{e}.$$

Hence, as $\theta$ varies from $\frac{\pi}{2}$, to $\alpha$, $r$ increases from $l$ to $\infty$.

Again as $\theta$ varies from $\alpha$ to $\pi$, $1 + e\cos\theta$ remains negative and changes from 0 to $1 - e$ and $r$, therefore changes from $-\infty$ to $\frac{l}{1-e}$.

The conic, a hyperbola, in this case, is as shown:

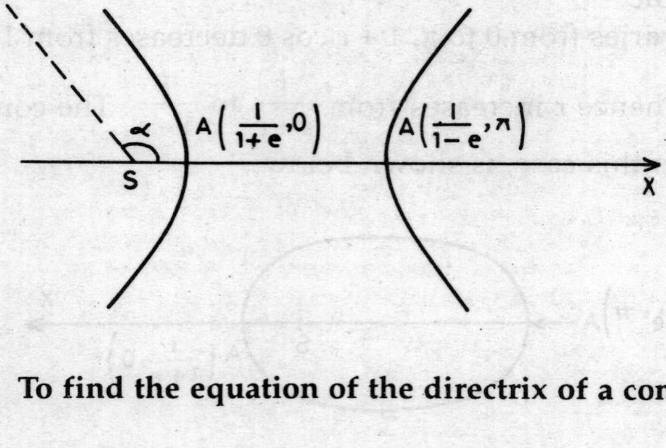


**14.4.2. To find the equation of the directrix of a conic.**

Lest $P(r, \theta)$ be any point on the directrix of the conic,

$$\frac{l}{r} = 1 + e\cos\theta,$$

whose focus, $S$, is the pole. Draw its latus rectum, $SA$; then

$$\overline{SA} = e.\overline{AT}$$

$$= e.\overline{SZ}$$

$$\Rightarrow \quad l = e.r\cos\theta$$

$$\Rightarrow \quad \frac{l}{r} = e\cos\theta,$$

which is the required equation of the corresponding directrix.

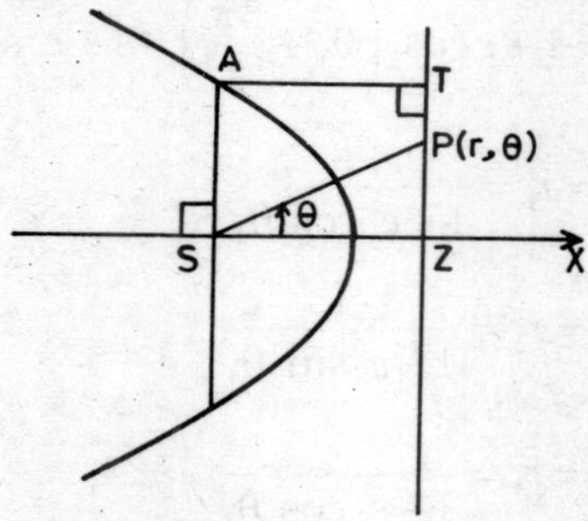


**Example 14.4**. If $PP'$, $QQ'$ be two perpendicular focal chords of the conic,

$$\frac{l}{r} = 1 + e\cos\theta, \text{ prove that}$$

(i) $$\frac{1}{PP'} + \frac{1}{QQ'} = \text{constant},$$

(ii) $$\frac{1}{SP.SP'} + \frac{1}{SQ.SQ'} = \text{constant}.$$

*Solution*. Let the co-ordinates of $P$ be $(r_1, \theta_1)$. Then as $PP'$ and $QQ'$ are perpendicular focal chords, the co-ordinates of $Q$, $P'$ and $Q'$ are $(r_2, \theta_1 + \pi/2)$, $(r_3, \theta + \pi)$, $(r_4, \theta_1 + 3\pi/2)$ respectively. Now $P$, $Q$, $P'$, and $Q'$ lie on the conic,

$$\frac{l}{r} = 1 + e\cos\theta$$

$$\therefore \quad \frac{l}{r_1} = 1 + e\cos\theta_1$$

$$\frac{l}{r_2} = 1 + e\cos\left(\theta_1 + \frac{\pi}{2}\right) = 1 - e\ \sin\ \theta_1,$$

$$\frac{l}{r_3} = 1 + e\cos\ (\theta_1 + \pi) = 1 - e\ \cos\ \theta_1,$$

and
$$\frac{l}{r_4} = 1 + e\cos\left(\theta_1 + \frac{3\pi}{2}\right) = 1 + e\ \sin\ \theta_1$$

Now
$$SP = r_1 = \frac{l}{1 + e\ \cos\ \theta_1}$$

$$SQ = r_2 = \frac{l}{1 - e\ \sin\ \theta_1}$$

$$SP' = r_3 = \frac{l}{1 - e\ \cos\ \theta_1},$$

and
$$SQ' = r_4 = \frac{l}{1 + e\ \sin\ \theta_1}$$

(i) Now
$$PP' = SP + SP' = \frac{2l}{1 - e^2\cos^2\theta}$$

and
$$QQ' = \frac{2l}{1 - e^2\sin^2\theta_1}$$

$$\Rightarrow \frac{1}{PP'} + \frac{1}{QQ'} = \frac{2 - e^2}{2l} \text{ which is a constant.}$$

(ii) Also
$$\frac{1}{SP.SP'} + \frac{1}{SQ.SQ'} = \frac{1}{r_1.r_3} + \frac{1}{r_2.r_4}$$

$$= \frac{1 - e^2\cos^2\theta_1}{l^2} + \frac{1 - e^2\sin^2\theta_1}{l^2}$$

$$= \frac{2-e^2}{l^2} \text{ which is again a constant.}$$

**Example 14.5.** Show that

$$\frac{l}{r} = 1 + e\cos\theta \text{ and } \frac{l}{r} = -1 + e\cos\theta$$

represent the same conic.

*Solution*. Let $r = f(\theta)$ be the equation of any curve. Let $(r, \theta)$ be any arbitrary point on it.

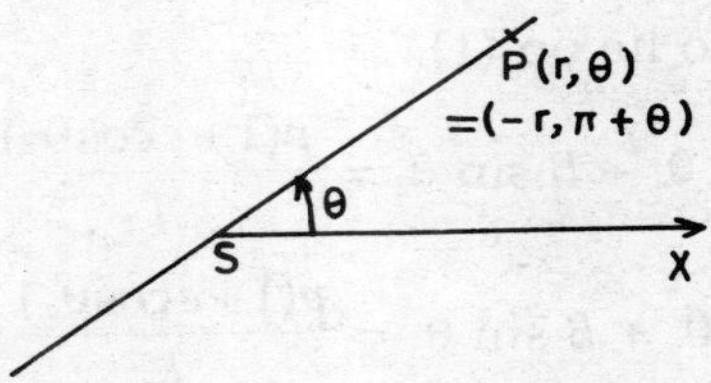


Now $(-r, \pi + \theta)$ is the same point as $(r, \theta)$

$$\therefore \quad -r = f(\pi + \theta)$$

$$\Rightarrow \quad r = -f(\pi + \theta)$$

is the same curve as $r = f(\theta)$. In the present case, the conic is given by

$$r = \frac{l}{1 + e\cos\theta}$$

$$\therefore \quad r = \frac{l}{1 + e\cos\theta} \text{ and } r = \frac{-l}{1 + e\cos(\pi + \theta)}$$

$$= \frac{-l}{1 - e\cos\theta)}$$

represent the same conic.

i.e. $\quad \frac{l}{r} = 1 + e\cos\theta$ and $\frac{l}{r} = -1 + e\cos\theta$

represent the same conic.

**14.4.3. To find the equation of the chord joining the points P ($r_1$, $\theta_1$) and Q ($r_2$, $\theta_2$) on the conic, $l/r = 1 + e\cos\theta$.**

By § 14.2.3, the general equation of a line is

$$A\cos\theta + B\sin\theta = \frac{p}{r} \tag{1}$$

Now $P(r_1, \theta_1)$ and $Q(r_2, \theta_2)$ lie on the conic,

$$\frac{l}{r} = 1 + e\cos\theta,$$

$$\therefore \quad \frac{l}{r_1} = 1 + e\cos\theta_1, \quad \frac{l}{r_2} = 1 + e\cos\theta_2.$$

These points also lie on (1),

$$\therefore \quad A\cos\theta_1 + B\sin\theta_1 = \frac{p(1 + e\cos\theta_1)}{l}$$

and $$A\cos\theta_2 + B\sin\theta_2 = \frac{p(1 + e\cos\theta_2)}{l}.$$

i.e. $$A\cos\theta_1 + B\sin\theta_1 - \frac{p}{l}(1 + e\cos\theta_1) = 0,$$

and $$A\cos\theta_2 + B\sin\theta_2 - \frac{p'}{l}(1 + e\cos\theta_2) = 0.$$

Solving these for $A$ and $B$, we get

$$A = \frac{\frac{p}{l}\left[(\sin\theta_1 - \sin\theta_2) + e\sin(\theta_1 - \theta_2)\right]}{\sin(\theta_1 - \theta_2)}$$

$$= \frac{p}{l}\left[\cos\left(\frac{\theta_1 + \theta_2}{2}\right)\sec\left(\frac{\theta_1 + \theta_2}{2}\right) + e\right] \tag{2}$$

$$B = \frac{\frac{p}{l}\left[\cos\theta_2 - \cos\theta_1\right]}{\sin(\theta_1 - \theta_2)}$$

$$= \frac{p}{l}\sin\left(\frac{\theta_1 + \theta_2}{2}\right)\sec\left(\frac{\theta_1 - \theta_2}{2}\right) \tag{3}$$

From (1), (2) and (3), the required equation of the chord PQ is

$$\frac{l}{r} = \sec\left(\frac{\theta_2 - \theta_1}{2}\right)\cos\left(\theta - \frac{\theta_2 + \theta_1}{2}\right) + e\cos\theta \qquad (4)$$

**Corollary**. If we take $\theta_1 = \alpha - \beta$ and $\theta_2 = \alpha + \beta$, the equation of the chord joining the points "$\alpha - \beta$" and "$\alpha + \beta$" is

$$\frac{l}{r} = \sec\beta\cos(\theta - \alpha) + e\cos\theta.$$

**14.4.4. To find the equation of the tangent at the point "$\theta_1$" of the conic.**

Put $\theta_2 = \theta_1$, in equation (4), of the above article. The equation of the tangent at "$\theta_1$" to the conic,

$$\frac{l}{r} = 1 + e\cos\theta \text{ is, } \frac{l}{r} = \cos(\theta - \theta_1) + e\cos\theta.$$

**Note.** The equation of the tangent to the conic,

$$\frac{l}{r} = 1 + e\cos(\theta - \gamma)$$

at the point "$\theta_1$" is

$$\frac{l}{r} = \cos(\theta - \theta_1) + e\cos(\theta - \gamma).$$

**14.4.5. To find the equation of the normal at the point "$\theta_1$" of the conic $\frac{l}{r} = 1 + e\cos\theta$.**

The equation of any line perpendicular to the tangent

$$\frac{l}{r} = \cos(\theta - \theta_1) + e\cos\theta, \text{ at "}\theta_1\text{" is}$$

$$\frac{L}{r} = \cos\left(\frac{\pi}{2} + \theta - \theta_1\right) + e\cos\left(\frac{\pi}{2} + \theta\right) \text{(by § 14.2.3.)}$$

$$\Rightarrow \qquad \frac{L}{r} = -\sin(\theta - \theta_1) - e\sin\theta$$

Now it passes through $(r_1, \theta_1)$,

where $\quad r_1 = \dfrac{l}{1+e\cos\theta_1}$

$$\therefore \quad \frac{L(1+e\cos\theta_1)}{l} = -\,e\sin\theta_1$$

$$\Rightarrow \quad L = \frac{-el\sin\theta_1}{1+e\cos\theta_1}.$$

Hence, the equation of the normal at "$\theta_1$" to the given conic is,

$$\left(\frac{el\sin\theta_1}{1+e\cos\theta_1}\right).\frac{1}{r} = \sin(\theta-\theta_1) + e\sin\theta$$

**14.4.6. To find the equation of the asymptotes of the conic, $\dfrac{l}{r} = 1 + e\cos\theta$ (e > 1)**

The equation of the tangent at "$\theta_1$" to this conic is

$$\frac{l}{r} = \cos(\theta-\theta_1) + e\cos\theta \qquad (1)$$

Here $e > 1$; therefore, the conic is a hyperbola and it has asymptotes.

Now (1) will be an asymptote to the conic if for $\theta = \theta_1$, $r \to \infty$,

whence $\quad 0 = 1 + e\cos\theta_1 \Rightarrow \cos\theta_1 = \dfrac{-1}{e}$

$$\Rightarrow \sin\theta_1 = \pm\sqrt{1-\cos^2\theta_1} = \pm\frac{\sqrt{e^2-1}}{e}$$

Now (1) $\Rightarrow \quad \dfrac{l}{r} = \cos\theta\cos\theta_1 + \sin\theta\sin\theta_1 + e\cos\theta$

$$= \cos\theta\left(-\frac{1}{e}\right) + \sin\theta\left(\pm\frac{\sqrt{e^2-1}}{e}\right) + e\cos\theta$$

$$\Rightarrow \frac{el}{r} = (e^2-1)\cos\theta \pm \sqrt{e^2-1}\,\sin\theta,$$

which are the required equations of the asymptotes to the given conic.

**14.4.7. To find the equation of the cnord of contact of the point (r′, α) w.r.t. the conic, $\frac{l}{r}=1+e\cos\theta$.**

Let the tangents from $P(r', \alpha)$ to the conic, touch it at $Q(r_1, \theta_1)$ and $R(r_2, \theta_2)$.

Now the equation of the chord joining $\theta_1$ and $\theta_2$ is

$$\frac{l}{r}=\sec\left(\frac{\theta_2-\theta_1}{2}\right)\cos\left(\theta-\left(\frac{\theta_2+\theta_1}{2}\right)\right)+e\cos\theta \quad (1)$$

Also the equations of the tangents at "$\theta_1$" and "$\theta_2$" are

$$\frac{l}{r}=\cos(\theta-\theta_1)+e\cos\theta$$

and $$\frac{l}{r}=\cos(\theta-\theta_2)+e\cos\theta.$$

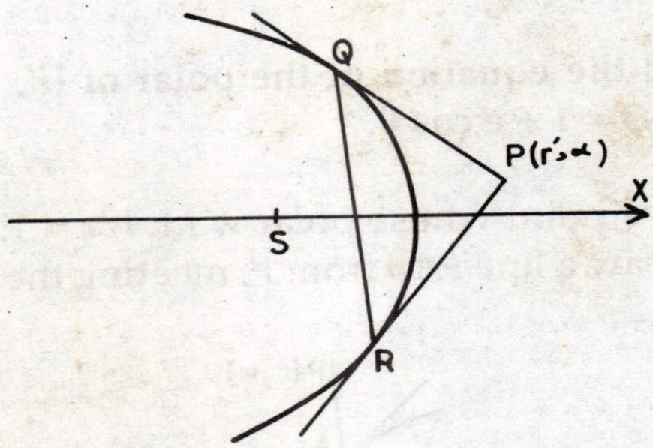


These tangents meet at $P(r', \alpha)$ which by Example 14.3, is given by

$$\left(\frac{l}{\cos\left(\frac{\theta_2-\theta_1}{2}\right)+e\cos\left(\frac{\theta_2+\theta_1}{2}\right)}, \frac{\theta_2+\theta_1}{2}\right)$$

$$\therefore \quad r'=\frac{l}{\cos\left(\frac{\theta_2-\theta_1}{2}\right)+e\cos\left(\frac{\theta_2+\theta_1}{2}\right)}$$

and $$\alpha = \frac{\theta_2 + \theta_1}{2} \qquad (2)$$

$$\Rightarrow \quad \frac{l}{r'} = \cos\left(\frac{\theta_2 - \theta_1}{2}\right) + e\cos\alpha$$

$$\Rightarrow \quad \cos\left(\frac{\theta_2 - \theta_1}{2}\right) = \frac{l}{r'} - e\cos\alpha \ldots \qquad (3)$$

Now, equation (1) can be rewritten as

$$\left(\frac{l}{r} - e\cos\theta\right)\cos\left(\frac{\theta_2 - \theta_1}{2}\right) = \cos\left(\theta - \left(\frac{\theta_2 + \theta_1}{2}\right)\right).$$

Using (2) and (3), this becomes

$$\left(\frac{l}{r} - e\cos\theta\right)\left(\frac{l}{r'} - e\cos\alpha\right) = \cos\,(\theta - \alpha) \qquad (4)$$

which is the required equation of the chord of contact of the point $(r', \alpha)$ w.r.t. the given conic.

### 14.4.8. To find the equation of the polar of (r′, α) w.r.t. the conic, l/r = 1 + e cos θ.

Let $P\,(r', \alpha)$ be a point whose polar w.r.t. $l/r = 1 + e\cos\theta$ is to be found. Draw a line $PAB$ from $P$, meeting the given conic at $A$ and $B$.

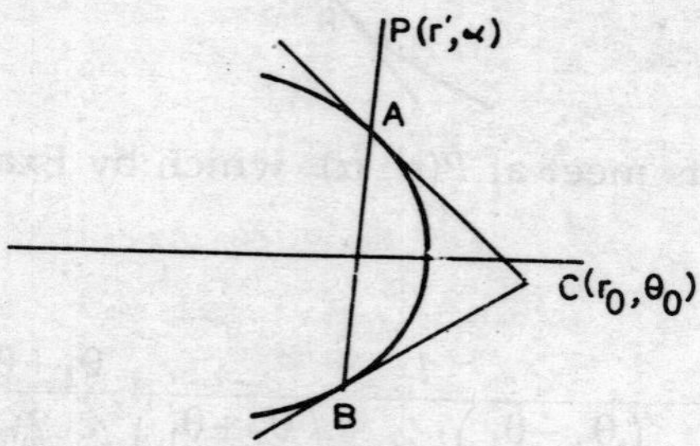


Let the tangents at $A$ and $B$ meet at $C(r_0, \theta_0)$. We have to find the locus of the point $C$.

The equation of $AB$, the chord of contact of the point $C$ w.r.t. the given conic, is,

$$\left(\frac{l}{r} - e\cos\theta\right)\left(\frac{l}{r_0} - e\cos\theta_0\right) = \cos(\theta - \theta_0).$$

It passes through $(r', \alpha)$.

$$\Rightarrow \quad \left(\frac{l}{r'} - e\cos\alpha\right)\left(\frac{l}{r_0} - e\cos\theta_0\right) = \cos(\alpha - \theta_0).$$

$\therefore$ The locus of $C(r_0, \theta_0)$ is,

$$\left(\frac{l}{r'} - e\cos\alpha\right)\left(\frac{l}{r} - e\cos\theta\right) = \cos(\alpha - \theta)$$

$$\Rightarrow \quad \left(\frac{l}{r} - e\cos\theta\right)\left(\frac{l}{r'} - e\cos\alpha\right) = \cos(\theta - \alpha),$$

which is the required equation of the polar of $P(r', \alpha)$ w.r.t. the given conic.

We note that this equation is same as the equation (4) of the previous article.

### 14.4.9 To find the equation of the director circle of the conic, l/r = 1 + e cos θ.

Let the tangents at "$\theta_1$" and "$\theta_2$" to the given conic be at right angles and let them meet at $(r', \alpha)$. We are required to find the locus of $(r', \alpha)$.
(see definition of the *director circle* of a conic in Question 5 of Exercise 8.7).

Now, the equations of the tangents at "$\theta_1$" and "$\theta_2$" to the conic, $\frac{l}{r} = 1 + e\cos\theta$; are

$$\frac{l}{r} = \cos(\theta - \theta_1) + e\cos\theta$$

and $\quad \frac{l}{r} = \cos(\theta - \theta_2) + e\cos\theta$, respectively.

As in § 14.4.7, $(r', \alpha)$, their point of intersection, is given by

$$\left.\begin{aligned} \frac{l}{r'} &= \cos\left(\frac{\theta_2 - \theta_1}{2}\right) + e\cos\alpha \\ \text{and} \quad \alpha &= \frac{\theta_2 + \theta_1}{2} \end{aligned}\right\} \quad (1)$$

Again the slopes of these tangents, by Example 14.2, are $\frac{-(e+\cos\,\theta_1)}{\sin\,\theta_1}$ and $\frac{-(e+\cos\,\theta_2)}{\sin\,\theta_2}$ respectively.

These tangents, being at right angles, we get

$$\frac{(e+\cos\,\theta_1)(e+\cos\,\theta_2)}{\sin\,\theta_1 \sin\,\theta_2} = -1$$

$$\Rightarrow \quad e^2 + (\cos\,\theta_2 + \cos\,\theta_1)e + \cos\,(\theta_2 - \theta_1) = 0$$

$$\Rightarrow \quad e^2 + 2\cos\left(\frac{\theta_2+\theta_1}{2}\right)\cos\left(\frac{\theta_2-\theta_1}{2}\right)e + 2\cos^2\left(\frac{\theta_2-\theta_1}{2}\right) - 1 = 0$$

$$\Rightarrow \quad e^2 + 2\cos\alpha\left(\frac{l}{r'} - e\cos\alpha\right)e + 2\left[\frac{l}{r'} - e\cos\alpha\right]^2 - 1 = 0$$

using equation (1).

$$\Rightarrow \quad e^2 + 2e\cos\alpha\left(\frac{l}{r'} - e\cos\alpha\right) + 2\left[\frac{l^2}{r'^2} - \frac{2le\cos\alpha}{r'} + e^2\cos^2\alpha\right] - 1 = 0$$

$$\Rightarrow \quad (e^2 - 1)r'^2 - 2\,ler'\cos\alpha + 2l^2 = 0.$$

$\therefore$ the locus of $(r', \alpha)$ is

$$(e^2 - 1)r^2 - 2ler\cos\theta + 2l^2 = 0,$$

which is the required equation of the director circle of the given conic.

**Example 14.6.** Find the condition that the line

$$\frac{l}{r} = A\cos\theta + B\sin\theta$$

may touch the conic

$$\frac{l}{r} = 1 + e\cos\theta$$

*Solution.* Suppose the given line touches the given conic at the point "$\theta_1$".

Then the equation of the tangent at "$\theta_1$" to $l/r = 1 + e\cos\theta$, is

$$\frac{l}{r} = \cos(\theta - \theta_1) + e\cos\theta$$

$$\Rightarrow \quad \frac{l}{r} = (e + \cos\theta_1)\cos\theta + \sin\theta_1 . \sin\theta$$

But this is the same as, $\frac{l}{r} = A\cos\theta + B\sin\theta$.

$\therefore \quad e + \cos\theta_1 = A$ and $\sin\theta_1 = B$.

Eliminating $\theta_1$, we get

$$(A - e)^2 + B^2 = 1,$$

which is the reuired condition.

**Example 14.7**. In a parabola with (focus $S$), the tangents at any points $P$ and $Q$ on it meet in $T$. Prove that

(i) $SP.SQ = ST^2$

(ii) The triangles $SPT$, $SQT$ are similar.

*Solution.* Let the points $P$ and $Q$ be "$\theta_1$" and "$\theta_2$" on the parabola, $\frac{l}{r} = 1 + \cos\theta$ (Here: $e = 1$)

If $T$ be $(r', \alpha)$, then

$$\alpha = \frac{\theta_2 + \theta_1}{2}$$

and $$\frac{l}{r'} = \cos\left(\frac{\theta_2 - \theta_1}{2}\right) + \cos\left(\frac{\theta_2 + \theta_1}{2}\right), \quad \text{by § 14.4.7.}$$

$$= 2\cos\frac{\theta_2}{2}.\cos\frac{\theta_1}{2} \qquad \text{(i)}$$

Now $P$ and $Q$ lie on the conic, $\frac{l}{r} = 1 + \cos\theta$

$$\therefore \quad \frac{l}{SP} = \frac{l}{r_1} = 1 + \cos\theta_1 = 2\cos^2\frac{\theta_1}{2}$$

and $\dfrac{l}{SQ} = \dfrac{l}{r_2} = 1 + \cos\theta_2 = 2\cos^2\dfrac{\theta_2}{2}$.

$$\therefore \quad \frac{l^2}{SP.SQ} = \frac{l}{r_1}.\frac{l}{r_2} = 4\cos^2\frac{\theta_1}{2}.\cos^2\frac{\theta_2}{2} = \frac{l^2}{r'^2} = \frac{l^2}{ST^2}$$

using (1)

$\Rightarrow$ $SP.SQ = ST^2$. This is part (i).

(ii) Now $SP.SQ = ST^2$ can be rewritten as

$$\frac{SP}{ST} = \frac{ST}{SQ}$$

$\Rightarrow$ $\Delta SPT$ and $\Delta SQT$ are similar.

**Example 14.8**. A circle, passing through the focus of a conic, whose latus rectum is $2l$, meets the conic in four points whose distances from the focus are $r_1$, $r_2$, $r_3$ and $r_4$. Show that

$$\frac{1}{r_1} + \frac{1}{r_2} + \frac{1}{r_3} + \frac{1}{r_4} = \frac{2}{l}$$

*Solution*. Let the equation of the conic be

$$\frac{l}{r} = 1 + e\cos\theta \tag{1}$$

By § 14.3, the equation of the circle passing through the pole (which is also the focus of the conic in this case) is

$r = 2a \cos(\theta - \theta_1)$

where $a$ is the radius of the circle.

$\Rightarrow r = 2a[\cos\theta\cos\theta_1 + \sin\theta\sin\theta_1)$ (2)

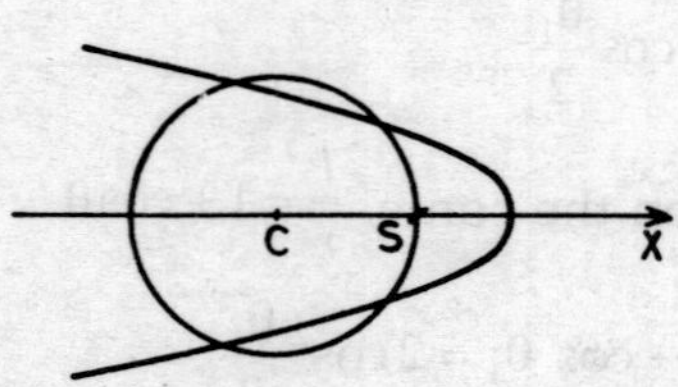


We eliminate $\theta$ from (1) and (2).

Now (2) can be rewritten as

$$[r - 2a \cos \theta_1 \cos \theta]^2 = 4a^2 \sin^2 \theta_1 . \sin^2 \theta \tag{3}$$

From (1), $\cos\theta = \dfrac{l-r}{er}$,

$\therefore$ (3) becomes

$$\left(r - 2a\cos\theta_1 . \frac{(l-r)}{er}\right)^2 - 4a^2 \sin^2\theta_1 \left(1 - \frac{(l-r)^2}{e^2r^2}\right) = 0$$

$\Rightarrow (er^2 - 2a \cos \theta_1 (l-r)^2 - 4a^2 \sin^2\theta . [e^2r^2 - (l-r)^2] = 0.$

$\Rightarrow e^2r^4 + 4ae \cos \theta_1 . r^3 + (4a^2 - 4ael \cos \theta_1 - 4a^2e^2 \sin^2 \theta_1)r^2$

$$- 8a^2lr + 4a^2l^2 = 0 \tag{4}$$

As $r_1, r_2, r_3, r_4$ are its roots,

$$\Sigma r_1r_2r_3 = \frac{8a^2l}{e^2} \tag{5}$$

and

$$r_1r_2r_3r_4 = \frac{4a^2l^2}{e^2} \tag{6}$$

Dividing (5) by (6), we obtain,

$$\frac{1}{r_1} + \frac{1}{r_2} + \frac{1}{r_3} + \frac{1}{r_4} = \frac{2}{l}$$

Hence the result.

**Example 14.9**. Prove that the locus of the foot of the perpendicular from the focus of the conic,

$l/r = 1 + e \cos \theta$, on a tangent to it is either a circle or a straight line.

*Solution*. Let the tangent at a point $P(r_1, \theta_1)$ to the given conic meet the perpendicular from the focus $S$ at $T$ and let $T$ be $(r, d)$. We want to find the locus of $T(r', \alpha)$.

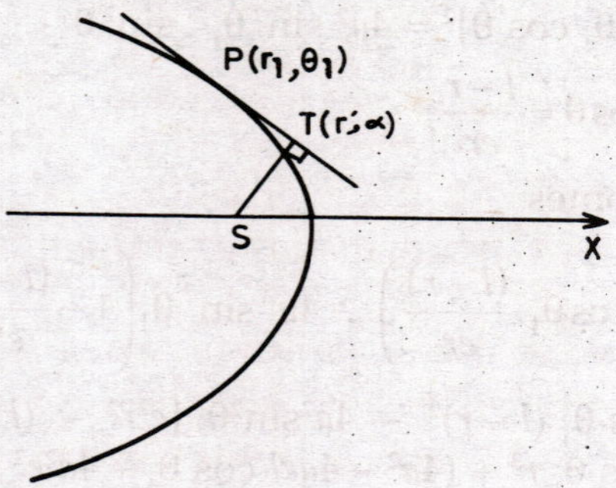


The equation of the tangent at $P(r_1, \theta_1)$ to the conic is

$$\frac{l}{r} = \cos(\theta - \theta_1) + e\cos\theta \qquad (1)$$

It passes through $T(r', \alpha)$

$$\therefore \quad \frac{l}{r'} = \cos(\alpha - \theta_1) + e\cos\alpha \qquad (2)$$

Also the equation of any line perpendicular to (1) is

$$\frac{L}{r} = \cos\left(\frac{\pi}{2} + \theta - \theta_1\right) + e\cos\left(\frac{\pi}{2} + \theta\right)$$

$$\Rightarrow \quad L = r[-\sin(\theta - \theta_1) - e\sin\theta]$$

It passes through the pole, $S(0, \theta)$

$$\therefore \quad L = 0.$$

Hence the equation of the line $ST$ is

$$\sin(\theta - \theta_1) + e\sin\theta = 0.$$

Again since $ST$ passes through $T(r', \alpha)$,

$$\therefore \sin(\alpha - \theta_1) + e\sin\alpha = 0 \qquad (3)$$

Eliminating $\theta_1$, from (2) and (3), we obtain,

$$\left(\frac{l^2}{r'} - e\cos\alpha\right)^2 + e^2\sin^2\alpha = 1$$

$$\Rightarrow \quad \frac{l^2}{r'^2} + e^2\cos^2\alpha - \frac{2el\cos\alpha}{r'} + e^2\sin^2\alpha = 1$$

$\therefore$ the locus of $(r', \alpha)$ is

$(e^2 - 1)r^2 - 2\,elr\cos\theta + l^2 = 0.$

On changing to Cartesian co-ordinates, this equation becomes

$$(e^2 - 1)(x^2 + y^2) - 2\,elx + l^2 = 0.$$

If the given conic is an ellipse or a hyperbola, $e \neq 1$; then the locus is a circle and if the given conic is a parabola, $e = 1$, then the locus is $x = l/2e$, which is a straight line.

**Example 14.10**. If $S$ be the focus and $P$ and $Q$ be two points on a conic such that angle $PSQ$ is constant, prove that the locus of the point of intersection of the tangents at $P$ and $Q$ is a conic section whose focus is $S$.

*Solution*. Let the equation of the conic with focus $S$ as pole be

$$\frac{l}{r} = 1 + e\cos\theta \qquad (1)$$

Let $P(r_1, \theta_1)$ and $Q(r_2, \theta_2)$ be points on (1).
Let the tangents at $P$ and $Q$ to (1) meet at $T(r', \alpha)$.

Then $\alpha$ and $r'$ are given by

$$\alpha = \frac{\theta_2 + \theta_1}{2}$$

and
$$\frac{l}{r'} = \cos\left(\frac{\theta_2 - \theta_1}{2}\right) + e\cos\alpha \qquad \text{(by § 14.4.7)}$$

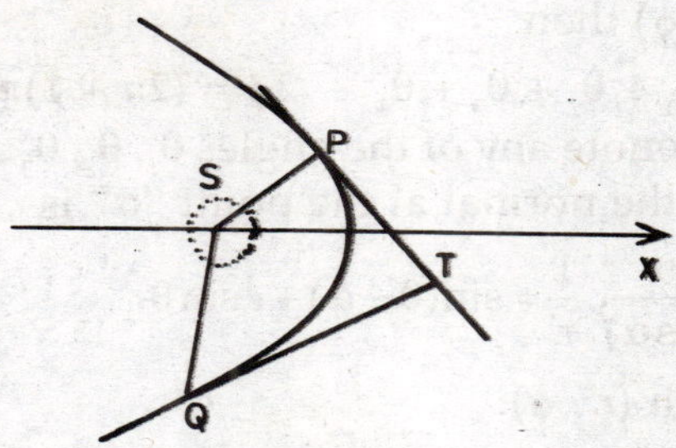


Let $\quad \angle PSQ = 2\beta$ (*a* constant)

Now $\quad \angle QSP + \angle PSQ = 2\pi$

$\Rightarrow \quad \theta_2 - \theta_1 + 2\beta = 2\pi$

$$\Rightarrow \quad \frac{\theta_2 - \theta_1}{2} = \pi - \beta$$

$$\therefore \quad \frac{l}{r'} = \cos(\pi - \beta) + e\cos\alpha = -\cos\beta + e\cos\alpha.$$

Hence, the locus of $(r', \alpha)$ is

$$\frac{l}{r} = -\cos\beta + e\cos\theta$$

$$\Rightarrow \quad \frac{l\sec\beta}{r} = -1 + (e\sec\beta)\cos\theta. \qquad (2)$$

But $(r, \theta)$ and $(-r, \pi + \theta)$ represent the same point.

$\therefore$ the equation (2) reduces to

$$\frac{l\sec\beta}{-r} = -1 + (e\sec\beta)\cos(\pi + \theta)$$

$$\Rightarrow \quad \frac{l\sec\beta}{r} = 1 + (e\sec\beta)\cos\theta$$

$$\Rightarrow \quad \frac{l'}{r} = 1 + e'\cos\theta$$

where $\quad l' = l\sec\beta$ and $e' = e\sec\beta$;

which is also a conic having $S$ as its focus.

**Example 14.11**. Show that if the normals at the points, whose vectorial angles are $\theta_1$, $\theta_2$, $\theta_3$ and $\theta_4$ on $l/r = 1 + e\cos\theta$, meet at the point $(r', \phi)$ then

$$\theta_1 + \theta_2 + \theta_3 + \theta_4 - 2\phi = (2n+1)\pi$$

*Solution*. Let $\alpha$ denote any of the angles $\theta_1$, $\theta_2$, $\theta_3$ and $\theta_4$. Then the equation of the normal at the point "$\alpha$" is

$$\frac{le\sin\alpha}{(1+e\cos\alpha)}\cdot\frac{1}{r} = \sin(\theta - \alpha) + e\sin\theta \qquad \text{by § 14.4.5.}$$

It passes through $(r', \phi)$.

$$\therefore \quad \frac{le\sin\alpha}{(1+e\cos\alpha)r'} = \sin(\phi - \alpha) + e\sin\phi$$

$$\Rightarrow \quad le\sin\alpha = [\sin(\phi - \alpha) + e\sin\phi](r' + r'e\cos\alpha)$$

$$\Rightarrow \quad le\sin\alpha - (\sin\phi\cos\alpha - \cos\phi\sin\alpha + e\sin\phi)$$

$$.(r' + r'e\cos\alpha) = 0. \qquad (1)$$

Let $$t = \tan\alpha/2.$$

Then $$\sin\alpha = \frac{2t}{1+t^2} \text{ and } \cos\alpha = \frac{1-t^2}{1+t^2}$$

Equation (1) then, becomes

$$le\frac{2t}{1+t^2} - \left[\sin\phi\frac{(1-t^2)}{(1+t^2)} - \cos\phi.\frac{2t}{1+t^2} + e\sin\phi\right]$$

$$\left[r' + r'\frac{e(1-t^2)}{(1+t^2)}\right] = 0$$

which simplifies to

$$r'\sin\phi\,(1-e)^2t^4 + 2[el + (1-e)r'\cos\phi]t^3$$
$$+ 2[el + (1+e)r'\cos\phi]t - r'\sin\phi\ (1+e)^2 = 0.$$

This is a fourth degree equation in $t = \tan\alpha/2$ and hence its roots are $\tan\theta_i/2$, $i = 1, 2, 3, 4$.

Now, from Trigonometry.

$$\tan\left(\frac{\theta_1}{2} + \frac{\theta_2}{2} + \frac{\theta_3}{2} + \frac{\theta_4}{2}\right) = \frac{s_1 - s_3}{1 - s_2 + s_4}$$

$$= \frac{\dfrac{-2[el + (1-e)r'\cos\phi] + 2\,[el + (1+e)r'\cos\phi]}{r'\sin\phi\,(1-e)^2}}{1 - 0 - \dfrac{(1+e)^2}{(1-e)^2}}$$

$$= \frac{4e\cot\phi}{-4e} = -\cot\phi = \tan\left(\frac{\pi}{2} + \phi\right)$$

$$\Rightarrow \quad \frac{\theta_1 + \theta_2 + \theta_3 + \theta_4}{2} = n\pi + \frac{\pi}{2} + \phi$$

$$\Rightarrow \quad \theta_1 + \theta_2 + \theta_3 + \theta_4 - 2\phi = (2n+1)\pi$$

where $n$ is any integer.

**Example 14.12**. If a normal is drawn at one extremity of the latus rectum, prove that the distance from the focus of the other point in which it meets the conic is

$$\frac{(1+3e^2+e^4)l}{1+e^2-e^4}.$$

*Solution*. The equation of the normal at $L(l,\pi/2)$ to the conic

$$\frac{l}{r}=1+e\cos\theta \qquad (1)$$

is
$$\frac{le}{r}=e\sin\theta-\cos\theta \qquad (2)$$

(see § 14.4.5)

We eliminate $r$ from (1) and (2) to get

$$e\sin\theta - \cos\theta = e\,[1+e\cos\theta]$$

$\Rightarrow$ $e\sin\theta = e+\cos\theta(e^2+1).$

$\Rightarrow$ $[e+\cos\theta(e^2+1)]^2 = e^2\sin^2\theta = e^2(1-\cos^2\theta)$

$\Rightarrow$ $(1+3e^2+e^4)\cos^2\theta + 2e(1+e^2)\cos\theta = 0.$

$\therefore$ $\cos\theta = 0$ or $\cos\theta = \dfrac{-2e(1+e^2)}{1+3e^2+e^4}$

Now $\cos\theta = 0 \Rightarrow \theta = \pi/2$ which corresponds to the point $L$.

$\therefore$ for the other end, $\cos\theta = \dfrac{-2e(1+e^2)}{1+3e^2+e^4}$

and then from (1)

$$\frac{l}{r}=1+e\left\{\frac{-2e(1+e^2)}{1+3e^2+e^4}\right\}=\frac{1+e^2-e^4}{1+3e^2+e^4}$$

$$\Rightarrow \qquad r=\frac{\left(1+3e^2+e^4\right)l}{1+e^2-e^4}$$

This is the distance from the focus of the other point in which the normal at one end of the latus rectum meets the conic.

## Exercise 14.3

1. In any conic, its semi-latus recturm is a harmonic mean between the segments of any focal chord.
2. A circle of given radius passing through the focus $S$ of a given conic, intersects it in $A$, $B$, $C$ and $D$. Show that $SA.SB.SC.SD$ is constant.

(*Hint*: See Example 14.8).

3. If the circle, $r + 2a \cos \theta = 0$, cuts the conic $l/r = 1 + e \cos(\theta - \alpha)$ in four real points, find the equation in $r$ which determines the distances of these points from the pole. Also show that if their algebraic sum equals $2a$, the eccentricity of the conic is, $2 \cos \alpha$.
4. Find the condition that the line, $l/r = A \cos \theta + B \sin \theta$ may touch the conic, $l/r = 1 + e \cos(\theta - \alpha)$.
(*Hint*: See Example 14.6)
5. Prove that the two conics $l_1/r = 1 + e_1 \cos \theta$ and $l_2/r = 1 + e_2 \cos(\theta - \alpha)$ will touch each other, if
$$l_1^2(1-e_2^2)+l_2^2(1-e_1^2)=2l_1l_2(1-e_1e_2\cos\alpha).$$
(*Hint*: They have a common tangent at "$\theta_1$" and eliminate $\theta_1$.)
6. Prove that the locus of the middle points of focal chords of a conic section is a conic section.
7. Chords of a conic substend a constant angle $2\alpha$ at the focus. Show that the locus of the points where the chords are met by the internal bisector of the angle that they substend at the focus is
$$\frac{l\cos\alpha}{r}=1+e\cos\alpha\cos\theta$$
8. In any conic, prove that the tangents at the ends of any focal chord meet on the directrix.
9. If $A, B, C$ be the points on a parabola, and the tangents at these points meet in $A', B', C$, show that
$$SA.SB.SC = SA'.SB'.SC',$$
$S$ being the focus of the parabola.
10. A conic is described having the same focus and eccentricity as the conic $l/r = 1 + e\cos\theta$ and the two conics touch at the point $\theta = \alpha$. Prove that the length of its latus rectum is
$$\frac{2l(1-e^2)}{e^2+2e\cos\alpha+1}$$
11. If $\alpha, \beta, \gamma$ be the vectorial angles of three points on
$$\frac{2a}{r}=1+\cos\theta,$$

the normals at which are concurrent, prove that

$$\tan\frac{\alpha}{2}+\tan\frac{\beta}{2}+\tan\frac{\gamma}{2}=0$$

12. If the normals at $\alpha$, $\beta$, $\gamma$ on $l/r = 1 + \cos\theta$ meet at the point $(r_1, \theta_1)$; then $2\theta_1 = \alpha + \beta + \gamma$.
13. Show that the equation of the pair of tangents from $P$ $(r', \theta')$ to the conic $l/r = 1 + e\cos\theta$ is

$$(S^2-1)(S'^2-1)=P^2,$$

where $S=\frac{l}{r}-e\cos\theta, S'=\frac{l}{r'}-e\cos\theta'$

and $P = SS' - \cos(\theta - \theta')$.

14. If the normal at $P$ to a conic cuts the axis at $G$, prove that

$$\overline{SG} = e\ \overline{SP}.$$

15. The tangents at the points $P$ and $Q$ on a conic, intersect in $T$ and the chord $PQ$ meets the directrix at $R$. Prove that the angle $TSR$ is a right angle.
16 Given the focus and directrix of a conic, show that the polar of a given point w.r.t. it, passes through a fixed point.
17. If $SM$, $SN$ be perpendiculars from the focus $S$ on the tangent and the normal at any point on the conic, $l/r = 1 + e\cos\theta$ and $ST$, the perpendicular on $MN$, show that the locus of $T$ is:

$$r(e^2 - 1) = el\cos\theta.$$

# ANSWERS

## CHAPTER 1

### Exercise 1.1

(2) 1, 5
(3) (7, 13), (23/9, 37/9)
(5) 0
(7) $|p-1|$

### Exercise 1.2

1. 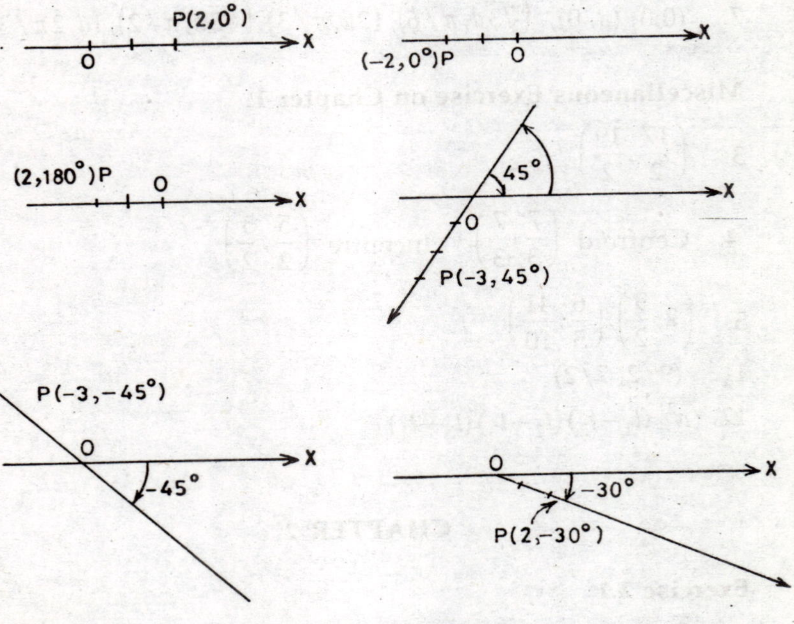


2. For example (− 3, − 45°) is same as (−3, − 45° + $2k\pi$) or (3, − 45° + π + $2k\pi$) where $k$ is any integer etc. etc.

3. (i) $(1, \sqrt{3})$, (−2, 0), (0, 3).

(ii) $\left(-\sqrt{3}, -1\right), \left(\frac{-3\sqrt{3}}{2}, \frac{3}{2}\right)$

4. (i) $\left(2\sqrt{2}, \frac{\pi}{4}\right), \left(2\sqrt{2}, \frac{7\pi}{4}\right), \left(2\sqrt{2}, \frac{3\pi}{4}\right), \left(2\sqrt{2}, \frac{5\pi}{4}\right)$

(ii) (2, 0°), (2, π/2).

5. (i) $r = \frac{-C}{A\cos\theta + B\sin\theta}$,

(ii) $r = a$.

(iii) $r^2 = a^2 \cos 2\theta$.

6. (i) $x^2 + y^2 - x = 0$,

(ii) $x^2 + y^2 - y = 0$,

(iii) $(x^2 + y^2)^2 = x^2 - y^2$,

(iv) $(x^2 + y^2)^2 = 2xy$.

7. (0,0) (a, 0), $\left(\sqrt{3}a, \pi/6\right), (2a, \pi/3), \left(\sqrt{3}a, \pi/2\right), (a, 2\pi/3)$

**Miscellaneous Exercise on Chapter 1.**

3. $\left(\frac{17}{2}, \frac{19}{2}\right)$.

4. Centroid $\left(\frac{7}{3}, \frac{7}{3}\right)$, Incentre $\left(\frac{5}{2}, \frac{5}{2}\right)$.

5. $\left(2, \frac{9}{2}\right), \left(\frac{6}{5}, \frac{41}{10}\right)$.

11. (9/2, 7/2)

12. $a^2 \left| (t_1 - t_2)(t_2 - t_3)(t_3 - t_1) \right|$

## CHAPTER 2

**Exercise 2.1.**

1. $|y| = 2|x|$ or $y^2 = 4x^2$.
3. $x^2 + y^2 + 10x + 9 = 0$.

4. (i) $5x - 2y - 29 = 0$ or $5x - 2y - 9 = 0$.
   (ii) $5x - 2y - 19 = 0$.
5. $y^2 = 6\,|x| + 9$ or $(y^2 - 9)^2 = 36x^2$.
6. $x^2 + y^2 - x(x_1 + x_2) - y(y_1 + y_2) + x_1x_2 + y_1y_2 = 0$
   where $A$ is $(x_1, y_1)$ and $B$ is $(x_2, y_2)$.
8. $(x + y)(x^2 + y^2) = axy$.
9. $\frac{x^2}{c^2} - \frac{y^2}{a^2 - c^2} = 1$.
10. $\frac{x^2}{a^2} + \frac{y^2}{b^2} = 1$ or $\frac{x^2}{b^2} + \frac{y^2}{a^2} = 1$.

**Exercise 2.2**

1. (i) $\{(2, 1)\}$,
   (ii) $\{(0, 0)\}$,
   (iii) $\{(x, y) \mid x + 2y = 5\}$,
   (iv) $\phi$ (empty set),
   (v) $\{(2, 1), (-2, -1)\}$.
2. (i) $a^2(1 + m^2) \geq c^2$, (a) $a^2(1 + m^2) = c^2$, (b) $a^2(1 + m^2) < c^2$.
   (ii) $b^2 + m^2a^2 \geq c^2$, (a) $b^2 + m^2a^2 = c^2$, (b) $b^2 + m^2a^2 < c^2$.
   (iii) $b^2 + c^2 \geq m^2a^2$, (a) $b^2 + c^2 = m^2a^2$, (b) $b^2 + c^2 < m^2a^2$.
   (iv) $a \geq mc$, (a) $a = mc$, (b) $a < mc$.

**Exercise 2.3**

1. (i) Any line through the origin.
   (ii) The lines $2x + 3y - 1 = 0$ and $x + 2y + 5 = 0$.
   (iii) The points (1, 2) and (–1, 2).
2. $T_2 \subset T_1 \subset T_3$.

**Miscellaneous Exercise on chapter 2.**

3. $a^2l^2 - b^2m^2 = n^2$
5. $20x^2 + 36y^2 = 45$
6. $(x^2 + y^2)^3 = 4a^2x^2y^2$

7. (i) $x + a = 0$

   (ii) $(6a - 2c)\,x = a^2 - c^2$

8. $x^2 + y^2 - 5x - 5y + 12 = 0$

9. $(x - \alpha)^2 + (y - \beta)^2 = k^2$

10. $y = m\,x + c/3$.

## CHAPTER 3

### Exercise 3.1.

1. $Y^2 = 4X$

2. (1, 0).

3. $3X^2 - 2XY + 4Y^2 = 1$.

4. $\dfrac{x^2}{2/3} + \dfrac{y^2}{2} = 1$

5. $\theta = \tan^{-1}\left(\dfrac{-A}{B}\right)$, given by $\sin\theta = \dfrac{-A}{\sqrt{A^2 + B^2}}$

   $\cos\theta = \dfrac{B}{\sqrt{A^2 + B^2}}$, $k = \dfrac{-c}{\sqrt{A^2 + B^2}}$

6. $X^2 - 14XY - 7Y^2 - 2 = 0$.

7. (i) $\theta = \dfrac{\pi}{4}, \left(\dfrac{1}{2}\right)y^2 - \left(\dfrac{1}{3}\right)x^2 = 1$

   (ii) $\theta = \tan^{-1}\left(\dfrac{1}{2}\right), \dfrac{x^2}{4} + \dfrac{y^2}{9} = 1$.

10. (–3/2, –5/2)

11. $3X^2 + 2\,Y^2 = 5$

12. $18X^2 - 7\,Y^2 + 54 = 0$.

13. $2XY + a^2 - b^2 - c^2 = 0$.

19. $\sqrt{5}\;Y^2 = 4X$

20. $Y^2 = 1$, $XY = 1$.

# CHAPTER 4

## Exercise 4.1.

1. (i) $y = -5.$

(ii) $x = 2.$

(iii) $9x + 4y - 17 = 0.$

(iv) $y - b = \left(\frac{d-b}{c-a}\right)(x-a)$

(v) $y - 3 = -5(x - 2).$

(vi) $\frac{x}{-4} + \frac{y}{5} = 1$

(vii) $x \pm y = \pm a.$

where $a > 0$ (e.g each of the lines

$x + y = 2,$

$x - y = 2,$

$x + y = 2,$

and $x - y = -2$

makes intercepts 2 on the axes).

(viii) $\frac{x+1}{\frac{1}{2}} = \frac{y-1}{\sqrt{\frac{3}{2}}} = r$

(ix) $-\frac{\sqrt{3}}{2}x + \frac{y}{2} = 5$

2. (i) $2x - (t_1 + t_2)y + 2at_1t_2 = 0,$

(ii) $\frac{x}{a}\cos\left(\frac{\theta_1 + \theta_2}{2}\right) + \frac{y}{b}\sin\left(\frac{\theta_1 + \theta_2}{2}\right) = \cos\left(\frac{\theta_1 - \theta_2}{2}\right)$

(iii) $\frac{x}{a}\cos\left(\frac{\theta_1 - \theta_2}{2}\right) - \frac{y}{b}\sin\left(\frac{\theta_1 + \theta_2}{2}\right) = \cos\left(\frac{\theta_1 + \theta_2}{2}\right)$

(iv) $x + t_1t_2y - a(t_1 + t_2) = 0,$

3. $\theta = \frac{\pi}{12}$

4. $x - y + 5 = 0.$

7. $bx = 2ay, \quad 2bx = ay$

8. $4\sqrt{2}$

11. $\alpha/x + \beta/y = 2$

**Exercise 4.2.**

1. (i) $\frac{2}{3}$

   (ii) $\frac{a}{b}$

   (iii) $-\cot\theta$;

   (iv) $m$.

2. (i) $\frac{-5}{2}, \frac{5}{3}$

   (ii) $\frac{-c}{a}, \frac{c}{b}$

   (iii) $p\sec\theta, p\operatorname{cosec}\theta$;

   (iv) $\frac{-c}{m}, c$

3. (i) $\frac{2}{\sqrt{13}}x + \frac{3}{\sqrt{13}}y = \frac{4}{\sqrt{13}}, p = \frac{4}{\sqrt{13}}$

   and $\theta = \tan^{-1}\left(\frac{3}{2}\right)$

   (ii) $\frac{-x}{\sqrt{2}} + \frac{-y}{\sqrt{2}} = \frac{1}{\sqrt{2}}, p = \frac{1}{\sqrt{2}}, \theta = \frac{5\pi}{4}$

   (iii) $\frac{y}{\sqrt{1+m^2}} - \frac{mx}{\sqrt{1+m^2}} = \frac{c}{\sqrt{1+m^2}}, p = \frac{c}{\sqrt{1+m^2}}$

   and $\theta = \tan^{-1}\left(\frac{-1}{m}\right)$

4. (i) $\frac{7}{13}$

(ii) $\dfrac{c}{\sqrt{1+m^2}}$;

(iii) $|p|$;

(iv) $\left|\dfrac{ab}{\sqrt{a^2+b^2}}\right|$.

5 (a) (i) $\dfrac{9}{13}$;

(ii) $\left|\dfrac{m+2-c}{\sqrt{1+m^2}}\right|$;

(iii) $|\cos\theta - 2\sin\theta + p|$;

(iv) $\left|\dfrac{2a-b-ab}{\sqrt{a^2+b^2}}\right|$.

(b) (i) $\left|\dfrac{12\alpha+5\beta-7}{13}\right|$;

(ii) $\left|\dfrac{m\alpha-\beta+c}{\sqrt{1+m^2}}\right|$;

(iii) $|\alpha\cos\theta + \beta\sin\theta - p|$;

(iv) $\left|\dfrac{\alpha b+\beta a-ab}{\sqrt{a^2+b^2}}\right|$.

6. $(m_1 + m_2)(x^2 - y^2) - 2xy(1 - m_1m_2) = 0$.

8. $y = 6x$, $2y = 3x$.

10. $(26/7, 26/7)$, $(-24/7, -24/7)$

**Exercise 4.3.**

1. (i) $\tan^{-1}\left(\dfrac{17}{6}\right)$;

(ii) $\frac{\pi}{2}$;

(iii) $\frac{\pi}{2}$.

2. (i) $3x - 4y + k = 0$, $4x + 3y + k' = 0$;
(ii) $3x + 4y + k = 0$, $4x - 3y + k' = 0$;
(iii) $lx + my + k = 0$, $mx - ly + k' = 0$;
(iv) $lx - my + k = 0$, $mx + ly + k' = 0$;
(v) $y = m_1x + k$, $my + x + k' = 0$.

3. (a) (i) $3x - 4y + 5 = 0$, $4x + 3y - 10 = 0$;
(ii) $3x + 4y - 11 = 0$, $4x - 3y + 2 = 0$;
(iii) $l(x - 1) + m(y - 2) = 0$,
$m(x - 1) - l(y - 2) = 0$;
(iv) $l(x - 1) - m(y - 2) = 0$,
$l(y - 2) + m(x - 1) = 0$;
(v) $y - 2 = m_1 (x - 1)$, $m_1(y - 2) + (x - 1) = 0$.
(b) (i) $3(x - x') - 4(y - y') = 0$,
$4(x - x') + 3(y - y') = 0$ etc. etc.

4. (i) $\frac{5}{13}$ (ii) $\frac{|c-d|}{\sqrt{1+m^2}}$.
5. $5x - 12y - 26 = 0$, $5x - 12y + 78 = 0$
6. $x + 11y + 9 = 0$, $11x - y - 23 = 0$
8. $(a - c)x + (b - d)y = 0$
9. $2x - y + 6 = 0$, $2x - y - 14 = 0$
10. $7x + 23y = 53$, $23x - 7y = 9$
11. $y + 1 = (2 \pm \sqrt{3})(x + 1)$
13. $(b/3, b/3)$

**Exercise 4.4.**

1. (i) $\left(0, \frac{1}{4}\right)$;

(ii) $(0, 0)$;

(iii) $\left(\frac{c_2 - c_1}{m_1 - m_2}, \frac{m_1c_2 - m_2c_1}{m_1 - m_2}\right)$

2. (i) $y = 3x$,
(ii) $38x + 5y + 43 = 0$,
(iii) $14x + 21y + 5 = 0$,
(iv) $105x - 70y + 44 = 0$,
(v) $x + y = 0$
(vi) $y = \pm x$

3. $x + 2y - 3 = 0$, $15x + 9y - 25 = 0$, $7x - 7y - 1 = 0$.

4. $\frac{x}{b} + \frac{y}{a} = \frac{a}{b}$.

5. $3x + 2y = 48$.

6. (i) $(1, -1)$ (ii) $(0, 3)$.

7. $3x - 2y + 17 = 0$.

8. $119x + 102y - 125 = 0$.

9. $y = 2$, $4x + 3y = 10$

10. $x = y$

13. $5/\sqrt{2}$, $5/\sqrt{13}$, $5/\sqrt{13}$

**Exercise 4.5.**

1. (i) Yes, $(11, 5)$,
(ii) No,
(iii) Yes, $\left(-\frac{1}{2}, -\frac{1}{2}\right)$.

3. (i) $\sum m_1(c_2 - c_3) = 0$,
(ii) $(a^3 + b^3 + c^3) - 3abc = 0$.

4. $k = -45$.

5. $\left(\frac{23}{21}, \frac{20}{21}\right)$.

8. $\left(\frac{11}{13}, \frac{-7}{13}\right)$.

9. $(-4, -7)$

**Exercise 4.6.**

1. (i) Collinear,
   (ii) Not collinear,
   (iii) Not collinear.
2. (i) $ad + be + cf = af + bc + de$,
   (ii) $t_2 t_1 = -1$.

**Exercise 4.7.**

1. $112x + 64y + 3 = 0$.
2. $y = 0, x = 0$.
3. $x = \pm y$.
4. $\frac{y - m_1 x}{\sqrt{1+m_1^2}} = \pm\left(\frac{y - m_2 x}{\sqrt{1+m_2^2}}\right)$.
6. $\left(\frac{23}{2}, 11\right)$.
8. $5x + 12y + 4 = 0$

**Miscellaneo 1s Exercise on Chapter 4.**

1. $bx + ay - 2ab = 0, 2a, 2b, \left|\frac{2ab}{\sqrt{a^2+b^2}}\right|$.
2. $3x - y - 7 = 0$, $x + 3y - 9 = 0$.
3. $x - y = 0$, $x + y - 2 = 0$.
4. (i) Passes through the point,
   (ii) Perpendicular,
   (iii) Parallel.
5. (iii).
6. (ii).
8. $(a - a')x + (b - b')y = 0$,
   $(a + a')x + (b + b')y + c + c' = 0$
11. (b) $56x + 32y + 39 = 0$.
12. $13(x^2 + y^2) - 20x + 30y - 14 = 0$.

13. 4 units.

24. $\frac{\pi}{4}$ or $\pi + \sin^{-1}\left(\frac{\sqrt{2}}{10}\right)$

29. $x^2 + y^2 \pm 2ay - a^2 = 0$

30. $x^2 \pm 2xy\cot\alpha - y^2 = a^2$

## CHAPTER 5

### Exercise 5.1.

1. (i) $x - 6y = 0, x + y = 0;$
   (ii) $x = 0, y = 0;$
   (iii) $x - y = 0, x - y = 0.$
   (iv) Imaginary lines.
   (v) $x + y = 0, x + 2y = 0, x + 3y = 0.$
3. $(ab' - a'b)^2 + 4(ah' - a'h)(bh' - b'h) = 0.$
7. $6x - 5y + 14 + 0, 6x + 5y - 56 = 0.$
8. $(x^2 + y^2)\cos 2\theta + 2xy = 0.$

10. $\left[\frac{2r(hq - bp)}{3(aq^2 - 2hpq + bp^2)}, \frac{2r(hp - aq)}{3(aq^2 - 2hpq + bp^2)}\right]$

### Exercise 5.2.

1. (i) $\theta = \tan^{-1}\left(\frac{5}{3}\right)$,

   (ii) $\theta = \tan^{-1}\left(\frac{9}{5}\right)$,

   (iii) $\frac{\pi}{2}$,

(iv) $\frac{\pi}{2}$.

3. (i) $x^2 + 4xy - y^2 = 0$,
(ii) $5x^2 - 22xy - 5y^2 = 0$,
(iii) $x = y$ and $x = -y$, i.e. $x^2 - y^2 = 0$,
(iv) $xy = 0$.

6. $m(a - b) = h(1 - m^2)$.

**Exercise 5.3.**

1. (i) Yes,
(ii) No.

2. $\theta = \tan^{-1}\left(\frac{5}{12}\right)$.

3. $x + 3y + 2 = 0$, $2x + y + 4 = 0$ $(-2, 0)$, $\frac{\pi}{4}$.

4. (i) $k = 0, \frac{15}{2}$;
(ii) $k = -5, \frac{-35}{4}$.

5. $\lambda = 2$.

7. (i) $2x^2 - xy - 6y^2 = 0$,
(ii) $2(x - x_1)^2 - (x - x_1)(y - y_1) - 6(y - y_1)^2 = 0$.

8. $(-2, 2)$.

9. $5\sqrt{2}a$

10. (i) $abc + 2fgh - af^2 - bg^2 - ch^2 = 0$,
(ii) $h = \sqrt{ab}$ and $g\sqrt{b} = f\sqrt{a}$ or $h = -\sqrt{ab}$ and $g\sqrt{b} = f\sqrt{a}$,
(iii) $h^2 - ab = 0, hf - bg = 0, f^2 - bc = 0$.

11. $a(x - 2\alpha)^2 + 2h(x - 2\alpha)(y - 2\beta) + b(y - 2\beta)^2 = 0$

12. 3, $2\sqrt{5}$.

14. $L - L' = 0$
$L + L' - c = 0$

**Exercise 5.4.**

2. $x^2 - xy - 2y^2 = 0.$

4. $\theta = \tan^{-1}\left(\frac{\sqrt{977}}{24}\right)$

5. $7x^2 - 2xy - y^2 = 0.$

7. $c^2 (x^2 + y^2) = a^2 (y - mx)^2$
$c^2 = a^2 (1 + m^2)$

**Miscellaneous Exercise on Chapter 5.**

**Section A**

3. (i) $(x - 1)^2 - 3(x - 1)(y - 4) + 2(y - 4)^2 = 0,$
(ii) $2(x - 1)^2 + 3(x - 1)(y - 4) + (y - 4)^2 = 0.$

4. $(-a, 0), (a, 0), x^2 + y^2 = a^2.$

14. $h(m^2 - l^2) + lm(a - b) = 0$

**Section B**

5. (i) $f_2(x, y) = hx^2 - (a - b)xy - hy^2 = 0,$
$f_3(x,y) = (a - b)x^2 + 4hxy - (a - b)y^2 = 0.$

9. $(b + d)(ad + be) + (e - a)^2 (a + c + e) = 0.$

17. $k^2$ does not lie between $a$ and $b$

19. $5x^2 - 24xy - 5y^2 = 0, \frac{36}{13}$

**CHAPTER 6**

**Exercise 6.1.**

1. (i) $(x + 1)^2 + (y - 2)^2 = 9;$
(ii) $(x - a)^2 + (y - b)^2 = (a + b)^2;$
(iii) $x^2 + y^2 - 4x - 6y + 8 = 0;$

(iv) $x^2 + y^2 \pm 2\sqrt{a^2 - d^2}\,.y - d^2 = 0.$

2. (i) real, (3, – 1), $\sqrt{13}$;

(ii) real, $\left(\frac{7}{4}, -2\right), \frac{\sqrt{145}}{4}$;

(iii) point circle, (–2, –3);

(iv) imaginary circle.

3. (i) $x^2 + y^2 - 5 = 0$,

(ii) $x^2 + y^2 - ax - by = 0$.

4. (i) $x^2 + y^2 - 5 = 0$,

(ii) $x^2 + y^2 + 34x + 26y - 62 = 0$.

5. (i) $x^2 + y^2 - 4x - 2y - 5 = 0$,

(ii) $x^2 + y^2 - ax - by = 0$,

(iii) $x^2 + y^2 - ax - by = 0$.

6. (i) $5x^2 + 5y^2 - 14x - 14y - 58 = 0$,

(ii) $x^2 + y^2 - 2x - 6y - 26 = 0$,

(iii) $x^2 + y^2 - y - 7 = 0$.

7. $b(x^2 + y^2) - (a^2 + b^2)\, y = 0$.

8. $y = x + 1$.

10. All these points lie on $x^2 + y^2 - 6x + 4y - 7 = 0$.

12. $13x^2 + 13y^2 - 20x + 30y - 14 = 0$. Centre $= \left(\frac{10}{13}, -\frac{15}{13}\right)$,

Radius $= \sqrt{3}$.

13. $(a^2 + b^2)\,(x^2 + y^2) + 2\,(ac + bd)\,x + 2\,(bc - ad)\,y + (c^2 + d^2 - r^2 a^2 - r^2 b^2) = 0$.

15. $y^2 + 2bx = a^2 + b^2$

18. (3, 0); (-4, 1)

19. $x^2 + y^2 - 2x - 2y + 1 = 0$

$x^2 + y^2 - 12x + 12y + 36 = 0$

$x^2 + y^2 - 6x - 6y + 9 = 0$

$x^2 + y + 4x + 4y + 4 = 0$

**Exercise 6.2.**

1. (i) inside,

(ii) outside,

(iii) on the circle.

2. (i) $x = 3 \cos \theta, y = 3 \sin \theta$;
   (ii) $x = 1 + 3 \cos \theta, y = 2 + 3 \sin \theta$.
   (iii) $x = a + r \cos \theta, y = b + r \sin \theta$.

3. (i) any point is $(2 \cos \theta, 2 \sin \theta)$.
   The points are (0, 2), (2,0), (-2, 0) $(\sqrt{2}, -\sqrt{2})$ respectively
   (ii) Any point is $(1 + 4 \cos \theta, -2 + 4 \sin \theta)$
   The points are (1, 2), (5, – 2), (–3. – 2) and $(1 + 2\sqrt{2}, -2 - 2\sqrt{2})$, respectively

4. (i) (0, 1), (3, -2) $3\sqrt{2}$;
   (ii) (1,2), (–2, 3), $\sqrt{10}$.

5. Same as 4 (i).

6. $3x^2 + 3y^2 - 2x + 30y + 26 = 0$.

7. $x = y, \sqrt{2[2c^2 - (a-b)^2]}$.

8. $9x^2 + 9y^2 + 31x - 28y + 4 = 0$.

9. (i) $7x^2 + 7y^2 - 16y - 1 = 0$;
   (ii) $11x^2 + 11y^2 - x - 25y - 2 = 0$.

10. (i) $6x^2 + 5y^2 - 11x - 13y + 22 = 0$,
    (ii) $(1 + m^2)(x^2 + y^2) - 2ax - 2amy = 0$,
    (iii) $(l^2 + m^2)(x^2 + y^2) + 2lnx + 2mny - a^2 (l^2 + m^2) + 2n^2 = 0$
    (iv) $x^2 + y^2 + 2gx + c = 0$.

11. (i) $2x^2 + 2y^2 + 2x + 6y + 1 = 0$,
    (ii) $x^2 + y^2 + x - y - 12 = 0$.

13. $\left(-\dfrac{ad + bc}{2ab}, -\dfrac{bd + ac}{2ab}\right)$

15. $2ab / \sqrt{a^2 + b^2}$

16. $x^2 + y^2 = a^2 - b^2$

18. $\dfrac{\alpha}{x} + \dfrac{\beta}{y} = 2$

**Exercise 6.3.**

1. (i) $2x + y = 5, x - 2y = 0$:

(ii) $x - 2y = 0, 2x + y = 0,$

(iii) Point $(0, -1)$, $x - 4y - 4 = 0$, $4x + y + 1 = 0$,

Point $(-2, -1)$, $x + 4y + 6 = 0$, $4x - y + 7 = 0$.

(iv) $x \cos\theta + y \sin\theta = a$; $x \sin\theta - y \cos\theta = 0$.

2. $x^2 + y^2 - 2\alpha x - 2\beta y = 0$.

4. (1, 1).

5. (i) $\frac{-c}{\sqrt{2}}, \frac{c}{\sqrt{2}}$

(ii) $\left(\frac{-amk}{\sqrt{1+m^2}}, \frac{ak}{\sqrt{1+m^2}}\right)$, where $k = \pm 1$.

6. (i) $a^2(l^2 + m^2) = n^2$,

(ii) $(gl + fm - n)^2 = (g^2 + f^2 - c)(l^2 + m^2)$,

(iii) $(cp + dq - a^2)^2 = b^2 (p^2 + q^2)$.

7. (i) $k = \pm\sqrt{50}$.

(ii) $4x + 3y + 19 + 0$, $4x + 3y - 31 = 0$; $(-1, -5)$, $(7, 1)$ respectively.

8. $(x - 3)^2 + (y - 4)^2 = 16$.

10. $x^2 + y^2 \pm 5x - 4y + 4 = 0$.

11. $x^2 + y^2 - 2x - 2y + 1 = 0$.

12. $x + y = \pm 2\sqrt{2}, x - y + 4 = \pm 2\sqrt{2}$

13. $13x^2 + 13y^2 - 13x - 61y + 70 = 0$.

14 $my + x \pm a\sqrt{1 + m^2} = 0$.

15. $lg + mf = n$

18. 1, 5.

21. $x^2 + y^2 - 4x + 4y = 0$

$x^2 + y^2 + 8x - 12y + 34 = 0$

23. $(x + 3)^2 + (y + 3)^2 = 72$

$(x + 12)^2 + (y - 2)^2 = 32$

$(x - 2)^2 + (y + 2)^2 = 32$

$(2x - 9)^2 + (2y - 9)^2 = 18$

**Exercise 6.4.**

1. (i) 1,

(ii) 2,

(iii) $\dfrac{\sqrt{a(\alpha^2+\beta^2)+2g\alpha+2f\beta+c}}{\sqrt{a}}$.

5. $10(x^2+y^2)-66x+11y+66=0$.
6. $(x^2+y^2-a^2)(p^2+q^2)=(px+qy+2)^2$
8. $(-1, 0), (3/2, 5/2)$.

**Exercise 6.5.**

2. $x^2+y^2-16x-18y-4=0$.
3. $19x^2+19y^2+14x-46y+40=0$.
4. $k(x^2+y^2)+(a-c)y-ck=0$.
5. $x^2+y^2-4x+2y+1=0$.
6. $x^2+y^2=1, 4x^2+4y^2-15x-4=0$.
7. (a) (i) $\tan^{-1}\left(\dfrac{1}{3}\right)$.

(ii) $\tan^{-1}\left(\dfrac{\sqrt{-c}(g-g')}{gg'-c}\right)$.

9. $x^2+y^2-6x-6y+9=0$
10. $x^2+y^2-cx-by+a^2=0$

**Exercise 6.6.**

1. (i) $2x-3y-5=0$,
   (ii) $11x+16y-1=0$.
2. $x^2+y^2-2ax=0$.
6. $(-4,3), 3\sqrt{3}$.
7. $(0, 0)$.
   $x^2+y^2=c$.
9. $\left(\dfrac{-17}{5}, \dfrac{11}{5}\right)$.
13. $3x\pm\sqrt{7}y=8$
    $x\pm\sqrt{15}y=8$
    $(8, 0), (8/3, 0)$

14. $5(x^2 + y^2) - 8x\ 14y - 32 = 0.$
15. $x^2 + y^2 - 18x - 16y + 120 = 0.$

**Exercise 6.7.**

1. $y$-axis and $x$-axis respectively.

**Exercise 6.8.**

1. (i) $(-1, 1), \left(\frac{1}{5}, \frac{8}{5}\right),$

   (ii) $(-1, 1), \left(\frac{1}{5}, \frac{8}{5}\right),$

   (iii) $\left(-3 - \frac{1}{\sqrt{2}}, -3 + \frac{1}{\sqrt{2}}\right), \left(-3 + \frac{1}{\sqrt{2}}, -3 - \frac{1}{\sqrt{2}}\right).$
2. $2x^2 + 2y^2 - x - 7y = 0.$
3. $9(x^2 + y^2) - 14x - 30y + 16 = 0.$
4. (i) $x^2 + y^2 + \lambda(4x + 2y + 5) = 0,$
   (ii) $x^2 + y^2 - 2x - 4y + 5 + \lambda(x - y - 5) = 0.$
5. $x^2 + y^2 - 2ax - 2by + a^2 + b^2 + \lambda[2x(a - c) + 2y\ (b - d) + (c^2 + d^2 - a^2 - b^2)] = 0.$
6. $x^2 + y^2 + 2x - 2y + 2 = 0$, and $x^2 + y^2 - 2x + 6y = 0.$
7. (i) $ax - by = 0,$

   (ii) $x^2 + y^2 + \frac{c}{2}\left(\frac{x}{a} + \frac{y}{b} + 2\right) = 0$, where $c$ is a parameter.
8. $\left(\frac{-cg}{f+g}, \frac{-cf}{f+g}\right)$

12. $x + y + 4 = 0$
    $(-5, -6)$

**Miscellaneous Exercise on Chapter 6.**

**Section A**

1. (b) (i) $x^2 + y^2 + 2x + 4y = 0,$

(ii) $x^2 + y^2 + 2x + 4y - 4 = 0$.

(c) $x^2 + y^2 - 8x - 10y + 1 = 0$.

3. $-\frac{\pi}{4}$.

10. $4x - 3y + 6 = 0$.

14. $25(x^2 + y^2) - 20x + 2y - 60 = 0$.

**Section B**

16. $x + 1 = 0,\ 4x - 3y = 5,\ y - 1 = 0,\ 3x + 4y = 5$.

18. $\frac{\pi}{6}, \frac{\pi}{6} \pm \cos^{-1}(-1/3)$

## CHAPTER 7

**Exercise 7.1.**

1. A straight line.
2. The pt. (0, 0).
3. The lines, $x + y = 0$ and $x - y = 0$.
4. The lines, $x = 0$ and $y = 0$.
5. Circle with centre at (0, 0) and radius = $\sqrt{2}$ .
6. Circle with centre at (3, -2) and radius = $2\sqrt{5}$
7. Ellipse with vertices at (± 3, 0) and (0, ± 2).
8. Ellipse with vertices at (± 2, 0) and (0, ± 3).
9. Parabola with $a = 2$ and vertex at (0, 0).
   Also focus = (2, 0).
10. Parabola with $a = -\frac{5}{2}$ and vertex at (0, 0).
   Also focus = $\left(-\frac{5}{2}, 0\right)$.
11. Parabola with $a = 1$, vertex at (0, 0) and focus at (0, 1).

12. Parabola with $a = -\frac{3}{4}$, vertex at (0, 0) and focus at $\left(0, -\frac{3}{4}\right)$
13. Hyperbola with vertices at (± 3, 0).
14. Hyperbola with vertices at (0, ± 2).
15. Rectangular hyperbola.

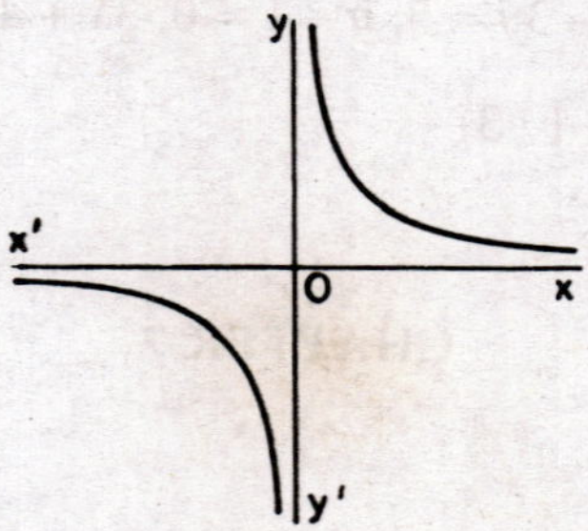


16. Rectangular hyperbola.

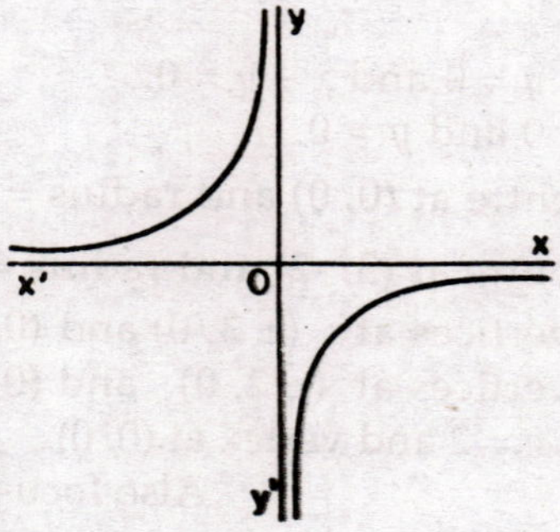


**(*Note*. See the later chapters for the curves in Questions 7 to 14).**

17. Hyperbola with vertices at (± 2, 0)
18. Hyperbola with vertices at (0, ± 3)

**Exercise 7.2.**

1. (i) Inside the ellipse,
(ii) on the parabola,
(iii) in the central region of the hyperbola,
(iv) outside the ellipse,
(v) outside the parabola,
(vi) in the central region of the hyperbola,
(vii) in the non-central region of the hyperbola,
(viii) in the central region of the hyperbola,
(ix) on the hyperbola,
(x) on the ellipse.
(xi) Inside.
(xii) Outside.

# CHAPTER 8

**Exercise 8.1.**

1. (i) (2, 2),
(ii) $(4a, 4a)$,
(iii) (0, 1).

2. (i) $4y = 5x \pm \dfrac{\sqrt{161}}{6}$
(ii) $y = 2x + 2,\ x + 2y + 16 = 0.$

3. (i) $n^2 = a^2 (l^2 + m^2)$,
(ii) $am^2 = nl$,
(iii) $n^2 = a^2l^2 - b^2m^2$,
(iv) $n^2 = 4lmc^2$.

Point of contact.

(i) $\left(\dfrac{-a^2 l}{n}, \dfrac{-a^2 m}{n}\right)$,

(ii) $\left(\dfrac{n}{l}, \dfrac{-2am}{l}\right)$,

(iii) $\left(\dfrac{-a^2l}{n}, \dfrac{b^2m}{n}\right)$

(iv) $\left(\dfrac{-n}{2l}, \dfrac{-n}{2m}\right)$

4. (i) $p = \pm a$,
(ii) $p = -a \tan \alpha \sin \alpha$,
(iii) $p^2 = a^2 \cos^2\alpha - b^2 \sin^2\alpha$,
(iv) $p^2 = 2c^2 \sin 2\alpha$,
(v) $p^2 = a^2 \cos^2\alpha + b^2 \sin^2\alpha$.

5. $x + a = 0$.

6. (i) $y = x \pm a\sqrt{2}, y = -\dfrac{1}{2}x \pm a\dfrac{\sqrt{5}}{2}, y = nx \pm a\sqrt{1+n^2}$

(ii) $y = x + a, y = -\dfrac{1}{2}x - 2a, y = nx + \dfrac{a}{n}$

(iii) $y = x \pm \sqrt{a^2 + b^2}, y = -\dfrac{1}{2}x \pm \dfrac{\sqrt{a^2 + 4b^2}}{2}$,

$y = +nx \pm \sqrt{a^2n^2 + b^2}$

(iv) $y = x \pm \sqrt{a^2 - b^2}$ $y = -\dfrac{1}{2}x \pm \dfrac{\sqrt{a^2 - 4b^2}}{2}$,

$y = nx \pm \sqrt{a^2n^2 - b^2}$

7. $x^2 + y^2 = \dfrac{1}{a^2}$

8. $y = x + a, \quad y = -\dfrac{1}{2}x - 2a$.

9. $y = x + 1, y = 3x + 5$.

10. $(-a, \pm a)$ $(2a, \pm 4a)$.

11. $y = \pm (x + 2)$.

12. $a^{1/3}x + b^{1/3}y + a^{2/3} b^{2/3} = 0$.

13. $mc = a(1 + m^2)$.

15. $\pm\sqrt{\dfrac{\sqrt{5}-1}{2}}$

**Exercise 8.2.**

1. (i) $xx' + yy' = a^2$,
   (ii) $xx' + yy' + g(x + x') + f(y + y') + c = 0$,
   (iii) $yy' = 2a(x + x')$,
   (iv) $\dfrac{xx'}{a^2} + \dfrac{yy'}{b^2} = 1$,
   (v) $\dfrac{xx'}{a^2} - \dfrac{yy'}{b^2} = 1$.
   (vi) $xy' + x'y = 2c^2$.
2. (i) $x - \sqrt{3}y + 4 = 0$
   (ii) $3x + y - 4 = 0$,
   (iii) $9x - 13y - 53 = 0$,
   (iv) $3x + 2y + 3 = 0$,
   (v) $x + 3y = 5$.
3. $(a, 2a), \left(\dfrac{9a}{4}, 3a\right)$, tangents: $y = x + a$, $6y = 4x + 9a$.
4. (*a*) (i) $x = a, x - y = a\sqrt{2}, x\cos\phi + y\sin\phi = a$.
   (ii) $x = a, \dfrac{x}{a} - \dfrac{y}{b} = \sqrt{2}, \dfrac{x}{a}\cos\phi + \dfrac{y}{b}\sin\phi = 1$
   (iii) $x = a, \dfrac{x\sqrt{2}}{a} + \dfrac{y}{b} = 1, \dfrac{x}{a}\sec\phi + \dfrac{y}{b}\tan\phi = 1$.
   (b) $x = 0, x + y + a = 0, py = x + ap^2$.
5. See Exercise 8.1 Question 3.
7. (a) (i) $x\cos\left(\dfrac{\theta_1 + \theta_2}{2}\right) + y\sin\left(\dfrac{\theta_1 + \theta_2}{2}\right) = a\cos\left(\dfrac{\theta_1 - \theta_2}{2}\right)$,
   (ii) $\dfrac{x}{a}\cos\left(\dfrac{\theta_1 + \theta_2}{2}\right) + \dfrac{y}{b}\sin\left(\dfrac{\theta_1 + \theta_2}{2}\right) = \cos\left(\dfrac{\theta_1 - \theta_2}{2}\right)$

(iii) $\frac{x}{a}\cos\left(\frac{\theta_1-\theta_2}{2}\right)-\frac{y}{b}\sin\left(\frac{\theta_1+\theta_2}{2}\right)=\cos\left(\frac{\theta_1+\theta_2}{2}\right)$

(b) (i) $y(t_1+t_2)=2[x+at_1t_2]$,

(ii) $x+t_1t_2y-c(t_1+t_2)=0$.

**Exercise 8.3.**

1. (i) $\sqrt{3x}+y=0$

(ii) $x-3y+2=0$,

(iii) $13x+9y-21=0$,

(iv) $2x-3y-11=0$,

(v) $9x-3y-5=0$.

3. (i) $2alm^2+al^3+m^2n=0$,

(ii) $\frac{(a^2+b^2)^2}{n^2}=\frac{a^2}{l^2}-\frac{b^2}{m^2}$

(iii) $lmn^2+c^2(l^2-m^2)^2=0$.

(iv) $gl+fm=n$.

4. $t_1+t_2+t_3=0$.

6. $(6,-4\sqrt{3})$.

7. $y=3(x-22)$, $(18,-12)$.

8. $(2a,-2\sqrt{2}a)$.

**Exercise 8.4.**

2. $gx+fy+c=0$,

3. $5x+9y+30=0$.

4. $\left(-\frac{a^2l}{n},-\frac{a^2m}{n}\right)$.

9. $x^2+y^2=\frac{a^4}{p^2}$.

12. $x^2+y^2-x$, $x-y$, $y=0$.

**Exercise 8.5.**

2. (i) $y = 3$,
   (ii) $5x - 8y + 2 = 0$,
   (iii) $x(4\alpha - 3\beta + 4) + y(-3\alpha + 10\beta + 7) + (4\alpha + 7\beta - 2) = 0$.

3. (i) $\left(\frac{5}{3}, \frac{5}{2}\right)$,
   (ii) $(1, 2)$.

4. (i) $\left(\frac{-a^2l}{n}, \frac{-a^2m}{n}\right)$,
   (ii) $\left(\frac{n}{l}, -\frac{2am}{l}\right)$,
   (iii) $\left(-\frac{a^2l}{n}, \frac{-b^2m}{n}\right)$
   (iv) $\left(\frac{-2c^2m}{n}, \frac{-2c^2l}{n}\right)$

5. (i) $\frac{a^6}{x^2} + \frac{b^6}{y^2} = (a^2 - b^2)^2$,
   (ii) $\frac{a^6}{x^2} - \frac{b^6}{y^2} = (a^2 + b^2)^2$
   (iii) $(x^2 - y^2)^2 + 4xyc^2 = 0$.

9. $\frac{x^2}{a^2} + \frac{a^2y^2}{b^4} = 1$

11. $x^2 + y^2 = ax$,

13. (i) $x^2 + y^2 - 3x + y - 4 = 0$.

15. $ly^2 - mxy + nx = 0$.

17. $\left(\frac{x^2}{a^4} + \frac{y^2}{b^4}\right)(a^2 + b^2) = 1$

18. (i) $x^2 + y^2 = 2a^2$,

(ii) $\frac{x^2}{a^4}+\frac{y^2}{b^4}=\frac{1}{a^2}+\frac{1}{b^2}$

(iii) $\frac{x^2}{a^4}+\frac{y^2}{b^4}=\frac{1}{a^2}-\frac{1}{b^2}$

19. $x + 4a = 0$.

27. (i) $a^2(ll' + mm') = nn'$,

(ii) $l'n + ln' = 2amm'$,

(iii) $a^2ll' + b^2mm' = nn'$,

(iv) $a^2ll' - b^2mm' = nn'$,

(v) $2c^2(lm' + l'm) = nn'$.

(vi) $\begin{vmatrix} a & h & g & l \\ h & b & f & m \\ g & f & c & n \\ l' & m' & n' & 0 \end{vmatrix} = 0$

32. $(ax - by)(ay - bx) + ab(a \pm b)(x \pm y) = 0$.

33. $4a^2 (y - b)^4 = (a^2 - b^2)[a^2 -(b - 2y)^2]x^2$

**Exercise 8.6.**

2. (i) $x - 2y - 5 = 0$,

(ii) $4x + 3y + 1 = 0$,

(iii) $x + y = 0$.

3. (i) $(2, 3)$,

(ii) $(2, -1)$,

(iii) $(4, 2)$.

(iv) $\left(\frac{ln}{l^2+m^2}, \frac{mn}{l^2+m^2}\right)$.

4. (i) $y(y - y') = 2a(x - x')$,

(ii) $\frac{x^2}{a^2}+\frac{y^2}{b^2}=\frac{xx'}{a^2}+\frac{yy'}{b^2}$

(iii) $\frac{x^2}{a^2}-\frac{y^2}{b^2}=\frac{xx'}{a^2}-\frac{yy'}{b^2}$

(iv) $xy' + yx' = 2xy$,

(v) $x^2 + y^2 = xx' + yy'$,

(vi) $axx' + h(xy' + yx') + byy' + gx' + fy'$
$= ax^2 + 2hxy + by^2 + gx + fy$.

5. (i) $\frac{y^2}{2a} + \frac{4a^3}{y^2} = x - 2a$

(ii) $\left(\frac{a^6}{x^2} - \frac{b^6}{y^2}\right)\left(\frac{x^2}{a^2} - \frac{y^2}{b^2}\right)^2 = (a^2 + b^2)^2$,

(iii) $(y^2 - x^2)^3 = 4ax^2y^2$,

(iv) $c^2(x^2 - y^2) + 4x^3y^3 = 0$.

6. (i) $2(x^2 + y^2) = a^2$,

(ii) $\frac{x^2}{a^4} + \frac{y^2}{b^4} = \left(\frac{a^2 + b^2}{a^2b^2}\right)\left(\frac{x^2}{a^2} + \frac{y^2}{b^2}\right)^2$,

(iii) $\frac{x^2}{a^4} + \frac{y^2}{b^4} = \left(\frac{b^2 - a^2}{a^2b^2}\right)\left(\frac{x^2}{a^2} - \frac{y^2}{b^2}\right)^2$,

(iv) $x^2 + y^2 = 0$.

7. $y^2 - 2ax + 8a^2 = 0$., $(4a, 0)$.

12. $(y^2 - 2ax)^2 = a^2(y^2 + 4a^2)$.

13. (i) $x + my = 0$,

(ii) $(ax + hy + g) + m(hx + by + f) = 0$,

(iii) $y = 2a/m$,

(iv) $ma^2y + xb^2 = 0$,

(v) $ma^2y - xb^2 = 0$,

(vi) $mx + y = 0$.

15. Assuming the circle as $x^2 + y^2 = a^2$, length of the chord as $2d$, the locus is $x^2 + y^2 = a^2 - d^2$.

16. $\left(\frac{pr}{p^2 + q^2}, \frac{qr}{p^2 + q^2}\right)$

18. $(x - \alpha)^2 + (y - \beta)^2 + x^2 + y^2 - a^2 = 0$

**Exercise 8.7.**

2. $x^2 + y^2 = 2c$
3. (i) $40(x^2 + y^2 - 1) = (4x - 5y - 1)^2$.
   (ii) $5(y^2 - 16x) = (y + 4x - 4)^2$.
4. (i) $(x^2 + y^2 - 2a^2)^2 \tan^2 \theta = 4a^2(x^2 + y^2 - a^2)$.
   (ii) $(x^2 + y^2 - a^2 - b^2)^2 \tan^2 \theta = 4(b^2x^2 + a^2y^2 - a^2 b^2)$.
   (iii) $(x^2 + y^2 - a^2 + b^2)^2 \tan^2 \theta = 4(a^2y^2 - b^2x^2 + a^2b^2)$.
   (iv) $(x^2 + y^2)^2 \tan^2 \theta = 16c^2 (c^2 - xy)$.
5. (i) $x^2 + y^2 = 2a^2$,
   (ii) $x^2 + y^2 = a^2 + b^2$,
   (iii) $x^2 + y^2 = a^2 - b^2$.
6. $x^2 + y^2 = (a^2 + b^2)\left(\dfrac{x^2}{a^2} + \dfrac{y^2}{b^2}\right)^2$.
9. $x^2 + y^2 + 2gx + 2fy + 4c = 3(g^2 + f^2)$.

**Miscellaneous Exercise on Chapter 8**

3. A circle.
10. $x^2 + y^2 = 2k^2$.

## CHAPTER 9

**Exercise 9.1.**

1. (i) $9x^2 - 12xy + 4y^2 - 126x + 110y + 532 = 0$, Parabola.
   (ii) $148x^2 + 40xy + 139y^2 - 16x + 20y - 4 = 0$, Ellipse.
   (iii) $3x^2 - y^2 - 40x - 2y + 99 = 0$, Hyperbola.
   (iv) $x^2 - 4x + 4y^2 = 0$, Parabola.
   (v) $7x^2 + 12xy - 2y^2 - 2x + 14y - 22 = 0$, Hyperbola.
2. (i) $\dfrac{x^2}{a^2} + \dfrac{y^2}{a^2(1-e^2)} = 1$ Ellipse.
   (ii) $y^2 = 4ax$, Parabola.
   (iii) $x^2 = -4ay$, Parabola.
   (iv) $\dfrac{x^2}{a^2} - \dfrac{y^2}{a^2(e^2-1)} = 1$ Hyperbola.

(v) $\dfrac{x^2}{a^2} - \dfrac{y^2}{a^2(e^2-1)} = 1$ Hyperbola.

3.

| | Eccentricity | Focus | Directrix | Conic |
|---|---|---|---|---|
| (i) | $\sqrt{3}$ | (1, 2), | $2x + 3y - 2 = 0$, | Hyperbola. |
| (ii) | 1, | (1, 0), | $4x - 3y + 1 = 0$, | Parabola. |
| (iii) | $\dfrac{1}{2}$ | (0, 0), | $5x + 3y + 7 = 0$, | Ellipse |

4. $x^2 + 2xy + y^2 - 2x + 2y - 5 = 0$.
5. $7x^2 + 2xy + 7y^2 + 10x - 10y + 7 = 0$.
6. $x^2 - 4xy + y^2 + 10x - 10y + 16 = 0$.
7. $(ax - by)^2 - 2a^3x - 2b^3y + (a^2 + b^2)\, b^2 = 0$.
8. $4x^2 + 9y^2 - 8x - 18y - 140 = 0$.
9. $16\,(x^2 + y^2) = 25\,(x \cos\alpha - y \sin\alpha - p)^2$.

# CHAPTER 10

## Exercise 10.1

1. L.R. Vertex Focus Axis Tangent at vertex Directrix Tracing

(i) 8 (0, 0) (2, 0) $y = 0$ $x = 0$ $x = -2$

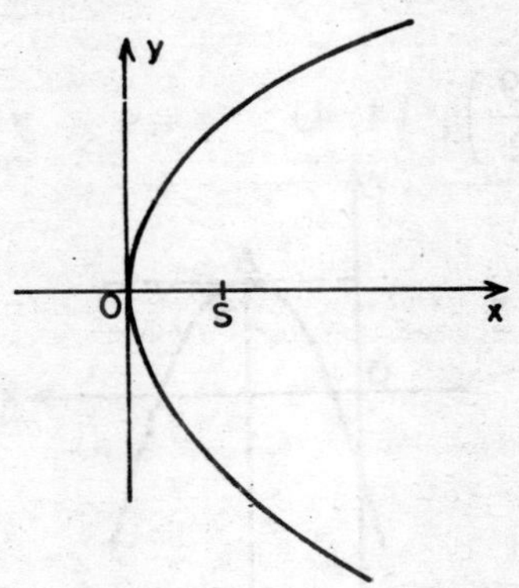

(ii) 4 (0, 0) (0, -1) $x = 0$ $y = 0$ $y = 1$

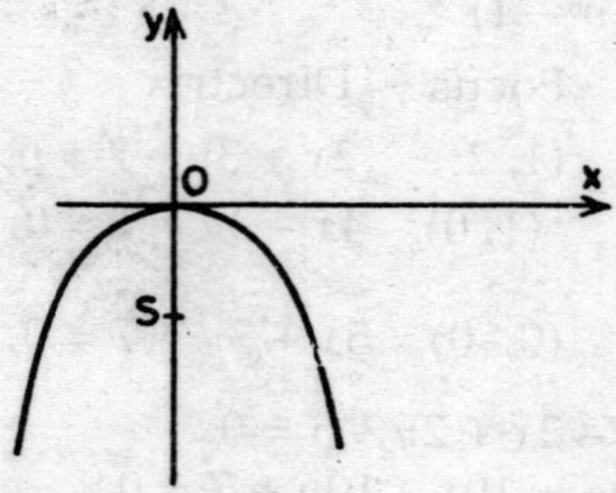


(iii) 2 $\left(\frac{19}{3}, 3\right)$ (9, 3) $y = 3$ $x = \frac{19}{2}$ $x = 10$

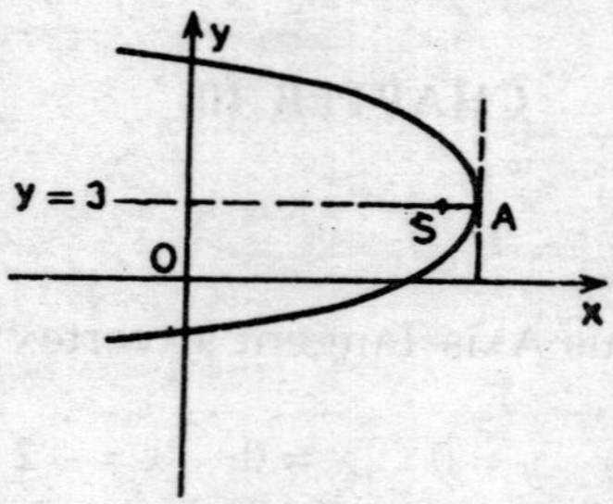


(iv) 2 $\left(4, \frac{19}{2}\right)$ (4, 4) $x = 4$ $y = \frac{9}{2}$ $y = 5$

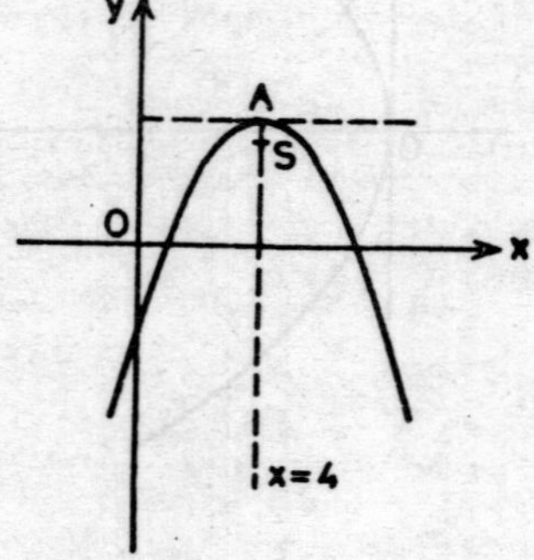

2. $x^2 = -4(y-1)$.
3. (a) $(y-3)^2 = 4(x-4)$,
(b) $(y-2)^2 = -16(x-2)$,
(c) $(y-10)^2 = 16(x-2)$.
4. $(x+h, k)$, $x = h - a$.
5. (0, 0) and (2, 6).
6. (a) $y^2 = 8x$,
(b) $y^2 = 16x$,
(c) $3y^2 = 25x$,
(d) $y^2 = 18x$.
9. (2, 4),
8,
$x - y + 2 = 0$
$x + y - 6 = 0$
10. $9x^2 - 24xy + 16y^2 - 88x - 91y + 704 = 0$
11. $\pm 4$.
14. $x^2 + 2xy + y^2 - 4ax + 2a^2 = 0$,
$x + y = a$

**Exercise 10.2.**

1. $\left(\frac{a}{t^2}, \frac{-2a}{t}\right)$.
6. 12.
8. $[at_1t_2, a(t_1 + t_2)]$.
10. $2\,l\operatorname{cosec}^2\theta$
18. (i) $y = cx$
(ii) $a = cx$
(iii) $y^2 - 2ax = cx^2$
(iv) $y^2 - 4ax = (x+a)^2 \tan c$.
21. (i) $(y-k)^2 = k^2 - 4ah$
(ii) $y^2 - k^2 = 4a(x-h)$.

**Exercise 10.4.**

2. $[a(2 + m_1^2 + m_2^2 + m_1m_2), am_1m_2(m_1 + m_2)]$.
4. $y = 0$; $(0, 0)$; $y = x - 3$, $(1, -2)$; $y = -x + 3$, $(1, 2)$.

10. $\left(3a, \pm 2\sqrt{3}a\right.$

20. $27a\,y^2 = (x - 5a)^2\,(2x - a)$.

**Miscellaneous Exercise on Chapter 10.**

14. $8a\sqrt{3}$

## CHAPTER 11

**Exercise 11.1.**

1.

| | Eccentricity | Foci | Directrices | Major axis | Minor axis | L.R. |
|---|---|---|---|---|---|---|
| (i) | $\frac{\sqrt{7}}{4}$, | $\left(\pm\sqrt{7}, 0\right)$ | $x = \pm\frac{16}{\sqrt{7}}$ | 8 | 6 | $\frac{9}{2}$ |
| (ii) | $\frac{4}{5}$, | $(\pm 4, 0)$ | $x = \pm\frac{25}{4}$ | 10 | 6 | $\frac{18}{5}$ |
| (iii) | $\frac{4}{5}$, | $(0, \pm 4)$ | $y = \pm\frac{25}{4}$ | 10 | 6 | $\frac{18}{5}$ |

2. (i) $\frac{\sqrt{3}}{2}$

(iii) $\frac{\sqrt{15}}{4}$.

3. (i) $\frac{x^2}{5} + \frac{y^2}{20} = 1$.

(ii) $\frac{x^2}{5} + \frac{y^2}{4} = 1$,

(iii) $\frac{x^2}{25} + \frac{y^2}{9} = 1$,

(iv) $\frac{x^2}{64} + \frac{y^2}{36} = 1$,

(v) $\frac{x^2}{48} + \frac{y^2}{36} = 1$.

4. $\frac{x^2}{81/4} + \frac{y^2}{45/4} = 1$.

5. $\frac{x^2}{9} + \frac{y^2}{(9/2)} = 1$.

6. $\frac{x^2}{100} + \frac{y^2}{50} = 1$.

7. $7x^2 - 2xy + 7y^2 - 32x - 32y + 64 = 0$
$(8/3, 8/3)$,
$(10/3, 10/3)$,
$3(x + y) = 32$.

8. (i) $\frac{(x-3)^2}{72} + \frac{(y-4)^2}{81} = 1$

(ii) $\frac{x^2}{16} + \frac{(y-1)^2}{20} = 1$.

9. $\frac{1}{\sqrt{2}}, \frac{x^2}{2} + \frac{y^2}{1} = 1$.

10. $(1, -3)$; $4/5$; $(5, -3)$, $(-3, -3)$
10, 6.

11. $\sqrt{3}/2$.

**Exercise 11.2.**

4. $\frac{xx_1}{a^2} + \frac{yy_1}{b^2} = 1$

6. $\frac{x\cos\alpha}{a} + \frac{y\sin\alpha}{b} = \cos\beta$

10. $x^2(y^2 - b^2)^2 + y^2(x^2 - b^2)^2 = 4x^2y^2(a^2 - y^2)$.

**Exercise 11.3.**

2. (i) $c(x^2 - a^2) = 2xy$
(ii) $y^2 - b^2 = c(x^2 - a^2)$.
(iii) $4x^2 y^2 - 2(x^2 - a^2)(y^2 - b^2) = c^2 (x^2 - a^2)^2$.

4. (i) $y = \pm x \pm \sqrt{a^2 + b^2}$

(ii) $\dfrac{\left[(r^2 - b^2)(a^2 - r^2)\right]^{1/2}}{r}$.

5. $\left(a\cos\dfrac{\alpha+\beta}{2}\sec\dfrac{\alpha-\beta}{2},\ b\sin\dfrac{\alpha+\beta}{2}\sec\dfrac{\alpha-\beta}{2}\right)$

14. (i) $2\pi/3$.

**Exercise 11.5.**

7. $a^2l^2 + b^2m^2 = 2n^2$.
9. $a^2l_1l_2 + b^2m_1m_2 = 0$.
11. $2(x^2 + y^2)^2 = a^2x^2 + b^2y^2$.
24. $a \cos \phi - b \sin \phi$
25. $y = \dfrac{bx}{a}\tan(\lambda/2)$

**Miscellaneous Exercise on Chapter 11**

7. (ii) $\dfrac{x^2}{a^2} + \dfrac{y^2}{b^2} = \dfrac{y}{b}$.
11. $a^2x^2 + b^2y^2 = a^4 + b^4$.
16. (i) $(a^2m^2 + b^2)(x^2 + y^2) + 2a^2 mcx - 2b^2cy + c^2(a^2 + b^2)$
$- a^2b^2(1 + m^2) = 0$.
23. $a^2lx + b^2 my = a^2 + b^2$.

# CHAPTER 12

## Exercise 12.1.

1. $45x^2 - 36y^2 = 80.$
2. $x^2 - 2y^2 = 2,\ 3x^2 - 2y^2 = 10.$
3. $x^2 - y^2 = \dfrac{1}{8}.$
4. (i) $\dfrac{x^2}{16} - \dfrac{y^2}{20} = 1$

   (ii) $16x^2 - 2y^2 = 1.$
5. $x^2 - 4xy - 2y^2 + 10x + 4y = 0$
6. (i) $(y - 3)^2 - 3x^2 = 3$

   (ii) $x^2 - 8y^2 = 32.$
7. $\dfrac{5}{4}, (\pm 5, 0), \dfrac{9}{2}$
8. (i) $e = -\dfrac{2}{\sqrt{3}}$, Foci $\left(0, \pm\dfrac{4}{\sqrt{3}}\right)$, directrices, $y = \pm\sqrt{3}$, equations of latera recta, $y = \pm\dfrac{4}{\sqrt{3}}$, length $= \dfrac{4}{3}$.

   (ii) $e = \sqrt{2}$ Foci $(\pm 2\sqrt{2}, 0)$, Directrices $x = \pm\sqrt{2}$,
9. $e = 2.$
10. (i) (1, 2); 6, 8; 5/3; (6, 2), (– 4, 2)

## Exercise 12.2.

2. (i) $\sqrt{2}$

   (ii) $x\alpha - y\beta = a^2,$

   (iii) $\left(-\dfrac{a^2A}{C}, \dfrac{a^2B}{C}\right)$
3. (i) $\sqrt{\dfrac{3}{2}},$

(ii) $\frac{1}{2}$.

5. $\frac{x}{a}\sec\alpha - \frac{y}{b}\tan\alpha = 1$

8. $\left(\frac{a\cos\frac{(\alpha-\beta)}{2}}{\cos\frac{(\alpha+\beta)}{2}}, \frac{b\sin\frac{(\alpha+\beta)}{2}}{\cos\frac{(\alpha+\beta)}{2}}\right)$

13. (i) $(x^2 + y^2)^2 = a^2x^2 - b^2y^2$
(ii) $a^2x^2 - b^2y^2 = (a^2 + b^2)^2$

**Exercise 12.3.**

1. $\pm x \mp y = \sqrt{a^2 - b^2}$

$(a^2 + b^2)$ $\sqrt{\frac{2}{a^2 - b^2}}$

3. $\sum \tan\frac{\theta_1}{2}\tan\frac{\theta_2}{2}$

$= \sum \frac{1}{\tan\frac{\theta_1}{2}\tan\frac{\theta_2}{2}}$.

**Exercise 12.4.**

1. (i) $\frac{x^2}{a^2} - \frac{y^2}{b^2} = k$
(ii) $(px + qy + r)(p'x + q'y + r') = 0$,
$(px + qy + r)(p'x + q'y + r') - A = 0$.

3. $9y^2 - 4x^2 = 0, \frac{36}{13}$.

7. (ii) $|\sec\lambda|$.

11. $\sqrt{3}$

16. $(a^2-b^2)^2\left(1-\frac{x^2}{a^2}+\frac{y^2}{b^2}\right) = (a^2-b^2-x^2-y^2)^2$.

**Exercise 12.5.**

1. $y=\frac{b^2}{ta^2}.x$

2. $y=\frac{32}{9}x$

3. $a^2l_1l_2 = b^2m_1m_2$.

5. $a^2A = b^2B$.

**Exercise 12.6.**

1. (i) $(c\sqrt{2}, c\sqrt{2})$

$(-c\sqrt{2}, -c\sqrt{2})$

$x+y=\pm c\sqrt{2}$

15. (iii) $\left(c\sum t_1, c\sum\frac{1}{t_1}\right)$

# CHAPTER 13

**Exercise 13.1.**

1. Centre (3, 2) $\begin{cases}\text{major axis} = 2\sqrt{6},\\ :x+2y=7\end{cases}$ $\begin{cases}\text{minor axis} = 4\\ 2x-y = 4\end{cases}$

2. Centre (2, – 2), major axis = $2\sqrt{2}$ minor axis = 2.

3. Ellipse Centre (2, – 3), major axis: $4x + 3y + 1 = 0$, length = 6; minor axis: $3x - 4y - 18 = 0$, length = 4.
4. Centre (1, -2), major axis = 6, minor axis = 4.
5. Centre , $\left(-\frac{1}{4}, -\frac{1}{4}\right)$ $13x^2 - 18xy + 37y^2 - 4 = 0$.
6. Major axis: $2x + y = 4$, length = 6;
minor axis: $x - 2y + 3 = 0$, length = 4.

**Exercise 13.2.**

1. Centre (2, 1), major axis: $3x + 2y - 8 = 0$, length = 4,

minor axis: $2x - 3y - 1 = 0$, length = 2, eccentricity $\frac{\sqrt{3}}{2}$,

Focus $\left(2 - 2\sqrt{\frac{3}{13}}, 1 + 3\sqrt{\frac{3}{13}}\right)$, Directrix: $2x - 3y = 1 + 4\sqrt{\frac{13}{3}}$,

Focus $\left(2 + 2\sqrt{\frac{3}{13}}, 1 - 3\sqrt{\frac{3}{13}}\right)$, Directrix: $2x - 3y = 1 - 4\sqrt{\frac{13}{3}}$,

L.R. = 1.
2. Ellipse, Centre (0, 1), major axis $x - y + 1 = 0$, length = $2\sqrt{3}$.
3. Ellipse, Centre (2, – 2), major axis: $3x - 2y - 10 = 0$, length = $2\sqrt{2}$; minor axis: $2x + 3y + 2 = 0$, length = 2.
4. Ellipse, Centre (2, 3) major axis: $2x - y - 1 = 0$, length = $2\sqrt{6}$; minor axis: $x + 2y - 8 = 0$, length = 4.
5. Ellipse, Centre (– 1, 1), major axis: $2x - y + 3 = 0$, length = 4; minor axis: $x + 2y - 1 = 0$, length = 2,

Directrices: $x + 2y - 1 \pm 4\sqrt{\frac{5}{3}} = 0$.
6. Eccentricity = 1/2, Foci (1, 1), (3, 1), directrices : $x = -2$ and $x = 6$.
7. Centre (2, – 3), foci $\left(2 \mp \sqrt{\frac{8}{3}}, -3 \pm \sqrt{\frac{8}{3}}\right)$,

major axis: $x + y + 1 = 0$
and minor axis: $x - y - 5 = 0$.

8. Ellipse $c\left(\frac{1}{\sqrt{2}}, \frac{1}{\sqrt{2}}\right)$
$a = \sqrt{2},\ b = 1$
foci $(0, \sqrt{2},), (\sqrt{2}, 0)$.

**Exercise 13.3.**

1. Vertex $\left(-\frac{1}{5}, \frac{2}{5}\right)$, directrix: $3x + 4y + 9 = 0$, focus $(1, 2)$.
2. Vertex $\left(\frac{-3}{25}, \frac{-4}{25}\right)$, directrix: $8x - 6y + 1 = 0$,
Focus $\left(\frac{-1}{25}, \frac{-11}{50}\right)$.
3. Axis: $3x + 4y - 2 = 0$, tangent at the vertex: $4x - 3y - 11 = 0$, vertex $(2, - 1)$, L.R. $= \frac{8}{5}$
4. axis: $2x - y - 1 = 0$, tangent at the vertex : $x + 2y - 1 = 0$, vertex $\left(\frac{3}{5}, \frac{1}{5}\right)$, L.R. $= \frac{4}{5}$.
5. Axis; $3x - 4y + 7 = 0$, tangent at the vertex: $4x + 3y + 2 = 0$, vertex $\left(-\frac{29}{33}, \frac{22}{25}\right)$, L.R. = 3.
6. Axis; $4x - 3y + 10 = 0$, tangent at the vertex: $3x + 4y + 5 = 0$, L.R. $= \frac{1}{5}$, vertex $\left(-\frac{11}{5}, \frac{2}{5}\right)$, Focus $\left(-\frac{217}{100}, \frac{11}{5}\right)$.
7. Axis: $2x + y + 1 = 0$, tangent at the vertex: $x - 2y = 0$, vertex $\left(-\frac{2}{5}, -\frac{1}{5}\right)$, L.R. $= \frac{1}{\sqrt{5}}$.

8. Axis: $\sqrt{a}x+\sqrt{b}y+\left(\frac{\sqrt{a}g+\sqrt{b}f}{a+b}\right)=0$.

10. $2|A|, 2|B|$ respectively.

11. Parabola, axis $x + y = 0$, Tangent at the vertex $y = x - 1$, L.R. = $2\sqrt{2}$.

12. $\left(\frac{a}{a^2+b^2}, \frac{b}{a^2+b^2}\right)$

$bx + ay = 0$

$ax - by = \frac{a^2 \quad b^2}{a^2 + b^2}$

13. $\alpha(bc' - b'c) + \beta(ca' - c'a) + \gamma(ab' - a'b) = 0$

**Exercise 13.4.**

1. Centre $(-2, 2)$, transverse axis: $x - y + 4 = 0$, length $= 2\sqrt{2}$, conjugate axis: $x + y = 0$, length $= \frac{2}{5}\sqrt{10}$.

2. Hyperbola with centre at $\left(\frac{5}{9}, \frac{2}{9}\right)$.

3. Eccentricity = 2, foci $\left(-1 \pm \sqrt{\frac{8}{3}}, 1 \pm \sqrt{\frac{8}{3}}\right)$.

4. Centre $(-1, 2)$, transverse axis: $x + 2y = 3$, length $= \sqrt{2}$, conjugate axis: $2x - y + 4 = 0$, length $= \frac{2}{3}\sqrt{3}$.

5. General equation :
$x^2 + xy - 2y^2 + 2x + y + C = 0$,
For asymptotes, $C = 1$.

6. Asymptotes: $3x^2 - 5xy - 2y^2 + 5x + 11y - 12 = 0$,
conjugate hyperbola: $3x^2 - 5xy - 2y^2 + 5x + 11y - 16 = 0$

7. (i) Asymptotes: $xy - 3x - 2y + 6 = 0$,
conjugate hyperbola: $xy - 3x - 2y + 12 = 0$.
(ii) $(x - 2)(y - 4) = 0$, $xy + 4x - 2y - 17 = 0$.

8. Asymptotes: $6x^2 - 7xy - 3y^2 - 2x - 8y - 4 = 0$, conjugate hyperbola: $6x^2 - 7xy - 3y^2 - 2x - 8y - 2 = 0$.
9. $(3x - 4y + 7)(4x + 3y + 1) - 7 = 0$.
10. $(2x + 3y - 8)(3x + 2y - 7) - 154 = 0$.
12. $(ax + by + c)(a'x + b'y + c') = 0$,
$(ax + by + c)(a'x + b'y + c') = \lambda^2$
14. Rectangular hyperbola, with asymptotes $3x - 4y + 1 = 0$, $4x + 3y - 7 = 0$.
16. (i) (1, 1); 16, 12, 5/4, (7, 9), (– 5, – 7)
(ii) (4, –4), 4, 4, $\sqrt{2}$ $(4 \pm 2\sqrt{2}, -4)$.

## CHAPTER 14

### Exercise 14.1.

1. (i) $a\sqrt{10}$
(ii) $b\sqrt{3}$

2. (i) $\dfrac{11\sqrt{3}\,b^2}{4}$
(ii) $\dfrac{\sqrt{3}}{4}$

### Exercise 14.2.

1. (i) $\dfrac{\pi}{6}$
(ii) $\dfrac{\pi}{2}$
(iii) $\dfrac{\pi}{6}$

2. $\left[\dfrac{\sqrt{p_1^2 + p_2^2 - 2p_1p_2 \cos(\theta_1 - \theta_2)}}{\sin(\theta_1 - \theta_2)},\right.$

$\left.\tan^{-1}\left(\dfrac{p_1 \cos\theta_2 - p_2 \cos\theta_1}{p_2 \sin\theta_1 - p_1 \sin\theta_2}\right)\right]$

3. $3r \sin\left(\theta - \dfrac{\pi}{6}\right) + 2 \sin\left(\dfrac{\pi}{6} - \theta\right) - 3 = 0$

4. $\dfrac{r_1[\sin(\theta_1 - \alpha) + e\sin\theta_1]}{r} = (\sin(\theta - \alpha) + e\sin\theta)$

6. $\sqrt{2}$

**Exercise 14.3**

4. $(A - e \cos \alpha)^2 + (B - e \sin \alpha)^2 = 1$

# INDEX


Abcissa, 5
Angle of intersection
—of two circles, 180
—of two lines, 71, 119, 494
Area of a triangle
—in Cartesian co-ordinates, 11
—in Polar co-ordinates, 490
Asymptotes, 414, 477, 512
Auxiliary circle, 334
Axes of a general conic, 452
Axis/axes
conjugate, 391
major, 327
minor, 327
principal, 326, 391
transverse, 391

Bisector of angles
between two lines, 95, 119

Central curve, 212, 452
Centre
—of a circle, 151
—of a conic, 212, 450
—of an ellipse, 327
—of a hyperbola, 391
Change of axes, 37
Chord joining two points, 238, 510
Chord of contact, 244, 513
Chord with a given middle point, 259
Circle, 151, 500
Co-axal system, 193
—limiting points of, 195
—simplified form of, 194
Collinear points, 92
Concentric circles, 203
Concurrence of three lines, 86
Condition of tangency
—in a circle, 170
—in a conic, 227, 516
Conic/conic section, 278, 281, 502
Conjugate axis, 391
Conjugate
—diameters of an ellipse, 361
—diameters of a hyperbola, 426
—lines, 251
—points, 251
Conjugate hyperbola, 393
Co-normal points
—of an ellipse, 340
—of a hyperbola, 407
—of a parabola, 305
Co-ordinate
—axes, 3
—plane, 3
Co-ordinates
—Cartesian, 5
—Polar, 17

Diameter
—form of a circle, 155
—of an ellipse, 361
—of a hyperbola, 426
—of a parabola, 320
Director circle, 273, 515
Directrix, 278, 284, 325, 389, 463
Discriminant of a general conic, 449
Distance between two points
—in Cartesian co-ordinates, 6
—in polar co-ordinates, 490
Distance between two parallel lines, 73

Eccentric angle, 335
Eccentricity, 278, 284, 325, 389, 455
Ellipse, 214, 278, 325

Equilateral hyperbola, 435
External division, 8

Focal
—chord, 285, 296, 339, 394
—distance, 296, 330, 394
Focus/Foci, 285, 328, 392, 462

General equation of second degree to represent a pair of straight lines, 123
Gradient of a line, 57

Homogeneous equation, 110
Hyperbola, 215, 278, 389

Intercept form, 56
Internal division, 7
Intersection
—of a line and a curve, 224
—of two lines, 76
Invariants, 49
Inverse points, 205

Joint equation of tangets from a point to a curve, 269

Latus rectum, 285, 327, 391
Length of the perpendicular, 67
Length of the tangent from a point to a circle, 177
Limiting points, 195
Line through the intersection of two lines, 77
Locus, 25

Major axis, 327
Mid-point formula, 8
Minor axis, 327

Normal, 169, 239, 511
—form of a line, 59

Oblique axes, 5
Ordinate, 5
—double, 285, 327
Origin, 1, 3
Orthogonal
—circles, 180
—co-axal system, 197

Parabola, 214, 278, 284
Parallel lines, 72
Parameter equation
—of a circle, 160
—of an ellipse, 219
—of a hyperbola, 219
—of a line, 58
—of a parabola, 219
—of a rectangular hyperbola, 219
Perpendicular distance formula
—in Cartesian co-ordinates, 67
—in Polar co-ordinates, 495
Perpendicular form of a line, 59
Perpendicular lines, 72
Point circle, 152
Point of intersection of two lines, 76
Point slope form, 58
Polar equation
—of a circle, 500
—of a conic, 502
—of a line, 498
Polar co-ordinates, 17
Pole and polar, 249, 514
Position of a point relative to
—a circle, 160
—an ellipse, 220
—a hyperbola, 222
—a parabola, 221
Principal axes, 327, 391

Quadrants, 3

Radical
—axis, 185
—centre, 187

Radius, 151
Radius vector, 17
Rectangular
—axes, 5
—hyperbola, 216, 434
Rotation of axes, 39

Section formula, 7
Self polar triangle, 258
Slope intercept form, 58
Slope of a line, 57
System of lines, 84
System of parallel chords
—of an ellipse, 360
—of a hyperbola, 426
—of a parabola, 319
Standard form
—of a circle, 152
—of an ellipse, 325
—of a hyperbola, 389
—of a parabola, 284

Tangent, 169, 225, 511
Tangents at/from a point, 233, 269
Touching circles, 187
Tracing
—of a circle, 213
—of an ellipse, 214, 464
—of a hyperbola, 215, 477
—of a parabola, 214, 468
—of a rectangular hyperbola, 216
Transformation, 37
Translation, 38
Transverse axis, 391
Two point form
—in Cartesian co-ordinates, 56
—in polar co-ordinates, 494
Two sides of a line, 94

Vertex/vertices
—of an ellipse, 327
—of a hyperbola, 391
—of a parabola, 285
Vectorial angle, 17